D1272287

PROTOCOLS AND ARCHITECTURES FOR WIRELESS SENSOR NETWORKS

PROTOCOLS AND ARCHITECTURES FOR WIRELESS SENSOR NETWORKS

Holger Karl
University of Paderborn, GERMANY

Andreas Willig
Hasso-Plattner-Institute at the University of Potsdam, GERMANY

John Wiley & Sons, Ltd

Telephone (+44) 1243 779777

Email (for orders and customer service enquiries): cs-books@wiley.co.uk
Visit our Home Page on www.wiley.com

Other Wiley Editorial Offices

John Wiley & Sons Inc., 111 River Street, Hoboken, NJ 07030, USA

Jossey-Bass, 989 Market Street, San Francisco, CA 94103-1741, USA

Wiley-VCH Verlag GmbH, Boschstr. 12, D-69469 Weinheim, Germany

John Wiley & Sons Australia Ltd, 33 Park Road, Milton, Queensland 4064, Australia

John Wiley & Sons (Asia) Pte Ltd, 2 Clementi Loop #02-01, Jin Xing Distripark, Singapore 129809

John Wiley & Sons Canada Ltd, 22 Worcester Road, Etobicoke, Ontario, Canada M9W 1L1

Wiley also publishes its books in a variety of electronic formats. Some content that appears in print may not be available in electronic books.

Library of Congress Cataloging-in-Publication Data

Karl, Holger, 1970-
Protocols and architectures for wireless sensor networks / Holger Karl, Andreas Willig.
p. cm.
Includes bibliographical references and index.
ISBN-13 978-0-470-09510-2 (cloth : alk. paper)
ISBN-10 0-470-09510-5 (cloth : alk. paper)
1. Sensor networks. 2. Wireless LANs. I. Willig, Andreas, 1968- II. Title.
TK7872.D48K37 2005
681′.2 – dc22

2005005800

British Library Cataloguing in Publication Data

A catalogue record for this book is available from the British Library

ISBN-13 978-0-470-09510-2 (HB)
ISBN-10 0-470-09510-5 (HB)

Typeset in 10/12 Times by Laserwords Private Limited, Chennai, India
Printed and bound in Great Britain by Antony Rowe Ltd, Chippenham, Wiltshire
This book is printed on acid-free paper responsibly manufactured from sustainable forestry in which at least two trees are planted for each one used for paper production.

Contents

Preface

Integrating simple processing, storage, sensing, and communication capabilities into small-scale, low-cost devices and joining them into so-called wireless sensor networks opens the door to a plethora of new applications – or so it is commonly believed. It is a struggle to find a business model that can turn the bright visions into a prosperous and actually useful undertaking. But this struggle can be won by applying creative ideas to the underlying technology, assuming that this technology and its abilities as well as shortcomings and limitations are properly understood. We have written this book in the hope of fostering this understanding.

Understanding (and presenting) this new type of networks is a formidable challenge. A key characteristic is the need to understand issues from many diverse areas, ranging from low-level aspects of hardware and radio communication to high-level concepts like databases or middleware and to the very applications themselves. Then, a joint optimization can be attempted, carefully tuning all system components, drawing upon knowledge from disciplines like electrical engineering, computer science and computer engineering, and mathematics. Such a complex optimization is necessary owing to the stringent resource restrictions – in particular, energy – by which these networks are constrained. As a consequence, a simple explanation along the lines of the ISO/OSI model or a similar layering model for communication networks fails. Nonetheless, we have attempted to guide the reader along the lines of such a model and tried to point out the points of interaction and interdependence between such different "layers".

In structuring the material and in the writing process, our goal was to explain the main problems at hand and principles and essential ideas for their solution. We usually did not go into the details of each of (usually many) several solution options; however, we did provide the required references for the readers to embark on a journey to the sources on their own. Nor did we attempt to go into any detail regarding performance characteristics of any described solution. The difficulty here lies in presenting such results in a comparable way – it is next to impossible to find generally comparable performance results in scientific publications on the topic of wireless sensor networks. What is perhaps missing is a suite of benchmarking applications, with clearly delimited rules and assumptions (the use of a prevalent simulator is no substitute here). Tracking might be one such application, but it clearly is not the only important application class to which wireless sensor networks can be applied.

Often, a choice had to be made whether to include a given idea, paper, or concept. Given the limited space in such a textbook, we preferred originality or an unusual but promising approach over papers that present solid but more technical work, albeit this type of work can make the difference whether a particular scheme is practicable at all.

We also tried to avoid, and explicitly argue against, ossification but rather tried to keep and promote an open mind-set about what wireless sensor networks are and what their crucial research topics entail. We feel that this still relatively young and immature field is sometimes inappropriately narrowed down to a few catchwords – energy efficiency being the most prominent example – which, although indubitably important, might prevent interesting ideas from forming and becoming publicly known. Here, we tried to give the benefit of the doubt and at least tried to include pointers and references to some "unusual" or odd approaches.

Nonetheless, we had to omit a considerable amount of material; areas like middleware, security, management, deployment, or modeling suffered heavily or were, in the end, entirely excluded. We also had to stop including new material at some point in time – at the rate of new publications appearing on this topic, this book would otherwise never be completed (if you feel that we have overlooked important work or misrepresented some aspects, we encourage you to contact us). We still hope that it can serve the reader as a first orientation in this young, vigorous, and fascinating research area. Visit the website accompanying this book, `www.wiley.com/go/wsn`, for a growing repository of lecture slides on ad hoc and sensor networks.

Audience and Prerequisites

The book is mainly targeted at senior undergraduate or graduate-level students, at academic and industrial researchers working in the field, and also at engineers developing actual solutions for wireless sensor networks. We consider this book as a good basis to teach a class on wireless sensor networks (e.g. for a lecture corresponding to three European Credit Transfer System points).

This book is not intended as a first textbook on wireless networking. While we do try to introduce most of the required background, it will certainly be helpful for the reader to have some prior knowledge of wireless communication already; some first contact with mobile ad hoc networking can be beneficial to understand the differences but is not essential. We do, however, assume general networking knowledge as a given.

Moreover, in several parts of the book, some concepts and results from discrete mathematics are used. It will certainly be useful for the reader to have some prior idea regarding optimization problems, NP completeness, and similar topics.

Acknowledgments

We are indebted to numerous people who have helped us in understanding this research field and in writing this book. A prominent place and heartfelt thanks are owed to our colleagues at the Telecommunication Networks Group at the Technische Universität Berlin, especially Prof. Adam Wolisz, Vlado Handziski, Jan-Hinrich Hauer, Andreas Köpke, Martin Kubisch, and Günther Schäfer. Also, we are grateful to many colleagues with whom we had the pleasure and the privilege to discuss WSN research issues – colleagues from different research projects like the EU IST project EYES and the German federal funded project AVM deserve a special mention here. Robert Mitschke from the Hasso Plattner Institute did an excellent job in proofreading and criticizing an intermediate version of this book. The anonymous reviewers provided us with many useful comments. The help of our editors and the support team at Wiley – in particular, Birgit Gruber, Julie Ward and Joanna Tootill – was very valuable.

We also want to express our deep gratitude to all the researchers in the field who have made their results and publications easily available over the World Wide Web. Without this help, collecting the material discussed in the present book alone would have been too big a challenge to embark on.

And last, but most importantly, both of us are very deeply indebted to our families for bearing with us during the year of writing, grumbling, hoping, and working.

Berlin & Paderborn
April 2005

List of abbreviations

ABR Associativity-Based Routing

ACPI Advanced Configuration and Power Interface

ACQUIRE ACtive QUery forwarding In sensoR nEtworks

ADC Analog/Digital Converter

AIDA Application-Independent Data Aggregation

ANDA Ad hoc Network Design Algorithm

AODV Ad hoc On-demand Distance Vector

APIT Approximate Point in Triangle

API Application Programming Interface

ARQ Automatic Repeat Request

ASCENT Adaptive Self-Configuring sEnsor Networks Topologies

ASIC Application-Specific Integrated Circuit

ASK Amplitude Shift Keying

AVO Attribute Value Operation

AWGN Additive White Gaussian Noise

BCH Bose–Chaudhuri–Hocquenghem

BER Bit-Error Rate

BIP Broadcast Incremental Power

BPSK Binary Phase Shift Keying

BSC Binary Symmetric Channel

CADR Constrained Anisotropic Diffusion Routing

CAMP Core-Assisted Mesh Protocol

CAP Contention Access Period

CCA Clear Channel Assessment

CCK Complementary Code Keying

CDMA Code Division Multiple Access

CDS Connected Dominating Set

CGSR Clusterhead Gateway Switch Routing

CIR Carrier to Interference Ratio

CMMBCR Conditional Max–Min Battery Capacity Routing

CODA COngestion Detection and Avoidance

CPU Central Processing Unit

CRC Cyclic Redundancy Check

CSD Cumulative Sensing Degree

CSIP Collaborative Signal and Information Processing

CSMA Carrier Sense Multiple Access

CTS Clear To Send

DAC Digital/Analog Converter

DAD Duplicate Address Detection

DAG Directed Acyclic Graph

DAML DARPA Agent Markup Language

DBPSK Differential Binary Phase Shift Keying

DCF Distributed Coordination Function

DCS Data-Centric Storage

DCS Dynamic Code Scaling

DHT Distributed Hash Table

DISCUS Distributed Source Coding Using Syndromes

DLL Data Link Layer

DMCS Dynamic Modulation-Code Scaling

DMS Dynamic Modulation Scaling

DPM Dynamic Power Management

DQPSK Differential Quaternary Phase Shift Keying

DREAM Distance Routing Effect Algorithm for Mobility

DSDV Destination-Sequenced Distance Vector

DSP Digital Signal Processor

DSR Dynamic Source Routing

DSSS Direct Sequence Spread Spectrum

DVS Dynamic Voltage Scaling

EEPROM Electrically Erasable Programmable Read-Only Memory

EHF Extremely High Frequency

ESRT Event-to-Sink Reliable Transport

FDMA Frequency Division Multiple Access

FEC Forward Error Correction

FFD Full Function Device

FFT Fast Fourier Transform

FHSS Frequency Hopping Spread Spectrum

FIFO First In First Out

FPGA Field-Programmable Gate Array

FSK Frequency Shift Keying

GAF Geographic Adaptive Fidelity

GAMER Geocast Adaptive Mesh Environment for Routing

GEAR Geographic and Energy Aware Routing

GEM Graph EMbedding

GHT Geographic Hash Table

GOAFR Greedy and (Other Adaptive) Face Routing

GPSR Greedy Perimeter Stateless Routing

GPS Global Positioning System

GRAB GRAdient Broadcast

GTS Guaranteed Time Slot

HHBA Hop-by-Hop Broadcast with Acknowledgments

HHB Hop-by-Hop Broadcast

HHRA Hop-by-Hop Reliability with Acknowledgments

HHR Hop-by-Hop Reliability

HMM Hidden Markov Model

HVAC Humidity, Ventilation, Air Conditioning

IDSQ Information-Driven Sensor Querying

IEEE Institute of Electrical and Electronics Engineers

IFS InterFrame Space

IF Intermediate Frequency

ISI InterSymbol Interference

ISM Industrial, Scientific, and Medical

LAR Location-Aided Routing

LBM Location-Based Multicast

LEACH Low-Energy Adaptive Clustering Hierarchy

LED Light-Emitting Diode

LNA Low Noise Amplifier

LOS Line Of Sight

MAC Medium Access Control

MANET Mobile Ad Hoc Network

MBCR Minimum Battery Cost Routing

MCDS Minimum Connected Dominating Set

MDS Minimum Dominating Set

MDS MultiDimensional Scaling

MEMS MicroElectroMechanical System

MIP Multicast Incremental Power

MLE Maximum Likelihood Estimation

MMBCR Min–Max Battery Cost Routing

MPDU MAC-layer Protocol Data Unit

MSE Mean Squared Error

MST Minimum Spanning Tree

MTPR Minimum Total Transmission Power Routing

MULE Mobile Ubiquitous LAN extension

MWIS Maximum Weight Independent Set

NAT Network Address Translation

NAV Network Allocation Vector

NLOS Non Line Of Sight

OOK On-Off-Keying

PAN Personal Area Network

PA Power Amplifier

PCF Point Coordination Function

PDA Personal Digital Assistant

PEGASIS Power-Efficient GAthering in Sensor Information Systems

PHY Physical Layer

PPDU Physical-layer Protocol Data Unit

PPM Pulse Position Modulation

PSD Power Spectral Density

PSFQ Pump Slowly Fetch Quickly

PSK Phase Shift Keying

PTAS Polynomial Time Approximation Scheme

QAM Quadrature Amplitude Modulation

QPSK Quaternary Phase Shift Keying

QoS Quality of Service

RAM Random Access Memory

RFD Reduced Function Device

RF ID Radio Frequency Identifier

RF Radio Frequency

RISC Reduced Instruction Set Computer

RMST Reliable Multisegment Transport

RNG Relative Neighborhood Graph

ROHC RObust Header Compression

ROM Read-Only Memory

RSSI Received Signal Strength Indicator

RS Reed–Solomon

RTS Request To Send

SAR Sequential Assignment Routing

SDMA Space Division Multiple Access

SFD Start Frame Delimiter

SINR Signal to Interference and Noise Ratio

SMACS Self-Organizing Medium Access Control for Sensor Networks

SNR Signal-to-Noise Ratio

SPIN Sensor Protocol for Information via Negotiation

SPT Shortest Path Tree

SQL Standard Query Language

SRM Scalable Reliable Multicast

SSR Signal Stability Routing

STEM Sparse Topology and Energy Management

TAG Tiny Aggregation

TBF Trajectory-Based Forwarding

TCP Transmission Control Protocol

TDMA Time Division Multiple Access

TDoA Time Difference of Arrival

TORA Temporally Ordered Routing Algorithm

TRAMA Traffic-Adaptive Medium Access

TTDD Two-Tier Data Dissemination

TTL Time To Live

ToA Time of Arrival

UML Unified Modeling Language

UTM Universal Transverse Mercator

UWB UltraWideBand

VCO Voltage-Controlled Oscillator

VLF Very Low Frequency

VOR VHF Omnidirectional Ranging

VPCR Virtual Polar Coordinate Routing

VPCS Virtual Polar Coordinate Space

WLAN Wireless Local Area Network

WPAN Wireless Personal Area Network

WRP Wireless Routing Protocol

WSDL Web Service Description Language

WSN Wireless Sensor Network

A guide to the book

The design and optimization of a wireless sensor network draws on knowledge and understanding of many different areas: properties of the radio front end determine what type of MAC protocols can be used, the type of application limits the options for routing protocols, and battery self-recharge characteristics influence sleeping patterns of a node. A book, on the other hand, is a linear entity. We are therefore forced to find a consecutive form of presenting an inherently nonconsecutive, but densely interwoven, topic.

To overcome this problem, we structured the book in two parts (Figure 1). The three chapters of the first part give a high-level overview of applications and problems, of hardware properties, and of the essential networking architecture. These first three chapters build a foundation upon which we build a detailed treatment of individual communication protocols in the second part of the book.

This second part is loosely oriented along the lines of the standard ISO/OSI layering model but, of course, focuses on algorithms and protocols relevant to wireless sensor networks. We start out by looking at the protocols needed between two neighboring nodes in the physical, link, and medium access layers. Then, a discussion about names and addresses in a wireless sensor network follows. The next three chapters – time synchronization, localization and positioning, and topology control – describe functionality that is important for the correct or efficient operation of a sensor network but that is not directly involved in the exchange of packets between neighboring nodes. In a sense, these are "helper protocols".

On the basis of this understanding of communication between neighbors and on essential helper functionality, the following three chapters treat networking functionality regarding routing protocols in various forms, transport layer functionality, and an appropriate notion of quality of service. The book is complemented by a final chapter on advanced application support. For extra learning materials in the form of lecture slides, go to the accompanying website, `www.wiley.com/go/wsn`, which is gradually being populated.

A Full Course

Selecting the material for a full course from this book should be relatively easy. Essentially, all topics should be covered, more or less in depth, using a variable number of the example protocols discussed in the book.

A Reduced Course

If time does not permit covering of all the topics, a selection has to be made. We consider the following material rather important and recommend to cover it, if at all possible.

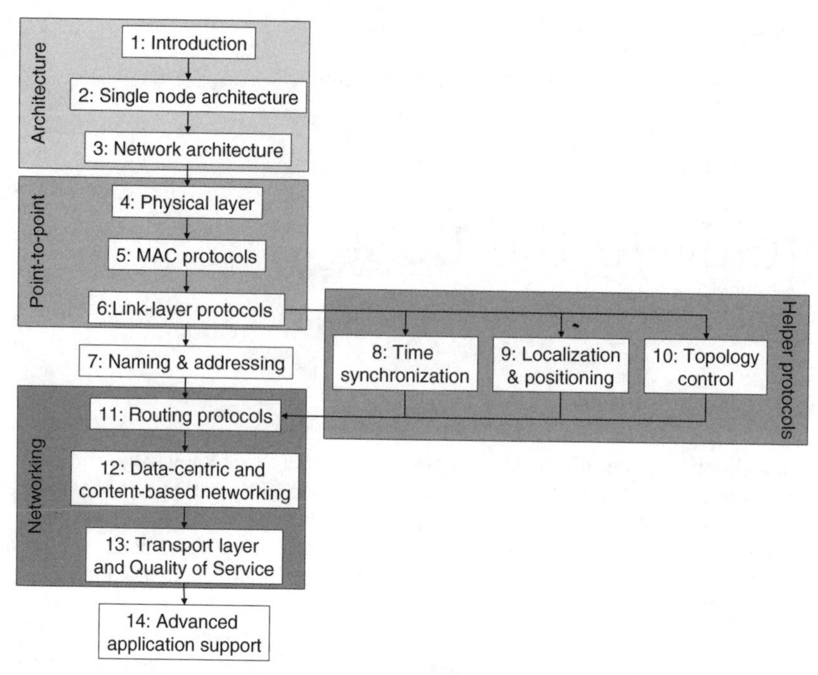

Figure 1 Structure of the book

Chapter 1: Introduction Completely.

Chapter 2: Single node architecture Treat at least Sections 2.1 and 2.2 to some level of detail. Section 2.3 on operating systems can be covered relatively briefly (depending on the focus of the course, this might not be very important material).

Chapter 3: Network architecture Cover Sections 3.1 to 3.3. The sections on service interface and gateways can be omitted for a first reading.

Chapter 4: Physical layer Depending on previous knowledge, this chapter can be skipped entirely. If possible, Section 4.3 should, however, be covered.

Chapter 5: MAC protocols An important chapter that should be covered, if possible, in its entirety. If time is short, some examples for each of different protocol classes can be curtailed.

Chapter 6: Link layer protocols Any of the three Sections 6.2, 6.3, or 6.4 can be selected for a more detailed treatment.

Chapter 7: Naming and addressing This chapter should be treated fairly extensively. Sections 7.3 and 7.4 can be omitted.

Chapter 8: Time synchronization This chapter can be skipped.

Chapter 9: Localization and positioning This chapter can be skipped.

Chapter 10: Topology control While this chapter can, in principle, be skipped as well, some of the basic ideas should be covered even in a condensed course. We would suggest to cover Section 10.1 and a single example from Sections 10.2 to 10.6 each.

Chapter 11: Routing protocols An important chapter. Sections 11.2 and 11.6 may be omitted.[1]

Chapter 12: Data-centric and content-based networking Quite important and characteristic for wireless sensor networks. Should receive extensive treatment in a lecture.

Chapter 13: Transport layer and Quality of Service This chapter also should be treated extensively.

Chapter 14: Advanced application support Much of this chapter can be skipped, but a few examples from Section 14.3 should make a nice conclusion for a lecture.

Evidently, the amount of detail and the focus of a lecture can be controlled by the number of examples discussed in class. It is probably infeasible to discuss the entire book in a lecture.

[1] We would like to make the reader aware of the Steiner tree problem described in Section 11.4.2. It did surprise us in preparing this book how often this problem has been "rediscovered" in the sensor network literature, often without recognizing it for what it is.

1

Introduction

Objectives of this Chapter

Applications should shape and form the technology for which they are intended. This holds true in particular for wireless sensor networks, which have, to some degree, been a technology-driven development. This chapter starts out by putting the idea of wireless sensor networks into a broader perspective and gives a number of application scenarios, which will later be used to motivate particular technical needs. It also generalizes from specific examples to types or classes of applications. Then, the specific challenges for these application types are discussed and why current technology is not up to meeting these challenges.

At the end of this chapter, the reader should have an appreciation for the types of applications for which wireless sensor networks are intended and a first intuition about the types of technical solutions that are required, both in hardware and in networking technologies.

Chapter Outline

1.1 The vision of Ambient Intelligence

The most common form of information processing has happened on large, general-purpose computational devices, ranging from old-fashioned mainframes to modern laptops or palmtops. In many applications, like office applications, these computational devices are mostly used to process information that is at its core centered around a human user of a system, but is at best indirectly related to the physical environment.

Protocols and Architectures for Wireless Sensor Networks H. Karl and A. Willig

In another class of applications, the physical environment is at the focus of attention. Computation is used to exert control over physical processes, for example, when controlling chemical processes in a factory for correct temperature and pressure. Here, the computation is integrated with the control; it is *embedded* into a physical system. Unlike the former class of systems, such **embedded systems** are usually not based on human interaction but are rather required to work without it; they are intimately tied to their control task in the context of a larger system.

Such embedded systems are a well-known and long-used concept in the engineering sciences (in fact, estimates say that up to 98 % of all computing devices are used in an embedded context [91]). Their impact on everyday life is also continuing to grow at a quick pace. Rare is the household where embedded computation is not present to control a washing machine, a video player, or a cell phone. In such applications, embedded systems meet human-interaction-based systems.

Technological progress is about to take this spreading of embedded control in our daily lives a step further. There is a tendency not only to equip larger objects like a washing machine with embedded computation and control, but also smaller, even dispensable goods like groceries; in addition, living and working spaces themselves can be endowed with such capabilities. Eventually, computation will surround us in our daily lives, realizing a vision of "**Ambient Intelligence**" where many different devices will gather and process information from many different sources to both control physical processes and to interact with human users. These technologies should be unobtrusive and be taken for granted – Marc Weiser, rightfully called the *father of ubiquitous computing*, called them *disappearing technologies* [867, 868]. By integrating computation and control in our physical environment, the well-known interaction paradigms of person-to-person, person-to-machine and machine-to-machine can be supplemented, in the end, by a notion of person-to-physical world [783]; the interaction with the physical world becomes more important than mere symbolic data manipulation [126].

To realize this vision, a crucial aspect is needed in addition to computation and control: communication. All these sources of information have to be able to transfer the information to the place where it is needed – an actuator or a user – and they should collaborate in providing as precise a picture of the real world as is required. For some application scenarios, such networks of sensors and actuators are easily built using existing, wired networking technologies. For many other application types, however, the need to wire together all these entities constitutes a considerable obstacle to success: Wiring is expensive (figures of up to US$200 per sensor can be found in the literature [667]), in particular, given the large number of devices that is imaginable in our environment; wires constitute a maintenance problem; wires prevent entities from being mobile; and wires can prevent sensors or actuators from being close to the phenomenon that they are supposed to control. Hence, *wireless communication* between such devices is, in many application scenarios, an inevitable requirement.

Therefore, a new class of networks has appeared in the last few years: the so-called Wireless Sensor Network (WSN) (see e.g. [17, 648]). These networks consist of individual nodes that are able to interact with their environment by sensing or controlling physical parameters; these nodes have to collaborate to fulfill their tasks as, usually, a single node is incapable of doing so; and they use wireless communication to enable this collaboration. In essence, the nodes without such a network contain at least some computation, wireless communication, and sensing or control functionalities. Despite the fact that these networks also often include actuators, the term wireless sensor network has become the commonly accepted name. Sometimes, other names like "wireless sensor and actuator networks" are also found.

These WSNs are powerful in that they are amenable to support a lot of very different real-world applications; they are also a challenging research and engineering problem because of this very flexibility. Accordingly, there is no single set of requirements that clearly classifies all WSNs, and there is also not a single technical solution that encompasses the entire design space. For example, in many WSN applications, individual nodes in the network cannot easily be connected to a wired power supply but rather have to rely on onboard batteries. In such an application, the energy efficiency of any proposed solution is hence a very important figure of merit as a long operation time is usually desirable. In other applications, power supply might not be an issue and hence other metrics, for example, the accuracy of the delivered results, can become more important. Also, the acceptable size and costs of an individual node can be relevant in many applications. Closely tied to the size is often the capacity of an onboard battery; the price often has a direct bearing on the quality of the node's sensors, influencing the accuracy of the result that can be obtained from a single node. Moreover, the number, price, and potentially low accuracy of individual nodes is relevant when comparing a distributed system of many sensor nodes to a more centralized version with fewer, more expensive nodes of higher accuracy. Simpler but numerous sensors that are close to the phenomenon under study can make the architecture of a system both simpler and more energy efficient as they facilitate distributed sampling – detecting objects, for example, requires a distributed system [17, 648].

Realizing such wireless sensor networks is a crucial step toward a deeply penetrating Ambient Intelligence concept as they provide, figuratively, the "last 100 meters" of **pervasive control**. To realize them, a better understanding of their potential applications and the ensuing requirements is necessary, as is an idea of the enabling technologies. These questions are answered in the following sections; a juxtaposition of wireless sensor networks and related networking concepts such as fieldbuses or mobile ad hoc network is provided as well.

1.2 Application examples

The claim of wireless sensor network proponents is that this technological vision will facilitate many existing application areas and bring into existence entirely new ones. This claim depends on many factors, but a couple of the envisioned application scenarios shall be highlighted.

Apart from the need to build cheap, simple to program and network, potentially long-lasting sensor nodes, a crucial and primary ingredient for developing actual applications is the actual sensing and actuating faculties with which a sensor node can be endowed. For many physical parameters, appropriate sensor technology exists that can be integrated in a node of a WSN. Some of the few popular ones are temperature, humidity, visual and infrared light (from simple luminance to cameras), acoustic, vibration (e.g. for detecting seismic disturbances), pressure, chemical sensors (for gases of different types or to judge soil composition), mechanical stress, magnetic sensors (to detect passing vehicles), potentially even radar (see references [245, 246] for examples). But even more sophisticated sensing capabilities are conceivable, for example, toys in a kindergarten might have tactile or motion sensors or be able to determine their own speed or location [783].

Actuators controlled by a node of a wireless sensor network are perhaps not quite as multifaceted. Typically, they control a mechanical device like a servo drive, or they might switch some electrical appliance by means of an electrical relay, like a lamp, a bullhorn, or a similar device.

On the basis of nodes that have such sensing and/or actuation faculties, in combination with computation and communication abilities, many different kinds of applications can be constructed, with very different types of nodes, even of different kinds within one application. A brief list of scenarios should make the vast design space and the very different requirements of various applications evident. Overviews of these and other applications are included in references [17, 26, 88, 91, 110, 126, 134, 245, 246, 351, 367, 392, 534, 648, 667, 783, 788, 803, 923].

Disaster relief applications One of the most often mentioned application types for WSN are disaster relief operations. A typical scenario is wildfire detection: Sensor nodes are equipped with thermometers and can determine their own location (relative to each other or in absolute coordinates). These sensors are deployed over a wildfire, for example, a forest, from an airplane. They collectively produce a "temperature map" of the area or determine the perimeter of areas with high temperature that can be accessed from the outside, for example, by firefighters equipped with Personal Digital Assistants (PDAs). Similar scenarios are possible for the control of accidents in chemical factories, for example.

Some of these disaster relief applications have commonalities with military applications, where sensors should detect, for example, enemy troops rather than wildfires. In such an application, sensors should be cheap enough to be considered disposable since a large number is necessary; lifetime requirements are not particularly high.

Environment control and biodiversity mapping WSNs can be used to control the environment, for example, with respect to chemical pollutants – a possible application is garbage dump sites. Another example is the surveillance of the marine ground floor; an understanding of its erosion processes is important for the construction of offshore wind farms. Closely related to environmental control is the use of WSNs to gain an understanding of the number of plant and animal species that live in a given habitat (biodiversity mapping).

The main advantages of WSNs here are the long-term, unattended, wirefree operation of sensors close to the objects that have to be observed; since sensors can be made small enough to be unobtrusive, they only negligibly disturb the observed animals and plants. Often, a large number of sensors is required with rather high requirements regarding lifetime.

Intelligent buildings Buildings waste vast amounts of energy by inefficient Humidity, Ventilation, Air Conditioning (HVAC) usage. A better, real-time, high-resolution monitoring of temperature, airflow, humidity, and other physical parameters in a building by means of a WSN can considerably increase the comfort level of inhabitants and reduce the energy consumption (potential savings of two quadrillion British Thermal Units in the US alone have been speculated about [667]). Improved energy efficiency as well as improved convenience are some goals of "intelligent buildings" [415], for which currently wired systems like BACnet, LonWorks, or KNX are under development or are already deployed [776]; these standards also include the development of wireless components or have already incorporated them in the standard.

In addition, such sensor nodes can be used to monitor mechanical stress levels of buildings in seismically active zones. By measuring mechanical parameters like the bending load of girders, it is possible to quickly ascertain via a WSN whether it is still safe to enter a given building after an earthquake or whether the building is on the brink of collapse – a considerable advantage for rescue personnel. Similar systems can be applied to bridges. Other types of sensors might be geared toward detecting people enclosed in a collapsed building and communicating such information to a rescue team.

The main advantage here is the collaborative mapping of physical parameters. Depending on the particular application, sensors can be retrofitted into existing buildings (for HVAC-type applications) or have to be incorporated into the building already under construction. If power supply is not available, lifetime requirements can be very high – up to several dozens of years – but the number of required nodes, and hence the cost, is relatively modest, given the costs of an entire building.

Facility management In the management of facilities larger than a single building, WSNs also have a wide range of possible applications. Simple examples include keyless entry applications where people wear badges that allow a WSN to check which person is allowed to enter which areas of a larger company site. This example can be extended to the detection of intruders, for example of vehicles that pass a street outside of normal business hours. A wide-area WSN could track such a vehicle's position and alert security personnel – this application shares many commonalities with corresponding military applications. Along another line, a WSN could be used in a chemical plant to scan for leaking chemicals.

These applications combine challenging requirements as the required number of sensors can be large, they have to collaborate (e.g. in the tracking example), and they should be able to operate a long time on batteries.

Machine surveillance and preventive maintenance One idea is to fix sensor nodes to difficult-to-reach areas of machinery where they can detect vibration patterns that indicate the need for maintenance. Examples for such machinery could be robotics or the axles of trains. Other applications in manufacturing are easily conceivable.

The main advantage of WSNs here is the cablefree operation, avoiding a maintenance problem in itself and allowing a cheap, often retrofitted installation of such sensors. Wired power supply may or may not be available depending on the scenario; if it is not available, sensors should last a long time on a finite supply of energy since exchanging batteries is usually impractical and costly. On the other hand, the size of nodes is often not a crucial issue, nor is the price very heavily constrained.

Precision agriculture Applying WSN to agriculture allows precise irrigation and fertilizing by placing humidity/soil composition sensors into the fields. A relatively small number is claimed to be sufficient, about one sensor per 100 m $\times$ 100 m area. Similarly, pest control can profit from a high-resolution surveillance of farm land. Also, livestock breeding can benefit from attaching a sensor to each pig or cow, which controls the health status of the animal (by checking body temperature, step counting, or similar means) and raises alarms if given thresholds are exceeded.

Medicine and health care Along somewhat similar lines, the use of WSN in health care applications is a potentially very beneficial, but also ethically controversial, application. Possibilities range from postoperative and intensive care, where sensors are directly attached to patients – the advantage of doing away with cables is considerable here – to the long-term surveillance of (typically elderly) patients and to automatic drug administration (embedding sensors into drug packaging, raising alarms when applied to the wrong patient, is conceivable). Also, patient and doctor tracking systems within hospitals can be literally life saving.

Logistics In several different logistics applications, it is conceivable to equip goods (individual parcels, for example) with simple sensors that allow a simple tracking of these objects during transportation or facilitate inventory tracking in stores or warehouses.

In these applications, there is often no need for a sensor node to *actively* communicate; passive readout of data is often sufficient, for example, when a suitcase is moved around on conveyor belts in an airport and passes certain checkpoints. Such passive readout is much simpler and cheaper than the active communication and information processing concept discussed in the other examples; it is realized by so-called Radio Frequency Identifier (RF ID) tags.

On the other hand, a simple RFID tag cannot support more advanced applications. It is very difficult to imagine how a passive system can be used to locate an item in a warehouse; it can also not easily store information about the history of its attached object – questions like "where has this parcel been?" are interesting in many applications but require some active participation of the sensor node [246, 392].

Telematics Partially related to logistics applications are applications for the telematics context, where sensors embedded in the streets or roadsides can gather information about traffic conditions at a much finer grained resolution than what is possible today [296]. Such a so-called "intelligent roadside" could also interact with the cars to exchange danger warnings about road conditions or traffic jams ahead.

In addition to these, other application types for WSNs that have been mentioned in the literature include airplane wings and support for smart spaces [245], applications in waste water treatment plants [367], instrumentation of semiconductor processing chambers and wind tunnels [392], in "smart kindergartens" where toys interact with children [783], the detection of floods [88], interactive museums [667], monitoring a bird habitat on a remote island [534], and implanting sensors into the human body (for glucose monitoring or as retina prosthesis) [745]

While most of these applications are, in some form or another, possible even with today's technologies and without wireless sensor networks, all current solutions are "sensor starved" [667]. Most applications would work much better with information at higher spatial and temporal resolution about their object of concern than can be provided with traditional sensor technology. wireless sensor networks are to a large extent about providing the required information at the required accuracy in time with as little resource consumption as possible.

1.3 Types of applications

Many of these applications share some basic characteristics. In most of them, there is a clear difference between **sources** of data – the actual nodes that sense data – and **sinks** – nodes where the data should be delivered to. These sinks sometimes are part of the sensor network itself; sometimes they are clearly systems "outside" the network (e.g. the firefighter's PDA communicating with a WSN). Also, there are usually, but not always, more sources than sinks and the sink is oblivious or not interested in the identity of the sources; the data itself is much more important.

The **interaction patterns** between sources and sinks show some typical patterns. The most relevant ones are:

Event detection Sensor nodes should report to the sink(s) once they have detected the occurrence of a specified event. The simplest events can be detected locally by a single sensor node in isolation (e.g. a temperature threshold is exceeded); more complicated types of events require the collaboration of nearby or even remote sensors to decide whether a (composite) event has occurred (e.g. a temperature gradient becomes too steep). If several different events can occur, **event classification** might be an additional issue.

Periodic measurements Sensors can be tasked with periodically reporting measured values. Often, these reports can be triggered by a detected event; the reporting period is application dependent.

Function approximation and edge detection The way a physical value like temperature changes from one place to another can be regarded as a function of location. A WSN can be used to approximate this unknown function (to extract its spatial characteristics), using a limited number of samples taken at each individual sensor node. This approximate mapping should be made available at the sink. How and when to update this mapping depends on the application's needs, as do the approximation accuracy and the inherent trade-off against energy consumption.

Similarly, a relevant problem can be to find areas or points of the same given value. An example is to find the isothermal points in a forest fire application to detect the border of the actual fire. This can be generalized to finding "edges" in such functions or to sending messages along the boundaries of patterns in both space and/or time [274].

Tracking The source of an event can be mobile (e.g. an intruder in surveillance scenarios). The WSN can be used to report updates on the event source's position to the sink(s), potentially with estimates about speed and direction as well. To do so, typically sensor nodes have to cooperate before updates can be reported to the sink.

These interactions can be scoped both in time and in space (reporting events only within a given time span, only from certain areas, and so on). These requirements can also change dynamically overtime; sinks have to have a means to inform the sensors of their requirements at runtime. Moreover, these interactions can take place only for one specific request of a sink (so-called "one-shot queries"), or they could be long-lasting relationships between many sensors and many sinks.

The examples also have shown a wide diversity in **deployment options**. They range from well-planned, fixed deployment of sensor nodes (e.g. in machinery maintenance applications) to random deployment by dropping a large number of nodes from an aircraft over a forest fire. In addition, sensor nodes can be mobile themselves and compensate for shortcomings in the deployment process by moving, in a postdeployment phase, to positions such that their sensing tasks can be better fulfilled [17]. They could also be mobile because they are attached to other objects (in the logistics applications, for example) and the network has to adapt itself to the location of nodes.

The applications also influence the available **maintenance options**: Is it feasible and practical to perform maintenance on such sensors – perhaps even required in the course of maintenance on associated machinery? Is maintenance irrelevant because these networks are only deployed in a strictly ad hoc, short-term manner with a clear delimitation of maximum mission time (like in disaster recovery operations)? Or do these sensors have to function unattended, for a long time, with no possibility for maintenance?

Closely related to the maintenance options are the **options for energy supply**. In some applications, wired power supply is possible and the question is mute. For self-sustained sensor nodes, depending on the required mission time, energy supply can be trivial (applications with a few days of usage only) or a challenging research problem, especially when no maintenance is possible but nodes have to work for years. Obviously, acceptable price and size per node play a crucial role in designing energy supply.

1.4 Challenges for WSNs

Handling such a wide range of application types will hardly be possible with any single realization of a WSN. Nonetheless, certain common traits appear, especially with respect to the characteristics and the required mechanisms of such systems. Realizing these characteristics with new mechanisms is the major challenge of the vision of wireless sensor networks.

1.4.1 Characteristic requirements

The following characteristics are shared among most of the application examples discussed above:

Type of service The service type rendered by a conventional communication network is evident – it moves bits from one place to another. For a WSN, moving bits is only a means to an end, but not the actual purpose. Rather, a WSN is expected to provide meaningful information and/or actions about a given task: "People want answers, not numbers" (Steven Glaser, UC Berkeley, in [367]). Additionally, concepts like *scoping* of interactions to specific geographic regions or to time intervals will become important. Hence, new paradigms of using such a network are required, along with new interfaces and new ways of thinking about the service of a network.

Quality of Service Closely related to the type of a network's service is the quality of that service. Traditional quality of service requirements – usually coming from multimedia-type applications – like bounded delay or minimum bandwidth are irrelevant when applications are tolerant to latency [26] or the bandwidth of the transmitted data is very small in the first place. In some cases, only occasional delivery of a packet can be more than enough; in other cases, very high reliability requirements exist. In yet other cases, delay *is* important when actuators are to be controlled in a real-time fashion by the sensor network. The packet delivery ratio is an insufficient metric; what is relevant is the amount and quality of information that can be extracted at given sinks about the observed objects or area.

Therefore, adapted quality concepts like reliable detection of events or the approximation quality of a, say, temperature map is important.

Fault tolerance Since nodes may run out of energy or might be damaged, or since the wireless communication between two nodes can be permanently interrupted, it is important that the WSN as a whole is able to tolerate such faults. To tolerate node failure, redundant deployment is necessary, using more nodes than would be strictly necessary if all nodes functioned correctly.

Lifetime In many scenarios, nodes will have to rely on a limited supply of energy (using batteries). Replacing these energy sources in the field is usually not practicable, and simultaneously, a WSN must operate at least for a given mission time or as long as possible. Hence, the **lifetime** of a WSN becomes a very important figure of merit. Evidently, an energy-efficient way of operation of the WSN is necessary.

As an alternative or supplement to energy supplies, a limited power source (via power sources like solar cells, for example) might also be available on a sensor node. Typically, these sources are not powerful enough to ensure continuous operation but can provide some recharging of batteries. Under such conditions, the lifetime of the network should ideally be infinite.

The lifetime of a network also has direct trade-offs against quality of service: investing more energy can increase quality but decrease lifetime. Concepts to harmonize these trade-offs are required.

The precise *definition of lifetime* depends on the application at hand. A simple option is to use the time until the first node fails (or runs out of energy) as the network lifetime. Other options include the time until the network is disconnected in two or more partitions, the time until 50 % (or some other fixed ratio) of nodes have failed, or the time when for the first time a point in the observed region is no longer covered by at least a single sensor node (when using redundant deployment, it is possible and beneficial to have each point in space covered by several sensor nodes initially).

Scalability Since a WSN might include a large number of nodes, the employed architectures and protocols must be able scale to these numbers.

Wide range of densities In a WSN, the number of nodes per unit area – the *density* of the network – can vary considerably. Different applications will have very different node densities. Even within a given application, density can vary over time and space because nodes fail or move; the density also does not have to homogeneous in the entire network (because of imperfect deployment, for example) and the network should adapt to such variations.

Programmability Not only will it be necessary for the nodes to process information, but also they will have to react flexibly on changes in their tasks. These nodes should be programmable, and their programming must be changeable during operation when new tasks become important. A fixed way of information processing is insufficient.

Maintainability As both the environment of a WSN and the WSN itself change (depleted batteries, failing nodes, new tasks), the system has to adapt. It has to monitor its own health and status to change operational parameters or to choose different trade-offs (e.g. to provide lower quality when energy resource become scarce). In this sense, the network has to maintain itself; it could also be able to interact with external maintenance mechanisms to ensure its extended operation at a required quality [534].

1.4.2 Required mechanisms

To realize these requirements, innovative mechanisms for a communication network have to be found, as well as new architectures, and protocol concepts. A particular challenge here is the need to find mechanisms that are sufficiently specific to the idiosyncrasies of a given application to support the specific quality of service, lifetime, and maintainability requirements [246]. On the other hand, these mechanisms also have to generalize to a wider range of applications lest a complete from-scratch development and implementation of a WSN becomes necessary for every individual application – this would likely render WSNs as a technological concept economically infeasible.

Some of the mechanisms that will form typical parts of WSNs are:

Multihop wireless communication While wireless communication will be a core technique, a direct communication between a sender and a receiver is faced with limitations. In particular, communication over long distances is only possible using prohibitively high transmission power. The use of intermediate nodes as relays can reduce the total required power. Hence, for many forms of WSNs, so-called *multihop communication* will be a necessary ingredient.

Energy-efficient operation To support long lifetimes, energy-efficient operation is a key technique. Options to look into include energy-efficient data transport between two nodes (measured in J/bit) or, more importantly, the energy-efficient determination of a requested information. Also, nonhomogeneous energy consumption – the forming of "hotspots" – is an issue.

Auto-configuration A WSN will have to configure most of its operational parameters autonomously, independent of external configuration – the sheer number of nodes and simplified deployment will require that capability in most applications. As an example, nodes should be able to determine their geographical positions only using other nodes of the network – so-called "self-location". Also, the network should be able to tolerate failing nodes (because of a depleted battery, for example) or to integrate new nodes (because of incremental deployment after failure, for example).

Collaboration and in-network processing In some applications, a single sensor is not able to decide whether an event has happened but several sensors have to collaborate to detect an event and only the joint data of many sensors provides enough information. Information is processed in the network itself in various forms to achieve this collaboration, as opposed to having every node transmit all data to an external network and process it "at the edge" of the network.

An example is to determine the highest or the average temperature within an area and to report that value to a sink. To solve such tasks efficiently, readings from individual sensors can be *aggregated* as they propagate through the network, reducing the amount of data to be transmitted and hence improving the energy efficiency. How to perform such aggregation is an open question.

Data centric Traditional communication networks are typically centered around the transfer of data between two specific devices, each equipped with (at least) one network address – the operation of such networks is thus **address-centric**. In a WSN, where nodes are typically deployed redundantly to protect against node failures or to compensate for the low quality of a single node's actual sensing equipment, the identity of the particular node supplying data becomes irrelevant. What is important are the answers and values themselves, not which node has provided them. Hence, switching from an address-centric paradigm to a **data-centric** paradigm in designing architecture and communication protocols is promising.

An example for such a data-centric interaction would be to request the average temperature in a given location area, as opposed to requiring temperature readings from individual nodes. Such a data-centric paradigm can also be used to set conditions for alerts or events ("raise an alarm if temperature exceeds a threshold"). In this sense, the data-centric approach is closely related to query concepts known from databases; it also combines well with collaboration, in-network processing, and aggregation.

Locality Rather a design guideline than a proper mechanism, the principle of locality will have to be embraced extensively to ensure, in particular, scalability. Nodes, which are very limited in resources like memory, should attempt to limit the state that they accumulate during protocol processing to only information about their direct neighbors. The hope is that this will allow the network to scale to large numbers of nodes without having to rely on powerful processing at each single node. How to combine the locality principle with efficient protocol designs is still an open research topic, however.

Exploit trade-offs Similar to the locality principle, WSNs will have to rely to a large degree on exploiting various inherent trade-offs between mutually contradictory goals, both during system/protocol design and at runtime. Examples for such trade-offs have been mentioned already: higher energy expenditure allows higher result accuracy, or a longer lifetime of the entire network trades off against lifetime of individual nodes. Another important trade-off is node density: depending on application, deployment, and node failures at runtime, the density of the network can change considerably – the protocols will have to handle very different situations, possibly present at different places of a single network. Again, not all the research questions are solved here.

Harnessing these mechanisms such that they are easy to use, yet sufficiently general, for an application programmer is a major challenge. Departing from an address-centric view of the network requires new programming interfaces that go beyond the simple semantics of the conventional socket interface and allow concepts like required accuracy, energy/accuracy trade-offs, or scoping.

1.5 Why are sensor networks different?

On the basis of these application examples and main challenges, two close relatives of WSNs become apparent: Mobile Ad Hoc Networks (MANETs) on the one hand and fieldbuses on the other hand.

1.5.1 Mobile ad hoc networks and wireless sensor networks

An ad hoc network is a network that is setup, literally, for a specific purpose, to meet a quickly appearing communication need. The simplest example of an ad hoc network is perhaps a set of computers connected together via cables to form a small network, like a few laptops in a meeting room. In this example, the aspect of *self-configuration* is crucial – the network is expected to work without manual management or configuration.

Usually, however, the notion of a MANET is associated with wireless communication and specifically *wireless* multihop communication; also, the name indicates the mobility of participating nodes as a typical ingredient. Examples for such networks are disaster relief operations – firefighters communicate with each other – or networks in difficult locations like large construction sites, where the deployment of wireless infrastructure (access points etc.), let alone cables, is not a feasible option. In such networks, the individual nodes together form a network that relays packets between nodes to extend the reach of a single node, allowing the network to span larger geographical areas than would be possible with direct sender – receiver communication. The two basic challenges in a MANET are the reorganization of the network as nodes move about and handling the problems of the limited reach of wireless communication. Literature on MANETs that summarize these problems and their solutions abound, as these networks are still a very active field of research; popular books include [635, 793, 827].

These general problems are shared between MANETs and WSNs. Nonetheless, there are some principal differences between the two concepts, warranting a distinction between them and regarding separate research efforts for each one.

Applications and equipment MANETs are associated with somewhat different applications as well as different user equipment than WSNs: in a MANET, the terminal can be fairly powerful (a laptop or a PDA) with a comparably large battery. This equipment is needed because in the typical MANET applications, there is usually a human in the loop: the MANET is used for voice communication between two distant peers, or it is used for access to a remote infrastructure like a Web server. Therefore, the equipment has to be powerful enough to support these applications.

Application specific Owing to the large number of conceivable combinations of sensing, computing, and communication technology, many different application scenarios for WSNs become possible. It is unlikely that there will be a "one-size-fits-all" solution for all these potentially very different possibilities. As one example, WSNs are conceivable with very different network densities, from very sparse to very dense deployments, which will require different or at least adaptive protocols. This diversity, although present, is not quite as large in MANETs.

Environment interaction Since WSNs have to interact with the environment, their traffic characteristics can be expected to be very different from other, human-driven forms of networks. A typical consequence is that WSNs are likely to exhibit very low data rates over a large timescale, but can have very bursty traffic when something happens (a phenomenon known from real-time systems as event showers or alarm storms). Long periods (months) of inactivity can alternate with short periods (seconds or minutes) of very high activity in the network, pushing its capacity to the limits. MANETs, on the other hand, are used to support more conventional applications (Web, voice, and so on) with their comparably well understood traffic characteristics.

Scale Potentially, WSNs have to scale to much larger numbers (thousands or perhaps hundreds of thousands) of entities than current ad hoc networks, requiring different, more scalable solutions. As a concrete case in point, endowing sensor nodes with a unique identifier is costly (either at production or at runtime) and might be an overhead that could be avoided – hence, protocols that work without such identifiers might become important in WSNs, whereas it is fair to assume such identifiers to exist in MANET nodes.

Energy In both WSNs and MANETs, energy is a scare resource. But WSNs have tighter requirements on network lifetime, and recharging or replacing WSN node batteries is much less an option than in MANETs. Owing to this, the impact of energy considerations on the entire system architecture is much deeper in WSNs than in MANETs.

Self configurability Similar to ad hoc networks, WSNs will most likely be required to self-configure into connected networks, but the difference in traffic, energy trade-offs, and so forth, could require new solutions. Nevertheless, it is in this respect that MANETs and WSNs are probably most similar.

Dependability and QoS The requirements regarding dependability and QoS are quite different. In a MANET, each individual node should be fairly reliable; in a WSN, an individual node is next to irrelevant. The quality of service issues in a MANET are dictated by traditional applications (low jitter for voice applications, for example); for WSNs, entirely new QoS concepts are required, which also take energy explicitly into account.

Data centric Redundant deployment will make data-centric protocols attractive in WSNs. This concept is alien to MANETs. Unless applications like file sharing are used in MANETs, which do bear some resemblance to data centric approaches, data-centric protocols are irrelevant to MANETs – but these applications do not represent the typically envisioned use case.

Simplicity and resource scarceness Since sensor nodes are simple and energy supply is scarce, the operating and networking software must be kept orders of magnitude simpler compared to today's desktop computers. This simplicity may also require breaking with conventional layering rules for networking software, since layering abstractions typically cost time and space. Also, resources like memory, which is relevant for comparably heavy-weight routing protocols as those used in MANETs, is not available in arbitrary quantities, requiring new, scalable, resource-efficient solutions.

Mobility The mobility problem in MANETs is caused by nodes moving around, changing multihop routes in the network that have to be handled. In a WSN, this problem can also exist if the sensor nodes are mobile in the given application. There are two additional aspects of mobility to be considered in WSNs.

First, the sensor network can be used to detect and observe a physical phenomenon (in the intrusion detection applications, for example). This phenomenon is the cause of events that happen in the network (like raising of alarms) and can also cause some local processing, for example, determining whether there really is an intruder. What happens if this phenomenon moves about? Ideally, data that has been gathered at one place should be available at the next one. Also, in tracking applications, it is the explicit task of the network to ensure that some form of activity happens in nodes that surround the phenomenon under observation.

Second, the sinks of information in the network (nodes where information should be delivered to) can be mobile as well. In principle, this is no different than node mobility in the general MANET sense, but can cause some difficulties for protocols that operate efficiently in fully static scenarios. Here, carefully observing trade-offs is necessary.

Furthermore, in both MANET and WSNs, mobility can be correlated – a group of nodes moving in a related, similar fashion. This correlation can be caused in a MANET by, for example, belonging to a group of people traveling together. In a WSN, the movement of nodes can be correlated because nodes are jointly carried by a storm, a river, or some other fluid.

In summary, there are commonalities, but the fact that WSNs have to support very different applications, that they have to interact with the physical environment, and that they have to carefully adjudicate various trade-offs justifies considering WSNs as a system concept distinct from MANETs.

1.5.2 Fieldbuses and wireless sensor networks

Fieldbuses are networks that are specifically designed for operation under hard real-time constraints and usually with inbuilt fault tolerance, to be used predominantly in control applications, that is, as part of a control loop. Examples include the Profibus and IEEE 802.4 Token Bus networks [372] for factory floor automation or the CAN bus for onboard networks in cars; some example summaries on the topic include [532, 644, 881]. Because of the stringent hard real-time requirements, these networks are usually wired and only the layers one (physical), two (link layer), and seven (application) of the OSI reference model are used, avoiding communication over multiple hops and associated queuing delays in intermediate nodes. Nevertheless, a number of research efforts deal with realizing fieldbus semantics on top of wireless communication, despite its inherently limited error rates that jeopardize real-time guarantees [200, 687, 878].

Since fieldbuses also have to deal with the physical environment for which they report sensing data and which they control, they are in this sense very similar to WSNs. With some justification, WSNs can be considered examples of wireless fieldbuses. Some differences do exist, however: WSNs do mostly not attempt to provide real-time guarantees in the range of (tens of) milliseconds but are rather focused on applications that can tolerate longer delays and some jitter (delay variability). Also, the adaptive trade-offs that WSNs are willing to make (accuracy against energy efficiency, for example) is a concept that is not commonly present in the fieldbus literature; specifically, fieldbuses make no attempt to conserve energy, and their protocols are not prepared to do so.

But these distinctions can only serve as a rough guideline; the borderline between these two research areas is certainly a blurry one.

1.6 Enabling technologies for wireless sensor networks

Building such wireless sensor networks has only become possible with some fundamental advances in enabling technologies. First and foremost among these technologies is the miniaturization of hardware. Smaller feature sizes in chips have driven down the power consumption of the basic components of a sensor node to a level that the constructions of WSNs can be contemplated. This is particularly relevant to microcontrollers and memory chips as such, but also, the radio modems, responsible for wireless communication, have become much more energy efficient. Reduced chip size and improved energy efficiency is accompanied by reduced cost, which is necessary to make redundant deployment of nodes affordable.

Next to processing and communication, the actual sensing equipment is the third relevant technology. Here, however, it is difficult to generalize because of the vast range of possible sensors – Chapter 2 will go more into details here.

These three basic parts of a sensor node have to accompanied by power supply. This requires, depending on application, high capacity batteries that last for long times, that is, have only a negligible self-discharge rate, and that can efficiently provide small amounts of current. Ideally, a sensor node also has a device for **energy scavenging**, recharging the battery with energy gathered from the environment – solar cells or vibration-based power generation are conceivable options. Such a concept requires the battery to be efficiently chargeable with small amounts of current, which is not a standard ability. Both batteries and energy scavenging are still objects of ongoing research.

The counterpart to the basic hardware technologies is software. The first question to answer here is the principal division of tasks and functionalities in a single node – the architecture of the operating system or runtime environment. This environment has to support simple retasking, cross-layer information exchange, and modularity to allow for simple maintenance. This software architecture on a single node has to be extended to a network architecture, where the division of tasks between nodes, not only on a single node, becomes the relevant question – for example, how to structure interfaces for application programmers. The third part to solve then is the question of how to design appropriate communication protocols.

This book only touches briefly on the hardware aspects of WSNs. It is also not much concerned with the questions of appropriate runtime environments. It focuses, rather, on the WSNs architecture and protocols to solve the communication questions as such.

Part I
Architectures

2

Single-node architecture

Objectives of this Chapter

This fairly long chapter explains the basic part of a wireless sensor network: the nodes as such. It discusses the principal tasks of a node – computation, storage, communication, and sensing/actuation – and which components are required to perform these tasks. Then, the energy consumption of these components is described: how energy can be stored, gathered from the environment, and saved by intelligently controlling the mode of operation of node components. This control has to be exerted by an operating system like execution environment, which is described in the last major section of this chapter. Finally, some examples of sensor nodes are given.

At the end of this chapter, the reader should have an understanding of the capabilities and limitations of the nodes in a sensor network. It lays the foundation for the following chapter, which discusses the principal options on how individual sensor nodes can be connected into a wireless sensor network.

Chapter Outline

Building a wireless sensor network first of all requires the constituting nodes to be developed and available. These nodes have to meet the requirements that come from the specific requirements of a given application: they might have to be small, cheap, or energy efficient, they have to be equipped with the right sensors, the necessary computation and memory resources, and they need adequate communication facilities. These hardware components and their composition into a functioning node are described in Section 2.1; the power consumption of these components and the ensuing trade-offs are discussed in Section 2.2. As this chapter only focuses onto an individual node, the consequences of choosing a particular communication technology for the architecture of a wireless sensor network as a whole are described in Chapter 3.

In addition to the hardware of sensor nodes, the operating system and programming model is an important consideration. Section 2.3 describes the tasks of such an operating system along with some examples as well as suitable programming interfaces.

2.1 Hardware components

2.1.1 Sensor node hardware overview

When choosing the hardware components for a wireless sensor node, evidently the application's requirements play a decisive factor with regard mostly to size, costs, and energy consumption of the nodes – communication and computation facilities as such are often considered to be of acceptable quality, but the trade-offs between features and costs is crucial. In some extreme cases, an entire sensor node should be smaller than 1 cc, weigh (considerably) less than 100 g, be substantially cheaper than US$1, and dissipate less than 100 μW [667]. In even more extreme visions, the nodes are sometimes claimed to have to be reduced to the size of grains of dust. In more realistic applications, the mere size of a node is not so important; rather, convenience, simple power supply, and cost are more important [126].

These diversities notwithstanding, a certain common trend is observable in the literature when looking at typical hardware platforms for wireless sensor nodes. While there is certainly not a single standard available, nor would such a standard necessarily be able to support all application types, this section will survey these typical sensor node architectures. In addition, there are a number of research projects that focus on shrinking any of the components in size, energy consumption, or costs, based on the fact that custom off-the-shelf components do currently not live up to some of the more stringent application requirements. But as this book focuses on the networking aspects of WSNs, these efforts are not discussed here.

A basic sensor node comprises five main components (Figure 2.1):

Controller A controller to process all the relevant data, capable of executing arbitrary code.

Memory Some memory to store programs and intermediate data; usually, different types of memory are used for programs and data.

Sensors and actuators The actual interface to the physical world: devices that can observe or control physical parameters of the environment.

Communication Turning nodes into a network requires a device for sending and receiving information over a wireless channel.

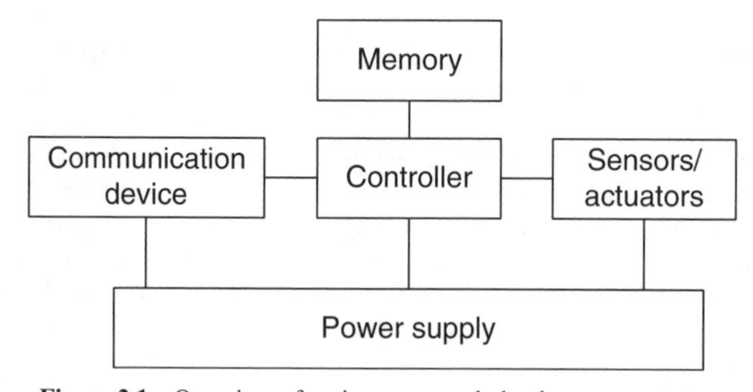

Figure 2.1 Overview of main sensor node hardware components

Power supply As usually no tethered power supply is available, some form of batteries are necessary to provide energy. Sometimes, some form of recharging by obtaining energy from the environment is available as well (e.g. solar cells).

Each of these components has to operate balancing the trade-off between as small an energy consumption as possible on the one hand and the need to fulfill their tasks on the other hand. For example, both the communication device and the controller should be turned off as long as possible. To wake up again, the controller could, for example, use a preprogrammed timer to be reactivated after some time. Alternatively, the sensors could be programmed to raise an interrupt if a given event occurs – say, a temperature value exceeds a given threshold or the communication device detects an incoming transmission.

Supporting such alert functions requires appropriate interconnection between individual components. Moreover, both control and data information have to be exchanged along these interconnections. This interconnection can be very simple – for example, a sensor could simply report an analog value to the controller – or it could be endowed with some intelligence of its own, preprocessing sensor data and only waking up the main controller if an actual event has been detected – for example, detecting a threshold crossing for a simple temperature sensor. Such preprocessing can be highly customized to the specific sensor yet remain simple enough to run continuously, resulting in improved energy efficiency [26].

2.1.2 Controller

Microcontrollers versus microprocessors, FPGAs, and ASICs

The controller is the core of a wireless sensor node. It collects data from the sensors, processes this data, decides when and where to send it, receives data from other sensor nodes, and decides on the actuator's behavior. It has to execute various programs, ranging from time-critical signal processing and communication protocols to application programs; it is the Central Processing Unit (CPU) of the node.

Such a variety of processing tasks can be performed on various controller architectures, representing trade-offs between flexibility, performance, energy efficiency, and costs.

One solution is to use general-purpose processors, like those known from desktop computers. These processors are highly overpowered, and their energy consumption is excessive. But simpler processors do exist, specifically geared toward usage in embedded systems. These processors are commonly referred as **microcontrollers**. Some of the key characteristics why these microcontrollers are particularly suited to embedded systems are their flexibility in connecting with other devices (like sensors), their instruction set amenable to time-critical signal processing, and their typically low power consumption; they are also convenient in that they often have memory built in. In addition, they are freely programmable and hence very flexible. Microcontrollers are also suitable for WSNs since they commonly have the possibility to reduce their power consumption by going into **sleep states** where only parts of the controller are active; details vary considerably between different controllers. Details regarding power consumption and energy efficiency are discussed in Section 2.2. One of the main differences to general-purpose systems is that microcontroller-based systems usually do not feature a memory management unit, somewhat limiting the functionality of memory – for example, protected or virtual memory is difficult, if not impossible, to achieve.

A specialized case of programmable processors are Digital Signal Processors (DSPs). They are specifically geared, with respect to their architecture and their instruction set, for processing large amounts of vectorial data, as is typically the case in signal processing applications. In a wireless sensor node, such a DSP could be used to process data coming from a simple analog, wireless communication device to extract a digital data stream. In broadband wireless communication, DSPs are an appropriate and successfully used platform. But in wireless sensor networks, the requirements on wireless communication are usually much more modest (e.g. simpler, easier to process modulations are used that can be efficiently handled in hardware by the communication device itself) and the signal processing tasks related to the actual sensing of data is also not overly complicated. Hence, these advantages of a DSP are typically not required in a WSN node and they are usually not used.

Another option for the controller is to depart from the high flexibility offered by a (fairly general-purpose) microcontroller and to use Field-Programmable Gate Arrays (FPGAs) or Application-Specific Integrated Circuits (ASICs) instead. An FPGA can be reprogrammed (or rather reconfigured) "in the field" to adapt to a changing set of requirements; however, this can take time and energy – it is not practical to reprogram an FPGA at the same frequency as a microcontroller could change between different programs. An ASIC is a specialized processor, custom designed for a given application such as, for example, high-speed routers and switches. The typical trade-off here is loss of flexibility in return for a considerably better energy efficiency and performance. On the other hand, where a microcontroller requires software development, ASICs provide the same functionality in hardware, resulting in potentially more costly hardware development.

For a dedicated WSN application, where the duties of a the sensor nodes do not change over lifetime and where the number of nodes is big enough to warrant the investment in ASIC development, they can be a superior solution. At the current stage of WSN technology, however, the bigger flexibility and simpler usage of microcontrollers makes them the generally preferred solution. However, this is not necessarily the final solution as "convenient programmability over several orders of energy consumption and data processing requirements is a worthy research goal" [648]. In addition, splitting processing tasks between some low-level, fixed functionality put into a very energy-efficient ASIC and high-level, flexible, relatively rarely invoked processing on a microcontroller is an attractive design and research option [26, 648].

For the remainder of this book, a microcontroller-based architecture is assumed.

Some examples for microcontrollers

Microcontrollers that are used in several wireless sensor node prototypes include the Atmel processor or Texas Instrument's MSP 430. In older prototypes, the Intel StrongArm processors have also been used, but this is no longer considered as a practical option; it is included here for the sake of completeness. Nonetheless, as the principal properties of these processors and controllers are quite similar, conclusions from these earlier research results still hold to a large degree.

Intel StrongARM

The Intel StrongARM [379] is, in WSN terms, a fairly high-end processor as it is mostly geared toward handheld devices like PDAs. The SA-1100 model has a 32-bit Reduced Instruction Set Computer (RISC) core, running at up to 206 MHz.

Texas Instruments MSP 430

Texas Instrument provides an entire family of microcontrollers under the family designation MSP 430 [814]. Unlike the StrongARM, it is explicitly intended for embedded applications. Accordingly, it runs a 16-bit RISC core at considerably lower clock frequencies (up to 4 MHz) but comes with a wide range of interconnection possibilities and an instruction set amenable to easy handling of peripherals of different kinds. It features a varying amount of on-chip RAM (sizes are 2–10 kB), several 12-bit analog/digital converters, and a real-time clock. It is certainly powerful enough to handle the typical computational tasks of a typical wireless sensor node (possibly with the exception of driving the radio front end, depending on how it is connected – bit or byte interface – to the controller).

Atmel ATmega

The Atmel ATmega 128L [28] is an 8-bit microcontroller, also intended for usage in embedded applications and equipped with relevant external interfaces for common peripherals.

2.1.3 Memory

The memory component is fairly straightforward. Evidently, there is a need for Random Access Memory (RAM) to store intermediate sensor readings, packets from other nodes, and so on. While RAM is fast, its main disadvantage is that it loses its content if power supply is interrupted. Program code can be stored in Read-Only Memory (ROM) or, more typically, in Electrically Erasable Programmable Read-Only Memory (EEPROM) or flash memory (the later being similar to EEPROM but allowing data to be erased or written in blocks instead of only a byte at a time). Flash memory can also serve as intermediate storage of data in case RAM is insufficient or when the power supply of RAM should be shut down for some time. The long read and write access delays of flash memory should be taken into account, as well as the high required energy.

Correctly dimensioning memory sizes, especially RAM, can be crucial with respect to manufacturing costs and power consumption. However, even general rules of thumbs are difficult to give as the memory requirements are very much application dependent.

2.1.4 Communication device

Choice of transmission medium

The communication device is used to exchange data between individual nodes. In some cases, wired communication can actually be the method of choice and is frequently applied in many sensor networklike settings (using field buses like Profibus, LON, CAN, or others). The communication devices for these networks are custom off-the-shelf components.

The case of wireless communication is considerably more interesting. The first choice to make is that of the transmission medium – the usual choices include radio frequencies, optical communication, and ultrasound; other media like magnetic inductance are only used in very specific cases. Of these choices, Radio Frequency (RF)-based communication is by far the most relevant one as it best fits the requirements of most WSN applications: It provides relatively long range and high data rates, acceptable error rates at reasonable energy expenditure, and does not require line of sight between sender and receiver. Thus, RF-based communication and transceiver will receive the lion share of attention here; other media are only treated briefly at the end of this section.

For a practical wireless, RF-based system, the carrier frequency has to be carefully chosen. Chapter 4 contains a detailed discussion; for the moment, suffice it to say that wireless sensor networks typically use communication frequencies between about 433 MHz and 2.4 GHz.

The reader is expected to be familiar with the basics of wireless communication; a survey is included in Chapter 4.

Transceivers

For actual communication, both a transmitter and a receiver are required in a sensor node. The essential task is to convert a bit stream coming from a microcontroller (or a sequence of bytes or frames) and convert them to and from radio waves. For practical purposes, it is usually convenient to use a device that combines these two tasks in a single entity. Such combined devices are called **transceivers**. Usually, half-duplex operation is realized since transmitting and receiving at the same time on a wireless medium is impractical in most cases (the receiver would only hear the own transmitter anyway).

A range of low-cost transceivers is commercially available that incorporate all the circuitry required for transmitting and receiving – modulation, demodulation, amplifiers, filters, mixers, and so on. For a judicious choice, the transceiver's tasks and its main characteristics have to be understood.

Transceiver tasks and characteristics

To select appropriate transceivers, a number of characteristics should be taken into account. The most important ones are:

Service to upper layer A receiver has to offer certain services to the upper layers, most notably to the Medium Access Control (MAC) layer. Sometimes, this service is **packet oriented**; sometimes, a transceiver only provides a **byte interface** or even only a **bit interface** to the microcontroller.

In any case, the transceiver must provide an interface that somehow allows the MAC layer to initiate frame transmissions and to hand over the packet from, say, the main memory of the sensor node into the transceiver (or a byte or a bit stream, with additional processing required on the microcontroller). In the other direction, incoming packets must be streamed into buffers accessible by the MAC protocol.

Power consumption and energy efficiency The simplest interpretation of energy efficiency is the energy required to transmit and receive a single bit. Also, to be suitable for use in WSNs, transceivers should be switchable between different states, for example, active and sleeping. The idle power consumption in each of these states and during switching between them is very important – details are discussed in Section 2.2.

Carrier frequency and multiple channels Transceivers are available for different carrier frequencies; evidently, it must match application requirements and regulatory restrictions. It is often useful if the transceiver provides several carrier frequencies ("channels") to choose from, helping to alleviate some congestion problems in dense networks. Such channels or "subbands" are relevant, for example, for certain MAC protocols (FDMA or multichannel CSMA/ALOHA techniques, see Chapter 5).

State change times and energy A transceiver can operate in different modes: sending or receiving, use different channels, or be in different power-safe states. In any case, the time and the energy required to change between two such states are important figures of merit. The turnaround time between sending and receiving, for example, is important for various medium access protocols (see Chapter 5).

Data rates Carrier frequency and used bandwidth together with modulation and coding determine the gross data rate. Typical values are a few tens of kilobits per second – considerably less than in broadband wireless communication, but usually sufficient for WSNs. Different data rates can be achieved, for example, by using different modulations or changing the symbol rate.

Modulations The transceivers typically support one or several of on/off-keying, ASK, FSK, or similar modulations. If several modulations are available, it is convenient for experiments if they are selectable at runtime even though, for real deployment, dynamic switching between modulations is not one of the most discussed options.

Coding Some transceivers allow various coding schemes to be selected.

Transmission power control Some transceivers can directly provide control over the transmission power to be used; some require some external circuitry for that purpose. Usually, only a discrete number of power levels are available from which the actual transmission power can be chosen. Maximum output power is usually determined by regulations.

Noise figure The **noise figure** NF of an element is defined as the ratio of the Signal-to-Noise Ratio (SNR) ratio SNR_I at the input of the element to the SNR ratio SNR_O at the element's output:

$$\mathrm{NF} = \frac{\mathrm{SNR}_I}{\mathrm{SNR}_O}$$

It describes the degradation of SNR due to the element's operation and is typically given in dB:

$$\mathrm{NF\,dB} = \mathrm{SNR}_I\,\mathrm{dB} - \mathrm{SNR}_O\,\mathrm{dB}$$

Gain The **gain** is the ratio of the output signal power to the input signal power and is typically given in dB. Amplifiers with high gain are desirable to achieve good energy efficiency.

Power efficiency The **efficiency** of the radio front end is given as the ratio of the radiated power to the overall power consumed by the front end; for a power amplifier, the efficiency describes the ratio of the output signal's power to the power consumed by the overall power amplifier.

Receiver sensitivity The **receiver sensitivity** (given in dBm) specifies the minimum signal power at the receiver needed to achieve a prescribed E_b/N_0 or a prescribed bit/packet error rate. Better sensitivity levels extend the possible range of a system.

Range While intuitively the range of a transmitter is clear, a formal definition requires some care. The range is considered in absence of interference; it evidently depends on the maximum transmission power, on the antenna characteristics, on the attenuation caused by the environment, which in turn depends on the used carrier frequency, on the modulation/coding scheme that is used, and on the bit error rate that one is willing to accept at the receiver. It also depends on the quality of the receiver, essentially captured by its sensitivity. Typical values are difficult to give here, but prototypes or products with ranges between a few meters and several hundreds of meters are available.

Blocking performance The blocking performance of a receiver is its achieved bit error rate in the presence of an interferer. More precisely, at what power level can an interferer (at a fixed distance) send at a given offset from the carrier frequency such that target BER can still be met? An interferer at higher frequency offsets can be tolerated at large power levels. Evidently, blocking performance can be improved by interposing a filter between antenna and transceiver.

An important special case is an adjacent channel interferer that transmits on neighboring frequencies. The adjacent channel suppression describes a transceiver's capability to filter out signals from adjacent frequency bands (and thus to reduce adjacent channel interference) has a direct impact on the observed Signal to Interference and Noise Ratio (SINR).

Out of band emission The inverse to adjacent channel suppression is the out of band emission of a transmitter. To limit disturbance of other systems, or of the WSN itself in a multichannel setup, the transmitter should produce as little as possible of transmission power outside of its prescribed bandwidth, centered around the carrier frequency.

Carrier sense and RSSI In many medium access control protocols, sensing whether the wireless channel, the carrier, is busy (another node is transmitting) is a critical information. The receiver has to be able to provide that information. The precise semantics of this carrier-sense signal depends on the implementation. For example, the IEEE 802.15.4 standard [468] distinguishes the following modes:

- The received energy is above threshold; however, the underlying signal does not need to comply with the modulation and spectral characteristics.
- A carrier has been detected, that is, some signal which complies with the modulation.
- Carrier detected and energy is present.

Also, the signal strength at which an incoming data packet has been received can provide useful information (e.g. a rough estimate about the distance from the transmitter assuming the transmission power is known); a receiver has to provide this information in the Received Signal Strength Indicator (RSSI).

Frequency stability The **frequency stability** denotes the degree of variation from nominal center frequencies when environmental conditions of oscillators like temperature or pressure change. In extreme cases, poor frequency stability can break down communication links, for example, when one node is placed in sunlight whereas its neighbor is currently in the shade.

Voltage range Transceivers should operate reliably over a range of supply voltages. Otherwise, inefficient voltage stabilization circuitry is required.

Transceivers appropriate for WSNs are available from many manufacturers. Usually, there is an entire family of devices to choose from, for example, customized to different regulatory restrictions on carrier frequency in Europe and North America. Currently popular product series include the RFM TR 1001, the Chipcon CC 1000 and CC 2420 (as one of the first IEEE 802.15.4 compliant models), and the Infineon TDA525x family, to name but a few. They are described in a bit more detail at the end of this section.

An important peculiarity and a key difference compared to other communication devices is the fact that these simple transceivers often lack a unique identifier: each Ethernet device, for example, has a MAC-level address that uniquely identifies this individual device. For simple transceivers, the additional cost of providing such an identifier is relatively high with respect to the device's total costs, and thus, unique identifiers cannot be relied upon to be present in all devices. The availability of such device identifiers is very useful in many communication protocols and their absence will have considerable consequences for protocol design.

Improving these commercial designs to provide better performance at lower energy consumption and reduced cost is an ongoing effort by a large research community, facing challenges such as low transistor transconductance or limitations of integrated passive RF components. As these hardware-related questions are not the main focus of this book, the reader is referred to other material [26, 134, 647].

Transceiver structure

A fairly common structure of transceivers is into the Radio Frequency (RF) front end and the baseband part:

- the **radio frequency front end** performs analog signal processing in the actual radio frequency band, whereas
- the **baseband processor** performs all signal processing in the digital domain and communicates with a sensor node's processor or other digital circuitry.

Between these two parts, a frequency conversion takes place, either directly or via one or several Intermediate Frequencys (IFs). The boundary between the analog and the digital domain is constituted by Digital/Analog Converters (DACs) and Analog/Digital Converters (ADCs).

A detailed discussion of the low-power design of RF front end and baseband circuitry is well beyond the scope of this book; one place to start with is reference [3].

The **RF front end** performs analog signal processing in the actual radio frequency band, for example in the 2.4 GHz Industrial, Scientific, and Medical (ISM) band; it is the first stage of the interface between the electromagnetic waves and the digital signal processing of the further transceiver stages [46, 470]. Some important elements of an RF front ends architecture are sketched in Figure 2.2:

- The Power Amplifier (PA) accepts upconverted signals from the IF or baseband part and amplifies them for transmission over the antenna.
- The Low Noise Amplifier (LNA) amplifies incoming signals up to levels suitable for further processing without significantly reducing the SNR [470]. The range of powers of the incoming signals varies from very weak signals from nodes close to the reception boundary to strong signals from nearby nodes; this range can be up to 100 dB. Without management actions, the LNA is active all the time and can consume a significant fraction of the transceiver's energy.
- Elements like local oscillators or voltage-controlled oscillators and mixers are used for frequency conversion from the RF spectrum to intermediate frequencies or to the baseband. The incoming signal at RF frequencies f_{RF} is multiplied in a mixer with a fixed-frequency signal from the local oscillator (frequency f_{LO}). The resulting intermediate-frequency signal has frequency $f_{LO} - f_{RF}$. Depending on the RF front end architecture, other elements like filters are also present.

The efficiency of RF front ends in wireless sensor networks is discussed in Section 4.3.

Transceiver operational states

Many transceivers can distinguish four operational states [670]:

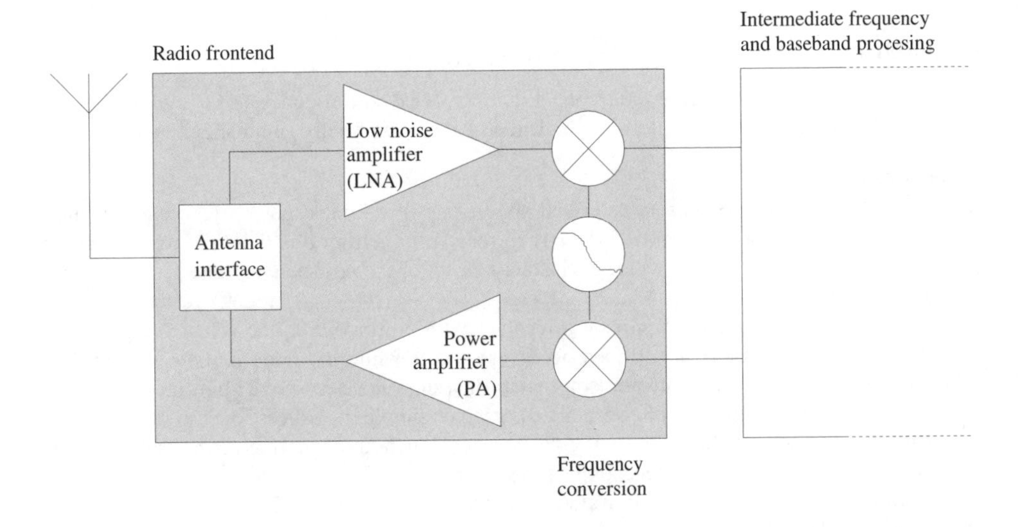

Figure 2.2 RF front end [46]

Transmit In the **transmit state**, the transmit part of the transceiver is active and the antenna radiates energy.

Receive In the **receive state** the receive part is active.

Idle A transceiver that is ready to receive but is not currently receiving anything is said to be in an **idle state**. In this idle state, many parts of the receive circuitry are active, and others can be switched off. For example, in the synchronization circuitry, some elements concerned with acquisition are active, while those concerned with tracking can be switched off and activated only when the acquisition has found something. MYERS et al. [580] also discuss techniques for switching off parts of the acquisition circuitry for IEEE 802.11 transceivers. A major source of power dissipation is **leakage**.

Sleep In the **sleep state**, significant parts of the transceiver are switched off. There are transceivers offering several different sleep states, see reference [580] for a discussion of sleep states for IEEE 802.11 transceivers. These sleep states differ in the amount of circuitry switched off and in the associated **recovery times** and **startup energy** [855]. For example, in a complete power down of the transceiver, the startup costs include a complete initialization as well as configuration of the radio, whereas in "lighter" sleep modes, the clock driving certain transceiver parts is throttled down while configuration and operational state is remembered.

The sensor node's protocol stack and operating software must decide into which state the transceiver is switched, according to the current and anticipated communications needs. One problem complicating this decision is that the operation of state changes also dissipate power [670]. For example, a transceiver waking up from the sleep mode to the transmit mode requires some startup time and startup energy, for example, to ramp up phase-locked loops or voltage-controlled oscillators. During this startup time, no transmission or reception of data is possible [762]. The problem of scheduling the node states (equivalently: switching on and off node/transceiver components) so as to minimize average power consumption (also called **power management**) is rather complex, an in-depth treatment can be found in reference [85], and a further reference is [741].

Advanced radio concepts

Apart from these basic transceiver concepts, a number of advanced concepts for radio communication are the objectives of current research. Three of them are briefly summarized here.

Wakeup radio

Looking at the transceiver concepts described above, one of the most power-intensive operations is waiting for a transmission to come in, ready to receive it. During this time, the receiver circuit must be powered up so that the wireless channel can be observed, spending energy without any immediate benefit.

While it seems unavoidable to provide a receiver with power during the actual reception of a packet, it would be desirable not to have to invest power while the node is only waiting for a packet to come in. A receiver structure is necessary that does not need power but can detect when a packet starts to arrive. To keep this specialized receiver simple, it suffices for it to raise an event to notify other components of an incoming packet; upon such an event, the main receiver can be turned on and perform the actual reception of the packet.

Such receiver concepts are called **wakeup receivers** [312, 667, 752, 931, 931]: Their only purpose is to wake up the main receiver without needing (a significant amount of) power to do so – ZHONG et al. [931] state a target power consumption of less than 1 μW. In the simplest case, this wakeup would happen for every packet; a more sophisticated version would be able to decide, using proper address information at the start of the packet, whether the incoming packet is actually destined for this node and only then wake up the main receiver.

Such wakeup receivers are tremendously attractive as they would do away with one of the main problems of WSNs: the need to be permanently able to receive in a network with low average traffic. It would considerably simplify a lot of the design problems of WSNs, in particular of the medium access control – Section 5.2.4 will discuss these aspects and some ensuing problems in more detail. Unfortunately, so far the realization of a reliable, well-performing wakeup receiver has not been achieved yet.

Spread-spectrum transceivers

Simple transceiver concepts, based on modulations like Amplitude Shift Keying (ASK) or Frequency Shift Keying (FSK), can suffer from limited performance, especially in scenarios with a lot of interference. To overcome this limitation, the use of spread-spectrum transceivers has been proposed by some researchers [155, 281]. These transceivers, however, suffer mostly from complex hardware and consequently higher prices, which has prevented them from becoming a mainstream concept for WSNs so far. Section 4.2.5 presents details.

Ultrawideband communication

UltraWideBand (UWB) communication is a fairly radical change from conventional wireless communication as outlined above. Instead of modulating a digital signal onto a carrier frequency, a very large bandwidth is used to directly transmit the digital sequence as very short impulses (to form nearly rectangular impulses requires considerable bandwidth, because of which this concept is not used traditionally) [44, 646, 866, 885].[1] Accordingly, these impulses occupy a large spectrum starting from a few Hertz up to the range of several GHz. The challenge is to synchronize sender and receiver sufficiently (to an accuracy of trillionth of seconds) so that the impulses can be correctly detected. A side effect of precisely timed impulses is that UWB is fairly resistant to multipath fading [181, 472], which can be a serious obstacle for carrier-based radio communication.

Using such a large bandwidth, an ultrawideband communication will overlap with the spectrum of a conventional radio system. But, because of the large spreading of the signal, a very small transmission power suffices. This power can be small enough so that it vanishes in the noise floor from the perspective of a traditional radio system.

As one concrete example, consider a time-hopping Pulse Position Modulation (PPM) proposed as combined modulation and multiple access scheme by Win and Scholtz [885]. For each symbol, a number of pulses are transmitted with almost periodic spacing. The deviations from the periodicity encode both the modulation as well as the transmitting user.

For a communication system, the effect is that a very high data rate can be realized over short distances; what is more, UWB communication can relatively easily penetrate obstacles such as doors, which are impermeable to narrowband radio waves. For a WSN, the high data rate is not strictly necessary but can be leveraged to reduce the on-time of the transceivers. The nature of UWB also allows to precisely measure distances (with claimed precision of centimeters).

These desirable features of UWB communication have to be balanced against the difficulties of building such transceivers at low-cost and low-power consumption. More precisely, an UWB transmitter is actually relatively simple since it does not need oscillators or related circuitry found in transmitters for a carrier-frequency-based transmitter. The receivers, on the other hand, require complex timing synchronization. As of this writing, UWB transceivers have not yet been used in prototypes for wireless sensor nodes.

[1] A more precise definition of an ultrawideband system is that it uses at least 500 MHz or a fractional spectrum of at least 20 % of the carrier frequency. This definition would theoretically encompass also spread-spectrum systems with high bandwidth; however, most people have the usage of short pulses in mind when speaking about UWB.

One of the best sources of information about UWB in WSN might be the documents of the IEEE 802.15.4a study group, which looks at UWB as an alternative physical layer for the IEEE 802.15.4 standard for short-range, low bitrate wireless communication. Some references to start from are [82, 187, 566, 603, 884]. A comparison between UWB and Direct Sequence Spread Spectrum (DSSS) technologies for sensor networks has been made in [939], under the assumption of an equal bandwidth for both types of systems.

Nonradio frequency wireless communication

While most of the wireless sensor network work has focused on the use of radio waves as communication media, other options exists. In particular, optical communication and ultrasound communication have been considered as alternatives.

Optical

Kahn et al. [392] and others have considered the use of optical links between sensor nodes. Its main advantage is the very small energy per bit required for both generating and detecting optical light – simple Light-Emitting Diodes (LEDs) are good examples for high-efficiency senders. The required circuitry for an optical transceiver is also simpler and the device as a whole can be smaller than the radio frequency counterpart. Also, communication can take place concurrently with only negligible interference. The evident disadvantage, however, is that communicating peers need to have a line of sight connection and that optical communication is more strongly influenced by weather conditions.

As a case in point, consider the so-called "corner-cube reflector": three mirrors placed at right angles to each other in a way that each beam of light directed at it is reflected back to its source (as long as it comes from a cone centered around the main diagonal of the cube) – an example for such a structure is shown in Figure 2.3. This reflection property holds only as long as the mirrors are exactly at right angles. When one the mirrors is slightly moved, a signal can be modulated onto an incoming ray of light, effectively transmitting information back to the sender. In fact, data rates up to 1 kb/s have been demonstrated using such a device. Its main advantage is that the mechanical movement of one such mirror only takes very little energy, compared to actually generating a beam of light or even a radio wave. Hence, a passive readout of sensor nodes can be done very energy

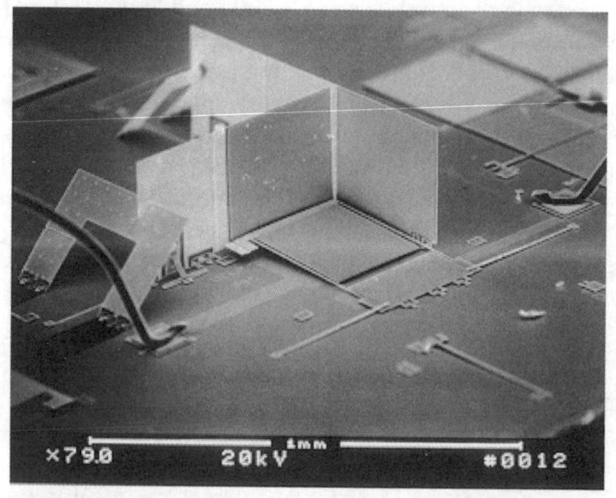

Figure 2.3 Example of a corner-cube reflector for optical communication [168]. Reproduced by permission of IEEE

efficiently over long distances as long as the reader has enough power to produce the laser beam (up to 150 m have been demonstrated using a 5 mW laser).

Ultrasound

Both radio frequency and optical communication are suitable for open-air environments. In some application scenarios, however, sensor nodes are used in environments where radio or optical communication is not applicable because these waves do not penetrate the surrounding medium. One such medium is water, and an application scenario is the surveillance of marine ground floor erosion to help in the construction of offshore wind farms. Sensors are deployed on the marine ground floor and have to communicate amongst themselves. In such an underwater environment, ultrasound is an attractive communication medium as it travels relatively long distances at comparably low power.

A further aspect of ultrasound is its use in location systems as a secondary means of communication with a different propagation speed. Details will be discussed in Chapter 9.

Some examples of radio transceivers

To complete this discussion of possible communication devices, a few examples of standard radio transceivers that are commonly used in various WSN prototype nodes should be briefly described. All these transceivers are in fact commodity, off-the-shelf items available via usual distributors. They are all single-chip solutions, integrating transmitter and receiver functionality, requiring only a small number of external parts and have a fairly low-power consumption. In principle, similar equipment is available from a number of manufacturers – as can be expected, there is not one "best product" available, but each of them has particular advantages and disadvantages.

RFM TR1000 family

The TR1000 family of radio transceivers from RF Monolithics[2] is available for the 916 MHz and 868 MHz frequency range. It works in a 400 kHz wide band centered at, for example, 916.50 MHz. It is intended for short-range radio communication with up to 115.2 kbps. The modulation is either on-off-keying (at a maximum rate of 30 kbps) or ASK; it also provides a dynamically tunable output power. The maximum radiated power is given in the data sheet [690] as 1.5 dBm, $\approx$ 1.4 mW, whereas in the Mica motes a number of 0.75 mW is given [351]. The transceiver offers received signal strength information. It is attractive because of its low-power consumption in both send and receive modes and especially in sleep mode. Details about parameters and configurations can be found in the data sheet [690].

Hardware accelerators (Mica motes)

The Mica motes use the RFM TR1000 transceiver and contain also a set of **hardware accelerators**. On the one hand, the transceiver offers a very low-level interface, giving the microcontroller tight control over frame formats, MAC protocols, and so forth. On the other hand, framing and MAC can be very computation intensive, for example, for computing checksums, for making bytes out of serially received bits or for detecting Start Frame Delimiters (SFDs) in a stream of symbols. The hardware accelerators offer some of these primitive computations in hardware, right at the disposal of the microcontroller.

Chipcon CC1000 and CC2420 family

Chipcon[3] offers a wide range of transceivers that are appealing for use in WSN hardware. To name but two examples: The CC1000 operates in a wider frequency range, between 300 and 1000 MHz,

[2] `http://www.rfm.com`
[3] `http://www.chipcon.com`

programmable in steps of 250 Hz. It uses FSK as modulation, provides RSSI, and has programmable output power. An interesting feature is the possibility to compensate for crystal temperature drift. It should also be possible to use it in frequency hopping protocols. Details can be found in the data sheet[157].

The CC2420 [158] is a more complicated device. It implements the physical layer as prescribed by the IEEE 802.15.4 standard with the required support for this standard's MAC protocol. In fact, the company claims that this is the first commercially available single-chip transceiver for IEEE 802.15.4. As a consequence of implementing this standard, the transceiver operates in the 2.4 GHz band and features the required DSSS modem, resulting in a data rate of 250 kbps. It achieves this at still relatively low-power consumption, although not quite on par with the simpler transceivers described so far.

Infineon TDA 525x family

The Infineon TDA 525x family provides flexible, single-chip, energy-efficient transceivers. The TDA 5250 [375], as an example, is a 868–870 MHztransceiver providing both ASK and FSK modulation, it has a highly efficient power amplifier, RSSI information, a tunable crystal oscillator, an onboard data filter, and an intelligent power-down feature. One of the interesting features is a self-polling mechanism, which can very quickly determine data rate. Compared to some other transceiver, it also has an excellent blocking performance that makes it quite resistant to interference.

IEEE 802.15.4/Ember EM2420 RF transceiver

The IEEE 802.15.4 low-rate Wireless Personal Area Network (WPAN) [468] works in three different frequency bands and employs a DSSS scheme. Some basic data can be found in Table 2.1. For one particular RF front-end design, the Ember[4] EM2420 RF Transceiver [240], some numbers on power dissipation are available. For a radiated power of -0.5 dBm (corresponding to ≈ 0.9 mW) and with a supply voltage of 3.3 V, the transmit mode draws a current of 22.7 mA, corresponding to ≈ 74.9 mW, whereas in the receive mode, 25.2 mA current are drawn, corresponding to ≈ 83.2 mW. In the sleep mode, only 12 μA are drawn.

In all bands, DSSS is used. In the 868 MHz band, only a single channel with a data rate of 20 kbps is available, in the 915 MHz band ten channels of 40kbps each and in the 2.4 GHz band 16 channels of 250 kbps are available. In the lower two bands, the chips are Binary Phase Shift Keying (BPSK)-modulated, and the data symbols are encoded differentially. A pseudonoise sequence of 15 chips is used for every bit. The modulation scheme in the 2.4 GHz band is a little

Table 2.1 The different PHY's of the IEEE 802.15.4 standard [468]. Reproduced by permission of IEEE

Band	868 MHz	915 MHz	2.4 GHz
Frequency [MHz]	868–868.6	902–928	2400–2483.5
Chip rate [kchips/s]	300	600	2000
# of channels	1	10	16
Modulation	BPSK	BPSK	O-QPSK
Data rate [kb/s]	20	40	250
Symbol rate [ksymbols/s]	20	40	62.5
Symbol type	binary	binary	16-ary orthogonal

[4] http://www.ember.com

bit more complicated. As can be observed from the table, a channel symbol consists of four user bits. These 16 different symbol values are distinguished by using 16 different nearly orthogonal pseudorandom chip sequences. The resulting chip sequence is then modulated using a modulation scheme called *offset*-Quaternary Phase Shift Keying (QPSK). Some of the design rationale for this modulation scheme is also given in reference [115, Chap. 3].

National Semiconductor LMX3162

The radio hardware of the μAMPS-1 node [563, 762, 872] consists of a digital baseband processor implemented on an FPGA, whereas for the RF front end, a (now obsolete) National Semiconductor LMX3162 transceiver [588] is used. The LMX3162 operates in the 2.4 GHz band and offers six different radiated power levels from 0 dBm up to 20 dBm. To transmit data, the baseband processor can control an externally controllable Voltage-Controlled Oscillator (VCO). The main components of the RF front end (phase-lock loop, transmit and receive circuitry) can be shut off. The baseband processor controls the VCO and also provides timing information to a TDMA-based MAC protocol (see Chapter 5). For data transmission, FSK with a data rate of 1 Mbps is used.

Conexant RDSSS9M

The WINS sensor node of Rockwell[5] carries a Conexant RDSSS9M transceiver, consisting of the RF part working in the ISM band between 902 and 928 MHzand a microcontroller (a 65C02) responsible for processing DSSS signals with a spreading factor of 12 bits per chip. The data rate is 100 kbps. The RF front end offers radiated power levels of 1 mW, 10 mW and 100 mW. A number of 40 sub-bands are available, which can be freely selected. The microcontroller implements portions of a MAC protocol also.

2.1.5 Sensors and actuators

Without the actual sensors and actuators, a wireless sensor network would be beside the point entirely. But as the discussion of possible application areas has already indicated, the possible range of sensors is vast. It is only possible to give a rough idea on which sensors and actuators can be used in a WSN.

Sensors

Sensors can be roughly categorized into three categories (following reference [670]):

Passive, omnidirectional sensors These sensors can measure a physical quantity at the point of the sensor node without actually manipulating the environment by active probing – in this sense, they are passive. Moreover, some of these sensors actually are self-powered in the sense that they obtain the energy they need from the environment – energy is only needed to amplify their analog signal. There is no notion of "direction" involved in these measurements. Typical examples for such sensors include thermometer, light sensors, vibration, microphones, humidity, mechanical stress or tension in materials, chemical sensors sensitive for given substances, smoke detectors, air pressure, and so on.

Passive, narrow-beam sensors These sensors are passive as well, but have a well-defined notion of direction of measurement. A typical example is a camera, which can "take measurements" in a given direction, but has to be rotated if need be.

Active sensors This last group of sensors actively probes the environment, for example, a sonar or radar sensor or some types of seismic sensors, which generate shock waves by small

[5] See `http://wins.rsc.rockwell.com/`.

explosions. These are quite specific – triggering an explosion is certainly not a lightly undertaken action – and require quite special attention.

In practice, sensors from all of these types are available in many different forms with many individual peculiarities. Obvious trade-offs include accuracy, dependability, energy consumption, cost, size, and so on – all this would make a detailed discussion of individual sensors quite ineffective.

Overall, most of the theoretical work on WSNs considers passive, omnidirectional sensors. Narrow-beam-type sensors like cameras are used in some practical testbeds, but there is no real systematic investigation on how to control and schedule the movement of such sensors. Active sensors are not treated in the literature to any noticeable extent.

An assumption occasionally made in the literature [128, 129] is that each sensor node has a certain **area of coverage** for which it can reliably and accurately report the particular quantity that it is observing. More elaborately, a sensor detection model is used, relating the distance between a sensor and the to-be-detected event or object to a detection probability; an example for such a detection model is contained in references [599, 944].

Strictly speaking, this assumption of a coverage area is difficult to justify in its simplest form. Nonetheless, it can be practically useful: It is often possible to postulate, on the basis of application-specific knowledge, some properties of the physical quantity under consideration, in particular, how quickly it can change with respect to distance. For example, temperature or air pressure are unlikely to vary very strongly within a few meters. Hence, allowing for some inevitable inaccuracies in the measurement, the maximum rate of changeover distance can be used to derive such a "coverage radius" within which the values of a single sensor node are considered "good enough". The precise mathematical tools for such a derivation are spatial versions of the sampling theorems.

Actuators

Actuators are just about as diverse as sensors, yet for the purposes of designing a WSN, they are a bit simpler to take account of: In principle, all that a sensor node can do is to open or close a switch or a relay or to set a value in some way. Whether this controls a motor, a light bulb, or some other physical object is not really of concern to the way communication protocols are designed. Hence, in this book, we shall treat actuators fairly summarily without distinguishing between different types.

In a real network, however, care has to be taken to properly account for the idiosyncrasies of different actuators. Also, it is good design practice in most embedded system applications to pair any actuator with a controlling sensor – following the principle to "never trust an actuator" [429].

2.1.6 Power supply of sensor nodes

For untethered wireless sensor nodes, the power supply is a crucial system component. There are essentially two aspects: First, storing energy and providing power in the required form; second, attempting to replenish consumed energy by "scavenging" it from some node-external power source over time.

Storing power is conventionally done using batteries. As a rough orientation, a normal AA battery stores about 2.2–2.5 Ah at 1.5 V. Battery design is a science and industry in itself, and energy scavenging has attracted a lot of attention in research. This section can only provide some small glimpses of this vast field; some papers that deal with these questions (and serve as the basis for this section) are references [134, 392, 667, 670] and, in particular, reference [703].

Storing energy: Batteries

Traditional batteries

The power source of a sensor node is a battery, either nonrechargeable ("primary batteries") or, if an energy scavenging device is present on the node, also rechargeable ("secondary batteries").

Table 2.2 Energy densities for various primary and secondary battery types [703]

Primary batteries			
Chemistry	Zinc-air	Lithium	Alkaline
Energy (J/cm^3)	3780	2880	1200
Secondary batteries			
Chemistry	Lithium	NiMHd	NiCd
Energy (J/cm^3)	1080	860	650

In some form or other, batteries are electro-chemical stores for energy – the chemicals being the main determining factor of battery technology.

Upon these batteries, very tough requirements are imposed:

Capacity They should have high capacity at a small weight, small volume, and low price. The main metric is energy per volume, J/cm^3. Table 2.2 shows some typical values of energy densities, using traditional, macroscale battery technologies. In addition, research on "microscale" batteries, for example, deposited directly onto a chip, is currently ongoing.

Capacity under load They should withstand various usage patterns as a sensor node can consume quite different levels of power over time and actually draw high current in certain operation modes.

Current numbers on power consumption of WSN nodes vary and are treated in detail in Section 2.2, so it is difficult to provide precise guidelines. But for most technologies, the larger the battery, the more power can be delivered instantaneously. In addition, the rated battery capacity specified by a manufacturer is only valid as long as maximum discharge currents are not exceeded, lest capacity drops or even premature battery failure occurs [670].[6]

Self-discharge Their self-discharge should be low; they might also have to last for a long time (using certain technologies, batteries are operational only for a few months, irrespective of whether power is drawn from them or not).

Zinc-air batteries, for example, have only a very short lifetime (on the order of weeks), which offsets their attractively high energy density.

Efficient recharging Recharging should be efficient even at low and intermittently available recharge power; consequently, the battery should also not exhibit any "memory effect".

Some of the energy-scavenging techniques described below are only able to produce current in the μA region (but possibly sustained) at only a few volts at best. Current battery technology would basically not recharge at such values.

Relaxation Their relaxation effect – the seeming self-recharging of an empty or almost empty battery when no current is drawn from it, based on chemical diffusion processes within the cell – should be clearly understood. Battery lifetime and usable capacity is considerably extended if this effect is leveraged. As but one example, it is possible to use multiple batteries in parallel and "schedule" the discharge from one battery to another, depending on relaxation properties and power requirements of the operations to be supported [153].

[6] This effect is due to the need for active material in a battery to be transported to the electrodes. If too much power is drawn, this transport is not fast enough and the battery fails even though energy is still stored in it.

Unconventional energy stores

Apart from traditional batteries, there are also other forms of energy reservoirs that can be contemplated. In a wider sense, fuel cells also qualify as an electro-chemical storage of energy, directly producing electrical energy by oxidizing hydrogen or hydrocarbon fuels. Fuel cells actually have excellent energy densities (e.g. methanol as a fuel stores 17.6 kJ/cm^3), but currently available systems still require a nonnegligible minimum size for pumps, valves, and so on. A slightly more traditional approach to using energy stored in hydrocarbons is to use miniature versions of heat engines, for example, a turbine [243]. Shrinking such heat engines to the desired sizes still requires a considerable research effort in MicroElectroMechanical Systems (MEMSs); predictions regarding power vary between 0.1–10 W at sizes of about 1 cc [703]. And lastly, even radioactive substances have been proposed as an energy store [463]. Another option are so-called "gold caps", high-quality and high-capacity capacitors, which can store relatively large amounts of energy, can be easily and quickly recharged, and do not wear out over time.

DC–DC Conversion

Unfortunately, batteries (or other forms of energy storage) alone are not sufficient as a direct power source for a sensor node. One typical problem is the reduction of a battery's voltage as its capacity drops. Consequently, less power is delivered to the sensor node's circuits, with immediate consequences for oscillator frequencies and transmission power – a node on a weak battery will have a smaller transmission range than one with a full battery, possibly throwing off any calibrations done for the range at full battery ranges.

A DC – DC converter can be used to overcome this problem by regulating the voltage delivered to the node's circuitry. To ensure a constant voltage even though the battery's supply voltage drops, the DC – DC converter has to draw increasingly higher current from the battery when the battery is already becoming weak, speeding up battery death (see Figure 3 in reference [670]). Also, the DC – DC converter does consume energy for its own operation, reducing overall efficiency. But the advantages of predictable operation during the entire life cycle can outweigh these disadvantages.

Energy scavenging

Some of the unconventional energy stores described above – fuel cells, micro heat engines, radioactivity – convert energy from some stored, secondary form into electricity in a less direct and easy to use way than a normal battery would do. The entire energy supply is stored on the node itself – once the fuel supply is exhausted, the node fails.

To ensure truly long-lasting nodes and wireless sensor networks, such a limited energy store is unacceptable. Rather, energy from a node's environment must be tapped into and made available to the node – **energy scavenging** should take place. Several approaches exist [667, 701, 703]:

Photovoltaics The well-known solar cells can be used to power sensor nodes. The available power depends on whether nodes are used outdoors or indoors, and on time of day and whether for outdoor usage. Different technologies are best suited for either outdoor or indoor usage. The resulting power is somewhere between 10 $\mu W/cm^2$ indoors and 15 mW/cm^2 outdoors. Single cells achieve a fairly stable output voltage of about 0.6 V (and have therefore to be used in series) as long as the drawn current does not exceed a critical threshold, which depends, among other factors, on the light intensity. Hence, solar cells are usually used to recharge secondary batteries. Best trade-offs between complexity of recharging circuitry, solar cell efficiency, and battery lifetime are still open questions.

Temperature gradients Differences in temperature can be directly converted to electrical energy. Theoretically, even small difference of, for example, 5 K can produce considerable power, but practical devices fall very short of theoretical upper limits (given by the Carnot efficiency).

Seebeck effect-based thermoelectric generators are commonly considered; one example is a generator, which will be commercially available soon, that achieves about 80 $\mu W/cm^2$ at about 1 V from a 5 Kelvin temperature difference.[7]

Vibrations One almost pervasive form of mechanical energy is vibrations: walls or windows in buildings are resonating with cars or trucks passing in the streets, machinery often has low-frequency vibrations, ventilations also cause it, and so on. The available energy depends on both amplitude and frequency of the vibration and ranges from about 0.1 $\mu W/cm^3$ up to $10,000$ $\mu W/cm^3$ for some extreme cases (typical upper limits are lower).

Converting vibrations to electrical energy can be undertaken by various means, based on electromagnetic, electrostatic, or piezoelectric principles. Figure 2.4 shows, as an example, a generator based on a variable capacitor [549]. Practical devices of 1 cm^3 can produce about 200 $\mu W/cm^3$ from 2.25 m/s^2, 120 Hz vibration sources, actually sufficient to power simple wireless transmitters [702].

Pressure variations Somewhat akin to vibrations, a variation of pressure can also be used as a power source. Such piezoelectric generators are in fact used already. One well-known example is the inclusion of a piezoelectric generator in the heel of a shoe, to generate power as a human walks about [759]. This device can produce, on average, 330 $\mu W/cm^2$. It is, however, not clear how such technologies can be applied to WSNs.

Flow of air/liquid Another often-used power source is the flow of air or liquid in wind mills or turbines. The challenge here is again the miniaturization, but some of the work on millimeter-scale MEMS gas turbines might be reusable [243]. However, this has so far not produced any notable results.

To summarize, Table 2.3 gives an overview of typical values of power and energy densities for different energy sources. The values in this table vary somewhat from those presented above as partially different technologies or environments were assumed; all these numbers can only serve as a general orientation but should always be taken with a grain of salt.

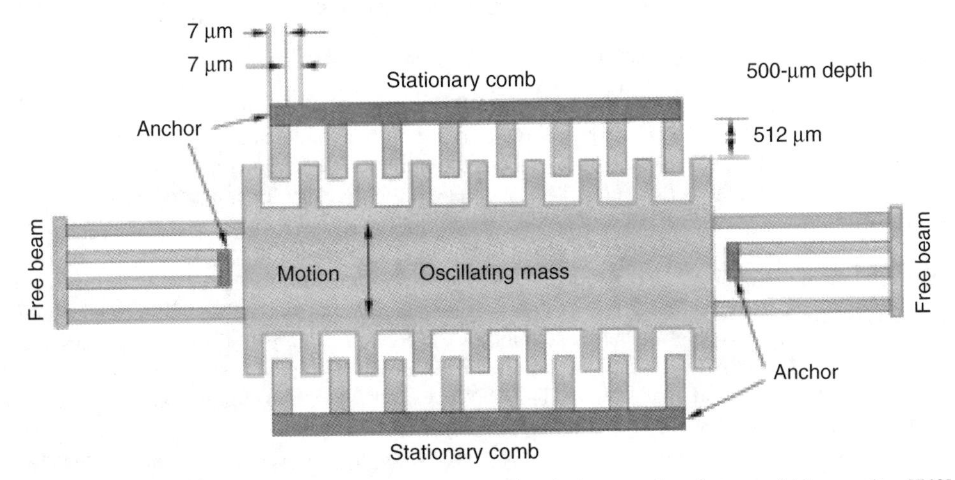

Figure 2.4 A MEMS device for converting vibrations to electrical energy, based on a variable capacitor [549]. Reproduced by permission of IEEE

[7] Compare `http://www.adsx.com`.

Table 2.3 Comparison of energy sources [667]

Energy source	Energy density
Batteries (zinc-air)	1050–1560 mWh/cm^3
Batteries (rechargeable lithium)	300 mWh/cm^3 (at 3–4 V)
Energy source	**Power density**
Solar (outdoors)	15 mW/cm^2 (direct sun)
	0.15 mW/cm^2 (cloudy day)
Solar (indoors)	0.006 mW/cm^2 (standard office desk)
	0.57 mW/cm^2 (<60 W desk lamp)
Vibrations	0.01–0.1 mW/cm^3
Acoustic noise	$3 \cdot 10^{-6}$ mW/cm^2 at 75 dB
	$9,6 \cdot 10^{-4}$ mW/cm^2 at 100 dB
Passive human-powered systems	1.8 mW (shoe inserts)
Nuclear reaction	80 mW/cm^3, 10^6 mWh/cm^3

As these examples show, energy scavenging usually has to be combined with secondary batteries as the actual power sources are not able to provide power consistently, uninterruptedly, at a required level; rather, they tend to fluctuate over time. This requires additional circuitry for recharging of batteries, possibly converting to higher power levels, and a battery technology that can be recharged at low currents. An alternative approach is to align the task execution pattern of the sensor network (which sensor is active when) with the characteristics of energy scavenging – Kansal and Srivastava [399] introduce this idea and describe some protocols and algorithms; they show that the network lifetime is extended by up to 200 % if these scavenging characteristics are taken into account in the task allocation.

2.2 Energy consumption of sensor nodes

2.2.1 Operation states with different power consumption

As the previous section has shown, energy supply for a sensor node is at a premium: batteries have small capacity, and recharging by energy scavenging is complicated and volatile. Hence, the energy consumption of a sensor node must be tightly controlled. The main consumers of energy are the controller, the radio front ends, to some degree the memory, and, depending on the type, the sensors.

To give an example, consider the energy consumed by a microcontroller per instruction. A typical ball park number is about 1 nJ per instruction [391]. To put this into perspective with the battery capacity numbers from Section 2.1.6, assume a battery volume of one cubic millimeter, which is about the maximum possible for the most ambitious visions of "smart dust". Such a battery could store about 1 J. To use such a battery to power a node even only a single day, the node must not consume continuously more than $1/(24 \cdot 60 \cdot 60)$ Ws/s ≈ 11.5 μW. No current controller, let alone an entire node, is able to work at such low-power levels.

One important contribution to reduce power consumption of these components comes from chip-level and lower technologies: Designing low-power chips is the best starting point for an energy-efficient sensor node. But this is only one half of the picture, as any advantages gained by such designs can easily be squandered when the components are improperly operated.

The crucial observation for proper operation is that most of the time a wireless sensor node has nothing to do. Hence, it is best to turn it off. Naturally, it should be able to wake up again, on the basis of external stimuli or on the basis of time. Therefore, completely turning off a node is not possible, but rather, its operational state can be adapted to the tasks at hand. Introducing and using multiple states of operation with reduced energy consumption in return for reduced functionality is the core technique for energy-efficient wireless sensor node. In fact, this approach is well known even from standard personal computer hardware, where, for example, the Advanced Configuration and Power Interface (ACPI) [8] introduces one state representing the fully operational machine and four sleep states of graded functionality/power consumption/wakeup time (time necessary to return to fully operational state). The term Dynamic Power Management (DPM) summarizes this field of work (see e.g. reference [63] for a slightly older, but quite a broad-range overview).

These modes can be introduced for all components of a sensor node, in particular, for controller, radio front end, memory, and sensors. Different models usually support different numbers of such sleep states with different characteristics; some examples are provided in the following sections. For a controller, typical states are "active", "idle", and "sleep"; a radio modem could turn transmitter, receiver, or both on or off; sensors and memory could also be turned on or off. The usual terminology is to speak of a "deeper" sleep state if less power is consumed.

While such a graded sleep state model is straightforward enough, it is complicated by the fact that transitions between states take both time and energy. The usual assumption is that the deeper the sleep state, the more time and energy it takes to wake up again to fully operational state (or to another, less deep sleep state). Hence, it may be worthwhile to remain in an idle state instead of going to deeper sleep states even from an energy consumption point of view.

Figure 2.5 illustrates this notion based on a commonly used model (used in, e.g. references [558, 769]). At time t_1, the decision whether or not a component (say, the microcontroller) is to be put into sleep mode should be taken to reduce power consumption from P_{active} to P_{sleep}. If it remains active and the next event occurs at time t_{event}, then a total energy of $E_{\text{active}} = P_{\text{active}}(t_{\text{event}} - t_1)$ has be spent uselessly idling. Putting the component into sleep mode, on the other hand, requires a time τ_{down} until sleep mode has been reached; as a simplification, assume that the average power consumption during this phase is $(P_{\text{active}} + P_{\text{sleep}})/2$. Then, P_{sleep} is consumed until t_{event}. In total, $\tau_{\text{down}}(P_{\text{active}} + P_{\text{sleep}})/2 + (t_{\text{event}} - t_1 - \tau_{\text{down}})P_{\text{sleep}}$ energy is required in sleep mode as opposed to $(t_{\text{event}} - t_1)P_{\text{active}}$ when remaining active. The energy saving is thus

$$\begin{aligned} E_{\text{saved}} = &(t_{\text{event}} - t_1)P_{\text{active}} - (\tau_{\text{down}}(P_{\text{active}} + P_{\text{sleep}})/2 + \\ &(t_{\text{event}} - t_1 - \tau_{\text{down}})P_{\text{sleep}}). \end{aligned} \tag{2.1}$$

Once the event to be processed occurs, however, an additional overhead of

$$E_{\text{overhead}} = \tau_{\text{up}}(P_{\text{active}} + P_{\text{sleep}})/2, \tag{2.2}$$

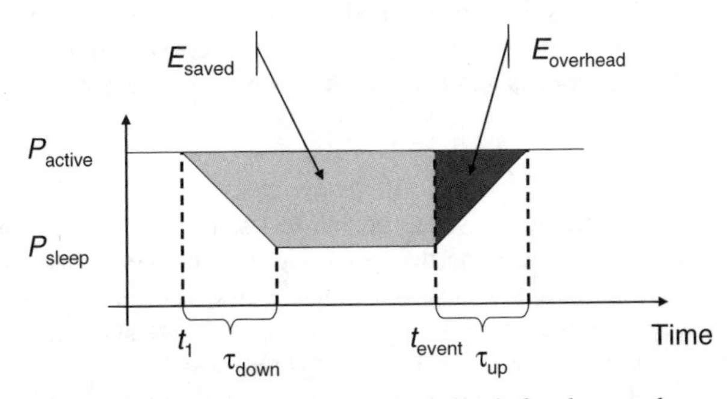

Figure 2.5 Energy savings and overheads for sleep modes

is incurred to come back to operational state before the event can be processed, again making a simplifying assumption about average power consumption during makeup. This energy is indeed an overhead since no useful activity can be undertaken during this time. Clearly, switching to a sleep mode is only beneficial if $E_{\text{overhead}} < E_{\text{saved}}$ or, equivalently, if the time to the next event is sufficiently large:

$$(t_{\text{event}} - t_1) > \frac{1}{2}\left(\tau_{\text{down}} + \frac{P_{\text{active}} + P_{\text{sleep}}}{P_{\text{active}} - P_{\text{sleep}}}\tau_{\text{up}}\right). \tag{2.3}$$

Careful scheduling of such transitions has been considered from several perspectives – reference [769], for example, gives a fairly abstract treatment – and in fact, a lot of medium access control research in wireless sensor networks can be regarded as the problem of when to turn off the receiver of a node.

2.2.2 Microcontroller energy consumption

Basic power consumption in discrete operation states

Embedded controllers commonly implement the concept of multiple operational states as outlined above; it is also fairly easy to control. Some examples probably best explain the idea.

Intel StrongARM

The Intel StrongARM [379] provides three sleep modes:

- In *normal mode*, all parts of the processor are fully powered. Power consumption is up to 400 mW.
- In *idle mode*, clocks to the CPU are stopped; clocks that pertain to peripherals are active. Any interrupt will cause return to normal mode. Power consumption is up to 100 mW.
- In *sleep mode*, only the real-time clock remains active. Wakeup occurs after a timer interrupt and takes up to 160 ms. Power consumption is up to 50 μW.

Texas Instruments MSP 430

The MSP430 family [814] features a wider range of operation modes: One fully operational mode, which consumes about 1.2 mW (all power values given at 1 MHz and 3 V). There are four sleep modes in total. The deepest sleep mode, LPM4, only consumes 0.3 μW, but the controller is only woken up by external interrupts in this mode. In the next higher mode, LPM3, a clock is also still running, which can be used for scheduled wake ups, and still consumes only about 6 μW.

Atmel ATmega

The Atmel ATmega 128L [28] has six different modes of power consumption, which are in principle similar to the MSP 430 but differ in some details. Its power consumption varies between 6 mW and 15 mW in idle and active modes and is about 75 μW in power-down modes.

Dynamic voltage scaling

A more sophisticated possibility than discrete operational states is to use a continuous notion of functionality/power adaptation by adapting the speed with which a controller operates. The idea is to choose the best possible speed with which to compute a task that has to be completed by a given deadline. One obvious solution is to switch the controller in full operation mode, compute the task at highest speed, and go back to a sleep mode as quickly as possible.

The alternative approach is to compute the task only at the speed that is required to finish it before the deadline. The rationale is the fact that a controller running at lower speed, that is, lower clock rates, consumes less power than at full speed. This is due to the fact that the supply voltage can be reduced at lower clock rates while still guaranteeing correct operation. This technique is called Dynamic Voltage Scaling (DVS) [133].

This technique is actually beneficial for CMOS chips: As the actual power consumption P depends quadratically on the supply voltage V_{DD} [649], reducing the voltage is a very efficient way to reduce power consumption. Power consumption also depends on the frequency f, hence $P \propto f \cdot V_{\mathrm{DD}}^2$.

Consequently, dynamic voltage scaling also reduces energy consumption. The Transmeta Crusoe processor, for example, can be scaled from 700 MHz at 1.65 V down to 200 MHz at 1.1 V [649]. This reduces the power consumption by a factor of $\frac{700 \cdot 1.65^2}{200 \cdot 1.1^2} = 7.875$, but the speed is only reduced by a factor of $700/200 = 3.5$. Hence, the energy required per instruction is reduced by $3.5/7.875 \approx 44\,\%$. Other processors and microcontrollers behave similarly, Figure 2.6 shows an example for the StrongARM SA-1100 [558]. The ultimate reason for this improvement is the convex shape of the function power against speed, caused by varying the supply voltage.

When applying dynamic voltage scaling, care has to be taken to operate the controller within its specifications. There are minimum and maximum clock rates for each device, and for each clock rate, there is a minimum and maximum threshold that must be obeyed. Hence, when there is nothing to process, going into sleep modes is still the only option. Also, using arbitrary voltages requires a quite efficient DC-DC converter to be used [134].

How to control DVS from an application or from the operating system is discussed in Section 2.3.4 on page 48.

2.2.3 Memory

From an energy perspective, the most relevant kinds of memory are on-chip memory of a microcontroller and FLASH memory – off-chip RAM is rarely if ever used. In fact, the power needed to drive on-chip memory is usually included in the power consumption numbers given for the controllers.

Hence, the most relevant part is FLASH memory – in fact, the construction and usage of FLASH memory can heavily influence node lifetime. The relevant metrics are the read and write times and

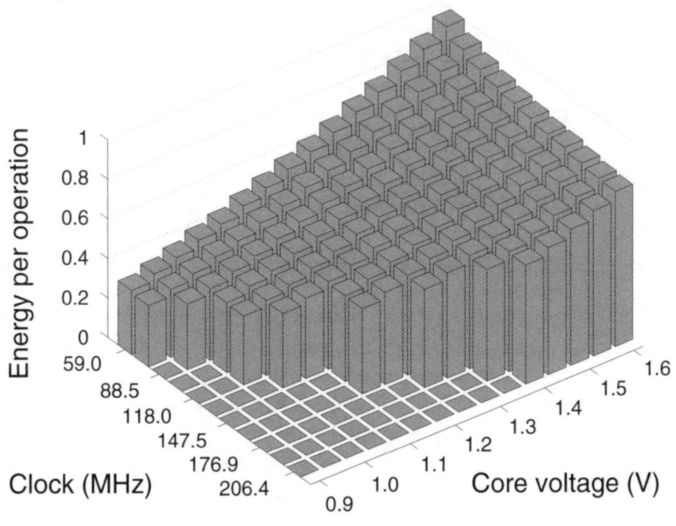

Figure 2.6 Energy per operation with dynamic power scaling on an Intel StrongARM SA-1100 [558]. Reproduced by permission of IEEE

energy consumption. All this information is readily available from manufacturers' data sheets and do vary depending on several factors. Read times and read energy consumption tend to be quite similar between different types of FLASH memory [329]. Writing is somewhat more complicated, as it depends on the granularity with which data can be accessed (individual bytes or only complete pages of various sizes). One means for comparability is to look at the numbers for overwriting the whole chip. Considerable differences in erase and write energy consumption exist, up to ratios of 900:1 between different types of memory [329].

To give a concrete example, consider the energy consumption necessary for reading and writing to the Flash memory used on the Mica nodes [534]. Reading data takes 1.111 nAh, writing requires 83.333 nAh.

Hence, writing to FLASH memory can be a time- and energy-consuming task that is best avoided if somehow possible. For detailed numbers, it is necessary to consult the documentation of the particular wireless sensor node and its FLASH memory under consideration.

2.2.4 Radio transceivers

A radio transceiver has essentially two tasks: transmitting and receiving data between a pair of nodes. Similar to microcontrollers, radio transceivers can operate in different modes, the simplest ones are being turned on or turned off. To accommodate the necessary low total energy consumption, the transceivers should be turned off most of the time and only be activated when necessary – they work at a low **duty cycle**. But this incurs additional complexity, time and power overhead that has to be taken into account.

To understand the energy consumption behavior of radio transceivers and their impact on the protocol design, models for the energy consumption per bit for both sending and receiving are required. Several such models of different accuracy and level of detail exist and are mostly textbook knowledge [14, 661, 682, 938] (for a research paper example see reference [762]); the presentation here mostly follows reference [559], in particular, with respect to concrete numbers.

Modeling energy consumption during transmission

In principle, the energy consumed by a transmitter is due to two sources [670]: one part is due to RF signal generation, which mostly depends on chosen modulation and target distance and hence on the transmission power P_{tx}, that is, the power radiated by the antenna. A second part is due to electronic components necessary for frequency synthesis, frequency conversion, filters, and so on. These costs are basically constant.

One of the most crucial decisions when transmitting a packet is thus the choice of P_{tx}. Chapter 4 will discuss some of the factors involved in such a decision; controlling the transmission power will also play a role in several other chapters of Part II. For the present discussion, let us assume that the desired transmission power P_{tx} is known – Chapter 4 will make it clear that P_{tx} is a function of system aspects like energy per bit over noise E_b/N_0, the bandwidth efficiency η_{BW}, the distance d and the path loss coefficient γ.

The transmitted power is generated by the amplifier of a transmitter. Its own power consumption P_{amp} depends on its architecture, but for most of them, their consumed power depends on the power they are to generate. In the most simplistic models, these two values are proportional to each other, but this is an oversimplification. A more realistic model assumes that a certain constant power level is always required irrespective of radiated power, plus a proportional offset:

$$P_{amp} = \alpha_{amp} + \beta_{amp} P_{tx}. \tag{2.4}$$

where α_{amp} and β_{amp} are constants depending on process technology and amplifier architecture [559].

As an example, MIN and CHANDRAKASAN [563] report, for the μAMPS-1 nodes, $\alpha_{amp} = 174\,\text{mW}$ and $\beta_{amp} = 5.0$. Accordingly, the **efficiency of the power amplifier** η_{PA} for $P_{tx} = 0$ dBm = 1 mW radiated power is given by

$$\eta_{PA} = \frac{P_{tx}}{P_{amp}} = \frac{1\,\text{mW}}{174\,\text{mW} + 5.0 \cdot 1\,\text{mW}} \approx 0.55\,\%.$$

This model implies that the amplifier's efficiency P_{tx}/P_{amp} is best at maximum output power. Maximum power is, however, not necessarily the common case and therefore such a design is not necessarily the most beneficial one – in cellular systems, for example, amplifiers often do not operate at their maximum output power. While it is not clear how this observation would translate to WSNs, it appears promising especially in dense networks to use amplifiers with different efficiency characteristics [447]. Nonetheless, here we shall restrict the attention to the model of Equation (2.4).

In addition to the amplifier, other circuitry has to be powered up during transmission as well, for example, baseband processors. This power is referred to as P_{txElec}.

The energy to transmit a packet n-bits long (including all headers) then depends on how long it takes to send the packet, determined by the nominal bit rate R and the coding rate R_{code}, and on the total consumed power during transmission. If, in addition, the transceiver has to be turned on before transmission, startup costs also are incurred (mostly to allow voltage-controlled oscillators and phase-locked loops to settle). Equation (2.5) summarizes these effects.

$$E_{tx}(n, R_{code}, P_{amp}) = T_{start}P_{start} + \frac{n}{R R_{code}}(P_{txElec} + P_{amp}). \qquad (2.5)$$

It should be pointed out that this equation does not depend on the modulation chosen for transmission (Section 4.3 will discuss in detail an example containing multiple modulations). Measurements based on IEEE 802.11 hardware [221] have shown that in fact there is a slight dependence on the modulation, but the difference between 1 Mbit/s and 11 Mbit/s is less than 10 % for all considered transmission power values, so this is an acceptable simplification. Moreover, it is assumed that the coding overhead only depends on the coding rate, which is an acceptable assumption. In this model, the antenna efficiency is missing as well, that is, it is assumed to have a perfect antenna. Otherwise, there would be further power losses between the output of the PA and the radiated power.

This model can be easily enhanced by the effects of Forward Error Correction (FEC) coding since, with respect to transmission, FEC just increases the number of bits approximately by a factor of one divided by the code rate (see Chapter 6), since the coding energy is negligible [563].

Disregarding the distance-independent terms in these energy costs and only assuming a simplified energy cost proportional to some power of the distance has been called "one of the top myths" of energy consumption in radio communication [560]. Clearly, choosing such an inappropriately simplified model would have considerable consequences on system design, for example, incorrectly favoring a multihop approach (see Chapter 3).

Modeling energy consumption during reception

Similar to the transmitter, the receiver can be either turned off or turned on. While being turned on, it can either actively receive a packet or can be idle, observing the channel and ready to receive. Evidently, the power consumption while it is turned off is negligible. Even the difference between idling and actually receiving is very small and can, for most purposes, be assumed to be zero.

To elucidate, the energy E_{rcvd} required to receive a packet has a startup component $T_{start}P_{start}$ similar to the transmission case when the receiver had been turned off (startup times are considered equal for transmission and receiving here); it also has a component that is proportional to the packet time $\frac{n}{RR_{\text{code}}}$. During this time of actual reception, receiver circuitry has to be powered up, requiring a (more or less constant) power of P_{rxElec} – for example, to drive the LNA in the RF front end. The last component is the decoding overhead, which is incurred for every bit – this decoding overhead can be substantial depending on the concrete FEC in use; Section 6.2.3 goes into details here. Equation (2.6) summarizes these components.

$$E_{\text{rcvd}} = T_{\text{start}} P_{\text{start}} + \frac{n}{R R_{\text{code}}} P_{\text{rxElec}} + n E_{\text{decBit}}. \tag{2.6}$$

The decoding energy is relatively complicated to model, as it depends on a number of hardware and system parameters – for example, is decoding done in dedicated hardware (by, for example, a dedicated Viterbi decoder for convolutional codes) or in software on a microcontroller; it also depends on supply voltage, decoding time per bit (which in turn depends on processing speed influenced by techniques like DVS), constraint length K of the used code, and other parameters. MIN and CHANDRAKASAN [559] give more details.

Again, it is worthwhile pointing out that different modulation schemes only implicitly affect this result via the increase in time to transmit the packet.

Some numbers

Providing concrete numbers for exemplary radio transceivers is even more difficult than it is for microcontrollers: The range of commercially available transceivers is vast, with many different characteristics. Transceivers that appear to have excellent energy characteristics might suffer from other shortcomings like poor frequency stability under temperature variations (leading to partitioning of a network when parts of the node are placed in the shade and others in sunlight), poor blocking performance, high susceptibility to interference on neighboring frequency channels, or undesirable error characteristics; they could also lack features that other transceivers have, like tunability to multiple frequencies. Hence, the numbers presented here should be considered very cautiously, even more so since they had been collected from different sources and were likely determined in noncomparable environments (and not all numbers are available for all examples). Still, they should serve to provide some impression of current performance figures for actual hardware.

Table 2.4 summarizes the parameters discussed here for a number of different nodes. These numbers have been collected from references [670] and [563];[8] the data sheets [588, 690] offer further information. Note that the way of reporting such figures in the literature is anything but uniform and that hence many of the numbers given here had to be calculated or estimated. The reader is encouraged to check with the original publications for full detail. In particular, the data about the WINS and MEDUSA-II node do not allow to distinguish between α_{amp} and P_{txElec} and β_{amp} is estimated by curve fitting. STEMM and KATZ [789] present additional data for some older hardware geared toward handheld devices. References [351, 353, 725, 769] also contain further examples for sensor nodes; FEENEY and NILSSON [254] and EBERT et al. [221] present actual measurement results for IEEE 802.11-based hardware. One useful reference number for rule-of-thumb estimations might be the 1 μJ required to transmit a single bit and 0.5 μJ to receive one for the RFM TR1000 transceiver [353].

Looking at the startup times in Table 2.4, we see that actually considerable time and energy can be spent to turn on a transceiver. CHANDRAKASAN et al. [134] argue therefore that architectures with short startup times are preferable and point out the impact of startup time on the energy per bit when using different modulations; they also propose an appropriate transceiver architecture with

[8] Since these numbers are likely obtained by different measurement methods, they are not directly comparable and the reader must be cautious. However, at least they give useful ballpark estimates.

Table 2.4 Some parameters of transceiver energy consumption

Symbol	Description	Example transceiver		
		μAMPS-1 [559]	WINS [670]	MEDUSA-II [670]
α_{amp}	Equation (2.4)	174 mW	N/A	N/A
β_{amp}	Equation (2.4)	5.0	8.9	7.43
P_{amp}	Amplifier pwr.	179–674 mW	N/A	N/A
P_{rxElec}	Reception pwr.	279 mW	368.3 mW	12.48 mW
P_{rxIdle}	Receive idle	N/A	344.2 mW	12.34 mW
P_{start}	Startup pwr.	58.7 mW	N/A	N/A
P_{txElec}	Transmit pwr.	151 mW	$\approx$ 386 mW	11.61 mW
R	Transmission rate	1 Mbps	100 kbps	OOK 30 kbps ASK 115.2 kbps
T_{start}	Startup time	466 μs	N/A	N/A

fast startup time. WANG et al. [855] also point this out and provide figures on how startup time influences the choice between modulations.

These startup costs motivate some considerations of the entire system architecture. One possible idea is to have only very simply functionalities on line that can handle most of the processing, for example, decide whether a packet is intended for a given node, and only startup other components, for example, the controller, if necessary [648]. Clearly, wakeup radios are the most advanced version of this concept. Naturally, startup costs also have to be taken into account during protocol design.

Another common observation based on these figures is that transmitting and receiving have comparable power consumption, at least for short-range communication [648]. Details differ, of course, but it is an acceptable approximation to assume $P_{txElec} = P_{rxElec}$ and even neglecting the amplifier part can be admissible as long as very low transmission powers are used. In fact, for some architectures, receiving consumes more power than transmitting.

CHANDRAKASAN et al. [132] summarize these numbers into an energy per bit versus bitrate figure, pointing out that energy efficiency improves as transmission rates go up if duty cycling is used on the radio.

Dynamic scaling of radio power consumption

Applying controller-based Dynamic Voltage Scaling (DVS) principles to radio transceivers as well is tempting, but nontrivial. Scaling down supply voltage or frequency to obtain lower power consumption in exchange for higher latency is only applicable to some of the electronic parts of a transceiver, but this would mean that the remainder of the circuitry – the amplifier, for instance, which cannot be scaled down as its radiated and hence its consumed power mostly depends on the communication distance – still has to be run at high power over an extended period of time [670].

However, the frequency/voltage versus performance trade-off exploited in DVS is not the only possible trade-off to exploit. Any such "parameter versus performance" trade-off that has a convex characteristic should be amenable to an analogous optimization technique. For radio communication, in particular, possible parameters include the choice of modulation and/or code, giving raise to Dynamic Modulation Scaling (DMS), Dynamic Code Scaling (DCS) and Dynamic Modulation-Code Scaling (DMCS) optimization techniques [449, 559, 650, 735, 738]. The claim that such trade-offs do not apply to communication is another one of the "myths" of energy consumption in communication [560].

The idea of these approaches is to dynamically adapt modulation, coding, or other parameters to maximize system metrics like throughput or, particularly relevant here, energy efficiency. It rests on the hardware's ability to actually perform such modulation adaptations, but this is a commonly found property of modern transceivers. In addition, delay constraints and time-varying radio channel properties have to be taken into account.

The details of these approaches are somewhat involved, and partially, complicated optimization problems have to be approximately solved. The required computational effort should not be underestimated and a combined analysis should be undertaken on how best to split up energy consumption. Nonetheless, these approaches are quite beneficial in energy efficiency terms.

2.2.5 Relationship between computation and communication

Looking at the energy consumption numbers for both microcontrollers and radio transceivers, an evident question to ask is which is the best way to invest the precious energy resources of a sensor node: Is it better to send data or to compute? What is the relation in energy consumption between sending data and computing?

Again, details about this relationship heavily depend on the particular hardware in use, but a few rule-of-thumb figures can be given here. Typically, computing a single instruction on a microcontroller requires about 1 nJ. Also, 1 nJ about suffices to take a single sample in a radio transceiver; Bluetooth transceivers could be expected to require roughly 100 nJ to transmit a single bit (disregarding issues like startup cost and packet lengths) [391]. For other hardware, the ratio of the energy consumption to send one bit compared to computing a single instruction is between 1500 to 2700 for Rockwell WINS nodes, between 220 to 2900 for MEDUSA II nodes, and about 1400 for WINS NG 2.0 nodes [670]. HILL et al. [353] notes, for the RFM TR1000 radio transceiver, 1 μJ to transmit a single bit and 0.5 μJ to receive one; their processor takes about 8 nJ per instruction. This results in a (actually quite good) ratio of about 190 for communication to computation costs. In a slightly different perspective, communicating 1 kB of data over 100 m consumes roughly the same amount of energy as computing three million instructions [648]. HILL and CULLER [351] give some more numbers for specific applications.

Disregarding the details, it is clear that communication is a considerably more expensive undertaking than computation. Still, energy required for computation cannot be simply ignored; depending on the computational task, it is usually still smaller than the energy for communication, but still noticeable. This basic observation motivates a number of approaches and design decisions for the networking architecture of wireless sensor networks. The core idea is to invest into computation within the network whenever possible to safe on communication costs, leading to the notion of *in-network processing* and *aggregation*. These ideas will be discussed in detail in Chapter 3.

2.2.6 Power consumption of sensor and actuators

Providing any guidelines about the power consumption of the actual sensors and actuators is next to impossible because of the wide diversity of these devices. For some of them – for example, passive light or temperature sensors – the power consumption can perhaps be ignored in comparison to other devices on a wireless node (although HILL et al. [353] report a power consumption of 0.6 to 1 mA for a temperature sensor). For others, in particular, active devices like sonar, power consumption can be quite considerable and must even be considered in the dimensioning of power sources on the sensor node, not to overstress batteries, for example. To derive any meaningful numbers, requires a look at the intended application scenarios and the intended sensors to be used. Some hints on power consumption of sensor/controller interfaces, namely, AD converters, can be found in reference [26].

In addition, the sampling rate evidently is quite important. Not only does more frequent sampling require more energy for the sensors as such but also the data has to processed and, possibly, communicated somewhere.

Table 2.5 Example characteristics of sensors. Reproduced from [534] by permission of ACM

Sensor	Accuracy	Interchangeability	Sample rate [Hz]	Startup [ms]	Current [mA]
Photoresistor	N/A	10 %	2000	10	1.235
I2C temperature	1 K	0.20 K	2	500	0.15
Barometric pressure	1.5 mbar	0.5 %	10	500	0.01
Bar. press. temp.	0.8 K	0.24 K	10	500	0.01
Humidity	2 %	3 %	500	500–3000	0.775
Thermopile	3 K	5 %	2000	200	0.17
Thermistor	5 K	10 %	2000	10	0.126

To give some quantitative ideas, Table 2.5 provides examples of various sensor characteristics.

2.3 Operating systems and execution environments

2.3.1 Embedded operating systems

The traditional tasks of an operating system are controlling and protecting the access to resources (including support for input/output) and managing their allocation to different users as well as the support for concurrent execution of several processes and communication between these processes [807]. These tasks are, however, only partially required in an embedded system as the executing code is much more restricted and usually much better harmonized than in a general-purpose system. Also, as the description of the microcontrollers has shown, these systems plainly do not have the required resources to support a full-blown operating system.

Rather, an operating system or an execution environment – perhaps the more modest term is the more appropriate one – for WSNs should support the specific needs of these systems. In particular, the need for energy-efficient execution requires support for energy management, for example, in the form of controlled shutdown of individual components or Dynamic Voltage Scaling (DVS) techniques. Also, external components – sensors, the radio modem, or timers – should be handled easily and efficiently, in particular, information that becomes available asynchronously (at any arbitrary point in time) must be handled.

All this requires an appropriate programming model, a clear way to structure a protocol stack, and explicit support for energy management – without imposing too heavy a burden on scarce system resources like memory or execution time. These three topics are treated in the following sections, with a case study completing the operating system considerations.

2.3.2 Programming paradigms and application programming interfaces

Concurrent Programming

One of the first questions for a programming paradigm is how to support concurrency. Such support for concurrent execution is crucial for WSN nodes, as they have to handle data communing from arbitrary sources – for example, multiple sensors or the radio transceiver – at arbitrary points in time. For example, a system could poll a sensor to decide whether data is available and process the data right away, then poll the transceiver to check whether a packet is available, and then immediately process the packet, and so on. (Figure 2.7). Such a simple sequential model would run the risk of missing data while a packet is processed or missing a packet when sensor information is processed. This risk is particularly large if the processing of sensor data or incoming packets takes substantial amounts of time, which can easily be the case. Hence, a simple, sequential programming model is clearly insufficient.

Process-based concurrency

Most modern, general-purpose operating systems support concurrent (seemingly parallel) execution of multiple processes on a single CPU. Hence, such a process-based approach would be a first candidate to support concurrency in a sensor node as well; it is illustrated in (b) of Figure 2.7. While indeed this approach works in principle, mapping such an execution model of concurrent processes to a sensor node shows, however, that there are some granularity mismatches [491]: Equating individual protocol functions or layers with individual processes would entail a high overhead in switching from one process to another. This problem is particularly severe if often tasks have to be executed that are small with respect to the overhead incurred for switching between tasks – which is typically the case in sensor networks. Also, each process requires its own stack space in memory, which fits ill with the stringent memory constraints of sensor nodes.

Event-based programming

For these reasons, a somewhat different programming model seems preferable. The idea is to embrace the reactive nature of a WSN node and integrate it into the design of the operating system. The system essentially waits for any event to happen, where an event typically can be the availability of data from a sensor, the arrival of a packet, or the expiration of a timer. Such an event is then handled by a short sequence of instructions that only stores the fact that this event has occurred and stores the necessary information – for example, a byte arriving for a packet or the sensor's value – somewhere. The actual processing of this information is not done in these event handler routines, but separately, decoupled from the actual appearance of events. This **event-based programming** [353] model is sketched in Figure 2.8.

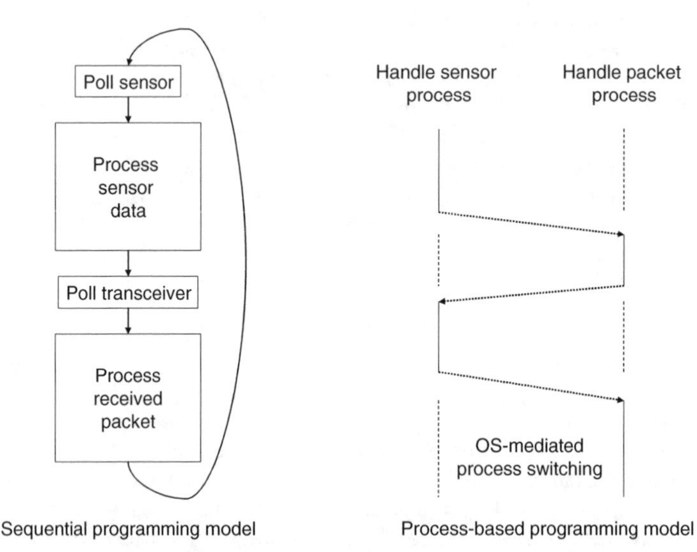

Figure 2.7 Two inadequate programming models for WSN operating systems: purely sequential execution (a) and process-based execution (b)

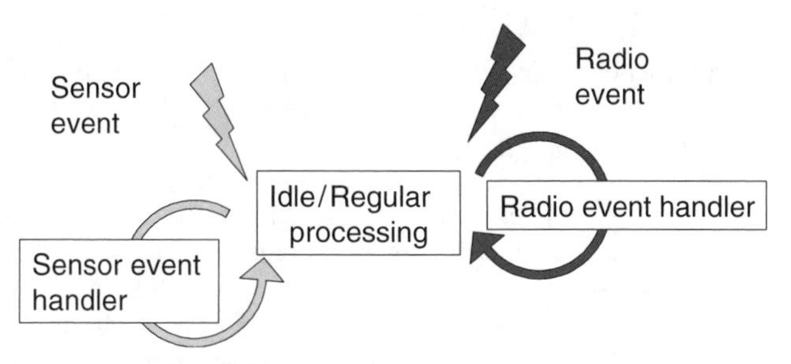

Figure 2.8 Event-based programming model

Such an event handler can interrupt the processing of any normal code, but as it is very simple and short, it can be required to **run to completion** in all circumstances without noticeably disturbing other code. Event handlers cannot interrupt each other (as this would in turn require complicated stack handling procedures) but are simply executed one after each other.

As a consequence, this event-based programming model distinguishes between two different "contexts": one for the time-critical event handlers, where execution cannot be interrupted and a second context for the processing of normal code, which is only triggered by the event handlers.

This event-based programming model is slightly different to what most programmers are used to and commonly requires some getting used to. It is actually comparable, on some levels, to communicating, extended finite state machines, which are used in protocol design formalisms as well as in some parallel programming paradigms. It does offer considerable advantages. Li et al. [491] compared the performance of a process-based and an event-based programming model (using TinyOS [353] described below) on the same hardware and found that performance improved by a factor of 8, instruction/data memory requirements were reduced by factors of 2 and 30, respectively, and power consumption was reduced by a factor of 12.

Interfaces to the operating system

In addition to the programming model that is stipulated, if not actually imposed, by the operating system, it is also necessary to specify some interfaces to how internal state of the system can be inquired and perhaps set. As the clear distinction between protocol stack and application programs vanishes somewhat in WSNs, such an interface should be accessible from protocol implementations and it should allow these implementations to access each other. This interface is also closely tied with the structure of protocol stacks discussed in the following section.

Such an Application Programming Interface (API) comprises, in general, a "functional interface, object abstractions, and detailed behavioral semantics" [558]. Abstractions are wireless links, nodes, and so on; possible functions include state inquiry and manipulation, sending and transmitting of data, access to hardware (sensors, actuators, transceivers), and setting of policies, for example, with respect to energy/quality trade-offs.

While such a general API would be extremely useful, there is currently no clear standard – or even an in-depth discussion – arising from the literature. Some first steps in this direction are more concerned with the networking architecture [751], not so much with accessing functionality on a single node. Until this changes, de facto standards will continue to be used and are likely to serve reasonably well. Section 2.3.5 describes one such de facto standard.

2.3.3 Structure of operating system and protocol stack

The traditional approach to communication protocol structuring is to use layering: individual protocols are stacked on top of each other, each layer only using functions of the layer directly below. This layered approach has great benefits in keeping the entire protocol stack manageable, in containing complexity, and in promoting modularity and reuse. For the purposes of a WSN, however, it is not clear whether such a strictly layered approach will suffice (the presentation here follows to some degree reference [431]).

As an example, consider the use of information about the strength of the signal received from a communication partner. This physical layer information can be used to assist in networking protocols to decide about routing changes (a signal becomes weaker if a node moves away and should perhaps no longer be used as a next hop), to compute location information by estimating distance from the signal strength, or to assist link layer protocols in channel-adaptive or hybrid FEC/ARQ schemes. Hence, one single source of information can be used to the advantage of many other protocols not directly associated with the source of this information.

Such **cross-layer information exchange** is but one way to loosen the strict confinements of the layered approach. Also, WSNs are not the only reason why such liberations are sought. Even in traditional network scenarios, efficiency considerations [170], the need to support wired networking protocols in wireless systems (e.g. TCP over wireless [42]), the need to migrate functionality into the backbone despite the prescriptions of Internet's end-to-end model [97], or the desire to support handover mechanisms by physical layer information in cellular networks [257] all have created a considerable pressure for a flexible, manageable, and efficient way of structuring and implementing communication protocols. HILL and CULLER [351] discuss some more examples in which cross-layer optimization is particularly useful in WSNs.

When departing from the layered architecture, the prevalent trend is to use a component model. Relatively large, monolithic layers are broken up into small, self-contained "**components**", "building blocks", or "modules" (the terminology varies). These components only fulfill one well-defined function each – for example, computation of a Cyclic Redundancy Check (CRC) – and interact with each other over clear interfaces. The main difference compared to the layered architecture is that these interactions are not confined to immediate neighbors in an up/down relationship, but can be with any other component.

This component model not only solves some of the structuring problems for protocol stacks, it also fits naturally with an event-based approach to programming wireless sensor nodes. Wrapping of hardware, communication primitives, in-network processing functionalities all can be conveniently designed and implemented as components.

One popular example for an operating system following this approach is TinyOS [353], described in detail later. It uses the notion of explicit wiring of components to allow event exchange to take place between them. While this is beneficial for "push" types of interactions (events are more or less immediately distributed to the receiving component), it does not serve well other cases where a "pull" type of information exchange is necessary. Looking at the case of the received signal strength information described above, the receiving component might not be interested in receiving all such events; rather, it might suffice to be informed asynchronously. A good solution for this is a blackboard, based on publish/subscribe principles [251], where information can be deposited and anonymously exchanged, allowing a looser coupling between components. This concept has been proposed in reference [431] and appears a promising add-on.

2.3.4 Dynamic energy and power management

Switching individual components into various sleep states or reducing their performance by scaling down frequency and supply voltage and selecting particular modulation and codings were the prominent examples discussed in Section 2.2 for improving energy efficiency. To control these possibilities, decisions have to be made by the operating system, by the protocol stack, or potentially by an application when to switch into one of these states. Dynamic Power Management (DPM) on a system level is the problem at hand.

One of the complicating factors to DPM is the energy and time required for the transition of a component between any two states. If these factors were negligible, clearly it would be optimal to always & immediately go into the mode with the lowest power consumption possible. As this is not the case, more advanced algorithms are required, taking into account these costs, the rate of updating power management decisions, the probability distribution of time until future events, and properties of the used algorithms. In fact, this field is very broad and only a few examples can be discussed here – for an overview, refer, for example, to reference [304] (especially parts III, V, and VI therein) or reference [62].

Probabilistic state transition policies

SINHA and CHANDRAKASAN [769] consider the problem of policies that regulate the transition between various sleep states. They start out by considering sensors randomly distributed over a fixed area and assume that events arrive with certain temporal distributions (Poisson process) and spatial distributions. This allows them to compute probabilities for the time to the next event, once an event has been processed (even for moving events). They use this probability to select the deepest sleep state out of several possible ones that still fulfill the threshold requirements of Equation (2.3).

In addition, they take into account the possibility of missing events when the sensor as such is also shut down in sleep mode. This can be acceptable for some applications, and SINHA and CHANDRAKASAN give some probabilistic rules on how to decide whether to go into such a deep sleep mode.

Other examples for state transition policies are discussed in references [611, 767, 784].

Controlling dynamic voltage scaling

To turn the possibilities of DVS into a technical solution also requires some further considerations. For example, it is the rare exception that there is only a single task to be run in an operating system; hence, a clever scheduler is required to decide which clock rate to use in each situation to meet all deadlines. This can require feedback from applications and has been mostly studied in "traditional" applications, for example, video playback in reference [649]. Another approach [259] incorporates dynamic voltage scaling control into the kernel of the operating system and achieves energy efficiency improvements in mixed workloads without modifications to user programs. Many other papers have considered DVS-based power management in various circumstances, often in the context of hard real-time systems, for example, references [109, 302, 307, 445, 537, 763, 869, 906] and the citations in reference [669]. Applying these results to the specific settings of a WSN is, however, still a research task as WSNs usually do not operate under similarly strict timing constraints, nor are the application profiles comparable.

Trading off fidelity against energy consumption

Most of the just described work on controlling DVS assumes hard deadlines for each task (the task has to be completed by a given time, otherwise its results are useless). In WSNs, such an assumption is often not appropriate. Rather, there are often tasks that can be computed with a higher or lower level of accuracy. The fidelity achieved by such tasks is a candidate for trading it off against other resources. When time is considered, the concept of "imprecise computation" results [515]. In a WSN, the natural trade-off is against energy required to compute a task. Essentially, the question arises again how best to invest a given amount of energy available for a given task [770]. *Deliberately embracing such inaccuracies in return for lower energy consumption is a characteristic feature of WSNs;* some examples will be discussed in various places in the book.

Some approaches to exploit such trade-offs have been described in the literature, for example, in references [260, 669], but mostly in the context of multimedia systems. SINHA et al. [770] discuss the energy-quality trade-off for algorithm design, especially for signal processing purposes (filtering, frequency domain transforms, and classification). The idea is to transform an algorithm such that it quickly approximates the final result and keeps computing as long as energy is available, producing incremental refinements (being a direct counterpart to imprecise computation [515], where computation can continue as long as time is available). As a simple example, the computation of a polynomial $f(x) = \sum_{i=0}^{N} k_i x^i$ is given: depending on whether $x < 1$ or $x \geq 1$, computation should start with the low-order or high-order terms for having the best possible approximation in case the computation has to be aborted because it exceeded its energy allocation. The performance of such (original or transformed) algorithms is studied using their $E - Q$ metric, indicating which (normalized) result quality can be achieved for how much (normalized) energy.

2.3.5 Case Study: TinyOS and nesC

Section 2.3.2 has advocated the use of an event-based programming model as the only feasible way to support the concurrency required for sensor node software while staying within the confined resources and running on top of the simple hardware provided by these nodes. The open question is how to harness the power of this programming model without getting lost in the complexity of many individual state machines sending each other events. In addition, modularity should be supported to easily exchange one state machine against another. The operating system TinyOS [353], along with the programming language nesC [285], addresses these challenges (the exposition here follows mainly these references).

TinyOS supports modularity and event-based programming by the concept of components. A component contains semantically related functionality, for example, for handling a radio interface or for computing routes. Such a component comprises the required state information in a *frame*, the program code for normal *tasks*, and handlers for *events* and *commands*. Both events and commands are exchanged between different components. Components are arranged hierarchically, from low-level components close to the hardware to high-level components making up the actual application. Events originate in the hardware and pass upward from low-level to high-level components; commands, on the other hand, are passed from high-level to low-level components.

Figure 2.9 shows a timer component that provides a more abstract version of a simple hardware time. It understands three commands ("init", "start", and "stop") and can handle one event ("fire") from another component, for example, a wrapper component around a hardware timer. It issues "setRate" commands to this component and can emit a "fired" event itself.

The important thing to note is that, in staying with the event-based paradigm, both command and event handlers must run to conclusion; they are only supposed to perform very simple triggering duties. In particular, commands must not block or wait for an indeterminate amount of time; they are simply a request upon which some task of the hierarchically lower component has to act. Similarly, an event handler only leaves information in its component's frame and arranges for a task to be executed later; it can also send commands to other components or directly report an event further up.

The actual computational work is done in the tasks. In TinyOS, they also have to run to completion, but can be interrupted by handlers. The advantage is twofold: there is no need for stack management and tasks are atomic with respect to each other. Still, by virtue of being triggered by handlers, tasks are seemingly concurrent to each other.

The arbitration between tasks – multiple can be triggered by several events and are ready to execute – is done by a simple, power-aware First In First Out (FIFO) scheduler, which shuts the node down when there is no task executing or waiting.

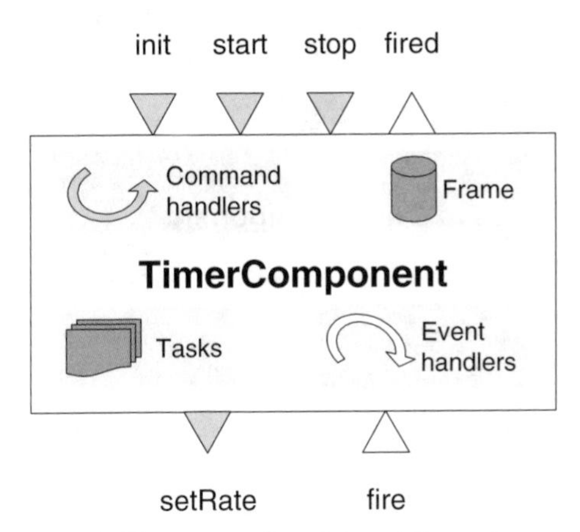

Figure 2.9 Example Timer component (adapted from references [285, 353])

With handlers and tasks all required to run to completion, it is not clear how a component could obtain feedback from another component about a command that it has invoked there – for example, how could an Automatic Repeat Request (ARQ) protocol learn from the MAC protocol whether a packet had been sent successfully or not? The idea is to split invoking such a request and the information about answers into two phases: The first phase is the sending of the command, the second is an explicit information about the outcome of the operation, delivered by a separate event. This **split-phase programming** approach requires for each command a matching event but enables concurrency under the constraints of run-to-completion semantics – if no confirmation for a command is required, no completion event is necessary.

Having commands and events as the only way of interaction between components (the frames of components are private data structures), and especially when using split-phase programming, a large number of commands and events add up in even a modestly large program. Hence, an abstraction is necessary to organize them. As a matter of fact, the set of commands that a component understands and the set of events that a component may emit are its interface to the components of a hierarchically higher layer; looked at it the other way around, a component can invoke certain commands at its lower component and receive certain events from it. Therefore, structuring commands and events that belong together forms an **interface** between two components.

The nesC language formalizes this intuition by allowing a programmer to define *interface types* that define commands and events that belong together. This allows to easily express split-phase programming style by putting commands and their corresponding completion events into the same interface. Components then *provide* certain interfaces to their users and in turn *use* other interfaces from underlying components.

Figure 2.10 shows how the Timer component of the previous example can be reorganized into using a *clock* interface and providing two interfaces *StdCtrl* and *Timer*. The corresponding nesC code is shown in Listing 2.1. Note that the component TimerComponent is defined here as a *module* since it is a primitive component, directly containing handlers and tasks.

Such primitive components or modules can be combined into larger *configurations* by simply "wiring" appropriate interfaces together. For this wiring to take place, only components that have the correct interface types can be plugged together (this is checked by the compiler). Figure 2.11 shows how the TimerComponent and an additional component HWClock can be wired together to form a new component CompleteTimer, exposing only the StdCtrl and Timer interfaces to the outside; Listing 2.2 shows the corresponding nesC code. Note that both modules and configurations are components.

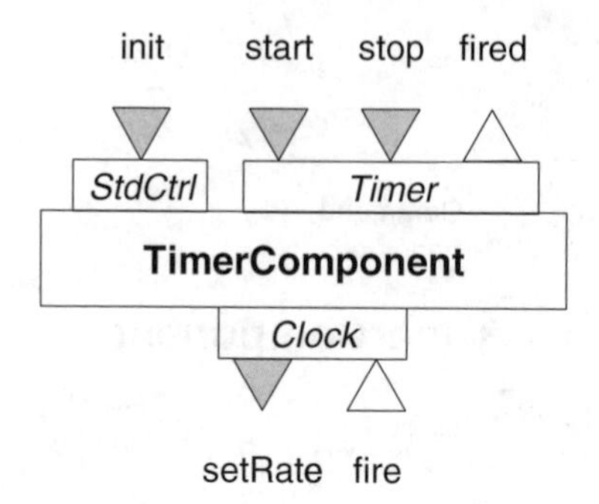

Figure 2.10 Organizing the Timer component using interfaces [285, 353]

Listing 2.1: Defining modules and interfaces [285]

```
interface StdCtrl {
  command result_t init();
}

interface Timer {
  command result_t start (char type, uint32_t interval);
  command result_t stop ();
  event result_t fired();
}

interface Clock {
  command result_t setRate (char interval, char scale);
  event result_t  fire ();
}

module TimerComponent {
  provides {
    interface StdCtrl;
    interface Timer;
  }
  uses interface Clock as Clk;
}
```

Using these component definition, implementation, and connection concepts, TinyOS and nesC together form a powerful and relatively easy to use basis to implement both core operating system functionalities as well as communication protocol stacks and application functions. Experience has shown [285] that in fact programmers do use these paradigms and arrive at relatively small, highly specialized components that are then combined as needed, proving the modularity claim. Also, code size and memory requirements are quite small.

Overall, TinyOS can currently be regarded as the standard implementation platform for WSNs. It is also becoming available for an increasing number of platforms other than the original "motes" on which it had been developed. For practical work, the project web page [820] provides a lot of valuable information along with a good tutorial [821].

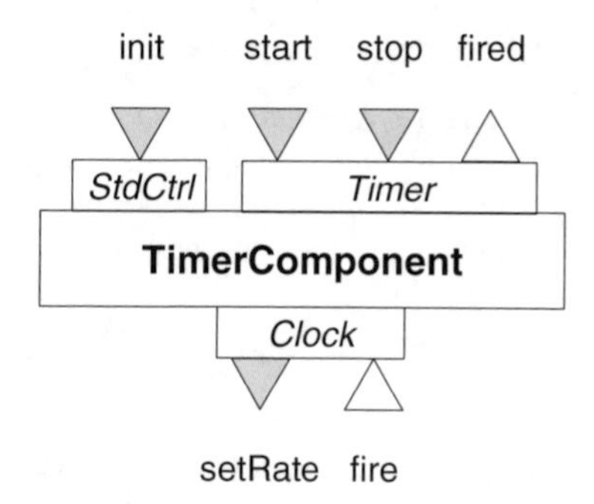

Figure 2.11 Building a larger configuration out of two components [285, 353]

Listing 2.2: Wiring components to form a configuration [285]

```
configuration CompleteTimer {
  provides {
    interface StdCtrl;
    interface Timer;
  }
  implementation {
    components TimerComponent , HWClock;
    StdCtrl = TimerComponent.HWClock;
    Timer = TimerComponent.Timer;
    TimerComponent.Clk = HWClock.Clock;
  }
}
```

On top of the TinyOS operating system, a vast range of extensions, protocols, and applications have been developed. Some brief examples must suffice here.[9] Levis and Culler [481] describe a virtual machine concept on top of TinyOS that provides a high-level interface to concisely represent programs; it is particularly beneficial for over-the-air reprogramming and retasking of an existing network. Conceiving of the sensor network as a relational database is made possible by the TinyDB project.

2.3.6 Other examples

Apart from TinyOS, there are a few other execution environments or operating systems for WSN nodes. One example is Contiki[10] [216], which has been ported to various hardware platforms and actually implements a TCP/IP stack on top of a platform with severely restricted resources [215]. Other examples are ecos [224] and the Mantis project [4].

[9] In February 2004, google found about 15.800 results when searching for "TinyOS"; in November 2004, already 123.000!

[10] `http://www.sics.se/~adam/contiki/`

2.4 Some examples of sensor nodes

There are quite a number of actual nodes available for use in wireless sensor network research and development. Again, depending on the intended application scenarios, they have to fulfill quite different requirements regarding battery life, mechanical robustness of the node's housing, size, and so on. A few examples shall highlight typical approaches; an overview of current developments can be found, for example, in reference [352].

2.4.1 The "Mica Mote" family

Starting in the late 1990s, an entire family of nodes has evolved out of research projects at the University of California at Berkeley, partially with the collaboration of Intel, over the years. They are commonly known as the *Mica motes*[11], with different versions (Mica, Mica2, Mica2Dot) having been designed [351, 353, 534]; references [285, 481] have an overview table of the family members; schematics for some of these designs are available from [822]. They are commercially available via the company Crossbow[12] in different versions and different kits. TinyOS is the usually used operating system for these nodes.

An early example for the schematics of such a node is shown in Figure 2.12 [353].

All these boards feature a microcontroller belonging to the Atmel family, a simple radio modem (usually a TR 1000 from RFM), and various connections to the outside. In addition, it is possible to connect additional "sensor boards" with, for example, barometric or humidity sensors, to the node as such, enabling a wider range of applications and experiments. Also, specialized enclosures have been built for use in rough environments, for example, for monitoring bird habitats [534]. Sensors are connected to the controller via an I2C bus or via SPI, depending on the version.

The MEDUSA-II nodes [670] share the basic components and are quite similar in design.

2.4.2 EYES nodes

The nodes developed by Infineon in the context of the European Union – sponsored project "Energy-efficient Sensor Networks" (EYES) [13] are another example of a typical sensor node (Figure 2.13). It is equipped with a Texas Instrument MSP 430 microcontroller, an Infineon radio modem TDA 5250, along with a SAW filter and transmission power control; the radio modem also reports the measured signal strength to the controller. The node has a USB interface to a PC and the possibility to add additional sensors/actuators.

2.4.3 BTnodes

The "Btnodes" [103] have been developed at the ETH Zürich out of several research projects (Figure 2.14). They feature an Atmel ATmega 128L microcontroller, 64 + 180 kB RAM, and 128 kB FLASH memory. Unlike most other sensor nodes (but similar to some nodes developed by Intel), they use Bluetooth as their radio technology in combination with a Chipcon CC1000 operating between 433 and 915 MHz.

2.4.4 Scatterweb

The ScatterWeb platform [694] was developed at the Computer Systems & Telematics group at the Freie Universität Berlin (Figure 2.15). This is an entire family of nodes, starting from a relatively

[11] A mote: a small particle, like a mote of dust.

[12] `http://www.xbow.com`

[13] `http://www.eyes.eu.org`

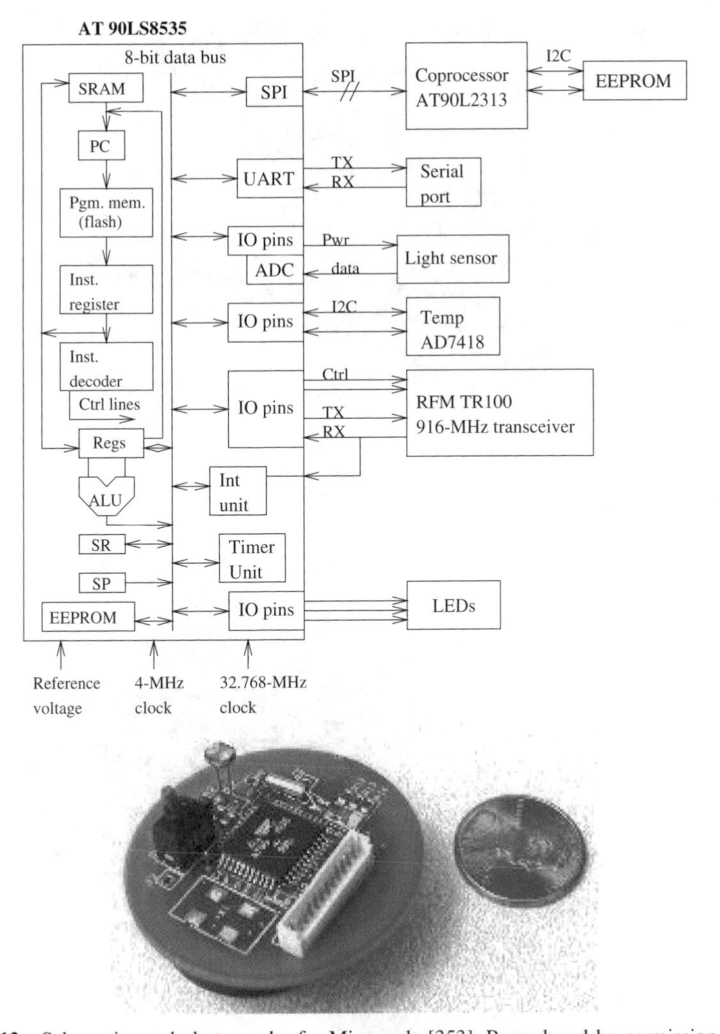

Figure 2.12 Schematics and photograph of a Mica node [353]. Reproduced by permission of ACM

standard sensor node (based on MSP 430 microcontroller) and ranges up to embedded web servers, which comes equipped with a wide range of interconnection possibilities – apart from Bluetooth and a low-power radio mode, connections for I^2C or CAN are available, for example.

2.4.5 Commercial solutions

Apart from these academic research prototypes, there are already a couple of sensor-node-type devices commercially available, including appropriate housing, certification, and so on. Some of these companies include "ember" (`www.ember.com`) or "Millenial" (`www.millenial.net`). The market here is more dynamic than can be reasonably reflected in a textbook and the reader is encouraged to watch for up-to-date developments.

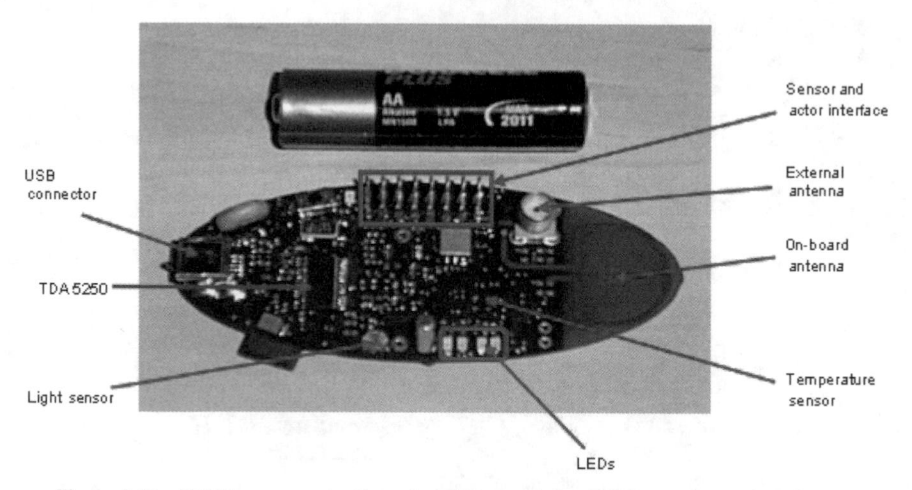

Figure 2.13 EYES sensor node. Reproduced by permission of Thomas Lentsch, Infineon

Figure 2.14 Btnode. Reproduced by permission of Jan Beutel, ETH Zurich

2.5 Conclusion

This chapter has introduced the necessary hardware prerequisites for building wireless sensor networks – the nodes as such. It has shown the principal ways of constructing such nodes and has shown some numbers on the performance and energy consumption of its main components – mainly the controller, the communication device, and the sensors. On the basis of these numbers, it will often be convenient to assume that a wireless sensor node consists of two separate parts [778]: One part that is continuously vigilant, can detect and report events, and has small or even negligible power consumption. This is complemented by a second part that performs actual processing and

Figure 2.15 A ScatterWeb embedded web server. Reproduced by permission of Prof. Dr.-Ing. J. Schiller, FU Berlin

communication, has higher, nonnegligible power consumption, and has therefore to be operated in a low duty cycle. This separation of functionalities is justified from the hardware properties as is it supported by operating systems like TinyOS.

Looking at the large variety of components to choose from, each with their own characteristic advantages and disadvantages, it is not surprising that there is not a single, "perfect" wireless sensor node – different application requirements will require different trade-offs to be made and different architectures to be used. As a consequence, there will be sensor networks that employ a heterogeneous mix of various node types to fulfill their tasks, for example, nodes with more or less computation power, different types of wireless communication, or different battery sizes. This can have consequences on how to design a wireless sensor network by exploiting this heterogeneity in hardware to assign different tasks to the best-suited nodes.

While much of the work described here is still on-going research or in its prototypical state, the emerging capabilities of future sensor nodes with respect to communication, computation, and storage as well as regarding their energy consumption trade-offs are quite apparent already. The absolute numbers are still subject to change, but it is unlikely that inherent trade-offs, for example, between the energy required for computation or communication, are going to change dramatically in the foreseeable future. These trade-offs form the basis for the construction of networking functionalities, geared toward the specific requirements of wireless sensor network applications.

3

Network architecture

Objectives of this Chapter

This chapter introduces the basic principles of turning individual sensor nodes into a wireless sensor network. On the basis of the high-level application scenarios of Chapter 1, more concrete scenarios and the resulting optimization goals of how a network should function are discussed. On the basis of these scenarios and goals, a few principles for the design of networking protocols in wireless sensor networks are derived – these principles and the resulting protocol mechanisms constitute the core differences of WSNs compared to other network types. To make the resulting capabilities of a WSN usable, a proper service interface is required, as is an integration of WSNs into larger network contexts.

At the end of this chapter, the reader should be able to appreciate the basic networking "philosophy" followed by wireless sensor network research. Upon this basis, the next part of the book will then discuss in detail individual networking functionalities.

Chapter Outline

The architecture of wireless sensor networks draws upon many sources. Historically, a lot of related work has been done in the context of self-organizing, mobile, ad hoc networks (references [635, 793, 827] provide some overview material). While these networks are intended for different purposes, they share the need for a decentralized, distributed form of organization. From a different perspective, sensor networks are related to real-time computing [429, 514] and even to some concepts from peer-to-peer computing [55, 480, 574, 608, 842], active networks [111], and mobile agents/swarm intelligence [86, 98, 176, 220, 892].

Protocols and Architectures for Wireless Sensor Networks H. Karl and A. Willig

Consequently, the number of ideas and publications on networking architectures for wireless sensor networks is vast, and it is often difficult to clearly attribute who first came up with a certain idea, especially since many of them are fairly obvious extrapolations of ideas from the areas just mentioned; also, similar concepts have often been proposed more or less concurrently by different authors. Nonetheless, proper attribution shall be given where possible. A (not necessarily complete) collection of important architectural papers on wireless sensor networks is [26, 88, 126, 134, 233, 245, 246, 274, 342, 344, 351, 353, 392, 433, 500, 534, 648, 653, 667, 758, 778, 788, 798, 921, 923]; pointers and discussion of architectural issues are also included in practically all overview papers, for example, [17, 367, 670, 699].

3.1 Sensor network scenarios

3.1.1 Types of sources and sinks

Section 1.3 has introduced several typical interaction patterns found in WSNs – event detection, periodic measurements, function approximation and edge detection, or tracking – it has also already briefly touched upon the definition of "sources" and "sinks". A source is any entity in the network that can provide information, that is, typically a sensor node; it could also be an actuator node that provides feedback about an operation.

A sink, on the other hand, is the entity where information is required. There are essentially three options for a sink: it could belong to the sensor network as such and be just another sensor/actuator node or it could be an entity outside this network. For this second case, the sink could be an actual device, for example, a handheld or PDA used to interact with the sensor network; it could also be merely a gateway to another larger network such as the Internet, where the actual request for the information comes from some node "far away" and only indirectly connected to such a sensor network. These main types of sinks are illustrated by Figure 3.1, showing sources and sinks in direct communication.

For much of the remaining discussion, this distinction between various types of sinks is actually fairly irrelevant. It is important, as discussed in Section 3.1.4, whether sources or sinks move, but what they do with the information is not a primary concern of the networking architecture. There are some consequences of a sink being a gateway node; they will be discussed in Section 3.5.

3.1.2 Single-hop versus multihop networks

From the basics of radio communication and the inherent power limitation of radio communication follows a limitation on the feasible distance between a sender and a receiver. Because of this

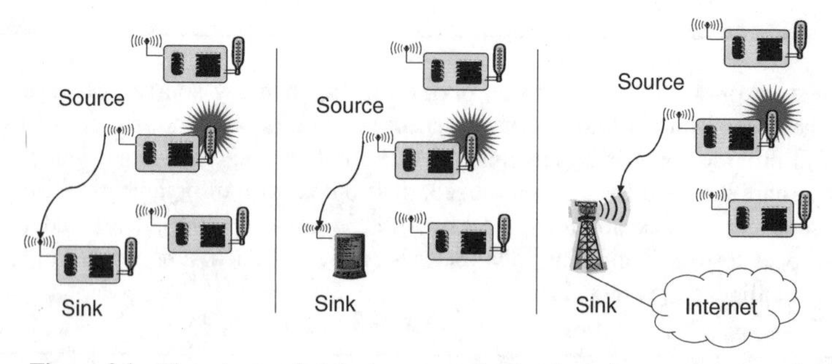

Figure 3.1 Three types of sinks in a very simple, single-hop sensor network

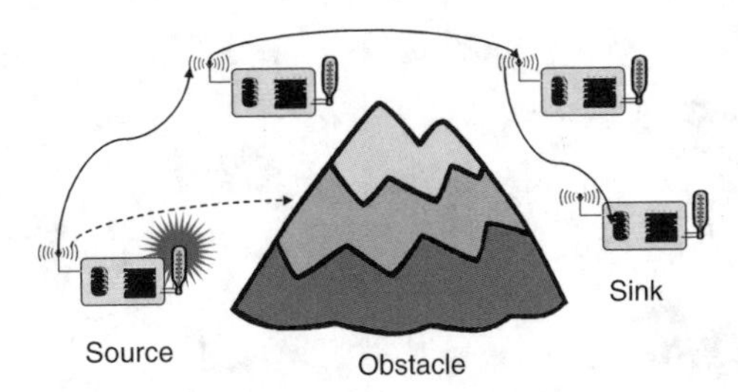

Figure 3.2 Multihop networks: As direct communication is impossible because of distance and/or obstacles, multihop communication can circumvent the problem

limited distance, the simple, direct communication between source and sink is not always possible, specifically in WSNs, which are intended to cover a lot of ground (e.g. in environmental or agriculture applications) or that operate in difficult radio environments with strong attenuation (e.g. in buildings).

To overcome such limited distances, an obvious way out is to use relay stations, with the data packets taking multi hops from the source to the sink. This concept of multihop networks (illustrated in Figure 3.2) is particularly attractive for WSNs as the sensor nodes themselves can act as such relay nodes, foregoing the need for additional equipment. Depending on the particular application, the likelihood of having an intermediate sensor node at the right place can actually be quite high – for example, when a given area has to be uniformly equipped with sensor nodes anyway – but nevertheless, there is not always a guarantee that such multihop routes from source to sink exist, nor that such a route is particularly short.

While multihopping is an evident and working solution to overcome problems with large distances or obstacles, it has also been claimed to improve the energy efficiency of communication. The intuition behind this claim is that, as attenuation of radio signals is at least quadratic in most environments (and usually larger), it consumes less energy to use relays instead of direct communication: When targeting for a constant SNR at all receivers (assuming for simplicity negligible error rates at this SNR), the *radiated* energy required for direct communication over a distance d is cd^α (c some constant, $\alpha \geq 2$ the path loss coefficient); using a relay at distance $d/2$ reduces this energy to $2c(d/2)^\alpha$.

But this calculation considers only the radiated energy, not the actually *consumed* energy – in particular, the energy consumed in the intermediate relay node. Even assuming that this relay belongs to the WSN and is willing to cooperate, when computing the total required energy it is necessary to take into account the complete power consumption of Section 2.2.4. It is an easy exercise to show that energy is actually wasted if intermediate relays are used for short distances d. Only for large d does the radiated energy dominate the fixed energy costs consumed in transmitter and receiver electronics – the concrete distance where direct and multihop communication are in balance depends on a lot of device-specific and environment-specific parameters. Nonetheless, this relationship is often not considered. In fact, Min and Chandrakasan [560] classify the misconception that multihopping saves energy as the number one myth about energy consumption in wireless communication. Great care should be taken when applying multihopping with the end of improved energy efficiency.

It should be pointed out that only multihop networks operating in a **store and forward** fashion are considered here. In such a network, a node has to correctly receive a packet before it can forward it somewhere. Alternative, innovative approaches attempt to exploit even erroneous reception of packets, for example, when multiple nodes send the same packet and each individual transmission could not be received, but collectively, a node can reconstruct the full packet. Such **cooperative relaying** techniques are not considered here.

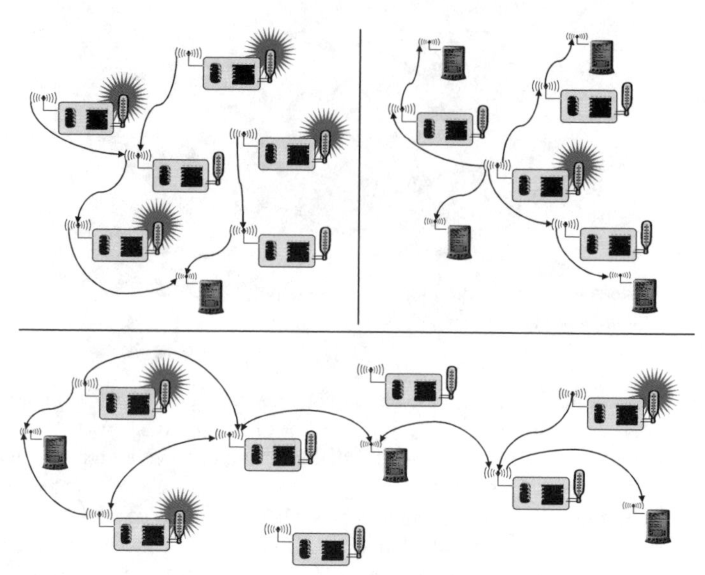

Figure 3.3 Multiple sources and/or multiple sinks. Note how in the scenario in the lower half, both sinks and active sources are used to forward data to the sinks at the left and right end of the network

3.1.3 Multiple sinks and sources

So far, only networks with a single source and a single sink have been illustrated. In many cases, there are multiple sources and/or multiple sinks present. In the most challenging case, multiple sources should send information to multiple sinks, where either all or some of the information has to reach all or some of the sinks. Figure 3.3 illustrates these combinations.

3.1.4 Three types of mobility

In the scenarios discussed above, all participants were stationary. But one of the main virtues of wireless communication is its ability to support mobile participants. In wireless sensor networks, mobility can appear in three main forms:

Node mobility The wireless sensor nodes themselves can be mobile. The meaning of such mobility is highly application dependent. In examples like environmental control, node mobility should not happen; in livestock surveillance (sensor nodes attached to cattle, for example), it is the common rule.

In the face of node mobility, the network has to reorganize itself frequently enough to be able to function correctly. It is clear that there are trade-offs between the frequency and speed of node movement on the one hand and the energy required to maintain a desired level of functionality in the network on the other hand.

Sink mobility The information sinks can be mobile (Figure 3.4). While this can be a special case of node mobility, the important aspect is the mobility of an information sink that is not part of the sensor network, for example, a human user requested information via a PDA while walking in an intelligent building.

In a simple case, such a requester can interact with the WSN at one point and complete its interactions before moving on. In many cases, consecutive interactions can be treated as

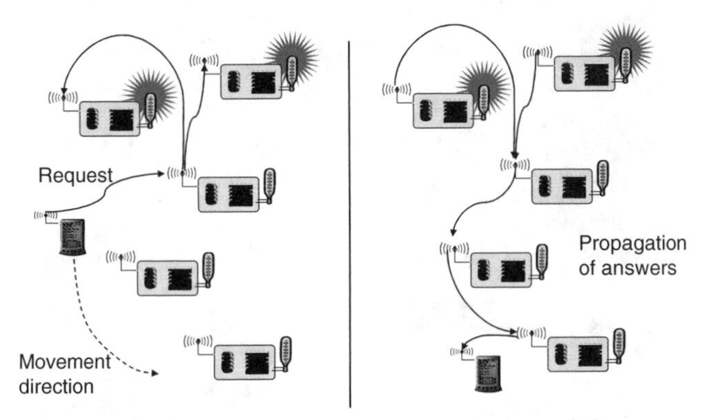

Figure 3.4 A mobile sink moves through a sensor network as information is being retrieved on its behalf

separate, unrelated requests. Whether the requester is allowed interactions with any node or only with specific nodes is a design choice for the appropriate protocol layers.

A mobile requester is particularly interesting, however, if the requested data is not locally available but must be retrieved from some remote part of the network. Hence, while the requester would likely communicate only with nodes in its vicinity, it might have moved to some other place. The network, possibly with the assistance of the mobile requester, must make provisions that the requested data actually follows and reaches the requester despite its movements [758].

Event mobility In applications like event detection and in particular in tracking applications, the cause of the events or the objects to be tracked can be mobile.

In such scenarios, it is (usually) important that the observed event is covered by a sufficient number of sensors at all time. Hence, sensors will wake up around the object, engaged in higher activity to observe the present object, and then go back to sleep. As the event source moves through the network, it is accompanied by an area of activity within the network – this has been called the *frisbee* model, introduced in reference [126] (which also describes algorithms for handling the "wakeup wavefront"). This notion is described by Figure 3.5, where the task is to detect a moving elephant and to observe it as it moves around. Nodes that do not actively detect anything are intended to switch to lower sleep states unless they are required to convey information from the zone of activity to some remote sink (not shown in Figure 3.5).

Communication protocols for WSNs will have to render appropriate support for these forms of mobility. In particular, event mobility is quite uncommon, compared to previous forms of mobile or wireless networks.

3.2 Optimization goals and figures of merit

For all these scenarios and application types, different forms of networking solutions can be found. The challenging question is how to optimize a network, how to compare these solutions, how to decide which approach better supports a given application, and how to turn relatively imprecise optimization goals into measurable figures of merit? While a general answer appears impossible considering the large variety of possible applications, a few aspects are fairly evident.

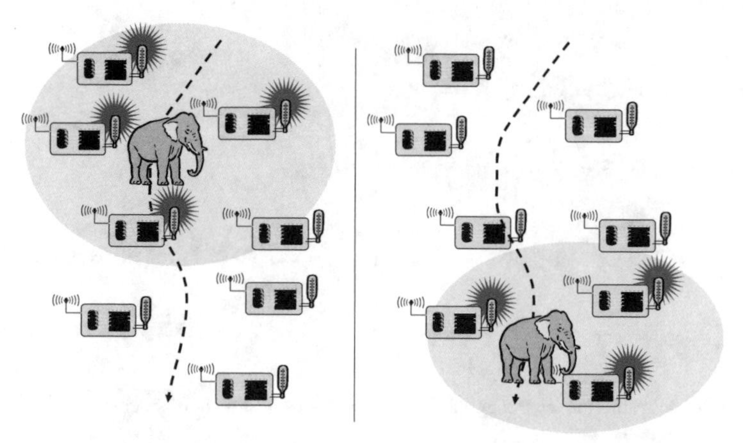

Figure 3.5 Area of sensor nodes detecting an event – an elephant [378] – that moves through the network along with the event source (dashed line indicate the elephant's trajectory; shaded ellipse the activity area following or even preceding the elephant)

3.2.1 Quality of service

WSNs differ from other conventional communication networks mainly in the type of service they offer. These networks essentially only move bits from one place to another. Possibly, additional requirements about the offered Quality of Service (QoS) are made, especially in the context of multimedia applications. Such QoS can be regarded as a low-level, networking-device-observable attribute – bandwidth, delay, jitter, packet loss rate – or as a high-level, user-observable, so-called subjective attribute like the perceived quality of a voice communication or a video transmission. While the first kind of attributes is applicable to a certain degree to WSNs as well (bandwidth, for example, is quite unimportant), the second one clearly is not, but is really the more important one to consider! Hence, high-level QoS attributes corresponding to the subjective QoS attributes in conventional networks are required.

But just like in traditional networks, high-level QoS attributes in WSN highly depend on the application. Some generic possibilities are:

Event detection/reporting probability What is the probability that an event that actually occurred is not detected or, more precisely, not reported to an information sink that is interested in such an event? For example, not reporting a fire alarm to a surveillance station would be a severe shortcoming.

Clearly, this probability can depend on/be traded off against the overhead spent in setting up structures in the network that support the reporting of such an event (e.g. routing tables) or against the run-time overhead (e.g. sampling frequencies).

Event classification error If events are not only to be detected but also to be classified, the error in classification must be small.

Event detection delay What is the delay between detecting an event and reporting it to any/all interested sinks?

Missing reports In applications that require periodic reporting, the probability of undelivered reports should be small.

Approximation accuracy For function approximation applications (e.g. approximating the temperature as a function of location for a given area), what is the average/maximum absolute or relative error with respect to the actual function?[1] Similarly, for edge detection applications, what is the accuracy of edge descriptions; are some missed at all?

Tracking accuracy Tracking applications must not miss an object to be tracked, the reported position should be as close to the real position as possible, and the error should be small. Other aspects of tracking accuracy are, for example, the sensitivity to sensing gaps [923].

3.2.2 Energy efficiency

Much of the discussion has already shown that energy is a precious resource in wireless sensor networks and that energy efficiency should therefore make an evident optimization goal. It is clear that with an arbitrary amount of energy, most of the QoS metrics defined above can be increased almost at will (approximation and tracking accuracy are notable exceptions as they also depend on the density of the network). Hence, putting the delivered QoS and the energy required to do so into perspective should give a first, reasonable understanding of the term energy efficiency.

The term "energy efficiency" is, in fact, rather an umbrella term for many different aspects of a system, which should be carefully distinguished to form actual, measurable figures of merit. The most commonly considered aspects are:

Energy per correctly received bit How much energy, counting all sources of energy consumption at all possible intermediate hops, is spent on average to transport one bit of information (payload) from the source to the destination? This is often a useful metric for periodic monitoring applications.

Energy per reported (unique) event Similarly, what is the average energy spent to report one event? Since the same event is sometimes reported from various sources, it is usual to normalize this metric to only the unique events (redundant information about an already known event does not provide additional information).

Delay/energy trade-offs Some applications have a notion of "urgent" events, which can justify an increased energy investment for a speedy reporting of such events. Here, the trade-off between delay and energy overhead is interesting.

Network lifetime The time for which the network is operational or, put another way, the time during which it is able to fulfill its tasks (starting from a given amount of stored energy). It is not quite clear, however, when this time ends. Possible definitions are:

Time to first node death When does the first node in the network run out of energy or fail and stop operating?

Network half-life When have 50 % of the nodes run out of energy and stopped operating? Any other fixed percentile is applicable as well.

Time to partition When does the first partition of the network in two (or more) disconnected parts occur? This can be as early as the death of the first node (if that was in a pivotal position) or occur very late if the network topology is robust.

[1] Clearly, this requires assumptions about the function to be approximated; discontinuous functions or functions with unlimited first derivative are impossible to approximate with a finite number of sensors.

Time to loss of coverage Usually, with redundant network deployment and sensors that can observe a region instead of just the very spot where the node is located, each point in the deployment region is observed by multiple sensor nodes. A possible figure of merit is thus the time when for the first time any spot in the deployment region is no longer covered by any node's observations.

If k redundant observations are necessary (for tracking applications, for example), the corresponding definition of loss of coverage would be the first time any spot in the deployment region is no longer covered by at least k different sensor nodes.

Time to failure of first event notification A network partition can be seen as irrelevant if the unreachable part of the network does not want to report any events in the first place. Hence, a possibly more application-specific interpretation of partition is the inability to deliver an event. This can be due to an event not being noticed because the responsible sensor is dead or because a partition between source and sink has occurred.

It should be noted that simulating network lifetimes can be a difficult statistical problem.

Obviously, the longer these times are, the better does a network perform. More generally, it is also possible to look at the (complementary) distribution of node lifetimes (with what probability does a node survive a given amount of time?) or at the relative survival times of a network (at what time are how many percent of the nodes still operational?). This latter function allows an intuition about many WSN-specific protocols in that they tend to sacrifice long lifetimes in return for an improvement in short lifetimes – they "sharpen the drop" (Figure 3.6).

All these metrics can of course only be evaluated under a clear set of assumptions about the energy consumption characteristics of a given node, about the actual "load" that the network has to deal with (e.g. when and where do events happen), and also about the behavior of the radio channel.

3.2.3 Scalability

The ability to maintain performance characteristics irrespective of the size of the network is referred to as scalability. With WSN potentially consisting of thousands of nodes, scalability is an evidently indispensable requirement. Scalability is ill served by any construct that requires globally consistent state, such as addresses or routing table entries that have to be maintained. Hence, the need to restrict such information is enforced by and goes hand in hand with the resource limitations of sensor nodes, especially with respect to memory.

The need for extreme scalability has direct consequences for the protocol design. Often, a penalty in performance or complexity has to be paid for small networks as discussed in the following Section 3.3.1. Architectures and protocols should implement *appropriate* scalability support rather than trying to be as scalable as possible. Applications with a few dozen nodes might admit more-efficient solutions than applications with thousands of nodes; these smaller applications might be

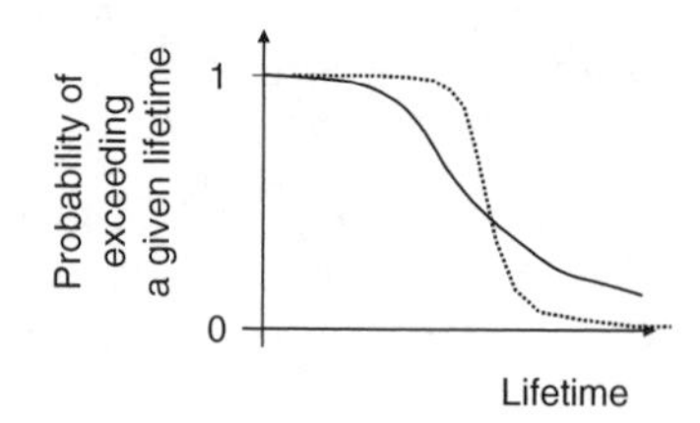

Figure 3.6 Two probability curves of a node exceeding a given lifetime – the dotted curve trades off better minimal lifetime against reduced maximum lifetime

more common in the first place. Nonetheless, a considerable amount of research has been invested into highly scalable architectures and protocols.

3.2.4 Robustness

Related to QoS and somewhat also to scalability requirements, wireless sensor networks should also exhibit an appropriate robustness. They should not fail just because a limited number of nodes run out of energy, or because their environment changes and severs existing radio links between two nodes – if possible, these failures have to be compensated for, for example, by finding other routes. A precise evaluation of robustness is difficult in practice and depends mostly on failure models for both nodes and communication links.

3.3 Design principles for WSNs

Appropriate QoS support, energy efficiency, and scalability are important design and optimization goals for wireless sensor networks. But these goals themselves do not provide many hints on how to structure a network such that they are achieved. A few basic principles have emerged, which can be useful when designing networking protocols; the description here follows partially references [246, 699]. Nonetheless, the general advice to always consider the needs of a concrete application holds here as well – for each of these basic principles, there are examples where following them would result in inferior solutions.

3.3.1 Distributed organization

Both the scalability and the robustness optimization goal, and to some degree also the other goals, make it imperative to organize the network in a distributed fashion. That means that there should be no centralized entity in charge – such an entity could, for example, control medium access or make routing decisions, similar to the tasks performed by a base station in cellular mobile networks. The disadvantages of such a centralized approach are obvious as it introduces exposed points of failure and is difficult to implement in a radio network, where participants only have a limited communication range. Rather, the WSNs nodes should cooperatively organize the network, using distributed algorithms and protocols. **Self-organization** is a commonly used term for this principle.

When organizing a network in a distributed fashion, it is necessary to be aware of potential shortcomings of this approach. In many circumstances, a centralized approach can produce solutions that perform better or require less resources (in particular, energy). To combine the advantages, one possibility is to use centralized principles in a localized fashion by dynamically electing, out of the set of equal nodes, specific nodes that assume the responsibilities of a centralized agent, for example, to organize medium access. Such elections result in a hierarchy, which has to be dynamic: The election process should be repeated continuously lest the resources of the elected nodes be overtaxed, the elected node runs out of energy, and the robustness disadvantages of such – even only localized – hierarchies manifest themselves. The particular election rules and triggering conditions for reelection vary considerably, depending on the purpose for which these hierarchies are used. Chapter 10 will, to a large degree, deal with the question of how to determine such hierarchies in a distributed fashion.

3.3.2 In-network processing

When organizing a network in a distributed fashion, the nodes in the network are not only passing on packets or executing application programs, they are also actively involved in taking decisions about how to operate the network. This is a specific form of information processing that happens in the network, but is limited to information about the network itself. It is possible to extend this concept by also taking the concrete data that is to be transported by the network into account in this information processing, making **in-network processing** a first-rank design principle.

Several techniques for in-network processing exist, and by definition, this approach is open to an arbitrary extension – any form of data processing that improves an application is applicable. A few example techniques are outlined here; they will reappear in various of the following chapters, especially in Chapter 12.

Aggregation

Perhaps the simplest in-network processing technique is aggregation. Suppose a sink is interested in obtaining periodic measurements from all sensors, but it is only relevant to check whether the average value has changed, or whether the difference between minimum and maximum value is too big. In such a case, it is evidently not necessary to transport are readings from all sensors to the sink, but rather, it suffices to send the average or the minimum and maximum value. Recalling from Section 2.3 that transmitting data is considerably more expensive than even complex computation shows the great energy-efficiency benefits of this approach. The name **aggregation** stems from the fact that in nodes intermediate between sources and sinks, information is aggregated into a condensed form out of information provided by nodes further away from the sink (and potentially, the aggregator's own readings).

Clearly, the aggregation function to be applied in the intermediate nodes must satisfy some conditions for the result to be meaningful; most importantly, this function should be **composable**. A further classification [528] of aggregate functions distinguishes *duplicate-sensitive* versus *insensitive*, *summary* versus *exemplary*, *monotone* versus *nonmonotone*, and *algebraic* versus *holistic* (a more detailed discussion can be found in Section 12.3). Functions like average, counting, or minimum can profit a lot from aggregation; holistic functions like the median are not amenable to aggregation at all.

Figure 3.7 illustrates the idea of aggregation. In the left half, a number of sensors transmit readings to a sink, using multihop communication. In total, 13 messages are required (the numbers in the figure indicate the number of messages traveling across a given link). When the highlighted nodes perform aggregation – for example, by computing average values (shown in the right half of the figure) – only 6 messages are necessary.

Challenges in this context include how to determine where to aggregate results from which nodes, how long to wait for such results, and determining the impact of lost packets.

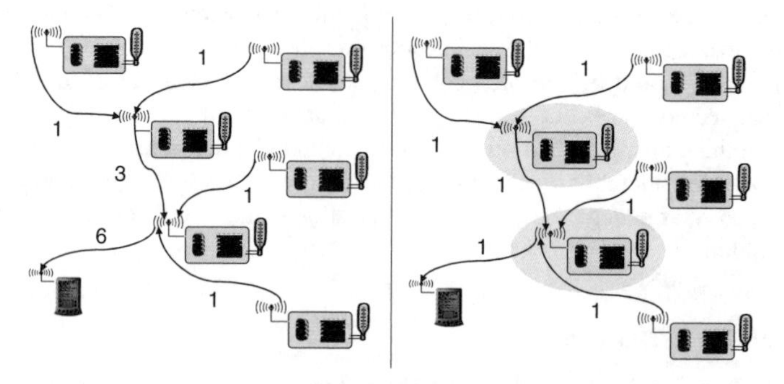

Figure 3.7 Aggregation example

Distributed source coding and distributed compression

Aggregation condenses and sacrifices information about the measured values in order not to have to transmit all bits of data from all sources to the sink. Is it possible to reduce the number of transmitted bits (compared to simply transmitting all bits) but still obtain the *full* information about all sensor readings at the sink?

While this question sounds surprising at first, it is indeed possible to give a positive answer. It is related to the coding and compression problems known from conventional networks, where a lot of effort is invested to encode, for example, a video sequence, to reduce the required bandwidth [901]. The problem here is slightly different, in that we are interested to encode the information provided by several sensors, not just by a single camera; moreover, traditional coding schemes tend to put effort into the encoding, which might be too computationally complex for simple sensor nodes.

How can the fact that information is provided by multiple sensors be exploited to help in coding? If the sensors were connected and could exchange their data, this would be conceivable (using relatively standard compression algorithms), but of course pointless. Hence, some implicit, joint information between two sensors is required. Recall here that these sensors are embedded in a physical environment – it is quite likely that the readings of adjacent sensors are going to be quite similar; they are *correlated*. Such **correlation** can indeed be exploited such that not simply the sum of the data must be transmitted but that overhead can be saved here. The theoretical basis is the theorem by SLEPIAN and WOLF [774], which carries their name. Good overview papers are references [653, 901].

Slepian-Wolf theorem–based work is an example of exploiting spatial correlation that is commonly present in sensor readings, as long as the network is sufficiently dense, compared to the derivate of the observed function and the degree of correlation between readings at two places. Similarly, **temporal correlation** can be exploited in sensor network protocols.

Distributed and collaborative signal processing

The in-networking processing approaches discussed so far have not really used the ability for *processing* in the sensor nodes, or have only used this for trivial operations like averaging or finding the maximum. When complex computations on a certain amount of data is to be done, it can still be more energy efficient to compute these functions on the sensor nodes despite their limited processing power, if in return the amount of data that has to be communicated can be reduced.

An example for this concept is the distributed computation of a Fast Fourier Transform (FFT) [152]. Depending on where the input data is located, there are different algorithms available to compute an FFT in a distributed fashion, with different trade-offs between local computation complexity and the need for communication. In principle, this is similar to algorithm design for parallel computers. However, here not only the latency of communication but also the energy consumption of communication and computation are relevant parameters to decide between various algorithms.

Such distributed computations are mostly applicable to signal processing type algorithms; typical examples are beamforming and target tracking applications. ZHAO and GUIBAS [924] provide a good overview of this topic.

Mobile code/Agent-based networking

With the possibility of executing programs in the network, other programming paradigms or computational models are feasible. One such model is the idea of **mobile code** or **agent-based networking**. The idea is to have a small, compact representation of program code that is small enough to be sent from node to node. This code is then executed locally, for example, collecting measurements, and then decides where to be sent next. This idea has been used in various environments; a classic example is that of a software agent that is sent out to collect the best possible travel itinerary by hopping from one travel agent's computer to another and eventually returning to the user who has posted this inquiry. There is a vast amount of literature available on mobile code/software agents in general, see, for example, references [98, 176, 892]. A newer take on this approach is to consider biologically inspired systems, in particular, the **swarm intelligence** of groups of simple entities, working together to reach a common goal [86, 220].

In wireless sensor networks, mobile agents and related concepts have been considered in various contexts, mostly with respect to routing of queries and for data fusion; see, for example, references [96, 99, 207, 663, 664, 665, 829]. Also, virtual machines for WSNs have been proposed that have a native language that admits a compact representation of the most typical operations that mobile code in a WSN would execute, allowing this code to be small [481].

3.3.3 Adaptive fidelity and accuracy

Section 2.3.4 has already discussed, in the context of a single node, the notion of making the fidelity of computation results contingent upon the amount of energy available for that particular computation. This notion can and should be extended from a single node to an entire network [246].

As an example, consider a function approximation application. Clearly, when more sensors participate in the approximation, the function is sampled at more points and the approximation is better. But in return for this, more energy has to be invested. Similar examples hold for event detection and tracking applications and in general for WSNs.

Hence, it is up to an application to somehow define the degree of accuracy of the results (assuming that it can live with imprecise, approximated results) and it is the task of the communication protocols to try to achieve at least this accuracy as energy efficiently as possible. Moreover, the application should be able to adapt its requirements to the current status of the network – how many nodes have already failed, how much energy could be scavenged from the environment, what are the operational conditions (have critical events happened recently), and so forth. Therefore, the application needs feedback from the network about its status to make such decisions.

But as already discussed in the context of WSN-specific QoS metrics, the large variety of WSN applications makes it quite challenging to come up with a uniform interface for expressing such requirements, let alone with communication protocols that implement these decisions. This is still one of the core research problems of WSN.

3.3.4 Data centricity

Address data, not nodes

In traditional communication networks, the focus of a communication relationship is usually the pair of communicating peers – the sender and the receiver of data. In a wireless sensor network, on the other hand, the interest of an application is not so much in the *identity* of a particular sensor node, it is much rather in the actual information reported about the physical environment. This is especially the case when a WSN is redundantly deployed such that any given event could be reported by multiple nodes – it is of no concern to the application precisely which of these nodes is providing data. This fact that not the identity of nodes but the data are at the center of attention is called **data-centric networking**. For an application, this essentially means that an interface is exposed by the network where data, not nodes, is addressed in requests. The set of nodes that is involved in such a *data-centric address* is implicitly defined by the property that a node can contribute data to such an address.

As an example, consider the elephant-tracking example from Figure 3.5. In a data-centric application, all the application would have to do is state its desire to be informed about events of a certain type – "presence of elephant" – and the nodes in the network that possess "elephant detectors" are implicitly informed about this request. In an identity-centric network, the requesting node would have to find out somehow all nodes that provide this capability and address them explicitly. As another example, it is useful to consider the location of nodes as a property that defines whether a node belongs to a certain group or not. The typical example here is the desire to communicate with all nodes in a given area, say, to retrieve the (average) temperature measured by all nodes in the living room of a given building.

Data-centric networking allows very different networking architectures compared to traditional, identity-centric networks. For one, it is the ultimate justification for some in-network processing techniques like data fusion and aggregation. Data-centric addressing also enables simple expressions of communication relationships – it is no longer necessary to distinguish between one-to-one, one-to-many, many-to-one, or many-to-many relationships as the set of participating nodes is only implicitly defined. In addition to this decoupling of identities, data-centric addressing also supports a decoupling in time as a request to provide data does not have to specify when the answer should happen – a property that is useful for event-detection applications, for example.

Apart from providing a more natural way for an application to express its requirements, data-centric networking and addressing is also claimed to improve performance and especially energy efficiency of a WSN. One reason is the hope that data-centric solutions scale better by being implementable using purely local information about direct neighbors. Another reason could be the easier integration of a notion of adaptive accuracy into a data-centric framework as the data as well as its desired accuracy can be explicitly expressed – it is not at all clear how stating accuracy requirements in an identity-centric network could even be formulated, let alone implemented. But this is still an objective of current research.

Implementation options for data-centric networking

There are several possible ways to make this abstract notion of data-centric networks more concrete. Each way implies a certain set of interfaces that would be usable by an application. The three most important ones are briefly sketched here and partially discussed in more detail in later chapters.

Overlay networks and distributed hash tables

There are some evident similarities between well-known peer-to-peer applications [55, 480, 574, 608, 842] like file sharing and WSN: In both cases, the user/requester is interested only in looking up and obtaining data, not in its source; the request for data and its availability can be decoupled in time; both types of networks should scale to large numbers.

In peer-to-peer networking, the solution for an efficient lookup of retrieval of data from an unknown source is usually to form an overlay network, implementing a Distributed Hash Table (DHT) [686, 704, 792, 922]. The desired data can be identified via a given key (a hash) and the DHT will provide one (or possibly several) sources for the data associated with this key. The crucial point is that this data source lookup can be performed efficiently, requiring $O(\log n)$ steps where n is the number of nodes, even with only distributed, localized information about where information is stored in the peer-to-peer network.

Despite these similarities, there are some crucial differences. First of all, it is not clear how the rather static key of a DHT would correspond to the more dynamic, parameterized requests in a WSN. Second, and more importantly, DHTs, coming from an IP-networking background, tend to ignore the distance/the hop count between two nodes and consider nodes as adjacent only on the basis of semantic information about their stored keys. This hop-count-agnostic behavior is unacceptable for WSNs where each hop incurs considerable communication overhead. There is some on-going work on taking the topology of the underlying network also into account [460, 683, 846] or the position of nodes [685, 760] when constructing the overlay network, but the applicability of this work to WSN is still open. Chapter 12 will deal with these approaches in more detail.

Publish/Subscribe

The required separation in both time and identity of a sink node asking for information and the act of providing this information is not well matched with the synchronous characteristics of a request/reply protocol. What is rather necessary is a means to express the need for certain data and the delivery of the data, where the data as such is specified and not the involved entities.

This behavior is realized by the **publish/subscribe** approach [251]: Any node interested in a given kind of data can *subscribe* to it, and any node can *publish* data, along with information about its kind as well. Upon a publication, all subscribers to this kind of data are notified of the new data. The elephant example is then easily expressed by sink nodes subscribing to the event "elephant detected"; any node that is detecting an elephant can then, at any later time, publish this event. If a subscriber is no longer interested, it can simply *unsubscribe* from any kind of event and will no longer be notified of such events. Evidently, subscription and publication can happen at different points in time and the identities of subscribers and publishers do not have to be known to each other.

Implementing this abstract concept of publishing and subscribing to information can be done in various ways. One possibility is to use a central entity where subscriptions and publications are matched to each other, but this is evidently inappropriate for WSNs. A distributed solution is preferable but considerably more complicated.

Also relevant is the expressiveness of the data descriptions (their "names") used to match publications and subscriptions. A first idea is to use explicit subjects or keywords as names, which have to be defined up front – published data only matches to subscriptions with the same keyword (like in the "elephant detected" example above). This subject-based approach can be extended into hierarchical schemes where subjects are arranged in a tree; a subscription to a given subject then also implies interest in any descendent subjects. A more general naming scheme allows to formulate the matching condition between subscriptions and publications as general predicates over the content of the publication and is hence referred to as **content-based publish/subscribe** approach (see e.g. reference [123] and the references therein for an introduction and overview).

In practice, general predicates on the content are somewhat clumsy to handle and restricted expressions (also called *filters*) of the form `(attribute, value, operator)` are preferable, where `attribute` corresponds to the subjects from above (e.g. `temperature`) and can assume values, `value` is a concrete value like `"25°C"` or a placeholder (`ALL` or `ANY`), and `operator` is a relational operator like "=", "<", "≤". Moreover, this formalism also lends itself very conveniently to the expression of accuracy requirements or periodic measurement support.

The question remains where to send publication and subscription messages if a decentralized approach is chosen – simply flooding all messages evidently defeats the purpose. MÜHL et al. [576] give an overview of various approaches, for example, flooding the subscriptions or exploiting information contained in the content-based filters to limit propagation of messages [121, 122, 575].

Publish/subscribe networking is a very popular approach for WSN. In fact, some of the most popular protocols are incarnations of this principle and are discussed in detail in Part II, in particular in Chapter 12.

Databases

A somewhat different view on WSN is to consider them as (dynamic) databases [269, 303, 374, 887]. This view matches very well with the idea of using a data-centric organization of the networking protocols. Being interested in certain aspects of the physical environment that is surveyed by a WSN is equivalent to formulating queries for a database.

To cast the sensor networks into the framework of relational databases, it is useful to regard the sensors as a virtual table to which relational operators can be applied. Then, extracting the average temperature reading from all sensors in a given room can be simply written as shown in Listing 3.1 [528] – it should come as no surprise to anybody acquainted with the Standard Query Language (SQL).

Listing 3.1: Example of an SQL-based request for sensor readings [528]

```
SELECT AVG(temperature)
FROM sensors
WHERE location = "Room 123"
```

Such SQL-based querying of a WSN can be extended to an easy-to-grasp interface to wireless sensor networks, being capable of expressing most salient interaction patterns with a WSN. It is, however, not quite as clear how to translate this interface into actual networking protocols that implement this interface and can provide the results for such queries. In a traditional relational database, this implementation of a query is done by determining an execution plan; the same is necessary here. Here, however, the execution plan has to be distributed and has to explicitly take communication costs into account.

3.3.5 Exploit location information

Another useful technique is to exploit location information in the communication protocols whenever such information is present. Since the location of an event is a crucial information for many applications, there have to be mechanisms that determine the location of sensor nodes (and possibly also that of observed events) – they are discussed in detail in Chapter 9. Once such information is available, it can simplify the design and operation of communication protocols and can improve their energy efficiency considerably. We shall see various examples in different protocols in Part II.

3.3.6 Exploit activity patterns

Activity patterns in a wireless sensor network tend to be quite different from traditional networks. While it is true that the data rate averaged over a long time can be very small when there is only very rarely an event to report, this can change dramatically when something does happen. Once an event has happened, it can be detected by a larger number of sensors, breaking into a frenzy of activity, causing a well-known event shower effect. Hence, the protocol design should be able to handle such bursts of traffic by being able to switch between modes of quiescence and of high activity.

3.3.7 Exploit heterogeneity

Related to the exploitation of activity patterns is the exploitation of heterogeneity in the network. Sensor nodes can be heterogenous by constructions, that is, some nodes have larger batteries, farther-reaching communication devices, or more processing power. They can also be heterogenous by evolution, that is, all nodes started from an equal state, but because some nodes had to perform more tasks during the operation of the network, they have depleted their energy resources or other nodes had better opportunities to scavenge energy from the environment (e.g. nodes in shade are at a disadvantage when solar cells are used).

Whether by construction or by evolution, heterogeneity in the network is both a burden and an opportunity. The opportunity is in an asymmetric assignment of tasks, giving nodes with more resources or more capabilities the more demanding tasks. For example, nodes with more memory or faster processors can be better suited for aggregation, nodes with more energy reserves for hierarchical coordination, or nodes with a farther-reaching radio device should invest their energy mostly for long-distance communication, whereas, shorter-distance communication can be undertaken by the other nodes. The burden is that these asymmetric task assignments cannot usually be static but have to be reevaluated as time passes and the node/network state evolves. Task reassignment in turn is an activity that requires resources and has to be balanced against the potential benefits.

3.3.8 Component-based protocol stacks and cross-layer optimization

Finally, a consideration about the implementation aspects of communication protocols in WSNs is necessary. Section 2.3.3 has already made the case for a component-based as opposed to a layering-based model of protocol implementation in WSN. What remains to be defined is mainly a default collection of components, not all of which have to be always available at all times on all sensor nodes, but which can form a basic "toolbox" of protocols and algorithms to build upon.

In fact, most of the chapters of Part II are about such building blocks. All wireless sensor networks will require some – even if only simple – form of physical, MAC and link layer[2] protocols; there will be wireless sensor networks that require routing and transport layer functionalities. Moreover, "helper modules" like time synchronization, topology control, or localization can be useful. On top of these "basic" components, more abstract functionalities can then be built. As a consequence, the set of components that is active on a sensor node can be complex, and will change from application to application.

Protocol components will also interact with each other in essentially two different ways [330]. One is the simple exchange of data packets as they are passed from one component to another as it is processed by different protocols. The other interaction type is the exchange of cross-layer information.

This possibility for cross-layer information exchange holds great promise for protocol optimization, but is also not without danger. Kawadia and Kumar [412], for example, argue that imprudent use of cross-layer designs can lead to feedback loops, endangering both functionality and performance of the entire system. Clearly, these concerns should not be easily disregarded and care has to be taken to avoid such unexpected feedback loops.

3.4 Service interfaces of WSNs

3.4.1 Structuring application/protocol stack interfaces

Looking at Section 2.3's discussion of a component-based operating system and protocol stack already enables one possibility to treat an application: It is just another component that can directly interact with other components using whatever interface specification exists between them (e.g. the command/event structure of TinyOS). The application could even consist of several components,

[2] While these components do not form a layer in the strict sense of the word, it will still be useful to refer to the corresponding functionality as that of a, say, "physical layer".

integrated at various places into the protocol stack. This approach has several advantages: It is streamlined with the overall protocol structure, makes it easy to introduce application-specific code into the WSN at various levels, and does not require the definition of an abstract, specific service interface. Moreover, such a tight integration allows the application programmer a very fine-grained control over which protocols (which components) are chosen for a specific task; for example, it is possible to select out of different routing protocols the one best suited for a given application by accessing this component's services.

But this generality and flexibility is also the potential downside of this approach. The allowing of the application programmer to mess with protocol stacks and operating system internals should not be undertaken carelessly. In traditional networks such as the Internet, the application programmer can access the services of the network via a commonly accepted interface: sockets [791]. This interface makes clear provisions on how to handle connections, how to send and receive packets, and how to inquire about state information of the network.[3] This clarity is owing to the evident tasks that this interface serves – the exchange of packets with one (sometimes, several) communication peers.

Therefore, there is the design choice between treating the application as just another component or designing a service interface that makes all components, in their entirety, accessible in a standardized fashion. These two options are outlined by Figure 3.8. A service interface would allow to raise the level of abstraction with which an application can interact with the WSN – instead of having to specify which value to read from which particular sensor, it might be desirable to provide an application with the possibility to express sensing tasks in terms that are close to the semantics of the application. In this sense, such a service interface can hide considerable complexity and is actually conceivable as a "middleware" in its own right.

Clearly, with a tighter integration of the application into the protocol stack, a broader optimization spectrum is open to the application programmer. On the downside, more experience will be necessary than when using a standardized service interface. The question is therefore on the one hand the price of standardization with respect to the potential loss of performance and on the other hand, the complexity of the service interface.

In fact, the much bigger complexity and variety of communication patterns in wireless sensor networks compared to Internet networks makes a more expressive and potentially complex service interface necessary. To better understand this trade-off, a clearer understanding of expressibility requirements of such an interface is necessary.

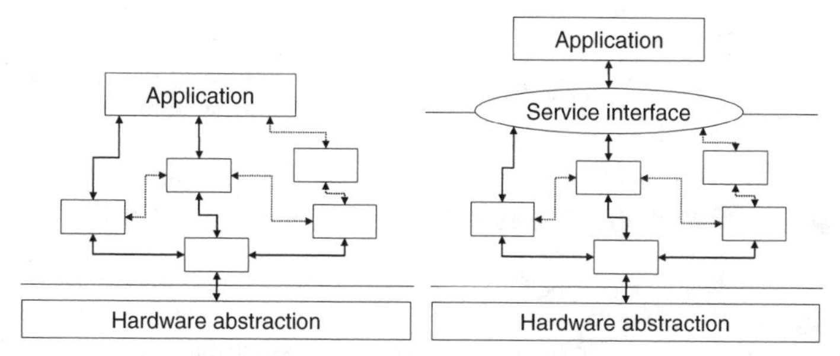

Figure 3.8 Two options for interfacing an application to a protocol stack: As just another component or via a deliberately designed, general service interface

[3] It is certainly correct to argue that the socket interface has its shortcomings and open issues, especially with regard to wireless communication. But these issues are mostly related to the wish to access lower-layer information, for example, received signal strength information, which is not directly exposed by the interface, but only via various, nonstandard workarounds.

3.4.2 Expressibility requirements for WSN service interfaces

The most important functionalities that a service interface should expose include:

- Support for simple request/response interactions: retrieving a measured value from some sensor or setting a parameter in some node. This is a synchronous interaction pattern in the sense that the result (or possibly the acknowledgment) is expected immediately. In addition, the responses can be required to be provided periodically, supporting periodic measurement-type applications.
- Support for asynchronous event notifications: a requesting node can require the network to inform it if a given condition becomes true, for example, if a certain event has happened. This is an asynchronous pattern in the sense that there is no a priori relationship between the time the request is made and the time the information is provided.
  This form of asynchronous requests should be accompanied by the possibility to cancel the request for information. It can be further refined by provisions about what should happen after the condition becomes true; a typical example is to request periodic reporting of measured values after an event.
- For both types of interactions, the addressees should be definable in several ways. The simplest option is an explicit enumeration of the single or multiple communication peers to whom a (synchronous or asynchronous) request is made – this corresponds to the peer address in a socket communication.
  More interesting is the question of how to express data centricity. One option, closely related to the publish/subscribe approach discussed in Section 3.3.4, is the implicit definition of peers by some form of a membership function of an abstract group of nodes. Possible examples for such membership functions include:
  - Location – all nodes that are in a given region of space belong to a group.
  - Observed value – all nodes that have observed values matching a given predicate belong to a group. An example would be to require the measured temperature to be larger than 20°C.

  Along with these groups, the usual set-theoretic operations of intersection, union, or difference between groups should be included in the service interface as well.
  Because of this natural need for a service interface semantics that corresponds to the publish/subscribe concept, this approach is a quite natural, but not the only possible, fit with WSNs.
- In-networking processing functionality has to be accessible. For an operation that accesses an entire group of nodes, especially when reading values from this group (either synchronously or asynchronously), it should be possible to specify what kind of in-network processing should be applied to it. In particular, processing that modifies the nature of the result (i.e., data fusion) must be explicitly allowed by the requesting application.
  In addition, it can be desirable for an application to be able to infuse its own in-network processing functions into the network. For example, a new aggregation function could be defined or a specific mobile agent has to be written by the application programmer anyway.
  In-network processing and application-specific code may also be useful to detect **complex events**: events that cannot be detected locally, by a single sensor, but for which data has to be exchanged between sensors.
- Related to the specification of aggregation functions is the specification of the required accuracy of a result. This can take on the form of specifying bounds on the number of group members that should contribute to a result, or the level of compression that should be applied. Hand in hand with required accuracy goes the acceptable energy expenditure to produce a given piece of information.
- Timeliness requirements about the delivery of data is a similar aspect. For example, it may be possible to provide a result quickly but at higher energy costs (e.g. by forcing nodes to wake up earlier than they would wake up anyway) or slowly but at reduced energy costs (e.g. by piggy-backing information on other data packets that have to exchanged anyway).

In general, any trade-offs regarding the energy consumption of any possible exchange of data packets should be made explicit as far as possible.

- The need to access location, timing, or network status information (e.g. energy reserves available in the nodes or the current rate of energy scavenging) via the service interface.
  It may also be useful to agglomerate location information into higher-level abstractions to be able to talk about objects that correspond to a human view of things, for example, "room 123". Similarly, facts like the administrative entity a sensor network belongs to can be practically important [751].
- To support the seamless connection of various nodes or entire networks as well as the simple access to services in an "unknown" network, there is a need for an explicit description of the set of available capabilities of the node/the network – for example, which physical parameters can be observed or which entities can be controlled. SGROI et al. [751] argue for a "concept repository" for this purpose.
- Security requirements as well as properties have be somehow expressed.
- While not a direct part of an actual service interface, additional management functionality, for example, for updating components, can be convenient to be present in the interface as well.

To avoid confusion, it is worthwhile to point out that the design of synchronous or asynchronous interface semantics has very little to do with a blocking or nonblocking design of the actual service invocation. It is, for example, easy to implement an asynchronous semantics with blocking invocations as long as the operating system provides threads. These really are separate issues.

3.4.3 Discussion

Evidently, the wealth of options that a general-purpose interface to WSNs would have to offer is vast. Looking at the overall picture, three key issues – data centricity, trade-offs against energy, and accuracy – make these networks quite different from all existing network types and how to offer them in a convenient service interface to an application programmer is anything but clear. It is hence perhaps not so surprising that there has only been relatively little work on a systematic approach to service interfaces for WSN.

One attempt has been undertaken by SGROI et al. [751], who start from a relatively conservative client/server interface paradigm and use it to arrange a "query manager" and a "command interface", embellished by additional sets of parameters. While their parameter sets are relatively extensive and can incorporate most of the issues above, it is not clear that this API can indeed support all types of programming models in WSNs. It is, in particular, unclear how to extend in-network processing functionalities (e.g. write new aggregation functions) based on their API, how to control energy trade-offs, or how to select from an application one out of several components that are suitable for a given tasks (e.g. select one of several routing protocols).

Some of the candidates for data centricity, in particular, publish/subscribe and databases, are relatively close to meeting all these requirements for a service interface, but all of them still need extensions. The publish/subscribe interface, for example, can be extended by subscriptions that express accuracy and tasking aspects. Akin to publish/subscribe is the notion of events where an application can express interests in single events or in certain complex events. One example is the DSWare system [490]. Also, the database approach appears promising.

On the basis of the idea of mobile agents, the SensorWare system [96] provides a set of simple commands, in particular, `query`, `send`, `wait`, `value`, and `replicate`. These commands allow mobile code to send itself to some other node, to replicate it into the network by sending itself to the "children" of a node, or to wait for the results returned from these children. This provides considerable flexibility, but is still a fairly low level of programming.

One example for a highly application-specific way of defining service requests is EnviroTrack [2], which is specialized to the tracking of mobile objects. It allows to define "contexts" for certain tracking tasks, which have activation functions and "reporting" objects, resulting in an extremely compact expression of the service request, which then has to be transformed into concrete interactions of sensor nodes.

As one example for another research approach, consider the attempt to model the behavior and characteristics of sensor networks as a set of Unified Modeling Language (UML) schemes, resulting in the "Sensor Modeling Language" (SensorML) [749]. As this effort is driven by specific application requirements (geosciences and earth-observing satellites), it concentrates mostly on the description of the capabilities of individual sensors, but makes provisions to express, for example, accuracy and data processing. The big advantage here is the potential to describe the "meaning" of measured parameters explicitly. Another example for such an approach is to use the DARPA Agent Markup Language (DAML) as an explicit description language for the capabilities of heterogeneous sensors [384]. How such UML-based concepts could be applied to entire networks is, however, completely open.

Looking at the high complexity of service interfaces necessary to harness all the possible options and requirements of how an application might want to interact with a protocol stack, it is rather questionable whether the existing, quite heavy-weight, but still limited, proposals for service interfaces are the last word on the topic. A better understanding in structuring this interaction is still necessary. Moreover, the price to pay in performance optimization when using a predefined service interface still has to be weighted against the danger of inexperienced application programmers messing with the protocol stack's internals.

3.5 Gateway concepts

3.5.1 The need for gateways

For practical deployment, a sensor network only concerned with itself is insufficient. The network rather has to be able to interact with other information devices, for example, a user equipped with a PDA moving in the coverage area of the network or with a remote user, trying to interact with the sensor network via the Internet (the standard example is to read the temperature sensors in one's home while traveling and accessing the Internet via a wireless connection). Figure 3.9 shows this networking scenario.

To this end, the WSN first of all has to be able to exchange data with such a mobile device or with some sort of gateway, which provides the physical connection to the Internet. This is relatively straightforward on the physical, MAC, and link layer – either the mobile device/the gateway is

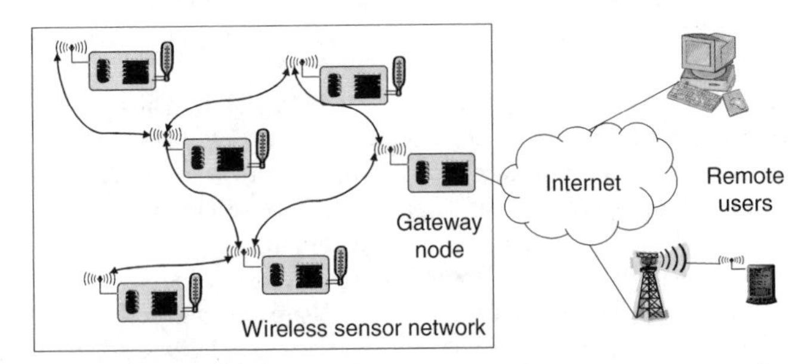

Figure 3.9 A wireless sensor network with gateway node, enabling access to remote clients via the Internet

equipped with a radio transceiver as used in the WSN, or some (probably not all) nodes in the WSN support standard wireless communication technologies such as IEEE 802.11. Either option can be advantageous, depending on the application and the typical use case. Possible trade-offs include the percentage of multitechnology sensor nodes that would be required to serve mobile users in comparison with the overhead and inconvenience to fit WSN transceivers to mobile devices like PDAs.

The design of gateways becomes much more challenging when considering their logical design. One option to ponder is to regard a gateway as a simple router between Internet and sensor network. This would entail the use of Internet protocols within the sensor network. While this option has been considered as well [215] and should not be disregarded lightly, it is the prevalent consensus that WSNs will require specific, heavily optimized protocols. Thus, a simple router will not suffice as a gateway.

The remaining possibility is therefore to design the gateway as an actual application-level gateway: on the basis of the application-level information, the gateway will have to decide its action. A rough distinction of the open problems can be made according to from where the communication is initiated.

3.5.2 WSN to Internet communication

Assume that the initiator of a WSN–Internet communication resides in the WSN (Figure 3.10) – for example, a sensor node wants to deliver an alarm message to some Internet host. The first problem to solve is akin to ad hoc networks, namely, how to find the gateway from within the network. Basically, a routing problem to a node that offers a specific service has to be solved, integrating routing and service discovery [139, 420, 435, 696, 799].

If several such gateways are available, how to choose between them? In particular, if not all Internet hosts are reachable via each gateway or at least if some gateway should be preferred for a given destination host? How to handle several gateways, each capable of IP networking, and the communication among them? One option is to build an IP overlay network on top of the sensor network [946].

How does a sensor node know to which Internet host to address such a message? Or even worse, how to map a semantic notion ("Alert Alice") to a concrete IP address? Even if the sensor node does not need to be able to process the IP protocol, it has to include sufficient information (IP address and port number, for example) in its own packets; the gateway then has to extract this information and translate it into IP packets. An ensuing question is which source address to use here – the gateway in a sense has to perform tasks similar to that of a Network Address Translation (NAT) device [225].

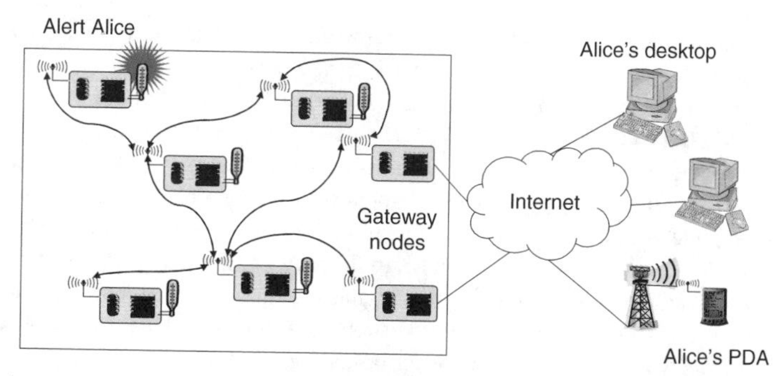

Figure 3.10 An event notification to "Alice" needs decisions about, among others, gateway choice, mapping "Alice" to a concrete IP address, and translating an intra-WSN event notification message to an Internet application message

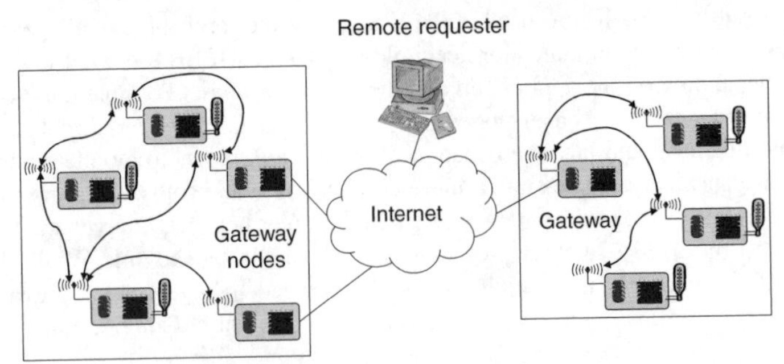

Figure 3.11 Requesting sensor network information from a remote terminal entails choices about which network to address, which gateway node of a given network, and how and where to adapt application-layer protocol in the Internet to WSN-specific protocols

3.5.3 Internet to WSN communication

The case of an Internet-based entity trying to access services of a WSN is even more challenging (Figure 3.11). This is fairly simple if this requesting terminal is able to directly communicate with the WSN, for example, a mobile requester equipped with a WSN transceiver, and also has all the necessary protocol components at its disposal. In this case, the requesting terminal can be a direct part of the WSN and no particular treatment is necessary.

The more general case is, however, a terminal "far away" requesting the service, not immediately able to communicate with any sensor node and thus requiring the assistance of a gateway node. First of all, again the question of service discovery presents itself – how to find out that there actually is a sensor network in the desired location, and how to find out about the existence of a gateway node?

Once the requesting terminal has obtained this information, how to access the actual services? Clearly, addressing an individual sensor (like addressing a communication peer in a traditional Internet application) both goes against the grain of the sensor network philosophy where an individual sensor node is irrelevant compared to the data that it provides and is impossible if a sensor node does not even have an IP address.

The requesting terminal can instead send a properly formatted request to this gateway, which acts as an application-level gateway or a proxy for the individual/set of sensor nodes that can answer this request; the gateway translates this request into the proper intrasensor network protocol interactions. This assumes that there is an application-level protocol that a remote requester and gateway can use and that is more suitable for communication over the Internet than the actual sensor network protocols and that is more convenient for the remote terminal to use. The gateway can then mask, for example, a data-centric data exchange within the network behind an identity-centric exchange used in the Internet.

It is by no means clear that such an application-level protocol exists that represents an actual simplification over just extending the actual sensor network protocols to the remote terminal, but there are some indications in this direction. For example, it is not necessary for the remote terminal to be concerned with maintaining multihop routes in the network nor should it be considered as "just another hop" as the characteristics of the Internet connection are quite different from a wireless hop.

In addition, there are some clear parallels for such an application-level protocol with so-called Web Service Protocols, which can explicitly describe services and the way they can be accessed. The Web Service Description Language (WSDL) [166], in particular, can be a promising starting point for extension with the required attributes for WSN service access – for example, required accuracy, energy trade-offs, or data-centric service descriptions. Moreover, the question arises as to how to

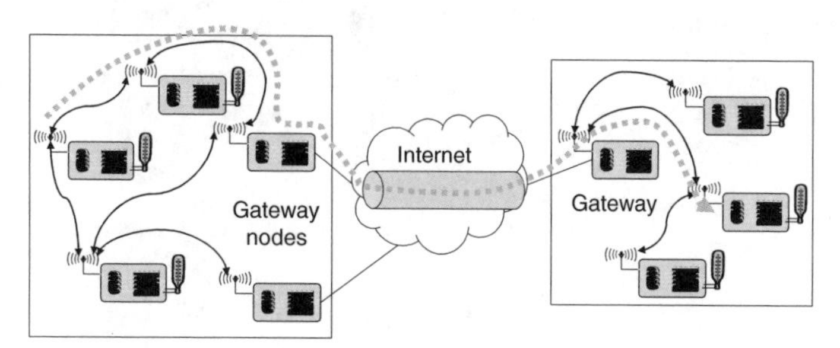

Figure 3.12 Connecting two WSNs with a tunnel over the Internet

integrate WSN with general middleware architectures [699] or how to make WSN services accessible from, say, a standard Web browser (which should be an almost automatic by-product of using WSDL and related standards in the gateway). However, research here is still in its early infancy [384, 508, 656]. Also, once a general-purpose service interface to WSNs is commonly accepted (such as [751]), this will have a clear impact on how to access WSN services from afar as well.

3.5.4 WSN tunneling

In addition to these scenarios describing actual interactions between a WSN and Internet terminals, the gateways can also act as simple extensions of one WSN to another WSN. The idea is to build a larger, "virtual" WSN out of separate parts, transparently "tunneling" all protocol messages between these two networks and simply using the Internet as a transport network (Figure 3.12) [751]. This can be attractive, but care has to be taken not to confuse the virtual link between two gateway nodes with a real link; otherwise, protocols that rely on physical properties of a communication link can get quite confused (e.g. time synchronization or localization protocols).

Such tunnels need not necessarily be in the form of fixed network connections; even mobile nodes carried by people can be considered as means for intermediate interconnection of WSNs [292]. FALL [252] also studies a similar problem in a more general setting.

3.6 Conclusion

The main conclusion to draw from this chapter is the fact that wireless sensor networks and their networking architecture will have many different guises and shapes. For many applications, but by no means all, multihop communication is the crucial enabling technology, and most of the WSN research as well as the following part of this book are focused on this particular form of wireless networking.

Four main optimization goals – WSN-specific forms of quality of service support, energy efficiency, scalability, and robustness – dominate the requirements for WSNs and have to be carefully arbitrated and balanced against each other. To do so, the design of WSNs departs in crucial aspects from that of traditional networks, resulting in a number of design principles. Most importantly, distributed organization of the network, the use of in-network processing, a data-centric view of the network, and the adaptation of result fidelity and accuracy to given circumstances are pivotal techniques to be considered for usage.

The large diversity of WSNs makes the design of a uniform, general-purpose service interface difficult; consequently, no final solutions to this problem are currently available. Similarly, the integration of WSNs in larger network contexts, for example, to allow Internet-based hosts a simple access to WSN services, is also still a fairly open problem.

Part II

Communication protocols

4

Physical layer

Objectives of this Chapter

This chapter is devoted to the physical layer, that is, those functions and components of a sensor node that mediate between the transmission and reception of wireless waveforms and the processing of digital data in the remaining node, including the higher-layer protocol processing.

It is a commonly acknowledged truth that the properties of the transmission channel and the physical-layer shape significant parts of the protocol stack. The first goal of this chapter is therefore to provide the reader with a basic understanding of some fundamental concepts related to digital communications over wireless channels.

The second important goal is to explain how the specific constraints of wireless sensor networks (regarding, for example, energy and node costs) in turn shape the design of modulation schemes and transceivers. The reader should get an understanding on some of the fundamental trade-offs regarding transmission robustness and energy consumption and how these are affected by the power-consumption properties of transceiver components.

Chapter Outline

4.1 Introduction

The physical layer is mostly concerned with modulation and demodulation of digital data; this task is carried out by so-called **transceivers**. In sensor networks, the challenge is to find modulation schemes and transceiver architectures that are simple, low cost, but still robust enough to provide the desired service.

Protocols and Architectures for Wireless Sensor Networks H. Karl and A. Willig

The first part of this chapter explains the most important concepts regarding wireless channels and digital communications (over wireless channels); its main purpose is to provide appropriate notions and to give an insight into the tasks involved in transmission and reception over wireless channels. We discuss some simple modulation schemes as well.

In the second part, we discuss the implications of the specific requirements of wireless sensor networks, most notably the scarcity of energy, for the design of transceivers and transmission schemes.

4.2 Wireless channel and communication fundamentals

This section provides the necessary background on wireless channels and digital communication over these. This is by no means an exhaustive discussion; it should just provide enough background and the most important notions to understand the energy aspects involved. Wireless channels are discussed in some more detail in references [124, 335, 620, 682, 744], some good introductory books on digital communication in general are references [772], [661], and more specific for wireless communications and systems are references [682, 848].

In wireless channels, electromagnetic waves propagate in (nearly) free space between a transmitter and a receiver. Wireless channels are therefore an **unguided medium**, meaning that signal propagation is not restricted to well-defined locations, as is the case in wired transmission with proper shielding.

4.2.1 Frequency allocation

For a practical wireless, RF-based system, the carrier frequency has to be carefully chosen. This carrier frequency determines the propagation characteristics – for example, how well are obstacles like walls penetrated – and the available capacity. Since a single frequency does not provide any capacity, for communication purposes always a finite portion of the electromagnetic spectrum, called a **frequency band**, is used. In radio-frequency (RF) communications, the range of usable radio frequencies in general starts at the Very Low Frequency (VLF) range and ends with the Extremely High Frequency (EHF) range (Figure 4.1). There is also the option of **infrared** or **optical** communications, used, for example, in the "Smart Dust" system [392]. The infrared spectrum is between wavelengths of 1 mm (corresponding to 300 GHz[1]) and 2.5 μm (120 THz), whereas the optical range ends at 780 nm (≈385 THz).

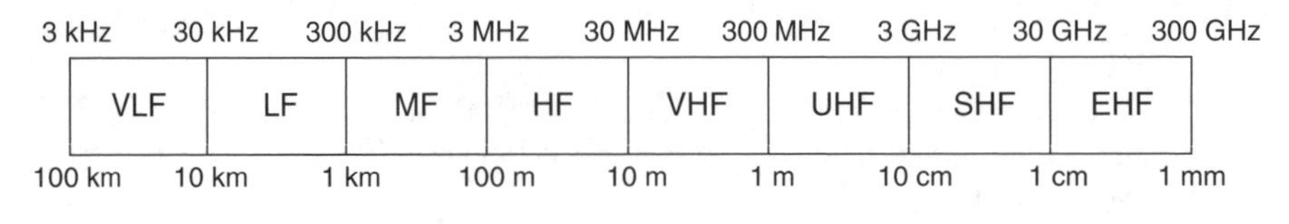

VLF = Very low frequency
LF = Low frequency
MF = Medium frequency
HF = High frequency
VHF = Very high frequency
UHF = Ultrahigh frequency
SHF = Super high frequency
EHF = Extremely high frequency

Figure 4.1 Electromagnetic spectrum – radio frequencies

[1] Assuming that the speed of light is 300,000,000 m/s.

Table 4.1 Some of the ISM bands

Frequency	Comment
13.553–13.567 MHz	
26.957–27.283 MHz	
40.66–40.70 MHz	
433–464 MHz	Europe
902–928 MHz	Only in the Americas
2.4–2.5 GHz	Used by WLAN/WPAN technologies
5.725–5.875 GHz	Used by WLAN technologies
24–24.25 GHz	

The choice of a frequency band is an important factor in system design. Except for ultrawideband technologies (see Section 2.1.4), most of today's RF-based systems work at frequencies below 6 GHz. The range of radio frequencies is subject to **regulation** to avoid unwanted interference between different users and systems. Some systems have special licenses for reserved bands; for example, in Europe, the GSM system can exclusively use the GSM 900 (880–915 MHz) and GSM 1800 (1710–1785 MHz) bands.[2] There are also licensefree bands, most notably the Industrial, Scientific, and Medical (ISM) bands, which are granted by the ITU for private and unlicensed use subject to certain restrictions regarding transmit power, power spectral density, or duty cycle. Table 4.1 lists some of the ISM frequency bands. Working in an unlicensed band means that one can just go to a shop, buy equipment, and start to transmit data without requiring any permission from the government/frequency allocation body. It is not surprising that these bands are rather popular, not only for sensor networks but also for/in other wireless technologies. For example, the 2.4-GHz ISM band is used for IEEE 802.11, Bluetooth, and IEEE 802.15.4.

Some considerations in the choice of frequency are the following:

- In the public ISM bands, any system has to live with interference created by other systems (using the same or different technologies) in the same frequency band, simply because there is no usage restriction. For example, many systems share the 2.4-GHz ISM band, including IEEE 802.11b [466, 467], Bluetooth [318, 319], and the IEEE 802.15.4 WPAN [468] – they **coexist** with each other in the same band. Therefore, all systems in these bands have to be robust against interference from other systems with which they cannot explicitly coordinate their operation. Coexistence needs to be approached both on the physical and the MAC layer [154, 359, 360, 469]. On the other hand, requesting allocation of some exclusive spectrum for a specific sensor network application from the competent regulatory organizations is a time consuming and likely futile endeavor.
- An important parameter in a transmission system is the **antenna efficiency**, which is defined as the ratio of the **radiated power** to the total input power to the antenna; the remaining power is dissipated as heat. The small form factor of wireless sensor nodes allows only small antennas. For example, radio waves at 2.4 GHz have a wave length of 12.5 cm, much longer than the intended dimensions of many sensor nodes. In general, it becomes more difficult to construct efficient antennas as the ratio of antenna dimension to wavelength decreases. As the efficiency decreases, more energy must be spent to achieve a fixed radiated power. These problems are discussed in some detail in reference [115, Chap. 8].

[2] `http://www.gsmworld.com/technology/spectrum/frequencies.shtml`

4.2.2 Modulation and demodulation

When digital computers communicate, they exchange **digital data**, which are essentially sequences of **symbols**, each symbol coming from a finite alphabet, the **channel alphabet**. In the process of **modulation**, (groups of) symbols from the channel alphabet are mapped to one of a finite number of **waveforms** of the same finite length; this length is called the **symbol duration**. With two different waveforms, a **binary modulation** results; if the size is $m \in \mathbb{N}, m > 2$, we talk about m-ary modulation. Some common cases for the symbol alphabet are binary data (the alphabet being $\{0, 1\}$) or bipolar data ($\{-1, 1\}$) in spread-spectrum systems.

When referring to the "speed" of data transmission/modulation, we have to distinguish between the following parameters:

Symbol rate The **symbol rate** is the inverse of the symbol duration; for binary modulation, it is also called **bit rate**.

Data rate The **data rate** is the rate in bit per second that the modulator can accept for transmission; it is thus the rate by which a user can transmit binary data. For binary modulation, bit rate and data rate are the same and often the term bit rate is (sloppily) used to denote the data rate.

For m-ary modulation, the data rate is actually given as the symbol rate times the number of bits encoded in a single waveform. For example, if we use 8-ary modulation, we can associate with each waveform one of eight possible groups of three bits and thus the bit rate is three times the symbol rate. The fundamentals of modulation and several modulation schemes are discussed in textbooks on digital communications, for example, references [78, 661, 772].

Modulation is carried out at the transmitter. The receiver ultimately wants to recover the transmitted symbols from a received waveform. The mapping from a received waveform to symbols is called **demodulation**. Because of noise, attenuation, or interference, the received waveform is a distorted version of the transmitted waveform and accordingly the receiver cannot determine the transmitted symbol with certainty. Instead, the receiver decides for the wrong symbol with some probability, called the **symbol error rate**. For digital data represented by bits, the notion of **bit error rate** (BER) is even more important: it describes the probability that a bit delivered to a higher layer is incorrect. If binary modulation is used, bit error probability and symbol error probability are the same; in case of m-ary modulation they can differ: even if a symbol is demodulated incorrectly, the delivered group of bits might be correct at some places (as long as the SNR is not too low, it is often acceptable to assume that an incorrect symbol maps to only a single incorrect bit). All upper layers are primarily interested in the bit error probability.

The most common form of modulation is the so-called **bandpass modulation**, where the information signal is modulated onto a periodic carrier wave of comparably high frequency [772, Chap. 3]. The spectrum used by bandpass modulation schemes is typically described by a **center frequency** f_c and a **bandwidth** B, and most of the signal energy can be found in the frequency range $\left[f_c - \frac{B}{2}, f_c + \frac{B}{2}\right]$.[3] The carrier is typically represented as a cosine wave, which is uniquely determined by amplitude, frequency, and phase shift.[4] Accordingly, the modulated signal $s(t)$ can, in general, be represented as:

$$s(t) = A(t) \cdot \cos(\omega(t) + \phi(t)),$$

[3] For theoretical reasons, it is not possible to have perfectly band-limited digital signals; there is always some minor signal energy leaking into neighboring frequency bands. For example, the spectrum occupied by a rectangular pulse can be described by a function similar to $\sin(x)/x$, which has nonzero values almost everywhere.

[4] There are three main advantages of bandpass modulation over digital baseband modulation like, for example, pulse modulation: it is technically comparably easy to generate sinusoids; one does not need to build huge antennas to transmit a 5-kHz data signal efficiently, and by choice of nonoverlapping bands, multiple users can transmit in parallel, which would not be possible in case of baseband modulation.

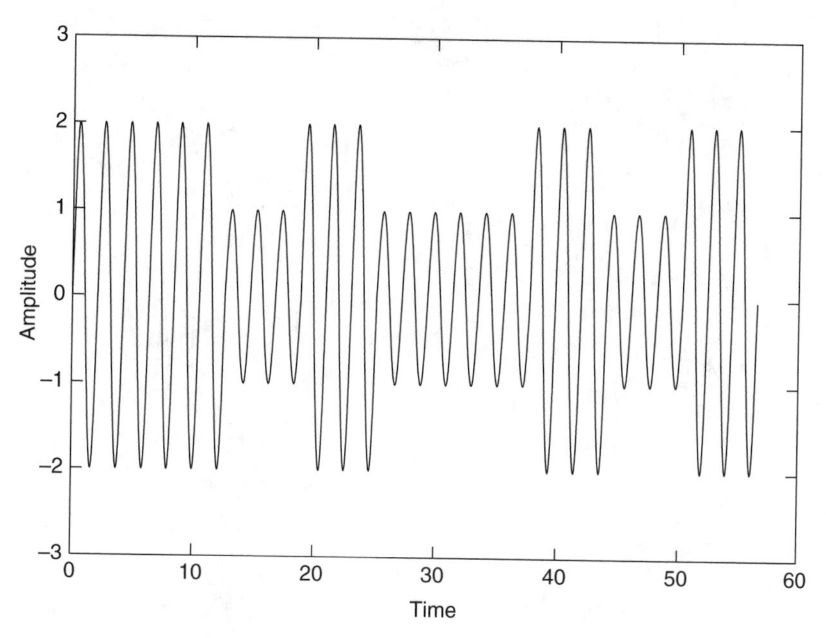

Figure 4.2 Amplitude shift keying (ASK) example

where $A(t)$ is the time-dependent amplitude, $\omega(t)$ is the time-dependent frequency, and $\phi(t)$ is the phase shift. Accordingly, there are three fundamental modulation types: Amplitude Shift Keying (ASK), Phase Shift Keying (PSK) and Frequency Shift Keying (FSK), which can be used as they are or in combination.

In ASK, the waveforms $s_i(\cdot)$ for the different symbols are chosen as:

$$s_i(t) = \sqrt{\frac{2E_i(t)}{T}} \cdot \cos\left[\omega_0 t + \phi\right],$$

where ω_0 is the center frequency, ϕ is an arbitrary constant initial phase, and $E_i(t)$ is constant over the symbol duration $[0, T]$ and assumes one of m different levels. The particular form of the amplitude $\sqrt{\frac{2E_i(t)}{T}}$ is a convention; it displays explicitly the **symbol energy** E. An example for ASK modulation is shown in Figure 4.2, where the binary data string 110100101 is modulated, using $E_0(t) = 1$ and $E_1(t) = 2$ for all t to represent logical zeros and ones. A special case of ASK modulation is a scheme with a binary channel alphabet where zeros are mapped to no signal at all, $E_0(t) = 0$, and $E_1(t) = 1$ for all t. Since it corresponds to switching off the transmitter, it is called On-Off-Keying (OOK).

In PSK, we have:

$$s_i(t) = \sqrt{\frac{2E}{T}} \cdot \cos\left[\omega_0 t + \phi_i(t)\right],$$

where ω_0 is the center frequency, E is the symbol energy, and $\phi_i(t)$ is one of m different constant values describing the phase shifts. The same binary data as in the ASK example is shown using PSK in Figure 4.3. Two popular PSK schemes are BPSK and QPSK; they are used, for example,

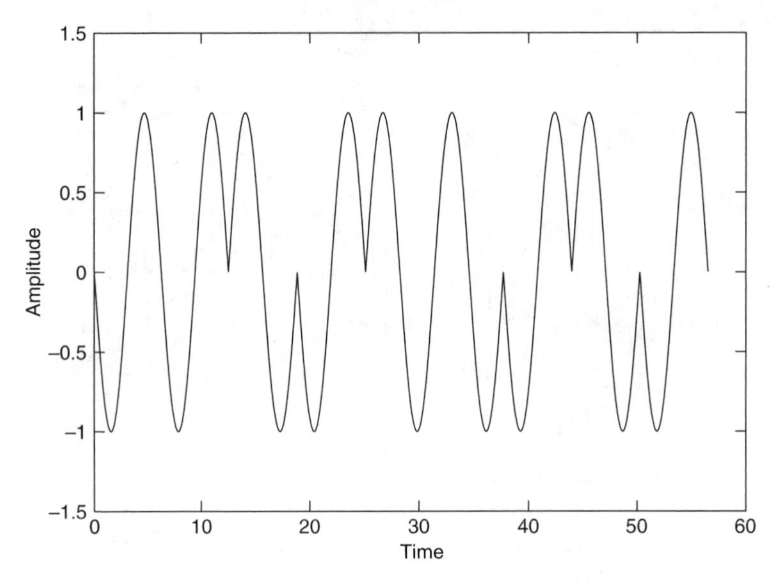

Figure 4.3 Phase shift keying (PSK) example

for the 1-Mbps and 2-Mbps modulations in IEEE 802.11 [467]. In BPSK, phase shifts of zero and π are used and in QPSK, four phase shifts of 0, $\frac{\pi}{2}$, π and $\frac{3\pi}{2}$ are used.[5]

In FSK, we have:

$$s_i(t) = \sqrt{\frac{2E}{T}} \cdot \cos\left[\omega_i(t) \cdot t + \phi\right],$$

where $\omega_i(t)$ is one of n different frequencies, E is the symbol energy, and ϕ is some constant initial phase. Figure 4.4 repeats the above example with FSK modulation.

Clearly, these basic types can be mixed. For example, Quadrature Amplitude Modulation (QAM) combines amplitude and phase modulation, using two different amplitudes and two different phases to represent two bits in one symbol.

4.2.3 Wave propagation effects and noise

Waveforms transmitted over wireless channels are subject to several physical phenomena that all *distort* the originally transmitted waveform at the receiver. This distortion introduces uncertainty at the receiver about the originally encoded and modulated data, resulting ultimately in bit errors.

Reflection, diffraction, scattering, doppler fading

The basic wave propagation phenomena [682, Chap. 3] are:

Reflection When a waveform propagating in medium A hits the boundary to another medium B and the boundary layer between them is smooth, one part of the waveform is reflected

[5] More precisely, IEEE 802.11 uses Differential Binary Phase Shift Keying (DBPSK) and Differential Quaternary Phase Shift Keying (DQPSK). In these differential versions, the information is not directly encoded in the phase of a symbol's waveform, but in the difference between phases of two subsequent symbols' waveforms.

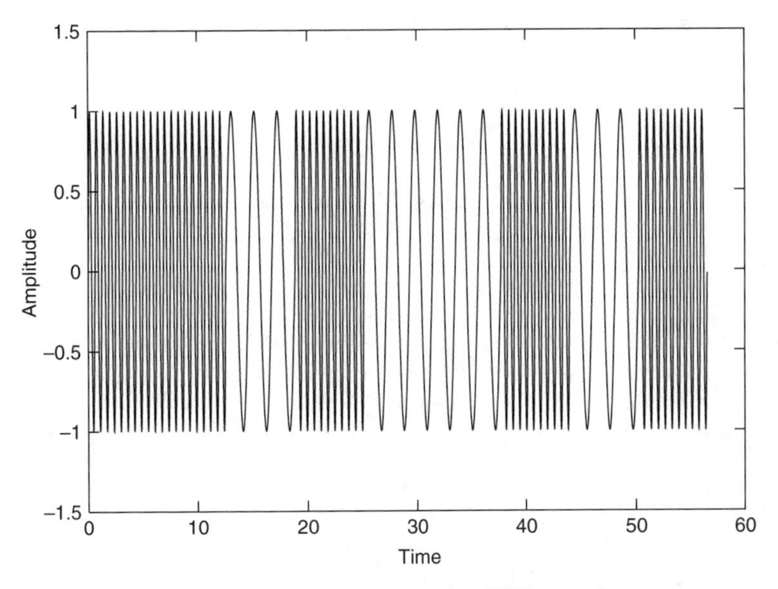

Figure 4.4 Frequency shift keying (FSK) example

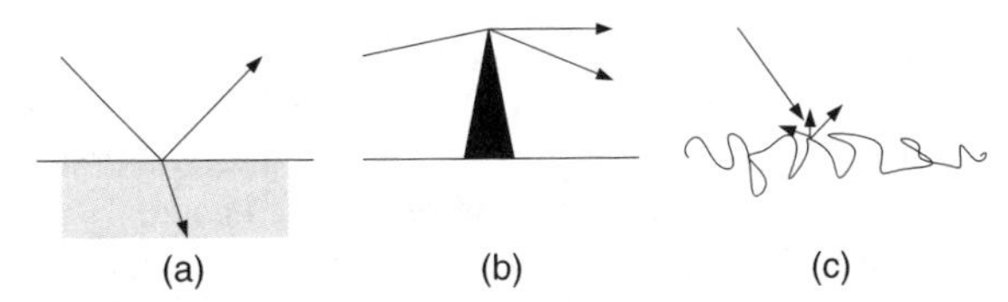

Figure 4.5 Illustration of wave propagation phenomena

back into medium A, another one is transmitted into medium B, and the rest is absorbed (Figure 4.5(a)). The amount of reflected/transmitted/absorbed energy depends on the materials and frequencies involved.

Diffraction By Huygen's principle, all points on a wavefront can be considered as sources of a new wavefront. If a waveform hits a sharp edge, it can by this token be propagated into a shadowed region (Figure 4.5(b)).

Scattering When a waveform hits a rough surface, it can be reflected multiple times and diffused into many directions (Figure 4.5(c)).

Doppler fading When a transmitter and receiver move relative to each other, the waveforms experience a shift in frequency, according to the Doppler effect. Too much of a shift can cause the receiver to sample signals at wrong frequencies.

Radio antennas radiate their signal into all directions at (nearly) the same strength, or they have a preferred direction characterized by a beam. In the first case, we have **omnidirectional antennas**,and in the second, we speak of **directed antennas**. In either case, it is likely that not only a single but multiple copies of the same signal would reach the receiver over different paths with different path lengths and attenuation (Figure 4.6), where a direct path or **Line Of Sight (LOS) path** and a reflected, or **Non line Of Sight (NLOS) path** are shown.

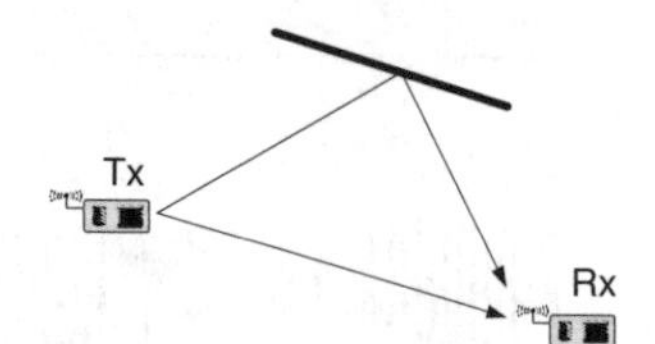

Figure 4.6 Multipath propagation

The signal at the receiver is therefore a superposition of multiple and delayed copies of the same signal. A signal actually occupies a certain spectrum, which can be represented by Fourier techniques. The different signal copies have different relative delays, which translate for each frequency component of the signal into different relative phase shifts at the receiver. Depending on the relative phase shift of the signal components, *destructive* or *constructive* interference can occur. If the channel treats all frequency components of a signal in "more or less the same way" (i.e., their amplitudes at the receiver are strongly correlated [682, Sec. 5.4]), we have **frequency-nonselective fading**, also often called **flat fading**; otherwise, we have a **frequency-selective channel**. The frequency (non-)selectivity of a channel is closely related to its **time dispersion** or **delay spread**, more exactly to the **RMS delay spread** value.[6] The **coherence bandwidth** captures, for a given propagation environment, the range of frequencies over which a channel can be considered flat; it is defined as the inverse of the RMS delay spread times a constant factor. A channel is a flat fading channel if the full signal bandwidth is smaller than the coherence bandwidth.

For wireless sensor networks with their small transmission ranges (leading to small RMS delay spread) and their comparably low symbol rates, it is reasonable to assume flat fading channels.

When transmitter and receiver move relatively to each other, the number and relative phase offset of the multiple paths changes over time and the received signal strength can fluctuate on the order of 30–40 dB within short time; this is called **fast fading** or **multipath fading**. Depending on the relative speed, the fluctuations occur at timescales of tens to hundreds of milliseconds.[7]

The importance of fading is its impact on the receiver. Since any receiver needs a minimum signal strength to have a chance for proper demodulation, a fade with its resulting drop in received signal strength is a source of errors. When the signal strength falls below this threshold because of fast fading, this is called a **deep fade**. When judging fast fading channels, specifically the rate at which the signal falls below this threshold (the **level-crossing rate**) and the duration of the deep fades are important. Qualitatively, fading channels tend to show **bursty errors**, that is, symbol errors tend to occur in clusters separated by errorfree periods.

Another source of errors (predominantly) caused by multipath propagation is InterSymbol Interference (ISI): When the transmitter transmits its symbols back-to-back, the presence of multiple paths with different delays can lead to a situation where waveforms belonging to some symbol s_t and reaching the receiver on an Line Of Sight (LOS) path overlap with delayed copies of previously sent symbols $s_{t-1}, s_{t-2}, \ldots$. The severity of ISI depends on the relationship between the symbol duration and the RMS delay spread.

[6] To characterize time dispersion of a multipath channel, the channel impulse response can be used: The transmitter emits a very short pulse and the receiver records the incoming pulses and their signal strength. The first received pulse corresponds to the shortest path and all subsequent pulses are from longer paths and likely attenuated. The time difference between the delayed pulses and the reference pulse are called **excess delays**, the **mean excess delay** is defined as the weighted average of the excess delays (using the pulse amplitudes as weights), and the **RMS delay spread** (root mean square) is the standard deviation of the weighted excess delays [682, Chap. 5].

[7] Example: For 2.4 GHz, the wavelength is 12.5 cm, and accordingly a change of 6.25 cm in the path length difference of two paths suffices to move from amplification (constructive interference) to cancellation (destructive interference) or vice versa.

Path loss and attenuation

Wireless waveforms propagating through free space are subject to a distance-dependent loss of power, called **path loss**. The received power at a distance of $d \geq d_0$ m between transmitter and receiver is described by the **Friis free-space equation** (compare reference [682, p.107], reflections are not considered):

$$
\begin{aligned}
P_{\text{rcvd}}(d) &= \frac{P_{\text{tx}} \cdot G_t \cdot G_r \cdot \lambda^2}{(4\pi)^2 \cdot d^2 \cdot L} \\
&= \frac{P_{\text{tx}} \cdot G_t \cdot G_r \cdot \lambda^2}{(4\pi)^2 \cdot d_0^2 \cdot L} \cdot \left(\frac{d_0}{d}\right)^2 = P_{\text{rcvd}}(d_0) \cdot \left(\frac{d_0}{d}\right)^2,
\end{aligned}
\tag{4.1}
$$

where P_{tx} is the transmission power, G_t and G_r are the **antenna gains**[8] of transmitter and receiver, d_0 is the so-called **far-field distance**, which is a reference distance[9] depending on the antenna technology, $d \geq d_0$ is the distance between transmitter and receiver, λ is the **wavelength** and $L \geq 1$ summarizes losses through transmit/receive circuitry. Note that this equation is only valid for $d \geq d_0$. For environments other than free space, the model is slightly generalized:

$$
P_{\text{rcvd}}(d) = P_{\text{rcvd}}(d_0) \cdot \left(\frac{d_0}{d}\right)^{\gamma}, \tag{4.2}
$$

where γ is the **path-loss exponent**, which typically varies between 2 (free-space path loss) and 5 to 6 (shadowed areas and obstructed in-building scenarios [682, Table 4.2]). However, even values $\gamma < 2$ are possible in case of constructive interference. The path loss is defined as the ratio of the radiated power to the received power $\frac{P_{\text{tx}}}{P_{\text{rcvd}}(d)}$ and, starting from Equation 4.2, can be expressed in decibel as:

$$
\text{PL}(d)[\text{dB}] = \text{PL}(d_0)[\text{dB}] + 10\gamma \log_{10}\left(\frac{d}{d_0}\right) \tag{4.3}
$$

This is the so-called **log-distance** path loss model. $\text{PL}(d_0)$[dB] is the known path loss at the reference distance.

We can draw some first conclusions from this equation. First, the received power depends on the frequency: the higher the frequency, the lower the received power. Second, the received power depends on the distance according to a power law. For example, assuming a path-loss exponent of 2, a node at a distance of $2d$ to some receiver must spent four times the energy of a node at distance d to the same receiver, to reach the same level of received power P_{rcvd}. Since, in general, the bit/symbol error rate at the receiver is a monotone function of the received power P_{rcvd}, higher frequencies or larger distances must be compensated by an appropriate increase in transmitted power to maintain a specified P_{rcvd} value. This will be elaborated further on in the following sections of this chapter.

An extension of the log-distance path-loss model takes the presence of obstacles into account. In the so-called **lognormal fading**, the deviations from the log-distance models due to obstacles

[8] Antenna gain: For directional antennas, this gives the ratio of the received power in the main direction to what would have been received from an isotropic/omnidirectional antenna (using the same transmit power).

[9] d_0 is for cellular systems with large coverage in the range of 1 km; for short range systems like WLANs, it is in the range of 1 m [682, p. 139].

are modeled as a multiplicative lognormal random variable. Equivalently, the received power can be expressed in dB as:

$$\mathrm{PL}(d)[\mathrm{dB}] = \mathrm{PL}(d_0)[\mathrm{dB}] + 10\gamma \log_{10}\left(\frac{d}{d_0}\right) + X_\sigma[\mathrm{dB}], \tag{4.4}$$

where X_σ is a zero-mean Gaussian random variable with variance σ^2, also called the **shadowing variance**.

Significant variations in the distance between transmitter and receiver or the movement beyond obstacles lead to variations of the long-term mean signal strength at the receiver. Movements and "distance hops" happen at timescales of (tens of) seconds to minutes and the variations are accordingly referred to as **slow fading**.

Besides path loss, there is often also **attenuation**. Most signals are not transmitted in a vacuum but in some media, for example, air, cables, liquids, and so on. In outdoor scenarios, there may also be fog or rain. These media types introduce additional, frequency-dependent signal attenuation. However, since attenuation obeys also a power law depending on the distance, it is only rarely modeled explicitly but accounted for in the path-loss exponent of the log-distance model.

Noise and interference

In general, **interference** refers to the presence of any unwanted signals from external (w.r.t. transmitter and receiver) sources, which obscure or mask a signal. These signals can come from other transmitters sending in the same band at the same time (**multiple access interference**) or from other devices like microwave ovens radiating in the same frequency band. In **co-channel interference**, the interference sources radiates in the same or in an overlapping frequency band as the transmitter and receiver node under consideration. In **adjacent-channel interference**, the interferer works in a neighboring band. Either the interferer leaks some signal energy into the band used by transmitter and receiver or the receiver has imperfect filters and captures signals from neighboring bands.

An important further phenomenon is **thermal noise** or simply **noise**. It is caused by thermal motions of electrons in any conducting media, for example, amplifiers and receiver/transmitter circuitry. Within the context of digital receivers, noise is typically measured by the single-sided noise Power Spectral Density (PSD)[10] N_0 given by [772, Sec. 4]:

$$N_0 = K \cdot T \left[\frac{\text{Watts}}{\text{Hertz}}\right]$$

where K is Boltzmanns constant ($\approx 1.38 \cdot 10^{-23}$ J/K) and T is the so-called system temperature in Kelvin. The thermal noise is **additive**, that is, the received signal $r(t)$ can be represented as a sum of the transmitted signal $s(t)$ (as it arrives at the receiver after path loss, attenuation, scattering, and so forth) and the noise signal $n(t)$:

$$r(t) = s(t) + n(t) \tag{4.5}$$

and furthermore this noise is **Gaussian**, that is, $n(t)$ has a Gaussian/normal distribution with zero mean and finite variance σ^2 for all t. A very important property of Gaussian noise is that its PSD can be assumed constant (with value $N_0/2$ over all frequencies of practical interest). A process with constant PSD is also called **white noise**. Hence, thermal noise is also often referred to as Additive White Gaussian Noise (AWGN).

[10] Technically, the PSD of a wide-sense-stationary random process $n(t)$ is the Fourier transform of the process's autocorrelation function; intuitively, the PSD describes the distribution of a signal's power in the frequency domain.

Symbols and bit errors

The symbol/bit error probability depends on the actual modulation scheme and on the ratio of the power of the received signal (P_{rcvd}) to the noise and interference power. When only AWGN is considered, this ratio is called Signal-to-Noise Ratio (SNR) and is given in decibel as:

$$\text{SNR} = 10 \log_{10} \left(\frac{P_{\text{rcvd}}}{N_0} \right)$$

where N_0 is the noise power and P_{rcvd} is the average received signal power. When other sources of interference are considered, too, often the Signal to Interference and Noise Ratio (SINR) is important:

$$\text{SINR} = 10 \log_{10} \left(\frac{P_{\text{rcvd}}}{N_0 + \sum_{i=1}^{k} I_i} \right)$$

where N_0 is the noise power and I_i is the power received from the i-th interferer.

The SINR describes the power that arrives at the receiver and is thus related to the symbols sent over the channel. In the end, the symbols are not relevant; the data bits are. To correctly demodulate and decode an arriving bit, the energy per such a bit E_b in relation to the noise energy N_0 is relevant. This ratio E_b/N_0 has a close relationship to the SNR (or SINR, when interference is treated as noise) [772, Sec. 3.7]:

$$\frac{E_b}{N_0} = \text{SNR} \cdot \frac{1}{R} = \frac{P_{\text{rcvd}}}{N_0} \cdot \frac{1}{R} \tag{4.6}$$

where R is the bit rate. It will be useful later on in this chapter to look also at the bandwidth W occupied by the modulated signal and to use the **bandwidth efficiency** $\eta_{\text{BW}} = \frac{R}{W}$ (in bit/s/Hz) as a measure of a modulation scheme's efficiency. This can be used to rewrite Equation 4.6 as:

$$\frac{E_b}{N_0} = \frac{P_{\text{rcvd}}}{N_0} \cdot \frac{1}{\eta_{\text{BW}} \cdot W}. \tag{4.7}$$

An important distinction not directly concerning modulation but concerning the receiver is the one between **coherent detection** and **noncoherent detection**. In coherent detection, the receiver has perfect phase and frequency information, for example, learned from preambles or synchronization sequences (see also Section 4.2.6). In general, coherent receivers are much more complex than noncoherent ones, but need lower signal-to-noise ratios to achieve a given target Bit-Error Rate (BER).

If we prescribe a desired maximum BER, we can, for many modulation schemes, determine some minimum SNR needed to achieve this BER on an AWGN channel. To illustrate this, we show in Figure 4.7 the BER versus the ratio E_b/N_0 given in decibel for coherently detected binary PSK and binary FSK. The qualitative behavior of such BER versus E_b/N_0 is the same for all popular modulation types. For example, with BPSK, the E_b/N_0 ratio must be larger than 4 dB to reach a BER of at least 10^{-3}. The noise power is fixed, so we have to tune the received power P_{rcvd} to achieve the desired SNR. For given antennas, this can only be achieved by increasing the radiated power at the transmitter P_{tx}; compare Equation 4.1. An alternative is clearly to use better modulation schemes.

The choice of modulation schemes for wireless sensor networks is discussed in Section 4.3.2.

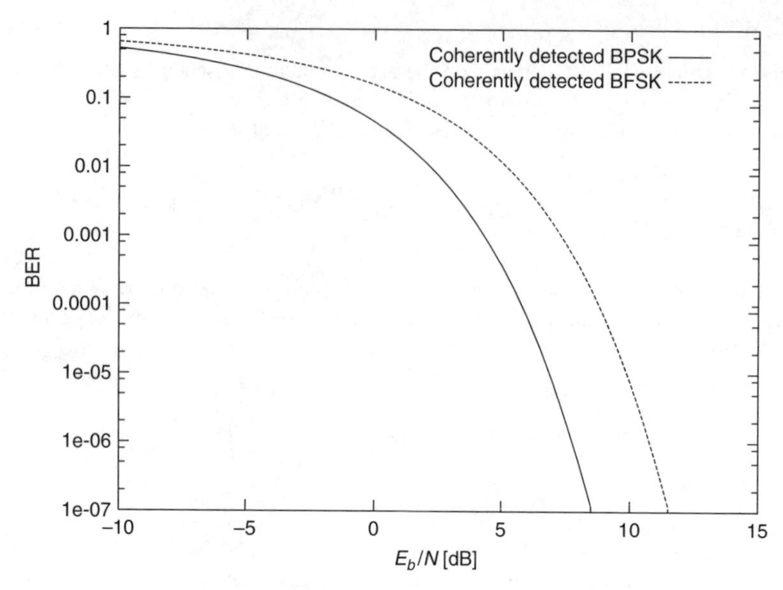

Figure 4.7 Bit error rate for coherently detected binary PSK and FSK

4.2.4 Channel models

For investigation of modulation or error control schemes, models for wireless channels are needed [36]. Because of the apparent complexity of real wireless channels, mostly *stochastic* models are used, which replace complex and tedious modeling of propagation environments by random variables. At the lowest level, such models work on the level of waveforms, describing the received signal. "Higher", more abstract models describe the statistics of symbol or bit errors or even of packet errors. These models are more amenable for investigation of network protocols, where often thousands or millions of packets are transmitted.

Signal models

We have already seen one waveform model, the AWGN model, having a constant SNR. As a reminder, this model expresses the received signal $r(t)$ as:

$$r(t) = s(t) + n(t),$$

where $s(t)$ is the transmitted signal and $n(t)$ is white Gaussian noise. One important property of this model is that the SNR is constant throughout. The simplicity of this model eases theoretical analysis; however, it is not appropriate to model time-varying channels like fading channels.

There are other popular models, specifically for frequency-nonselective fading channels [80]. These models assume that the SNR is a random variable, fluctuating from symbol to symbol or from block to block [79]. In the **Rayleigh fading** model, it is assumed that there is no LOS path. Instead, a large number of signal copies with stochastically independent signal amplitudes of the same mean value overlap at the receiver. By virtue of the central limit theorem, it can be shown that the amplitude of the resulting signal has a Rayleigh distribution, whereas the phase is uniformly distributed in $[0, 2\pi]$. A second popular model is the **Rice fading** model, which makes the same assumptions as the Rayleigh fading model, but additionally a strong LOS component is present. Such a fading channel together with AWGN can be represented as:

$$r(t) = R \cdot e^{i\theta} \cdot s(t) + n(t)$$

where again $n(t)$ is white Gaussian noise and $R \cdot e^{i\theta}$ is a Gaussian random variable such that R has a Rice or Rayleigh probability density function.

Digital models

In the AWGN channel, each transmitted symbol is erroneous with a certain fixed error probability, and errors of subsequent symbols are independent. If these two conditions hold true and if in addition the error probability does not depend on the symbol value, we have a Binary Symmetric Channel (BSC) [180].

There have been several efforts to find good stochastic models for (Rayleigh) fading channels on the bit/symbol level. These models try to capture the tendency of fading channels to have bursty errors. Often, such channels are modeled as **Markov chains** with the states of the chain corresponding to different channel "quality levels". For example, the popular two-state **Gilbert–Elliot model** [231, 290] describes the alternation between deep fades and good periods in a fading channel. WANG and MOAYERI [858] discuss how the parameters of an N-state Markov chain describing the received signal level can be derived under Rayleigh fading assumptions from simple physical parameters like wavelength, relative speed of the nodes, and others. A more general class of models, which has also often been used, are Hidden Markov Models (HMMs); see, for instance, reference [241, 834].

WSN-specific channel models

One design constraint of wireless sensor networks is the intention to use small transmission power (and consequently the radiated power) – on the order of 1 dBm [855] – with the hope to save energy by leveraging multihop communication. The choice of a small transmit power has several consequences for the channel characteristics:

- By the Friis equation (Equation 4.1), a small transmit power implies a small range.
- Having a small transmission range means that the rms delay spread will be in the range of nanoseconds [682, Table 5.1], which is small compared to symbol durations in the order of milli- or microseconds. Since in addition the data rates are moderate, it is reasonable to expect frequency nonselective fading channels with noise [762] and a low-to-negligible degree of ISI. Accordingly, no special provisions against ISI like equalizers are needed.

SOHRABI et al. [779] present measurements of the near-ground propagation conditions for a 200-MHz frequency band between 800 MHz and 1000 MHz in various environments. These measurements comprise the path-loss exponents γ, shadowing variance σ^2, the reference path loss $\mathrm{PL}(d_0)$[dB] at $d_0 = 1$ m and the coherence bandwidth. The measurement sites under consideration include parking lots, hallways, engineering buildings and plant fences, covering distances between 1 and 30 m. Mobility was not considered. The average path-loss exponents (the average is formed over the range of frequencies), the average shadowing variance, and the ranges of the reference path loss $\mathrm{PL}(d_0)$[dB] are quoted in Table 4.2. It is interesting to note that the average path-loss exponents can range from $\gamma = 1.9$ up to $\gamma = 5$. It is also interesting to note that already at a distance of 1 m the signal has lost between 30 and 50 dB. The coherence bandwidth depends strongly on the environment as well as on the distance; with increasing distance, the coherence

Table 4.2 Average path-loss exponents, shadowing variance, and range of path loss at reference distances for near-ground measurements in 800–1000 MHz [779]

Location	Average of γ	Average of σ^2 [dB]	Range of PL(1m)[dB]
Engineering building	1.9	5.7	[−50.5, −39.0]
Apartment hallway	2.0	8.0	[−38.2, −35.0]
Parking structure	3.0	7.9	[−36.0, −32.7]
One-sided corridor	1.9	8.0	[−44.2, −33.5]
One-sided patio	3.2	3.7	[−39.0, −34.2]
Concrete canyon	2.7	10.2	[−48.7, −44.0]
Plant fence	4.9	9.4	[−38.2, −34.5]
Small boulders	3.5	12.8	[−41.5, −37.2]
Sandy flat beach	4.2	4.0	[−40.8, −37.5]
Dense bamboo	5.0	11.6	[−38.2, −35.2]
Dry tall underbrush	3.6	8.4	[−36.4, −33.2]

bandwidth decreases, but is for many scenarios in the range of 50 MHz and beyond. Accordingly, low-bandwidth channels in this frequency range can be considered as frequency nonselective. Other references propose path-loss values in the range of $\gamma = 4$ [245, 648]. In reference [563], the parameters PL(1m)[dB] $= -30$ and $\gamma = 3.5$ are used to model transmission using the μAMPS-1 nodes (2.4 GHz, 1 Mbps FSK transceiver).

4.2.5 Spread-spectrum communications

In spread-spectrum systems [293, 297, 557], the bandwidth occupied by the transmitted waveforms is much larger than what would be really needed to transmit the given user data.[11] The user signal is *spreaded* at the transmitter and *despreaded* at the receiver. By using a wideband signal, the effects of narrowband noise/interference are reduced. Spread-spectrum systems offer an increased robustness against multipath effects but pay the price of a more complex receiver operation compared to conventional modulation schemes.

The two most popular kinds of spread-spectrum communications are Direct Sequence Spread Spectrum (DSSS) and Frequency Hopping Spread Spectrum (FHSS).

Direct sequence spread spectrum

In Direct Sequence Spread Spectrum (DSSS), the transmission of a data bit of duration t_b is replaced by transmission of a finite **chip sequence** $\mathbf{c} = c_1c_2 \dots c_n$ with $c_i \in \{0, 1\}$ if the user bit is a logical one, or $\overline{c_1}\overline{c_2}\dots\overline{c_n}$ if it is a logical zero ($\overline{c_i}$ is the logical inverse of c_i). Each chip c_i has duration $t_c = t_b/n$, where n is called the **spreading factor** or **gain**. Each chip is then modulated with a digital modulation scheme like BPSK or QPSK. Since the spectrum occupied by a digital signal is roughly inverse of the symbol duration, the spectrum of the chip sequence is much wider than the spectrum the user data signal would require in case of direct modulation. The intention is that the chip duration becomes smaller than the average or RMS delay spread value and the channel becomes, thus, frequency selective. Therefore, when multipath fading is present, a chip sequence **c** coming from an LOS and a delayed copy **c** (of the same chip sequence overlap, and the delay

[11] Information theorists would say that the Fourier bandwidth (describing the occupied spectrum) is much larger than the Shannon bandwidth (describing the number of dimensions of the signal space used per second) [542].

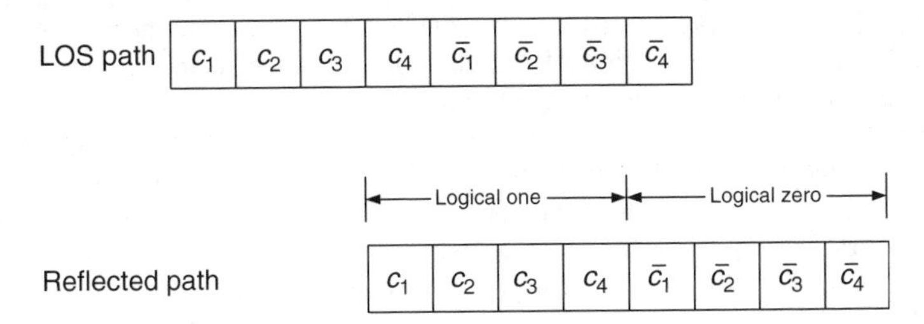

Figure 4.8 Direct sequence spread-spectrum example

difference (the **lag**) between these amounts to more than one chip duration. This is exploited by proper design of the chip sequences: these are **pseudorandom sequences** chosen such that the autocorrelation between a chip sequence and a lagged version of itself has a peak for lag zero and almost vanishes for all nonzero lags.

To explain this, consider the example shown in Figure 4.8. Both a direct LOS path and a reflected path are present, with the lag corresponding to three chip durations. The direct LOS chip sequence is given by $\mathbf{c} = c_1 c_2 \ldots c_n$ followed by $\mathbf{c}^{-1} = \overline{c_1} \overline{c_2} \ldots \overline{c_n}$, whereas the chip sequence from the reflected path starts with a lag of three chips. Somewhat simplified, the operation of the receiver can be described as follows (coherent matched filter receiver): Let us assume that the receiver is synchronized to the direct LOS path. It compares the incoming chip sequence with the well-known reference sequence **c** by computing the inner product (term-wise multiplication and final summation in terms of modulo-2 operations). If the received sequence is the same as **c**, then this operation yields the value n, if the incoming chip sequence is $\mathbf{c}^{-1}$, then the result is $-n$. By proper choice of the chip sequence, it can be achieved that the inner product formed between the chip sequence and a shifted/lagged version of it assumes absolute values smaller than n. For example, the 11-chip **Barker sequence** used in IEEE 802.11 [467] assumes for all shifted versions only the values -1, 0, or 1.[12] Delayed copies distort the direct signal in the same way as AWGN does. Thus, DSSS increases robustness against multipath effects.

However, there are also downsides. First, receivers must be properly synchronized with the transmitter, and second, there is the issue of management of chip sequences. In systems like IEEE 802.11 with DSSS Physical Layer (PHY) or IEEE 802.15.4, there is only a single chip sequence used by all nodes. Proper measures at the MAC level must be taken to avoid collisions. It is also possible to assign different chip sequences or **codes** to different users, which then can transmit in parallel and create only minor distortion to each other. Such an approach is called Code Division Multiple Access (CDMA) and is used, for example, in UMTS [847]. However, immediately the question how codes are assigned to nodes ("code management") comes up.

Frequency hopping spread spectrum

In Frequency Hopping Spread Spectrum (FHSS) systems like Bluetooth [318, 319] and the (outdated) FHSS version of IEEE 802.11, the available spectrum is subdivided into a number of equal-sized **subbands** or **channels** (not to be confused with the physical channels discussed above); Bluetooth and IEEE 802.11 divide their spectrum in the 2.4-GHz range into 78 subbands 1-MHz wide. The user data is always transmitted within one channel at a time; its bandwidth is thus limited. All nodes in a network hop synchronously through the channels according to a prespecified

[12] When the inner product of a chip sequence with a shifted version of itself assumes "large" values only for lag zero, but comparably small values for all other lags, it is also called **nearly orthogonal**.

schedule. This way, a channel currently in a deep fade is left at some point in time and the nodes switch to another, hopefully, good channel. Different networks can share the same geographic area by using (mostly) nonoverlapping hopping schedules.

As an example, the FHSS version of IEEE 802.11 hops with 2.5 Hz and many packets can be transmitted before the next hop. In Bluetooth, the hopping frequency is 1.6 kHz and at most one packet can be transmitted before the next hop. Packets can have lengths corresponding to one, three, or five hops. During a longer packet, hopping is suppressed – the packet is transmitted at the same frequency. Once a packet is finished, the systems continues with the frequency it would have reached if the long packet had been absent.

4.2.6 Packet transmission and synchronization

The MAC layer above the physical layer uses **packets** or **frames** as the basic unit of transmission.[13] From the perspective of the MAC layer, such a frame has structure; for the transceiver, however, it is just a block of bits. Transceivers perform the functions of modulation and demodulation along with associated high- and intermediate-frequency processing, typically in hardware, and provide an interface to the physical layer. They are discussed in Section 2.2.4.

The receiver must know certain properties of an incoming waveform to make any sense of it and to detect a frame, including its frequency, phase, start and end of bits/symbols, and start and end of frames [772, Chap. 8], [286]. What is the root of this **synchronization problem**? The generation of sinusoidal carriers and of local clocks (with respect to which symbol times are expressed) involves **oscillators** of a certain **nominal frequency**. However, because of production inaccuracies, temperature differences, aging effects, or any of several other reasons, the *actual frequency* of oscillators deviates from the nominal frequency. This **drift** is often expressed in **parts per million (ppm)** and gives the number of additional or missing oscillations a clock makes in the amount of time needed for one million oscillations at the nominal rate. As a rule of thumb, the cheaper the oscillator, the more likely are larger drifts.

To compensate this drift, the receiver has to *learn* about the frequency or time base of the transmitter. The receiver has to extract synchronization information from the incoming waveform. An often-found theme for such approaches is the distinction between **training** (or **acquisition**) and **tracking** phases. Frames are equipped with a well-known **training sequence** that allows the receiver to learn about the detailed parameters of the transmitter, for example, its clock rate – the receive can "train" its parameters. This training sequence is often placed at the beginning of frames (for example, in IEEE 802.11 [467] or IEEE 802.15.4 [468]), but sometimes it is placed in the middle (e.g. in GSM [848]). In the first case we speak of a **preamble**, and in the second case of a midamble. In either case, the training sequence imposes some overhead. As an example, in IEEE 802.15.4, the preamble consists of 32 zero bits.

After the receiver has successfully acquired initial synchronization from the training sequence, it enters a tracking mode, continuously readjusting its local oscillator.

Important synchronization problems are:

Carrier synchronization The receiver has to learn the frequency and, for coherent detection schemes, also the phase of the signal. A frequency drift can be caused by oscillators or by Doppler shift in case of mobile nodes. One way to achieve frequency synchronization is to let the transmitter occasionally send packets with known spectral shape and to let the receiver scan some portion of the spectrum around the nominal frequency band for this shape; for example, in the GSM system, special **frequency correction bursts** are used to

[13] In OSI terminology, this would be MAC PDUs. In fact, packets and frames are two words for the same thing; however, the word frame tends to be used more often when discussing lower layers.

this end [848, Chap. 3]. The phase varies typically much faster than the frequency; accordingly, phase synchronization must be done more often than frequency synchronization [286]. Phase synchronization can be avoided in noncoherent detection schemes but at the price of a higher BER at the same transmit power.

Bit/symbol synchronization Having acquired carrier synchronization, the receiver must determine both the symbol duration as well as the start and end of symbols to demodulate them successfully. The continuous readjustment in the tracking phase requires sufficient "stimuli" indicating symbol bounds. This can be explained with the example of OOK, where logical zeros are modulated as the absence of any carrier. If a long run of zeros occurs in the data, the receiver clock gets no stimulus for readjustment and may drift away from the transmitter clock, this way adding spurious symbols or skipping symbols. For example, for the RF Monolithics TR1000 transceiver used in the Mica motes, more than four consecutive zero or one bits should be avoided [351]. This situation can be avoided by choosing coding schemes with a sufficient number of logical ones, by bit-stuffing techniques, or by **scrambling** where the data stream is shifted through a linear-feedback shift register. The scrambling technique is, for example, applied in IEEE 802.11 and no extra symbols have to be sent. The other schemes incur some overhead symbols.

Frame synchronization The receiver of a frame must be able to detect where the frame starts and where it ends, that is, the frame bounds. Frame synchronization assumes that bit/symbol synchronization is already acquired. There are several techniques known for framing [327], including time gaps, length fields, usage of special flag sequences along with bit-stuffing techniques to avoid the occurrence of these sequences in the packet data, and others. One technique to mark the start of a frame is the approach of IEEE 802.15.4, where the preamble is immediately followed by a well-known Start Frame Delimiter (SFD). This SFD is part of the physical layer header, not of the data part, and thus no measures to avoid the SFD pattern in the data part have to be taken.

Let us discuss a simple example (Figure 4.9). In the Mica motes [351], one option for modulation is OOK. Accordingly, bits are represented by two transmission power levels: a power level of zero corresponds to a logical zero, whereas a nonzero power level corresponds to a logical one (ignoring the noise floor). A packet consists of a preamble, a start frame delimiter, and a data part. A long idle period on the medium is interpreted as boundary between packets. Within such a long idle period, the receiver of a packet needs to sample the medium for activity only occasionally. The time between samples must be smaller than the preamble length not to miss it, but large enough

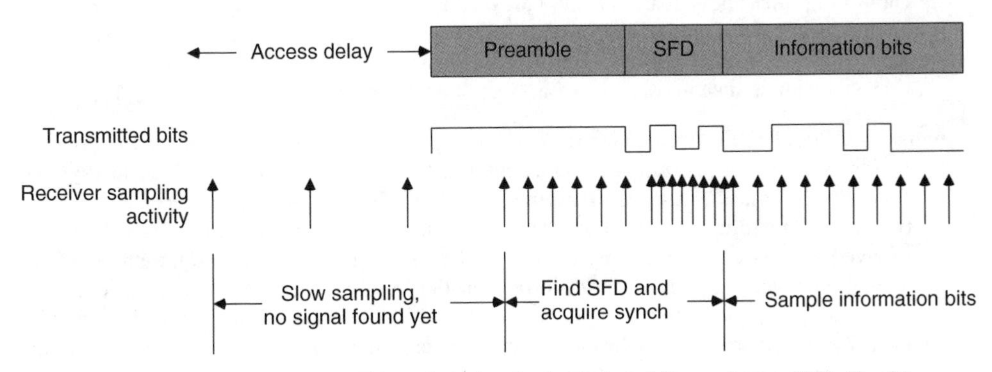

Figure 4.9 Example for sampling and synchronization (adapted from reference [351, Fig. 5])

to keep the energy costs induced by sampling. When sampling reveals activity in the channel, its frequency is increased to find the end of the preamble and to derive the length of a transmitted bit from the SFD. Once this information is determined, the receiver samples the medium in the mid of the data bits. To avoid the presence of long idle periods in the data part and misinterpretation as packet boundary, the length of runs of zeros (and ones) must be bounded, for example, by four. This has to be achieved by proper transformation of the user data.

4.2.7 Quality of wireless channels and measures for improvement

As opposed to wired channels, wireless channels often have a poorer quality in terms of bit/symbol error rate. The actual channel quality depends on many factors, including frequency, distance between transmitter and receiver, and their relative speed, propagation environment (number of paths and their respective attenuation), technology, and much more. Consequently, there is no such thing as "the" wireless channel. Many measurements of error rates have appeared in the literature; two of them are references [13, 223].

A great deal of work has been devoted to improve transmission quality on wireless channels, working on the physical as well as on higher layers and in many cases not taking energy concerns or other constraints specific for wireless sensor networks into account. Some of the mechanisms developed are the following:

Optimization of transmission parameters The choice of modulation scheme as well as the choice of radiated power (within legal constraints) can influence the BER significantly. Another control knob is the choice of packet sizes and the structure of packets. This is discussed in Chapter 6.

Diversity mechanisms All **diversity techniques** [682, Chap. 7], [625] seek to obtain and exploit statistically independent (or at least uncorrelated) replicas of the same signal. Simply speaking, it is hoped that even if one replica is in a deep fade and delivers symbol errors, another replica is currently good. The receiver tries to pick the best of all replicas. In **explicit diversity** schemes, the multiple copies are explicitly created by the transmitter, by sending the same packet over another frequency, during another time slot, or sending it into another spatial direction. In **implicit diversity** schemes, the signal is sent only once, but multiple copies are created *in the channel* through multipath propagation. In either case, the receiver needs mechanisms to take advantage of the multiple copies. One simple example is the so-called **receive diversity**, where the receiver is equipped with two or more appropriately spaced antennas and the receiver combines the different signals (e.g. by so-called **selection combining**: pick the signal with the best quality; or by **maximum ratio combining**: sum up all signals, weighted by their quality). Receive diversity works best when the signals at the two antennas are independent or at least uncorrelated. As a rule of thumb, this can be achieved with an antenna spacing of at least 40–50 % of the wavelength [682, Chap. 5].

Equalization **Equalization** techniques [682, Chap. 7], [660] are useful to combat InterSymbol Interference (ISI). Equalization works as follows: The transmitter sends a well-known symbol pattern/waveform, the so-called **training sequence**. The equalizer at the receiver works in two modes: **training** and **tracking**. During the training phase, the equalizer analyzes the received version of the well-known pattern, learns the mode of distortion, and computes an algorithm for "inverting" the distortion. In the tracking phase, the remaining packet is analyzed by applying the inversion algorithm to it and the equalizer continually readjusts the inversion algorithm. Equalization requires some signal processing at the receiver and the channel is assumed to be stationary during the packet transmission time. As a side effect, the training sequence can also be used to acquire **bit synchronization**.

Forward error correction (FEC) The transmitter accepts a stream or a block of user data bits or source bits, adds suitable redundancy, and transmits the result to the receiver. Depending on the amount and structure of the redundancy, the receiver might be able to correct some bit/symbol errors. It is known that AWGN channels have a higher capacity than Rayleigh fading channels and many coding schemes achieve better BER performance on AWGN than on fading channels with their bursty errors [79]. The operation of **interleaving** applies a permutation operation to a block of bits, hoping to distribute bursty errors smoothly and letting the channel "look" like an AWGN channel. FEC is discussed in some more detail in Section 6.2.3.

ARQ The basic idea of ARQ protocols [322, 511] can be described as follows: The transmitter prepends a header and appends a checksum to a data block. The resulting packet is then transmitted. The receiver checks the packet's integrity with the help of the checksum and provides some feedback to the transmitter regarding the success of packet transmission. On receiving negative feedback, the transmitter performs a retransmission. ARQ protocols are discussed in Section 6.2.2.

4.3 Physical layer and transceiver design considerations in WSNs

So far, we have discussed the basics of the PHY without specific reference to wireless sensor networks. Some of the most crucial points influencing PHY design in wireless sensor networks are:

- Low power consumption.
- As one consequence: small transmit power and thus a small transmission range.
- As a further consequence: low duty cycle. Most hardware should be switched off or operated in a low-power standby mode most of the time.
- Comparably low data rates, on the order of tens to hundreds kilobits per second, required.
- Low implementation complexity and costs.
- Low degree of mobility.
- A small form factor for the overall node.

In this section, we discuss some of the implications of these requirements.

In general, in sensor networks, the challenge is to find modulation schemes and transceiver architectures that are simple, low-cost but still robust enough to provide the desired service.

4.3.1 Energy usage profile

The choice of a small transmit power leads to an energy consumption profile different from other wireless devices like cell phones. These pivotal differences have been discussed in various places already but deserve a brief summary here.

First, the radiated energy is small, typically on the order of 0 dBm (corresponding to 1 mW). On the other hand, the overall transceiver (RF front end and baseband part) consumes much more energy than is actually radiated; Wang et al. [855] estimate that a transceiver working at frequencies beyond 1 GHz takes 10 to 100 mW of power to radiate 1 mW. In reference [115, Chap. 3], similar numbers are given for 2.4-GHz CMOS transceivers: For a radiated power of 0 dBm, the transmitter uses actually 32 mW, whereas the receiver uses even more, 38 mW. For the Mica motes, 21 mW are consumed in transmit mode and 15 mW in receive mode [351]. These numbers coincide well with the observation that many practical transmitter designs have efficiencies below 10 % [46] at low radiated power.

A second key observation is that for small transmit powers the transmit and receive modes consume more or less the same power; it is even possible that reception requires more power than transmission [670, 762]; depending on the transceiver architecture, the idle mode's power consumption can be less or in the same range as the receive power [670]. To reduce average power consumption in a low-traffic wireless sensor network, keeping the transceiver in idle mode all the time would consume significant amounts of energy. Therefore, it is important to put the transceiver into sleep state instead of just idling. It is also important to explicitly include the received power into energy dissipation models, since the traditional assumption that receive energy is negligible is no longer true.

However, there is the problem of the **startup energy/startup time**, which a transceiver has to spend upon waking up from sleep mode, for example, to ramp up phase-locked loops or voltage-controlled oscillators. During this startup time, no transmission or reception of data is possible [762]. For example, the μAMPS-1 transceiver needs a startup time of 466 µs and a power dissipation of 58 mW [561, 563]. Therefore, going into sleep mode is unfavorable when the next wakeup comes fast. It depends on the traffic patterns and the behavior of the MAC protocol to schedule the transceiver operational state properly. If possible, not only a single but multiple packets should be sent during a wakeup period, to distribute the startup costs over more packets. Clearly, one can attack this problem also by devising transmitter architectures with faster startup times. One such architecture is presented in reference [855].

A third key observation is the relative costs of communications versus computation in a sensor node. Clearly, a comparison of these costs depends for the communication part on the BER requirements, range, transceiver type, and so forth, and for the computation part on the processor type, the instruction mix, and so on. However, in [670], a range of energy consumptions is given for Rockwell's WIN nodes, UCLA's WINS NG 2.0 nodes, and the MEDUSA II nodes. For the WIN nodes, 1500 to 2700 instructions can be executed per transmitted bit, for the MEDUSA II nodes this ratio ranges from 220:1 up to 2900:1, and for the WINS NG nodes, it is around 1400:1. The bottom line is that computation is cheaper than communication!

4.3.2 Choice of modulation scheme

A crucial point is the choice of modulation scheme. Several factors have to be balanced here: the required and desirable data rate and symbol rate, the implementation complexity, the relationship between radiated power and target BER, and the expected channel characteristics.

To maximize the time a transceiver can spend in sleep mode, the transmit times should be minimized. The higher the data rate offered by a transceiver/modulation, the smaller the time needed to transmit a given amount of data and, consequently, the smaller the energy consumption.

A second important observation is that the power consumption of a modulation scheme depends much more on the symbol rate than on the data rate [115, Chap. 3]. For example, power consumption measurements of an IEEE 802.11b Wireless Local Area Network (WLAN) card showed that the power consumption depends on the modulation scheme, with the faster Complementary Code Keying (CCK) modes consuming more energy than DBPSK and DQPSK; however, the relative differences are below 10 % and all these schemes have the same symbol rate. It has also been found that for the μAMPS-1 nodes the power consumption is insensitive to the data rate [762].

Obviously, the desire for "high" data rates at "low" symbol rates calls for m-ary modulation schemes. However, there are trade-offs:

- m-ary modulation requires more complex digital and analog circuitry than 2-ary modulation [762], for example, to parallelize user bits into m-ary symbols.

Table 4.3 Bandwidth efficiency η_{BW} and E_b/N_0[dB] required at the receiver to reach a BER of 10^{-6} over an AWGN channel for m-ary orthogonal FSK and PSK (adapted from reference [682, Chap. 6])

m	2	4	8	16	32	64
m-ary PSK:η_{BW}	0.5	1.0	1.5	2.0	2.5	3.0
m-ary PSK:E_b/N_0	10.5	10.5	14.0	18.5	23.4	28.5
m-ary FSK:η_{BW}	0.40	0.57	0.55	0.42	0.29	0.18
m-ary FSK:E_b/N_0	13.5	10.8	9.3	8.2	7.5	6.9

- Many m-ary modulation schemes require for increasing m an increased E_b/N_0 ratio and consequently an increased radiated power to achieve the same target BER; others become less and less bandwidth efficient. This is exemplarily shown for coherently detected m-ary FSK and PSK in Table 4.3, where for different values of m, the achieved bandwidth efficiencies and the E_b/N_0 required to achieve a target BER of 10^{-6} are displayed. However, in wireless sensor network applications with only low to moderate bandwidth requirements, a loss in bandwidth efficiency can be more tolerable than an increased radiated power to compensate E_b/N_0 losses.
- It is expected that in many wireless sensor network applications most packets will be short, on the order of tens to hundreds of bits. For such packets, the startup time easily dominates overall energy consumption, rendering any efforts in reducing the transmission time by choosing m-ary modulation schemes irrelevant.

Let us explore the involved trade-offs a bit further with the help of an example.

Example 4.1 (Energy efficiency of m-ary modulation schemes) Our goal is to transmit data over a distance of $d = 10\,\mathrm{m}$ at a target BER of 10^{-6} over an AWGN channel having a path-loss exponent of $\gamma = 3.5$ (corresponding to the value determined in reference [563]). We compare two families of modulations: coherently detected m-ary PSK and coherently detected orthogonal m-ary orthogonal FSK. For these two families we display in Table 4.3, the bandwidth efficiencies η_{BW} and the E_b/N_0 in dB required at the receiver to reach a BER of 10^{-6} over an AWGN channel.

From the discussion in Section 4.2.3, the relationship between E_b/N_0 and the received power at a distance d is given as:

$$\begin{aligned} \frac{E_b}{N_0} &= \mathrm{SNR} \cdot \frac{1}{R} = \frac{P_{\mathrm{rcvd}}(d)}{N_0} \cdot \frac{1}{R} \\ &= \frac{1}{N_0 \cdot R} \cdot \frac{P_{\mathrm{tx}} \cdot G_t \cdot G_r \cdot \lambda^2}{(4\pi)^2 \cdot d_0^{\gamma} \cdot L} \cdot \left(\frac{d_0}{d}\right)^{\gamma}, \end{aligned} \tag{4.8}$$

which can be easily solved for P_{tx} given a required E_b/N_0 value and data rate R. We denote the solution as $P_{\mathrm{tx}}\left(\frac{E_b}{N_0}, R\right)$. One example: From Table 4.3 we obtain that 16-PSK requires an E_b/N_0 of 18.5 dB to reach the target BER. When fixing the parameters $G_t = G_r = L = 1$, $\lambda = 12.5\,\mathrm{cm}$ (according to a 2.4 GHz transceiver), reference distance $d_0 = 1\,\mathrm{m}$, distance $d = 10\,\mathrm{m}$, a data rate of $R = 1\,\mathrm{Mbps}$, and a noise level of $N_0 = -180\,\mathrm{dB}$ this corresponds to $P_{\mathrm{tx}}(18.5\,\mathrm{dB}, R) \approx 2.26\,\mathrm{mW}$.

We next utilize a transceiver energy consumption model developed in references [762, 855] that incorporates startup energy and transmit energy. In this model, it is assumed that during the startup time mainly a frequency synthesizer is active, consuming energy P_{FS}, while during the actual waveform transmission power is consumed by the frequency synthesizer, the modulator (using P_{MOD}), and the radiated energy $P_{\mathrm{tx}}(\cdot,\cdot)$. The power amplifier is not explicitly considered. Using reference [855], we assume $P_{\mathrm{FS}} = 10\,\mathrm{mW}$, $P_{\mathrm{MOD}} = 2\,\mathrm{mW}$ and a symbol rate of $B = 1$ M symbols/sec. The duration of the startup time is T_{start}. For the case of binary modulation, we assume the following energy model:

$$E_{\mathrm{binary}}\left(\frac{E_b}{N_0}, B\right) = P_{\mathrm{FS}} \cdot T_{\mathrm{start}} + \left(P_{\mathrm{MOD}} + P_{\mathrm{FS}} + P_{\mathrm{tx}}\left(\frac{E_b}{N_0}, B\right)\right) \cdot \frac{n}{B},$$

where n is the number of data bits to transmit in a packet. For the case of m-ary modulation, it is assumed that the power consumption of the modulator and the frequency synthesizer are increased by some factors $\alpha \geq 1$, $\beta \geq 1$, such that the overall energy expenditure is:

$$E_{m\text{-ary}}\left(\frac{E_b}{N_0}, B \cdot \log_2 m\right) = \beta \cdot P_{\mathrm{FS}} \cdot T_{\mathrm{start}} + \left(\alpha \cdot P_{\mathrm{MOD}} + \beta \cdot P_{\mathrm{FS}} + P_{\mathrm{tx}}\left(\frac{E_b}{N_0}, B \cdot \log_2 m\right)\right) \cdot \frac{n}{B \cdot \log_2(m)}.$$

Accepting the value $\beta = 1.75$ from reference [855] for both PSK and FSK modulation, one can evaluate the ratio $\frac{E_{m\text{-ary}}(\cdot,\cdot)}{E_{\mathrm{binary}}(\cdot,\cdot)}$ to measure the energy advantage or disadvantage of m-ary modulation over binary modulation. As an example, we show this ratio in Figure 4.10 for varying $m \in \{4, 8, 16, 32, 64\}$, with $\alpha = 2.0$, a startup time of 466 μs, and two different packet sizes, 100 bits and 2000 bits. The two upper curves correspond to a packet size of 100 bits; the two lower curves correspond to the packet size of 2000 bits. Other results obtained with a shorter startup time of 100 μs or $\alpha = 3.0$ look very similar. One can see that for large packet sizes m-ary FSK modulation is favorable, since the actual packet transmission times are shortened and furthermore the required E_b/N_0 *decreases* for increasing m, at the expense of a reduced bandwidth efficiency, which translates into a wider required spectrum (the FSK scheme is orthogonal FSK). For m-ary PSK, only certain values of m give an energy advantage; for larger m the increased E_b/N_0 requirements outweigh the gains due to reduced transmit times. For small packet sizes, the binary modulation schemes are more energy efficient for both PSK and FSK, because the energy costs are dominated by the startup time. If one reduces β to $\beta = 1$ (assuming no extra energy consumption of the frequency synthesizer due to m-ary modulation), then m-ary modulation would, for all parameters under consideration, be truly better than binary modulation. The results presented in reference [855] indicate that the advantage of m-ary modulation increases as the startup time decreases. For shorter startup times also the packet lengths required to make m-ary modulation pay out are smaller.

Can we conclude from this that it is favorable to use large packets? Unfortunately, the answer is: it depends. As we will see in Chapter 6, longer packets at the same bit error rate and without employing error-correction mechanisms lead to higher packet error rates, which in turn lead to retransmitted packets, easily nullifying the energy gains of choosing m-ary modulation. A careful joint consideration of modulation and other schemes for increasing transmission robustness (FEC or ARQ schemes) is needed.

But it can be beneficial to transmit multiple short packets during a single wakeup period, thus achieving a lower relative influence of the startup costs per packet [562].

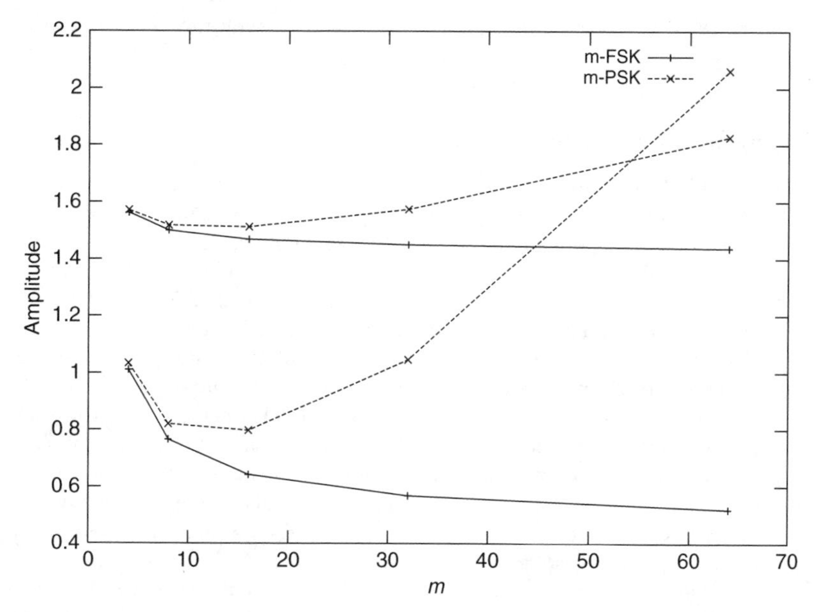

Figure 4.10 Comparison of the energy consumption of m-ary FSK/PSK to binary FSK/PSK for $\alpha = 2.0$ and startup time of 466 μs.

Clearly, this example provides only a single point in the whole design space. The bottom line here is that the choice of modulation scheme depends on several interacting aspects, including technological factors (in the example: α, β), packet size, target error rate, and channel error model (in reference [855], a similar example is carried out for the case of Rayleigh fading). The optimal decision would have to properly balance the modulation scheme and other measures to increase transmission robustness, since these also have energy costs:

- With retransmissions, entire packets have to be transmitted again.
- With FEC coding, more bits have to be sent and there is additional energy consumption for coding and decoding. While coding energy can be neglected, and the receiver needs significant energy for the decoding process [563]. This is especially cumbersome if the receiver is a power-constrained node. Coding and retransmission schemes are discussed in more detail in Chapter 6.
- The cost of increasing the radiated power [855] depends on the efficiency of the power amplifier (compare Section 2.2.4), but the radiated power is often small compared to the overall power dissipated by the transceiver, and additionally this drives the PA into a more efficient regime.[14]

In [670], a similar analysis as in our example has been carried out for m-ary QAM. Specifically, the energy-per-bit consumption (defined as the overall energy consumption for transmitting a packet of n bits divided by n) of different m-ary QAM modulation schemes has been investigated for different packet sizes, taking startup energy and the energy costs of power amplifiers as well as PHY and MAC packet overheads explicitly into account. For the particular setup used in this

[14] Of course, one disadvantage of using an increased transmit power is an increased interference for other transmissions and thus a decreased overall network capacity. However, this plays no role during low-load situations, which prevail in wireless sensor networks – unless event storms or other correlated traffic models are present.

investigation, 16-QAM seems to be the optimum modulation schemes for all different sizes of the user data.

4.3.3 Dynamic modulation scaling

Even if it is possible to determine the optimal scheme for a given combination of BER target, range, packet sizes and so forth, such an optimum is only valid for short time; as soon as one of the constraints changes, the optimum can change, too. In addition, other constraints like delay or the desire to achieve high throughput can dictate to choose higher modulation schemes.

Therefore, it is interesting to consider methods to *adapt* the modulation scheme to the current situation. Such an approach, called **dynamic modulation scaling**, is discussed in reference [738]. In particular, for the case of m-ary QAM and a target BER of 10^{-5}, a model has been developed that uses the symbol rate B and the number of levels per symbol m as parameters. This model expresses the energy required per bit and also the achieved delay per bit (the inverse of the data rate), taking into account that higher modulation levels need higher radiated energy. Extra startup costs are not considered. Clearly, the bit delay decreases for increasing B and m. The energy per bit depends much more on m than on B. In fact, for the particular parameters chosen, it is shown that both energy per bit and delay per bit are minimized for the maximum symbol rate. With modulation scaling, a packet is equipped with a delay constraint, from which directly a minimal required data rate can be derived. Since the symbol rate is kept fixed, the approach is to choose the smallest m that satisfies the required data rate and which thus minimizes the required energy per bit. Such delay constraints can be assigned either explicitly or implicitly. One approach explored in the paper is to make the delay constraint depend on the packet backlog (number of queued packets) in a sensor node: When there are no packets present, a small value for m can be used, having low energy consumption. As backlog increases, m is increased as well to reduce the backlog quickly and switch back to lower values of m. This modulation scaling approach has some similarities to the concept of **dynamic voltage scaling** discussed in Section 2.2.2.

4.3.4 Antenna considerations

The desired small form factor of the overall sensor nodes restricts the size and the number of antennas. As explained above, if the antenna is much smaller than the carrier's wavelength, it is hard to achieve good antenna efficiency, that is, with ill-sized antennas one must spend more transmit energy to obtain the same radiated energy.

Secondly, with small sensor node cases, it will be hard to place two antennas with suitable distance to achieve receive diversity. As discussed in Section 4.2.7, the antennas should be spaced apart at least 40–50 % of the wavelength used to achieve good effects from diversity. For 2.4 GHz, this corresponds to a spacing of between 5 and 6 cm between the antennas, which is hard to achieve with smaller cases.

In addition, radio waves emitted from an antenna close to the ground – typical in some applications – are faced with higher path-loss coefficients than the common value $\alpha = 2$ for free-space communication. Typical attenuation values in such environments, which are also normally characterized by obstacles (buildings, walls, and so forth), are about $\alpha = 4$ [245, 648].

Moreover, depending on the application, antennas must not protrude from the casing of a node, to avoid possible damage to it. These restrictions, in general, limit the achievable quality and characteristics of an antenna for wireless sensor nodes.

Nodes randomly scattered on the ground, for example, deployed from an aircraft, will land in random orientations, with the antennas facing the ground or being otherwise obstructed. This can lead to nonisotropic propagation of the radio wave, with considerable differences in the strength of the emitted signal in different directions. This effect can also be caused by the design of an antenna, which often results in considerable differences in the spatial propagation characteristics (so-called lobes of an antenna).

Antenna design is an issue in itself and is well beyond the scope of this book. Some specific considerations on antenna design for wireless sensor nodes are discussed in [115, Chap. 8].

4.4 Further reading

Jointly optimizing coding and modulation Biglieri et al. [79] consider coding and modulation from an information-theoretic perspective for different channel models, including the AWGN, flat fading channels and block fading channels. Specifically, the influence of symbol-by-symbol power control at the transmitter in the presence of channel-state information such that deep fades are answered with higher output powers ("channel inversion"), of receiver diversity and interleaving and of coding schemes with unequal protection (i.e., user bits of different importance are encoded differently) on the channel capacity are discussed. One particularly interesting result is that the capacity of a Rayleigh fading channel with power control can be higher than the capacity of an AWGN channel with the same average radiated power.

DSSS in WSN Some efforts toward the construction of DSSS transceivers for wireless sensor networks with their space and power constraints are described in references [155, 280, 281]. In addition, Myers et al. [580] discuss low-power spread-spectrum transceivers for IEEE 802.11.

Energy efficiency in GSM Reducing energy consumption is an issue not only in wireless sensor networks but also in other types of systems, for example, cellular systems. For the interested: advanced signal processing algorithms for reducing power consumption of GSM transceivers are discussed in references [525].

5

MAC protocols

Objectives of this Chapter

Medium Access Control (MAC) protocols solve a seemingly simple task: they coordinate the times where a number of nodes access a shared communication medium. An "unoverseeable" number of protocols have emerged in more than thirty years of research in this area. They differ, among others, in the types of media they use and in the performance requirements for which they are optimized.

This chapter presents the fundamentals of MAC protocols and explains the specific requirements and problems these protocols have to face in wireless sensor networks. The single most important requirement is energy efficiency and there are different MAC-specific sources of energy waste to consider: overhearing, collisions, overhead, and idle listening. We discuss protocols addressing one or more of these issues. One important approach is to switch the wireless transceiver into a sleep mode. Therefore, there are trade-offs between a sensor network's energy expenditure and traditional performance measures like delay and throughput.

Chapter Outline

Medium Access Control (MAC) protocols is the first protocol layer above the Physical Layer (PHY) and consequently MAC protocols are heavily influenced by its properties. The fundamental task of any MAC protocol is to regulate the access of a number of nodes to a shared medium in such a way that certain application-dependent performance requirements are satisfied. Some of the traditional performance criteria are delay, throughput, and fairness, whereas in WSNs, the issue of energy conservation becomes important.

Within the OSI reference model, the MAC is considered as a part of the Data Link Layer (DLL), but there is a clear division of work between the MAC and the remaining parts of the DLL. The MAC protocol determines for a node the points in time when it accesses the medium to try to transmit a data, control, or management packet to another node (unicast) or to a set of nodes (multicast, broadcast). Two important responsibilities of the remaining parts of the DLL are error control and flow control. Error control is used to ensure correctness of transmission and to take appropriate actions in case of transmission errors and flow control regulates the rate of transmission to protect a slow receiver from being overwhelmed with data. The link layer is discussed in Chapter 6.

In this chapter, we first give a brief introduction to MAC protocols in general and to the particular requirements and challenges found in wireless sensor networks (Section 5.1). Most notably, the issue of energy efficiency is the prime consideration in WSN MAC protocols, and therefore, we concentrate on schemes that explicitly try to reduce overall energy consumption. One of the main approaches to conserve energy is to put nodes into sleep state whenever possible. Protocols striving for low duty cycle or wakeup concepts (Section 5.2) are designed to accomplish this. Other important classes of useful MAC protocols are contention-based (Section 5.3) and schedule-based protocols (Section 5.4). The IEEE 802.15.4 protocol combines contention- and schedule-based elements and can be expected to achieve significant commercial impact; it is discussed in Section 5.5. The question why other commercially successful protocols like IEEE 802.11 and Bluetooth are not the primary choice in wireless sensor networks is touched in Section 5.6. The final Section 5.8 contains some concluding remarks and a comparison of the different protocols discussed in this chapter.

5.1 Fundamentals of (wireless) MAC protocols

In this section, we discuss some fundamental aspects and important examples of wireless MAC protocols, since the protocols used in wireless sensor networks inherit many of the problems and approaches already existing for this more general field.

MAC protocols are an active research area for more than 30 years now [5], and there exists a huge body of literature. Some survey papers covering MAC protocols in general as well as wireless MAC protocols can be found in references [7, 18, 23, 143, 311, 390, 579]. General introductions into MAC protocols can be found in references [6, 68, 709, 808]. Energy aspects were not one of the top priorities in earlier research on MAC protocols (this is not to say they have not been addressed [886]), but with the advent of wireless sensor networks, energy has been established as one of the primary design concerns.

5.1.1 Requirements and design constraints for wireless MAC protocols

Traditionally, the most important **performance requirements** for MAC protocols are throughput efficiency, stability, fairness, low access delay (time between packet arrival and first attempt to transmit it), and low transmission delay (time between packet arrival and successful delivery), as well as a low overhead. The overhead in MAC protocols can result from per-packet overhead (MAC headers and trailers), collisions, or from exchange of extra control packets. Collisions can happen if the MAC protocol allows two or more nodes to send packets at the same time. Collisions can result in the inability of the receiver to decode a packet correctly, causing the upper layers to perform a retransmission. For time-critical applications, it is important to provide deterministic or stochastic guarantees on delivery time or minimal available data rate. Sometimes, preferred treatment of important packets over unimportant ones is required, leading to the concept of **priorities**.

The operation and performance of MAC protocols is heavily influenced by the properties of the underlying physical layer. Since WSNs use a wireless medium, they inherit all the well-known problems of wireless transmission. One problem is time-variable, and sometimes quite high, error rates, which is caused by physical phenomena like slow and fast fading, path loss, attenuation, and man-made or thermal noise (see Chapter 4 and [682, Chapters 4 & 5]). Depending on modulation schemes, frequencies, distance between transmitter and receiver, and the propagation environment, instantaneous bit error rates in the range of $10^{-3} \dots 10^{-2}$ can easily be observed [213, 223, 594, 882].

As explained in Chapter 4, the received power P_{rcvd} decreases with the distance between transmitting and receiving node. This path loss combined with the fact that any transceiver needs a minimum signal strength to demodulate signals successfully leads to a maximum range that a sensor node can reach with a given transmit power. If two nodes are out of reach, they cannot hear each other. This gives rise to the well-known hidden-terminal/exposed-terminal problems [823]. The **hidden-terminal problem** occurs specifically for the class of Carrier Sense Multiple Access (CSMA) protocols, where a node senses the medium before starting to transmit a packet. If the medium is found to be busy, the node defers its packet to avoid a collision and a subsequent retransmission. Consider the example in Figure 5.1. Here, we have three nodes A, B, and C that are arranged such that A and B are in mutual range, B and C are in mutual range, but A and C cannot hear each other. Assume that A starts to transmit a packet to B and some time later node C also decides to start a packet transmission. A carrier-sensing operation by C shows an idle medium since C cannot hear A's signals. When C starts its packet, the signals collide at B and both packets are useless. Using simple CSMA in a hidden-terminal scenario thus leads to needless collisions.

In the **exposed-terminal scenario**, B transmits a packet to A, and some moment later, C wants to transmit a packet to D. Although this would be theoretically possible since both A and D would receive their packets without distortions, the carrier-sense operation performed by C suppresses C's transmission and bandwidth is wasted. Using simple CSMA in an exposed terminal scenario thus leads to needless waiting.

Two solutions to the hidden-terminal and exposed-terminal problems are busy-tone solutions [823] and the RTS/CTS handshake used in the IEEE 802.11 WLAN standard [815] and first presented in the MACA [407]/MACAW [75] protocols. These will be described in Section 5.1.2 in the context of CSMA protocols.

On wired media, it is often possible *for the transmitter* to detect a collision *at the receiver* immediately and to abort packet transmission. This feature is called *collision detection* (CD) and is used in Ethernet's CSMA/CD protocol to increase throughput efficiency. Such a collision detection works because of the low attenuation in a wired medium, resulting in similar SNRs at transmitter and receiver. Consequently, when the transmitter reads back the channel signal during transmission and observes a collision, it can infer that there must have been a collision at the receiver too. More importantly, the *absence* of a collision at the transmitter allows to conclude that there has been no

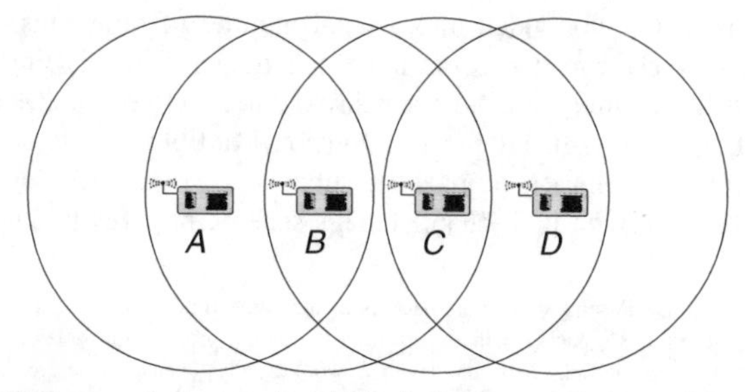

Figure 5.1 Hidden-terminal scenario (circles indicate transmission & interference range)

collision at the receiver during the packet transmission.[1] In a wireless medium, neither of these two conclusions holds true – the interference situation at the transmitter does not tell much about the interference situation at the receiver. Furthermore, simple wireless transceivers work only in a half-duplex mode, meaning that at any given time either the transmit or the receive circuitry is active but not both.[2] Therefore, collision detection protocols are usually not applicable to wireless media.

Another important problem arises when there is no dedicated frequency band allocated to a wireless sensor network and the WSN has to share its spectrum with other systems. Because of license-free operations, many wireless systems use the so-called ISM bands, with the 2.4 GHz ISM band being a prime example. This specific band is used by several systems, for example, the IEEE 802.11/IEEE 802.11b WLANs [466, 815], Bluetooth [318], and the IEEE 802.15.4 WPAN. Therefore, the issue of coexistence of these systems arises [154, 359, 360, 469].

Finally, the design of MAC protocols depends on the expected traffic load patterns. If a WSN is deployed to continuously observe a physical phenomenon, for example, the time-dependent temperature distribution in a forest, a continuous and low load with a significant fraction of periodic traffic can be expected. On the other hand, if the goal is to wait for the occurrence of an important event and upon its occurrence to report as much data as possible, the network is close to idle for a long time and then is faced with a bulk of packets that are to be delivered quickly. A high MAC efficiency is desirable during these overload phases. An example for this class of applications is wildfire observation [742].

5.1.2 Important classes of MAC protocols

A huge number of (wireless) MAC protocols have been devised during the last thirty years. They can be roughly classified into the following classes [311]: fixed assignment protocols, demand assignment protocols, and random access protocols.

Fixed assignment protocols

In this class of protocols, the available resources are divided between the nodes such that the resource assignment is long term and each node can use its resources exclusively without the risk of collisions. Long term means that the assignment is for durations of minutes, hours, or even longer, as opposed to the short-term case where assignments have a scope of a data burst, corresponding to a time horizon of perhaps (tens of) milliseconds. To account for changes in the topology – for example, due to nodes dying or new nodes being deployed, mobility, or changes in the load patterns – signaling mechanisms are needed in fixed assignment protocols to renegotiate the assignment of resources to nodes. This poses questions about the scalability of these protocols.

Typical protocols of this class are TDMA, FDMA, CDMA, and SDMA. The Time Division Multiple Access (TDMA) scheme [708] subdivides the time axis into fixed-length superframes and each superframe is again subdivided into a fixed number of time slots. These time slots are assigned to nodes exclusively and hence the node can transmit in this time slot periodically in every superframe. TDMA requires tight time synchronization between nodes to avoid overlapping of signals in adjacent time slots. In Frequency Division Multiple Access (FDMA), the available frequency band is subdivided into a number of subchannels and these are assigned to nodes, which can transmit exclusively on their channel. This scheme requires frequency synchronization,

[1] When two distant nodes A and B send very short packets at the same time, it may happen that A finishes its packet transmission before the signal from B's packet actually arrives (due to the propagation delay). In this case, neither A nor B would see any collision but nodes halfway between A and B would. Only when packets are long enough or the distance between nodes is suitably bounded, nodes A and B have a chance to detect collisions and react upon them.

[2] This way, transmit and receive circuitry can share components, leading to reduced transceiver complexity.

relatively narrowband filters, and the ability of a receiver to tune to the channel used by a transmitter. Accordingly, an FDMA transceiver tends to be more complex than a TDMA transceiver. In Code Division Multiple Access (CDMA) schemes [293, 297, 700], the nodes spread their signals over a much larger bandwidth than needed, using different codes to separate their transmissions. The receiver has to know the code used by the transmitter; all parallel transmissions using other codes appear as noise. Crucial to CDMA is the code management. Finally, in Space Division Multiple Access (SDMA), the spatial separation of nodes is used to separate their transmissions. SDMA requires arrays of antennas and sophisticated signal processing techniques [476] and cannot be considered a candidate technology for WSNs.

Demand assignment protocols

In demand assignment protocols, the exclusive allocation of resources to nodes is made on a short-term basis, typically the duration of a data burst. This class of protocols can be broadly subdivided into centralized and distributed protocols. In central control protocols (examples are the HIPERLAN/2 protocol [209, 247, 248, 249, 250], DQRUMA [408], or the MASCARA protocol [621]; polling schemes [757, 805, 824] can also be subsumed under this class), the nodes send out requests for bandwidth allocation to a central node that either accepts or rejects the requests. In case of successful allocation, a confirmation is transmitted back to the requesting node along with a description of the allocated resource, for example, the numbers and positions of assigned time slots in a TDMA system and the duration of allocation. The node can use these resources exclusively. The submission of requests from nodes to the central station is often donecontention based, that is, using a random access protocol on a dedicated (logical) signaling channel. Another option is to let the central station poll its associated nodes. In addition, the nodes often **piggyback** requests onto data packets transmitted in their exclusive data slots, thus avoiding transmission of separate request packets. The central node needs to be switched on all the time and is responsible for resource allocation. Resource deallocation is often done implicitly: when a node does not use its time slots any more, the central node can allocate these to other nodes. This way, nodes do not need to send extra deallocation packets. Summarizing, the central node performs a lot of activities, it must be constantly awake, and thus needs lots of energy. This class of protocols is a good choice if a sufficient number of energy-unconstrained nodes are present and the duties of the central station can be moved to these. An example is the IEEE 802.15.4 protocol discussed in Section 5.5. If there are no unconstrained nodes, a suitable approach is to rotate the central station duties among the nodes like, for example, in the LEACH protocol described in Section 5.4.1.

An example of *distributed* demand assignment protocols are token-passing protocols like IEEE 802.4 Token Bus [372]. The right to initiate transmissions is tied to reception of a small special **token frame**. The token frame is rotated among nodes organized in a **logical ring** on top of a broadcast medium. Special ring management procedures are needed to include and exclude nodes from the ring or to correct failures like lost tokens. Token-passing protocols have also been considered for wireless or error-prone media [387, 535, 883], but they tend to have problems with the maintenance of the logical ring in the presence of significant channel errors [883]. In addition, since token circulation times are variable, a node must always be able to receive the token to avoid breaking the logical ring. Hence, a nodes transceiver must be switched on most of the time. In addition, maintaining a logical ring in face of frequent topology changes is not an easy task and involves significant signaling traffic besides the token frames themselves.

Random access protocols

The nodes are uncoordinated, and the protocols operate in a fully distributed manner. Random access protocols often incorporate a random element, for example, by exploiting random packet arrival times, setting timers to random values, and so on. One of the first and still very important random access protocols is the ALOHA or slotted ALOHA protocol, developed at the University of Hawaii [5]. In the pure ALOHA protocol, a node wanting to transmit a new packet transmits it *immediately*. There is no coordination with other nodes and the protocol thus accepts the risk of collisions at the receiver. To detect this, the receiver is required to send an immediate acknowledgment for a properly received packet. The transmitter interprets the lack of an acknowledgment frame as a sign of a collision, backs off for a random time, and starts the next trial. ALOHA provides short access and transmission delays under light loads; under heavier loads, the number of collisions increases, which in turn decreases the throughput efficiency and increases the transmission delays. In slotted ALOHA, the time is subdivided into time slots and a node is allowed to start a packet transmission only at the beginning of a slot. A slot is large enough to accommodate a maximum-length packet. Accordingly, only contenders starting their packet transmission *in the same slot* can destroy a node's packet. If any node wants to start later, it has to wait for the beginning of the next time slot and has thus no chance to destroy the node's packet. In short, the synchronization reduces the probability of collisions and slotted ALOHA has a higher throughput than pure ALOHA.

In the class of **CSMA protocols** [422], a transmitting node tries to be respectful to ongoing transmissions. First, the node is required to listen to the medium; this is called **carrier sensing**. If the medium is found to be idle, the node starts transmission. If the medium is found busy, the node defers its transmission for an amount of time determined by one of several possible algorithms. For example, in **nonpersistent CSMA**, the node draws a random waiting time, after which the medium is sensed again. Before this time, the node does not care about the state of the medium. In different **persistent CSMA** variants, after sensing that the medium is busy, the node awaits the end of the ongoing transmission and then behaves according to a **backoff algorithm**. In many of these backoff algorithms, the time after the end of the previous frame is subdivided into time slots. In p-persistent CSMA, a node starts transmission in a time slot with some probability p and with probability $1 - p$ it waits for another slot.[3] If some other node starts to transmit in the meantime, the node defers and repeats the whole procedure after the end of the new frame. A small value of p makes collisions unlikely, but at the cost of high access delays. The converse is true for a large value of p.

In the backoff algorithm executed by the IEEE 802.11 Distributed Coordination Function (DCF), a node transmitting a new frame picks a random value from the current **contention window** and starts a timer with this value. The timer is decremented after each slot. If another node starts in the meantime, the timer is suspended and resumed after the next frame ends and contention continues. If the timer decrements to zero, the node transmits its frame. When a transmission error occurs (indicated, for example, by a missing acknowledgment frame), the size of the contention window is increased according to a modified binary exponential backoff procedure.[4] While CSMA protocols are still susceptible to collisions, they have a higher throughput efficiency than ALOHA protocols, since ongoing packets are not destroyed when potential competitors hear them on the medium.

As explained above, carrier-sense protocols are susceptible to the hidden-terminal problem since interference at the receiver cannot be detected by the transmitter. This problem may cause packet collisions. The energy spent on collided packets is wasted and the packets have to be retransmitted. Several approaches have appeared [268] to solve or at least to reduce the hidden-terminal problem; we present two important ones: the busy-tone solution and the RTS/CTS handshake.

In the original **busy-tone solution** [823], two different frequency channels are used, one for the data packets and the other one as a control channel. As soon as a node starts to receive a packet destined to it, it emits an unmodulated wave on the control channel and ends this when packet

[3] The special case $p = 1$ amounts to a node that always starts transmission when the preceding packet ends, surely creating collisions when two or more nodes want to transmit. This choice is best accompanied by a collision detection and resolution facility, like, for example, in Ethernet.

[4] In the binary exponential backoff procedure, the contention window is doubled after each collision/transmission error as indicated by lack of an immediate acknowledgment. In the truncated binary exponential backoff procedure, the contention window is doubled until an upper bound is reached. Afterward it stays constant.

reception is finished. A node that wishes to transmit a packet first senses the control channel for the presence of a busy tone. If it hears something, the node backs off according to some algorithm, for example similar to nonpersistent CSMA. If it hears nothing, the node starts packet transmission on the data channel. This protocol solves both the hidden- and exposed-terminal problem, given that the busy-tone signal can be heard over the same distance as the data signal. If the busy tone is too weak, a node within radio range of the receiver might start data transmission and destroy the receiver's signal. If the busy tone is too strong, more nodes than necessary suppress their transmissions. The control channel does not need much bandwidth but a narrow bandwidth channel requires good frequency synchronization. A solution with two busy tones, one sent by the receiver and the other by the transmitter node, is discussed in [203, 321]. Another variant of the busy-tone approach is used by PAMAS, discussed in Section 5.3.2.

The **RTS/CTS handshake** as used in IEEE 802.11 [815] is based on the MACAW protocol [75] and is illustrated in Figure 5.2. It uses only a single channel and two special control packets. Suppose that node *B* wants to transmit a data packet to node *C*. After *B* has obtained channel access (for example after sensing the channel as idle), it sends a Request To Send (RTS) packet to *C*, which includes a duration field indicating the remaining length of the overall transaction (i.e., until the point where *B* would receive the acknowledgment for its data packet). If *C* has properly received the RTS packet, it sends a Clear To Send (CTS) packet, which again contains a duration field. When *B* receives the CTS packet, it starts transmission of the data packet and finally *C* answers with an acknowledgment packet. The acknowledgment is used to tell *B* about the success of the transmission; lack of acknowledgment is interpreted as collision (the older MACA protocol [407] lacks the acknowledgment). Any other station *A* or *D* hearing either the RTS, CTS, data or acknowledgment packet sets an internal timer called Network Allocation Vector (NAV) to the remaining duration indicated in the respective frame and avoids sending any packet as long as this

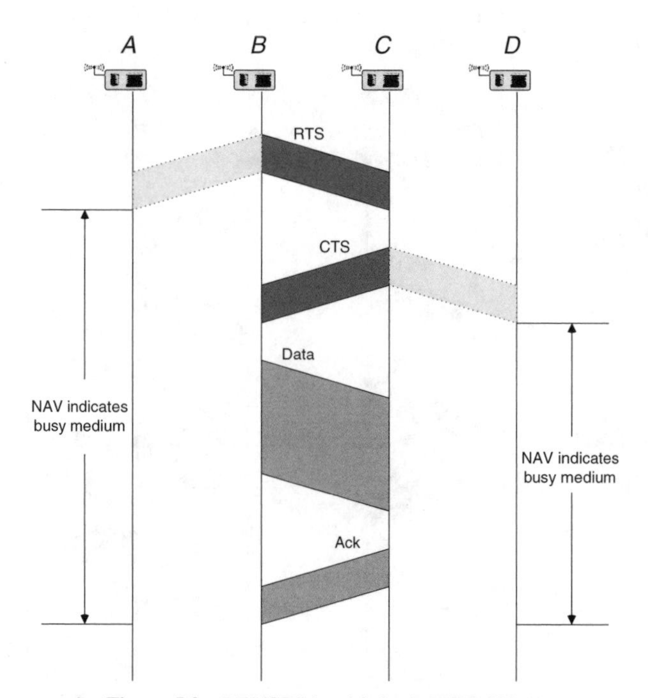

Figure 5.2 RTS/CTS handshake in IEEE 802.11

timer is not expired. Specifically, nodes A and D send no CTS answer packets even when they have received a RTS packet correctly. This way, the ongoing transmission is not distorted.

Does this scheme eliminate collisions completely? No, there still exist some collision scenarios. First, in the scenario described above, nodes A and C can issue RTS packets to B simultaneously. However, in this case, only the RTS packets are lost and no long data frame has been transmitted. Two further problems are illustrated in Figure 5.3 [668]: In the left part of the figure, nodes A and B run the RTS-CTS-Data-Ack sequence, and B's CTS packet also reaches node C. However, at almost the same time, node D sends an RTS packet to C, which collides at node C with B's CTS packet. This way, C has no chance to decode the duration field of the CTS packet and to set its NAV variable accordingly. After its failed RTS packet, D sends the RTS packet again to C and C answers with a CTS packet. Node C is doing so because it cannot hear A's ongoing transmission and has no proper NAV entry. C's CTS packet and A's data packet collide at B. In the figure's right part, the problem is created by C starting its RTS packet to D immediately before it can sense B's CTS packet, which C consequently cannot decode properly. One solution approach [668] is to ensure that CTS packets are longer than RTS packets. For an explanation, consider the right part of Figure 5.3. Here, even if B's CTS arrives at C immediately after C starts its RTS, it lasts long enough that C has a chance to turn its transceiver into receive mode and to sense B's signal. An additional protocol rule states that in such a case node C has to defer any further transmission for a sufficiently long time to accommodate one maximum-length data packet. Hence, the data packet between A and B can be transmitted without distortion. It is not hard to convince oneself that the problem in the left half of Figure 5.3 is eliminated too.

A further problem of the RTS/CTS handshake is its significant overhead of two control packets per data packet, not counting the acknowledgment packet. If the data packet is small, this overhead might not pay off and it may be simpler to use some plain CSMA variant. For long packets, the overhead of the RTS/CTS handshake can be neglected, but long packets are more likely to be hit by channel errors and must be retransmitted entirely, wasting precious energy (channel errors often hit only a few bits). A good compromise is to fragment a large packet like, for example, in

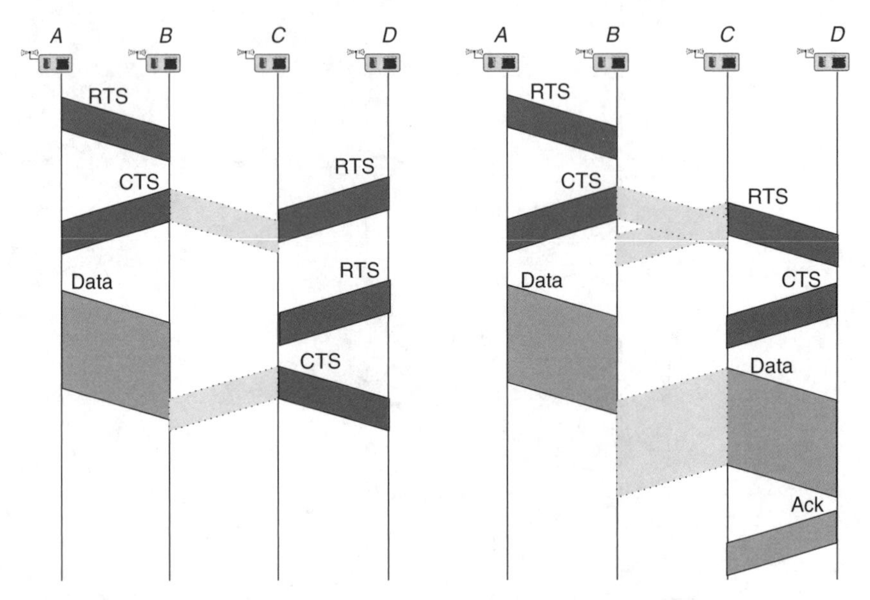

Figure 5.3 Two problems in RTS/CTS handshake [668]

IEEE 802.11 or in the S-MAC protocol discussed in Section 5.2.2 and to use the RTS/CTS only once for the whole set of fragments.

5.1.3 MAC protocols for wireless sensor networks

In this section, we narrow down the specific requirements and design considerations for MAC protocols in wireless sensor networks.

Balance of requirements

For the case of WSNs, the balance of requirements is different from traditional (wireless) networks. Additional requirements come up, first and foremost, the need to conserve energy. The importance of energy efficiency for the design of MAC protocols is relatively new and many of the "classical" protocols like ALOHA and CSMA contain no provisions toward this goal. Some papers covering energy aspects in MAC protocols are references [143, 299, 886]. Other typical performance figures like fairness, throughput, or delay tend to play a minor role in sensor networks. Fairness is not important since the nodes in a WSN do not represent individuals competing for bandwidth, but they collaborate to achieve a common goal. The access/transmission delay performance is traded against energy conservation, and throughput is mostly not an issue either.

Further important requirements for MAC protocols are scalability and robustness against frequent topology changes, as caused for example by nodes powering down temporarily to replenish their batteries by energy scavenging, mobility, deployment of new nodes, or death of existing nodes. The need for scalability is evident when considering very dense sensor networks with dozens or hundreds of nodes in mutual range.

Energy problems on the MAC layer

As we have discussed in Chapters 2 and 4, a nodes transceiver consumes a significant share of energy. Recall that a transceiver can be in one of the four main states (Section 2.1.4): transmitting, receiving, idling, or sleeping. Section 2.2.4 has discussed the energy-consumption properties of some transceiver designs in the different operational states. In a nutshell, the lessons are: Transmitting is costly, receive costs often have the same order of magnitude as transmit costs, idling can be significantly cheaper but also about as expensive as receiving, and sleeping costs almost nothing but results in a "deaf" node. Applying these lessons to the operations of a MAC protocol, we can derive the following **energy problems** and design goals [915]:

Collisions collisions incur useless receive costs at the destination node, useless transmit costs at the source node, and the prospect to expend further energy upon packet retransmission. Hence, collisions should be avoided, either by design (fixed assignment/TDMA or demand assignment protocols) or by appropriate collision avoidance/hidden-terminal procedures in CSMA protocols. However, if it can be guaranteed for the particular sensor network application at hand that the load is always sufficiently low, collisions are no problem.

Overhearing Unicast frames have one source and one destination node. However, the wireless medium is a broadcast medium and all the source's neighbors that are in receive state hear a packet and drop it when it is not destined to them; these nodes **overhear** the packet. References [668, 915] show that for higher node densities overhearing avoidance can save significant amounts of energy. On the other hand, overhearing is sometimes desirable, for example, when collecting neighborhood information or estimating the current traffic load for management purposes.

Protocol overhead Protocol overhead is induced by MAC-related control frames like, for example, RTS and CTS packets or request packets in demand assignment protocols, and furthermore by per-packet overhead like packet headers and trailers.

Idle listening A node being in idle state is ready to receive a packet but is not currently receiving anything. This readiness is costly and useless in case of low network loads; for many radio modems, the idle state still consumes significant energy. Switching off the transceiver is a solution; however, since mode changes also cost energy, their frequency should be kept at "reasonable" levels. TDMA-based protocols offer an implicit solution to this problem, since a node having assigned a time slot and exchanging (transmitting/receiving) data *only* during this slot can safely switch off its transceiver in all other time slots.

Most of the MAC protocols developed for wireless sensor networks attack one or more of these problems to reduce energy consumption, as we will see in the next sections.

A design constraint somewhat related to energy concerns is the requirement for **low complexity operation**. Sensor nodes shall be simple and cheap and cannot offer plentiful resources in terms of processing power, memory, or energy. Therefore, computationally expensive operations like complex scheduling algorithms should be avoided. The desire to use cheap node hardware includes components like oscillators and clocks. Consequently, the designer of MAC protocols should bear in mind that very tight time synchronization (as needed for TDMA with small time slots) would require frequent resynchronization of neighboring nodes, which can consume significant energy. Time synchronization issues are discussed in Chapter 8.

Structure of the following discussion

In the following sections, we discuss a number of different MAC protocols proposed for wireless sensor networks because of their ability to conserve energy. The presentation in the following sections is *not* structured according to the above discussed classes of MAC protocols (fixed assignment, demand assignment, random access) but instead it is according to the way they attack one or more of the energy problems.

In Section 5.2, we discuss protocols that explicitly attack the idle listening problem by applying periodic sleeping or even wakeup radio concepts.

Some other protocols are classified into either **contention-based** or **schedule-based** protocols. This distinction is to be understood by the number of possible contenders upon a transmit opportunity toward a receiver node:

- In contention-based protocols (Section 5.3), any of the receiver's neighbors might try its luck at the risk of collisions. Accordingly, those protocols contain mechanisms to avoid collisions or to reduce their probability.
- In schedule-based protocols (Section 5.4), only one neighbor gets an opportunity and collisions are avoided. These protocols have a TDMA component, which provides also an implicit idle listening avoidance mechanism: when a node knows its allocated slots and can be sure that it communicates (transmits/receives) *only* in these slots, it can safely switch off its receiver at all other times.

In Section 5.5, we discuss the IEEE 802.15.4 protocol, which combines elements of schedule- and contention-based protocols and can be expected to achieve some commercial impact.

5.2 Low duty cycle protocols and wakeup concepts

Low duty cycle protocols try to avoid spending (much) time in the idle state and to reduce the communication activities of a sensor node to a minimum. In an ideal case, the sleep state is left

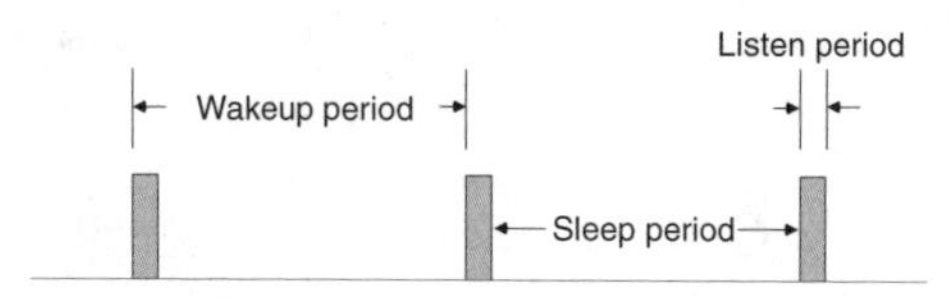

Figure 5.4 Periodic wakeup scheme

only when a node is about to transmit or receive packets. A concept for achieving this, the wakeup radio, is discussed in Section 5.2.4. However, such a system has not been built yet, and thus, there is significant interest to find alternative approaches.

In several protocols, a **periodic wakeup** scheme is used. Such schemes exist in different flavors. One is the **cycled receiver** approach [503], illustrated in Figure 5.4. In this approach, nodes spend most of their time in the sleep mode and wake up periodically to *receive* packets from other nodes. Specifically, a node A listens onto the channel during its **listen period** and goes back into sleep mode when no other node takes the opportunity to direct a packet to A. A potential transmitter B must acquire knowledge about A's listen periods to send its packet at the right time – this task corresponds to a *rendezvous* [503]. This rendezvous can, for example, be accomplished by letting node A transmit a short beacon at the beginning of its listen period to indicate its willingness to receive packets. Another method is to let node B send frequent request packets until one of them hits A's listen period and is really answered by A. However, in either case, node A only *receives* packets during its listen period. If node A itself wants to transmit packets, it must acquire the target's listen period. A whole cycle consisting of sleep period and listen period is also called a **wakeup period**. The ratio of the listen period length to the wakeup period length is also called the node's **duty cycle**. From this discussion, we already can make some important observations:

- By choosing a small duty cycle, the transceiver is in sleep mode most of the time, avoiding idle listening and conserving energy.
- By choosing a small duty cycle, the traffic directed from neighboring nodes to a given node concentrates on a small time window (the listen period) and in heavy load situations significant competition can occur.
- Choosing a long sleep period induces a significant **per-hop latency**, since a prospective transmitter node has to wait an average of half a sleep period before the receiver can accept packets. In the multihop case, the per-hop latencies add up and create significant end-to-end latencies.
- Sleep phases should not be too short lest the start-up costs outweigh the benefits.

In other protocols like, for example, S-MAC (Section 5.2.2), there is also a periodic wakeup but nodes can both *transmit and receive* during their wakeup phases. When nodes have their wakeup phases at the same time, there is no necessity for a node wanting to transmit a packet to be awake *outside* these phases to rendezvous its receiver.

Subsequently, we discuss some variations of this approach. They differ in various aspects, for example, the number of channels required or in the methods by which prospective transmitters can learn the listen periods of the intended receivers.

5.2.1 Sparse topology and energy management (STEM)

The Sparse Topology and Energy Management (STEM) protocol does not cover all aspects of a MAC protocol but provides a solution for the idle listening problem [742]. STEM targets networks that are deployed to wait for and report on the behaviour of a certain event, for example, when studying the paths of elephants in a habitat. From the perspective of a single sensor, most of

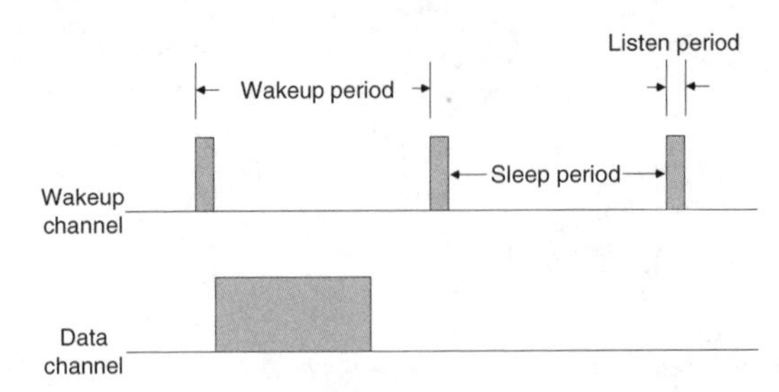

Figure 5.5 STEM duty cycle for a single node [742, Fig. 3]

the time there are no elephants and the sensor has nothing to report. However, once an elephant appears, the sensor reports its readings periodically. More abstractly, the network has a **monitor state**, where the nodes idle and do nothing, and also a **transfer state**, where the nodes exhibit significant sensing and communication activity. STEM tries to eliminate idle listening in the monitor state and to provide a fast transition into the transfer state, if required. In the transfer state, different MAC protocols can be employed. The term "topology" in STEMs name comes from the observation that as nodes enter and leave the sleep mode network topology changes. An important requirement for such topology-management schemes is that the network stays connected (or bi-connected or fulfills even higher connectivity requirements) even if a subset of nodes is in the sleep mode.

For an explanation of STEM, please refer to Figure 5.5. Two different channels are used, requiring two transceivers in each node: the **wakeup channel** and the **data channel**. The data channel is always in sleep mode, except when transmitting or receiving data packets. The underlying MAC protocol is executed solely on the data channel during the transfer states. On the wakeup channel the time is divided into fixed-length **wakeup periods** of length T. A wakeup period is subdivided into a **listen period** of length $T_{Rx} \ll T$ and a sleep period, where the wakeup channel transceiver enters sleep mode, too. If a node enters the listen period, it simply switches on its receiver for the wakeup channel and waits for incoming signals. If nothing is received during time T_{Rx}, the node returns into sleep mode. Otherwise the transmitter and receiver start a packet transfer on the data channel. There are two different variants for the transmitter to acquire the receiver's attention:

- In STEM-B, the transmitter issues so-called **beacons** on the wakeup channel periodically and without prior carrier sensing. Such a beacon indicates the MAC addresses of transmitter and receiver. As soon as the receiver picks up the beacon, it sends an acknowledgment frame back on the wakeup channel (causing the transmitter to stop beacon transmission), switches on the transceiver for the data channel, and both nodes can proceed to execute the regular MAC protocol on the data channel, like for example an RTS/CTS handshake. Any other node receiving the beacon on the wakeup channel recognizes that the packet is not destined for it and goes back to sleep mode. The transmitter sends these beacons at least for one full wakeup period to be sure to hit the receivers listen period.
- In STEM-T, the transmitter sends out a simple busy tone on the control channel (the T in STEM-T comes from "tone") for a time long enough to hit the receiver's listen period. Since the busy tone carries no address information, all the transmitter's neighbors (the receiver as well as other nodes) will sense the busy tone and switch on their data channel, without sending an acknowledgment packet. The other nodes can go back to sleep when they can deduce from the packet exchange on the data channel that they are not involved in the data transfer. A transceiver capable of generating and sensing busy tones can be significantly cheaper and less energy-consuming than a transceiver usable for data transmission but requires proper frequency synchronization.

In STEM-B, several transmitters might transmit their beacons simultaneously, leading to beacon collisions. A node waking up and receiving just some energy on the wakeup channel without being able to decode it behaves exactly as in STEM-T: It sends no acknowledgment, switches on its data channel, and waits what happens. The transmitter in this case transmits the beacons for the maximum time (since it hears no acknowledgment), then switches to the data channel, and tries to start the conversation with the receiver node.

It is noteworthy that in STEM a node entering the listen period remains silent, that is, transmits no packet. The opposite approach has been chosen, for example, in the mediation device protocol discussed in Section 5.2.3 or in the Piconet system [64], where node just waking up announces its willingness to receive a packet by transmitting a query beacon packet. In the approach taken by STEM, the transmitter has to send beacons or a busy tone for an average time of $\approx \frac{T}{2}$ and in the worst case for a maximum of $\approx T$. If packet transmissions are a rare event, it pays off to avoid the frequent (and mostly useless) query beacons and to put some extra burden on the transmitter to reach its receiver. Therefore, in low load situations, STEM-T is preferable over STEM-B.

The wakeup latency achievable with STEM-T or STEM-B is intimately related to the wakeup period T. Indeed, the results presented by SCHURGERS et al. [742] confirm the guess of a linear dependence of the mean wakeup latency on the wakeup period. STEM-B can achieve half the wakeup latency of STEM-T if no collisions occur on the wakeup channel. The reason for STEM-B's advantage under the assumption of no collisions is that in STEM-T (as well as in STEM-B with collisions) the busy tone/beacon frames are sent for the maximum time, while in STEM-B likely an acknowledgment frame will be received much earlier. With respect to energy expenditure, STEM-T can have advantages, since the acknowledgment packet is saved and the length of the listen period T_{Rx} can be significantly shorter in STEM-T than for STEM-B, since it suffices to detect energy, whereas in STEM-B this time has to accommodate at least one full beacon packet.

5.2.2 S-MAC

The S-MAC (Sensor-MAC) protocol [914, 915] provides mechanisms to circumvent idle listening, collisions, and overhearing. As opposed to STEM, it does not require two different channels.

S-MAC adopts a periodic wakeup scheme, that is, each node alternates between a fixed-length listen period and a fixed-length sleep period according to its **schedule**, compare Figure 5.6. However, as opposed to STEM, the listen period of S-MAC can be used to receive *and transmit* packets. S-MAC attempts to coordinate the schedules of neighboring nodes such that their listen periods start at the same time. A node x's listen period is subdivided into three different phases:

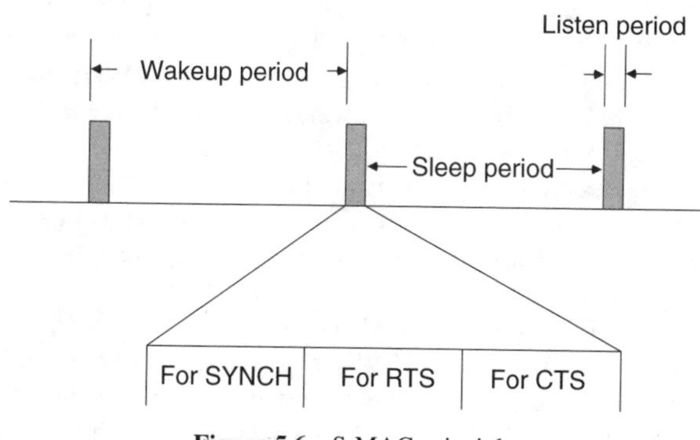

Figure 5.6 S-MAC principle

- In the first phase (**SYNCH phase**), node x accepts SYNCH packets from its neighbors. In these packets, the neighbors describe their own schedule and x stores their schedule in a table (the **schedule table**). Node x's SYNCH phase is subdivided into time slots and x's neighbors contend according to a CSMA scheme with additional backoff, that is, each neighbor y wishing to transmit a SYNCH packet picks one of the time slots randomly and starts to transmit if no signal was received in any of the previous slots. In the other case, y goes back into sleep mode and waits for x's next wakeup. In the other direction, since x knows a neighbor y's schedule, x can wake at appropriate times and send its own SYNCH packet to y (in broadcast mode). It is not required that x broadcasts its schedule in every of y's wakeup periods. However, for reasons of time synchronization and to allow new nodes to learn their local network topology, x should send SYNCH packets periodically. The according period is called **synchronization period**.
- In the second phase (**RTS phase**), x listens for RTS packets from neighboring nodes. In S-MAC, the RTS/CTS handshake described in Section 5.1.2 is used to reduce collisions of data packets due to hidden-terminal situations. Again, interested neighbors contend in this phase according to a CSMA scheme with additional backoff.
- In the third phase (**CTS phase**), node x transmits a CTS packet if an RTS packet was received in the previous phase. After this, the packet exchange continues, extending into x's nominal sleep time.

In general, when competing for the medium, the nodes use the RTS/CTS handshake, including the virtual carrier-sense mechanism, whereby a node maintains a NAV variable. The NAV mechanism can be readily used to switch off the node during ongoing transmissions to avoid overhearing. When transmitting in a broadcast mode (for example SYNCH packets), the RTS and CTS packets are dropped and the nodes use CSMA with backoff.

If we can arrange that the schedules of node x and its neighbors are synchronized, node x and all its neighbors wake up at the same time and x can reach all of them with a single SYNCH packet. The S-MAC protocol allows neighboring nodes to agree on the same schedule and to create **virtual clusters**. The clustering structure refers solely to the exchange of schedules; the transfer of data packets is not influenced by virtual clustering.

The S-MAC protocol proceeds as follows to form the virtual clusters: A node x, newly switched on, listens for a time of at least the (globally known) synchronization period. If x receives any SYNCH packet from a neighbor, it adopts the announced schedule and broadcasts it in one of the neighbors' next listen periods. In the other case, node x picks a schedule and broadcasts it. If x receives another node's schedule during the broadcast packet's contention period, it drops its own schedule and follows the other one. It might also happen that a node x receives a different schedule after it already has chosen one, for example, because bit errors destroyed previous SYNCH packets. If node x already knows about the existence of neighbors who adopted its own schedule, it keeps its schedule and in the future has to transmit its SYNCH and data packets according to both schedules. On the other hand, if x has no neighbor sharing its schedule, it drops its own and adopts the other one. Since there is always a chance to receive SYNCH packets in error, node x periodically listens for a whole synchronization period to relearn its neighborhood. This makes the virtual cluster formation fairly robust.

By this approach, a large multihop network is partitioned into "islands of schedule synchronization". Border nodes have to follow two or more different schedules for broadcasting their SYNCH packets and for forwarding data packets. Thus, they expend more energy than nodes only having neighbors of the same "schedule regime".

The periodic wakeup scheme adopted by S-MAC allows nodes to spend much time in the sleep mode, but there is also a price to pay in terms of latency. Without further modifications, the per-hop latency of S-MAC will be approximately equal to the sleep period on average when all nodes follow the same schedule. Ye et al. [915] describe the **adaptive-listening** scheme, which roughly halves the per-hop latency. Consider the following situation: Node x receives during its listen period an

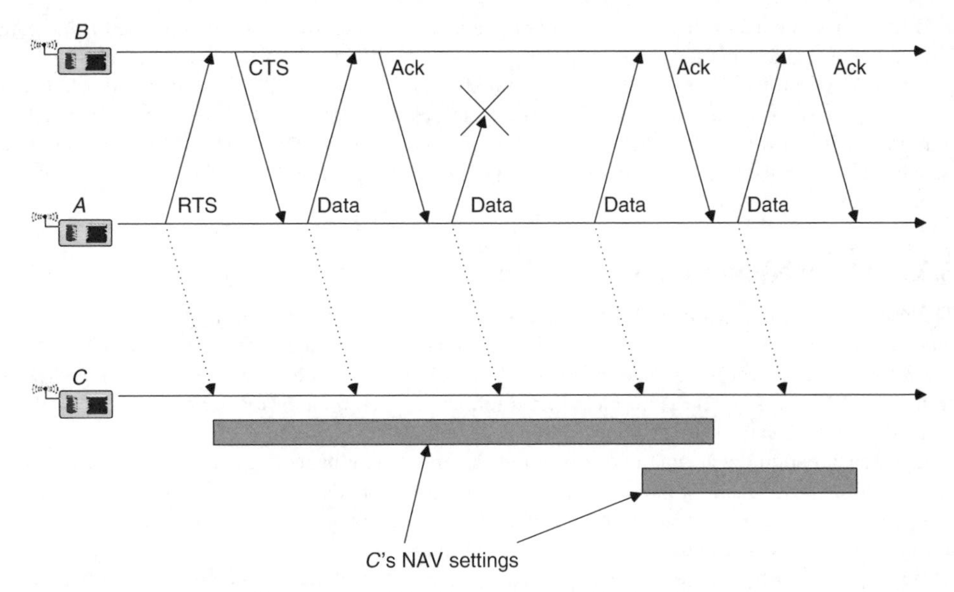

Figure 5.7 S-MAC fragmentation and NAV setting

RTS or CTS packet belonging to a packet exchange from neighbor node y to node z. From the duration field of these packets, x can infer the time t_0 when the packet exchange ends. Since it might happen that x is the next hop for z's packet, node x schedules an extra listen period around time t_0 and z tries to send an extra RTS at time t_0, ignoring x's normal wakeup cycle. Under ideal circumstances, x is awake when z sends the RTS and the packet can take the next hop quickly.

S-MAC also adopts a message-passing approach (illustrated in Figure 5.7), where a message is a larger data item meaningful to the application. In-network processing usually requires the aggregating node to receive a message completely. On the other hand, on wireless media, it is advisable to break a longer packet into several shorter ones (fragmentation, see also Chapter 6). S-MAC includes a fragmentation scheme working as follows. A series of fragments is transmitted with only one RTS/CTS exchange between the transmitting node *A* and receiving node *B*. After each fragment, *B* has to answer with an acknowledgment packet. All the packets (data, ack, RTS, CTS) have a duration field and a neighboring node *C* is required to set its NAV field accordingly. In S-MAC, the duration field of all packets carries the remaining length of the whole transaction, including all fragments and their acknowledgments. Therefore, the whole message shall be passed at once. If one fragment needs to be retransmitted, the remaining duration is incremented by the length of a data plus ack packet, and the medium is reserved for this prolonged time. However, there is the problem of how a nonparticipating node shall learn about the elongation of the transaction when he has only heard the initial RTS or CTS packets.

This scheme has some similarities to the fragmentation scheme used in IEEE 802.11 but there are important differences. In IEEE 802.11, the RTS and CTS frame reserve the medium only for the time of the first fragment, and any fragment reserves only for the next fragment. If one packet needs to be retransmitted, the initiating node has to give up the channel and recontend for it in the same way as for a new packet. The approach taken by S-MAC reduces the latency of complete messages by suppressing intertwined transmissions of other packets. Therefore, in a sense, this protocol is unfair because single nodes can block the medium for long time. However, as explained in Section 5.1.3, the fairness requirement has a different weight in a wireless sensor network than it has in a data network where users want to have fair medium access.

S-MAC has one major drawback: it is hard to adapt the length of the wakeup period to changing load situations, since this length is essentially fixed, as is the length of the listen period.

The T-MAC protocol presented by VAN DAM and LANGENDOEN [838] is similar to S-MAC but adaptively shortens the listen period. If a node x senses no activity on the medium for a specified duration, it is allowed to go back into sleep mode prematurely. Therefore, if no node wants to transmit to x, the listen period can be ended quickly, whereas in S-MAC, the listen period has a fixed length.

5.2.3 The mediation device protocol

The **mediation device protocol** [115, Chap. 4] is compatible with the peer-to-peer communication mode of the IEEE 802.15.4 low-rate WPAN standard [114, 115, 468, 521]. It allows each node in a WSN to go into sleep mode periodically and to wake up only for short times to receive packets from neighbor nodes. There is no global time reference, each node has its own sleeping schedule, and does not take care of its neighbors sleep schedules.

Upon each periodic wakeup, a node transmits a short **query beacon**, indicating its node address and its willingness to accept packets from other nodes. The node stays awake for some short time following the query beacon, to open up a window for incoming packets. If no packet is received during this window, the node goes back into sleep mode.

When a node wants to transmit a packet to a neighbor, it has to synchronize with it. One option would be to have the sender actively waiting for query beacon, but this wastes considerable energy for synchronization purposes only. The **dynamic synchronization** approach achieves this synchronization without requiring the transmitter to be awake permanently to detect the destinations query beacon. To achieve this, a **mediation device** (MD) is used. We first discuss the case where the mediation device is not energy constrained and can be active all the time; this scenario is illustrated in Figure 5.8. Because of its full duty cycle, the mediation device can receive the query beacons from all nodes in its vicinity and learn their wakeup periods.

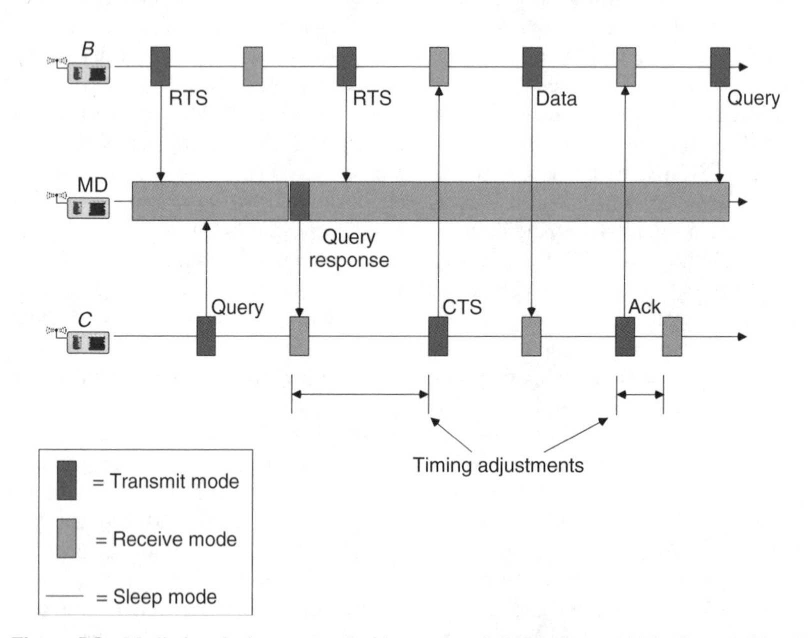

Figure 5.8 Mediation device protocol with unconstrained mediators [115, Chap. 4, Fig. 3]

Suppose that node A wants to transmit a packet to node B. Node A announces this to the mediation device by sending periodically **request to send** (RTS) packets, which the MD captures. Node A sends its RTS packets instead of its query beacons and thus they have the same period. Again, there is a short answer window after the RTS packets, where A listens for answers. After the MD has received A's RTS packet, it waits for B's next query beacon. The MD answers this with a **query response** packet, indicating A's address and a timing offset, which lets B know when to send the answering **clear to send** (CTS) to A such that the CTS packet hits the short answer window after A's next RTS packet. Therefore, B has learned A's period. After A has received the CTS packet, it can send its data packet and wait for B's immediate acknowledgment. After the transaction has finished, A restores its periodic wakeup cycle and starts to emit query beacons again. Node B also restores its own periodic cycle and thus *decouples* from A's period.

This protocol has some advantages. First, it does not require any time synchronization between the nodes, only the mediation device has to learn the periods of the nodes. Second, the protocol is asymmetric in the sense that most of the energy burden is shifted to the mediation device, which so far is assumed to be power unconstrained. The other nodes can be in the sleep state most of the time and have to spend energy only for the periodic beacons. Even when a transmitter wants to synchronize with the receiver, it does not have to wait actively for the query beacon, but can go back to sleep and wait for the mediation device to do the synchronization work. This way very low duty cycles can be supported. This protocol has also some drawbacks: The nodes transmit their query beacons without checking for ongoing transmissions and, thus, the beacons of different nodes may collide repeatedly when nodes have the same period and their wakeup periods overlap. If the wakeup periods are properly randomized and the node density is sufficiently low, this collision probability can be low too. However, in case of higher node densities or unwanted synchronization between the nodes, the number of collisions can be significant. A possible solution to this is the following: When the MD registers collisions, it might start to emit a dedicated **reschedule** control frame. All colliding nodes can hear this frame as long as the MD repeats it often enough. Reception of this frame causes each node to randomly pick a new period from a certain interval $[a, b]$ indicated in the reschedule frame. If the MD continues to perceive collisions, it can enlarge the interval accordingly.

The main drawbacks, however, are the assumptions that: (i) the mediation device is energy unconstrained, which does not conform to the idea of a "simply thrown out" wireless sensor network, and (ii) there are sufficient mediation devices to cover all nodes. The **distributed mediation device protocol** deals with these problems in a probabilistic manner. It lets nodes randomly wake up and serve as MD for a certain time and afterward lets them go back to their regular periodic wakeup behavior. The service time must be chosen to be at least as long as the maximum period of all neighbors plus the length of a query beacon. However, under these assumptions, it cannot be expected that a temporary MD knows all its neighbors' periods and wakeup times. If we assume that all nodes have the same period, then it suffices for the temporary MD to hear an RTS from the transmitter node A and a query beacon from receiving node B in order to compute their time offsets and to instruct node B accordingly in the MD's query response. A problem with this approach is that nodes A and B may have two or more MD devices in their vicinity, causing a collision of several query responses. By properly randomizing the times where nodes decide to serve as MD, the probability of this can be kept low.

5.2.4 Wakeup radio concepts

The ideal situation would be if a node were always in the receiving state when a packet is transmitted to it, in the transmitting state when it transmits a packet, and in the sleep state at all other times; the idle state should be avoided. The **wakeup radio** concept strives to achieve this goal by a simple, "powerless" receiver that can trigger a main receiver if necessary (see Section 2.1.4 for details).

One proposed wakeup MAC protocol [931] assumes the presence of several parallel data channels, separated either in frequency (FDMA) or by choosing different codes in a CDMA schemes. A node wishing to transmit a data packet randomly picks one of the channels and performs a carrier-sensing operation. If the channel is busy, the node makes another random channel choice and repeats the carrier-sensing operation. After a certain number of unsuccessful trials, the node backs off for a random time and starts again. If the channel is idle, the node sends a wakeup signal to the intended receiver, indicating both the receiver identification and the channel to use. The receiver wakes up its data transceiver, tunes to the indicated channel, and the data packet transmission can proceed. Afterward, the receiver can switch its data transceiver back into sleep mode. This wakeup radio concept has the significant advantage that only the low-power wakeup transceiver has to be switched on all the time while the much more energy consuming data transceiver is nonsleeping if and only if the node is involved in data transmissions. Furthermore, this scheme is naturally **traffic adaptive**, that is, the MAC becomes more and more active as the traffic load increases. Periodic wakeup schemes do not have this property.

However, there are also some drawbacks. First, to our knowledge, there is no real hardware yet for such an ultralow power wakeup transceiver. Second, the range of the wakeup radio and the data radio should be the same. If the range of the wakeup radio is smaller than the range of the data radio, possibly not all neighbor nodes can be woken up. On the other hand, if the range of the wakeup radio is significantly larger, there can be a problem with local addressing schemes (compare Chapter 7): These schemes do not use globally or networkwide-unique addresses but only locally unique addresses, such that no node has two or more one-hop neighbors with the same address. Put differently: A node's MAC address should be unique within its two-hop neighborhood. Since the packets exchanged in the neighbor discovery phase have to use the data channel, the two-hop neighborhood as seen on the data channel might be different from the two-hop neighborhood on the wakeup channel. Third, this scheme critically relies on the wakeup channel's ability to transport useful information like node addresses and channel identifications; this might not always be feasible for transceiver complexity reasons and additionally requires methods to handle collisions or transmission errors on the wakeup channel. If the wakeup channel does not support this feature, the transmitter wakes up *all* its neighbors when it emits a wakeup signal, creating an overhearing situation for most of them. If the transmitting node is about to transmit a long data packet, it might be worthwhile to prepend the data packet with a short **filter packet** [552] announcing the receiving node's address. All the other nodes can go back to sleep mode after receiving the filter packet. Instead of using an extra packet, all nodes can read the bits of the data packet until the destination address appeared. If the packet's address is not identical to its own address, the node can go back into sleep mode.

5.2.5 Further reading

The protocol described by Miller and Vaidya [552] has some similarities to STEM. It uses two different channels, one for wakeup and the other one for data transmission. However, the nodes try to adapt their wakeup period to the observed periodicity of the traffic destined to them. Each node has buffer space for L packets. If the buffer is full, the node wakes up its whole neighborhood and transmits all packets ("full wakeup"). However, if the traffic is sufficiently regular, the source and destination node agree on a time where only the destination wakes up and the nodes can exchange their packet ("triggered wakeup"). In this case, the transmitter can empty its buffer and the costly full wakeups arise less frequently.

In the DMAC protocol proposed by Lu et al. [520], the following problem is attacked. If all the nodes on the way from some source node to a sink node have their individual wakeup schedules, the accumulated latency from source to sink can be significant. Especially when the schedules are unsynchronized the per-hop latency can be enormous. DMAC attacks this problem by carefully arranging the wakeup schedules according to the distance of a node from the sink. In the best case, as soon as one node has received a packet, the listen period of its upstream neighbor starts.

In references [226, 227, 228], the preamble sampling technique is applied to both ALOHA and CSMA protocols. A node wakes up periodically and listens for a short time to the medium to check whether there is any signal. If so, the node tries to receive the packet. If a node wants to transmit a packet, it prepends the packet with a preamble, long enough to let the intended receiver pick it up. The remaining packet is transmitted after the preamble. One problem for this protocol can be the start-up energy needed to switch the transceiver from sleep into idle or receive mode (see Section 4.3.1). El-Hoiydi et al. [226] let a node try to learn its neighbors' wakeup periods and phases and to start the preamble only immediately before the receiving node wakes up, in order to keep the preamble length at minimum.

Lin et al. [503] consider different variants of cycled receiver schemes in which nodes periodically wake up to receive packets. They consider the problem of how transmitter and receiver find each other. One option is to let the transmitter send short packets frequently until they hit the receiver's listen period and trigger an answer of the receiver, or the receiver could send a short beacon packet at the start of its listen period. The trade-offs are investigated under a Gilbert–Elliot channel model.

5.3 Contention-based protocols

In contention-based protocols, a given transmit opportunity toward a receiver node can in principle be taken by any of its neighbors. If only one neighbor tries its luck, the packet goes through the channel. If two or more neighbors try their luck, these have to compete with each other and in unlucky cases, for example, due to hidden-terminal situations, a collision might occur, wasting energy for both transmitter and receiver. Section 5.1.2 briefly presented two important contention-based protocols: (slotted) ALOHA and CSMA, along with mechanisms to solve the hidden-terminal problem. In the following sections, we discuss variations of these protocols with the goal to conserve energy.

As opposed to some of the contention-based protocols having a periodic wakeup scheme (for example S-MAC, see Section 5.2.2), the protocols described in this section have no idle listening avoidance and make no restrictions as to when a node can receive a packet.

5.3.1 CSMA protocols

Woo and Culler [888] investigate several CSMA variants for their inherent energy costs and their fairness, without specifying any measures for idle listening avoidance or overhearing avoidance. "Inherent cost" subsumes mainly the energy spent on transmitting and receiving.

The authors consider a multihop network with a single or only a few sinks and the same traffic pattern as already envisioned for STEM (Section 5.2.1): A network that is idle for long times and starts to become active when triggered by an important external event. Upon the triggering event, all nodes wish to transmit simultaneously, potentially creating lots of collisions. In the case that the nodes want to send their packets periodically, the danger of collisions is repeated if no special measures are taken. The nodes are assumed to know an upstream neighbor to which they have to forward packets destined for the sink. This upstream neighbor is also called the *parent node*. Each node generates local sensor traffic and additionally works as a forwarder for downstream nodes. Only a single channel is required.

We briefly discuss the skeleton of the different CSMA protocols. Figure 5.9 shows the several steps a node passes through in case of a transmission as a finite state automaton. After a node gets a new packet for transmission from its upper layers, it starts with a random delay and initializes its trial counter `num_retries` with zero. The purpose of the random delay is to desynchronize

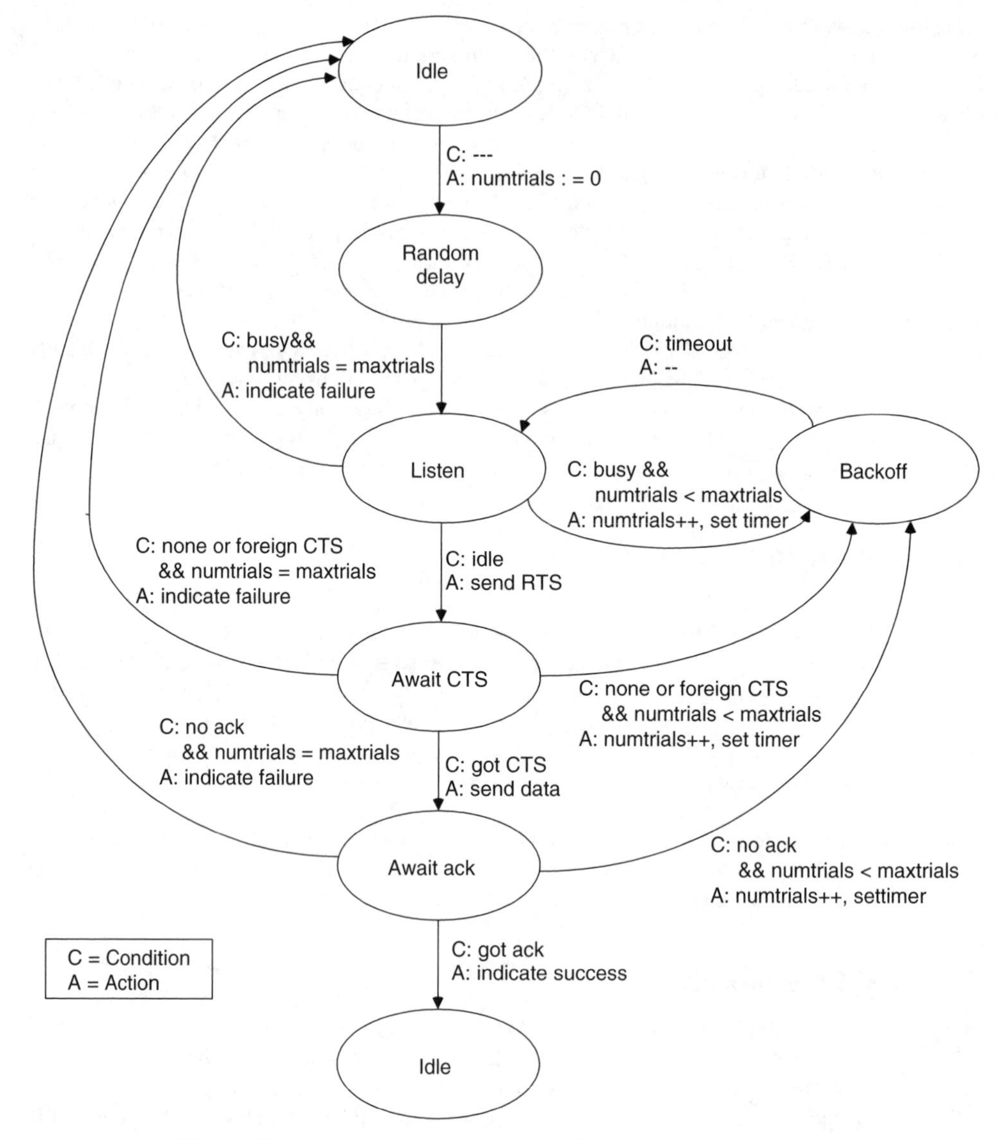

Figure 5.9 Schematic of the CSMA protocol presented in reference [888]

nodes that are initially synchronized by the external event. During this random delay, the node's transceiver can be put into sleep mode. During the following listen period, the node performs carrier sensing. If the medium is found to be busy and the number of trials so far is smaller than the maximum number, the node goes into the backoff mode. In the backoff mode, the node waits a random amount of time, which can depend on the number of trials and during which the node can sleep (the protocol is thus a nonpersistent CSMA variant). The backoff mode can also be used by the application layer to initiate a "phase change" for its locally generated periodic traffic. This phase change aims to desynchronize correlated or periodic traffic of different nodes. After the backoff mode finishes, the node listens again. If the medium is busy and the node has exhausted its maximum number of trials, the packet is dropped. If the medium is idle, the node transmits an RTS packet and enters the "Await CTS" state, where it waits for the corresponding CTS packet (this step can be skipped if the node knows that there is currently a low load situation). In case no CTS packet arrives or a CTS packet for another transaction is received, the node either enters the backoff mode or drops the packet, depending on the value of `num_retries`. If the CTS packet arrives, the node sends its data packet and waits for an acknowledgment. This acknowledgment can be either an explicit acknowledgment packet, or the parent node piggybacks the acknowledgment on a packet that it forwards to the node's grandparent. However, for such a piggybacked acknowledgment, it is not an easy task to determine an appropriate waiting time until the acknowledgment must arrive at the child.

Several variants of this skeleton (no random delay vs. random delay, random listening time vs. constant listening time, fixed window backoff vs. exponentially increasing backoff vs. exponentially decreasing backoff vs. no backoff) have been investigated in a single-hop scenario with a triggering event and it turns out that protocols with random delay, fixed listening time, and a backoff algorithm with sleeping radio transceiver give the best throughput as well as lowest aggregate energy consumption, when compared with other CSMA variants, including IEEE 802.11.

5.3.2 PAMAS

The PAMAS protocol (Power Aware Multiaccess with Signaling) presented by Raghavendra and Singh [668] is originally designed for ad hoc networks. It provides a detailed overhearing avoidance mechanism while it does not consider the idle listening problem. The protocol combines the busy-tone solution and RTS/CTS handshake similar to the MACA protocol [407] (MACA uses no final acknowledgment packet). A distinctive feature of PAMAS is that it uses two channels: a **data channel** and a **control channel**. All the signaling packets (RTS, CTS, busy tones) are transmitted on the control channel, while the data channel is reserved for data packets. We follow Raghavendra and Singh [668] in first describing the main protocol operation and then discussing the power-conservation enhancements.

Let us consider an idle node x to which a new packet destined to a neighboring node y arrives. First, x sends an RTS packet on the control channel without doing any carrier sensing. This packet carries both x's and y's MAC addresses. If y receives this packet, it answers with a CTS packet if y does not know of any ongoing transmission in its vicinity. Upon receiving the CTS, x starts to transmit the packet to y on the data channel. When y starts to receive the data, it sends out a **busy-tone** packet on the control channel. If x fails to receive a CTS packet within some time window, it enters the backoff mode, where a binary exponential backoff scheme is used (i.e., the backoff time is uniformly chosen from a time interval that is doubled after each failure to receive a CTS).

Now, let us look at the nodes receiving x's RTS packet on the control channel. There is the intended receiver y and there are other nodes; let z be one of them. If z is currently receiving a packet, it reacts by sending a busy-tone packet, which overlaps with y's CTS at node x and effectively destroys the CTS. Therefore, x cannot start transmission and z's packet reception is not disturbed. Since the busy-tone packet is longer than the CTS, we can be sure that the CTS is really destroyed. Next, we consider the intended receiver y. If y knows about an ongoing transmission in its vicinity, it suppresses its CTS, causing x to back off. Node y can obtain this knowledge by either sensing the data channel or by checking whether there was some noise on the control channel immediately after receiving the RTS. This noise can be an RTS or CTS of another node colliding at y. In the other case, y answers with a CTS packet and starts to send out a busy-tone packet as soon as x's transmission has started. Furthermore, y sends out busy-tone packets each time it receives some noise or a valid packet on the control channel, to prevent its neighborhood from any activities.

A node that receives an RTS packet while being in the backoff state starts its packet reception procedure, that is, it checks the conditions for sending a CTS.

When can a node put its transceivers (control and data) into sleep mode? Roughly speaking, any time a node knows that it cannot transmit or receive packets because some other node in its vicinity is already doing so. However, the decision to go into sleep mode raises an important question: when to wake up again? This decision is easy if a node x knows about the length of an ongoing transmission, for example from overhearing the RTS or CTS packets or the header of the data packets on the data channel. However, often this length is unknown to x, for example, because these packets are corrupted or a foreign data transmission cycle starts when x is just sleeping. Additional procedures are needed to resolve this.

Suppose that x wakes up and finds the data channel busy. There are two cases to distinguish: either x has no own packet to send or x wants to transmit. In the first case, x desires to go back into sleep mode and to wake up exactly when the ongoing transmission ends to be able to receive an immediately following packet. Waking up at the earliest possible time has the advantage of avoiding unwanted delays. However, since x may not have overheard the RTS, CTS, or data packet header belonging to the ongoing transmission, it runs a **probing protocol** on the control channel to inquire the length of the ongoing packet. This probing protocol works similar to a binary search algorithm. Let l be the maximal packet length in seconds. First, x sends a $t\_probe(l/2, l)$ packet, and any transmitter node who finishes in the time interval $[l/2, l]$ answers with a $t\_probe\_response(t)$ packet, indicating the time t where its transmission ends. If x manages to receive $t\_probe\_response(t)$ packet, it knows exactly when this single ongoing transmission ends and when to wake up the next time. If x receives only noise in response, several $t\_probe\_response(t)$ may have collided at x and x starts to search in the subinterval $[3l/4, l]$, again hoping for a single answer only. If no answer arrives at all upon $t\_probe(3l/4, l)$, x next checks the interval $[l/2, 3l/4]$, and so on.

In the other case, x wakes up during an ongoing transmission and wants to transmit a packet. Therefore, x has not only to take care of ongoing transmissions but also of ongoing receptions in its vicinity. To find the time for the next wakeup, x runs the described probing protocol for the set of transmitters, giving a time t when the longest ongoing transmission ends. In addition, x runs a similar probing protocol for the set of receivers in its neighborhood, indicating the time r when the longest ongoing reception ends. Finally, x schedules its wakeup for time $\min\{r, t\}$. The rationale for this choice is: If $t < r$, waking up at t might give another node y a chance to transmit a packet to x without any additional delay. On the other hand, if $r < t$, there is some chance that x can start its own transmission.

Raghavendra and Singh [668] compare the power-saving performance of PAMAS with overhearing avoidance against PAMAS without this feature. Analytical and simulation results are presented for several network topologies, node densities, and load scenarios. For the case of random networks, the power savings for low load situations depend on the average node degree, that is, the average number of neighbors that a node has. Clearly, the more neighbors a node x has, the more can switch their transceivers off when x actually transmits. For low loads also, the number of control packets is smaller than for high loads. This is particularly true for the busy-tone packets. PAMAS saves up to 60 % of energy for low loads and a high node degree, and still between 20 and 30 % are reached for low node degrees under a low load. In high load situations, between ≈10 and ≈40 % of energy savings can be achieved, with higher savings for higher node degrees.

5.3.3 Further solutions

Adireddy and Tong [9] take a more information-theoretic view on wireless sensor networks and investigate the **asymptotic stable throughput**, which is defined as the maximum stable throughput achievable in a high density wireless sensor network as the number of nodes goes to infinity but the overall load is kept fixed. Specifically, they investigate a variation of slotted ALOHA taking channel state information into account.

TSENG et al. [832] propose three different mechanisms that enhance IEEE 802.11's power-saving capabilities by letting nodes go into sleep mode periodically. The nodes are assumed to have independent clocks and independent sleep schedules. By different choices of the nodes' activity periods, it can be guaranteed that two nodes can reach each other eventually. Further changes to the original IEEE 802.11 power-saving protocol are proposed, for example, to let each node send a beacon irrespective of whether any of its neighbors sent one before; this allows a node to learn its neighborhood more quickly. Three different wakeup schemes (dominating-awake interval, periodically fully awake interval, and quorum-based interval) are proposed and investigated. These protocols offer different points in the energy-consumption/network adaption space, where network adaption refers to the network's ability to accommodate topological changes.

5.4 Schedule-based protocols

We discuss some schedule-based protocols that do not explicitly address idle listening avoidance but do so implicitly, for example, by employing TDMA schemes, which explicitly assign transmission and reception opportunities to nodes and let them sleep at all other times. A second fundamental advantage of schedule-based protocols is that transmission schedules can be computed such that no collisions occur at receivers and hence no special mechanisms are needed to avoid hidden-terminal situations.

However, these schemes also have downsides. First, the setup and maintenance of schedules involves signaling traffic, especially when faced to variable topologies. Second, if a TDMA variant is employed, time is divided into comparably small slots, and both transmitter and receiver have to agree to slot boundaries to actually meet each other and to avoid overlaps with other slots, which would lead to collisions. However, maintaining time synchronization involves some extra signaling traffic. For cheap sensor nodes with cheap oscillators, one can expect the clocks of different nodes to drift comparably quickly and resynchronization is required frequently (see Chapter 8). A third drawback is that such schedules are not easily adapted to different load situations on small timescales. Specifically, in TDMA, it is difficult for a node to give up unused time slots to its neighbors. A further disadvantage is that the schedule of a node (and possibly those of its neighbors) may require a significant amount of memory, which is a scarce resource in several sensor node designs. Finally, distributed assignment of conflict-free TDMA schedules is a difficult problem in itself (for example [169]).

5.4.1 LEACH

The LEACH protocol (Low-energy Adaptive Clustering Hierarchy) presented by HEINZELMAN et al. [344] assumes a dense sensor network of homogeneous, energy-constrained nodes, which shall report their data to a sink node. In LEACH, a TDMA-based MAC protocol is integrated with clustering and a simple "routing" protocol.

LEACH partitions the nodes into **clusters** and in each cluster a dedicated node, the **clusterhead**, is responsible for creating and maintaining a TDMA schedule; all the other nodes of a cluster are **member nodes**. To all member nodes, TDMA slots are assigned, which can be used to exchange data between the member and the clusterhead; there is no peer-to-peer communication. With the exception of their time slots, the members can spend their time in sleep state. The clusterhead aggregates the data of its members and transmits it to the sink node or to other nodes for further relaying. Since the sink is often far away, the clusterhead must spend significant energy for this transmission. For a member, it is typically much cheaper to reach the clusterhead than to transmit directly to the sink. The clusterheads role is energy consuming since it is always switched on and is responsible for the long-range transmissions. If a fixed node has this role, it would burn its energy quickly, and after it died, all its members would be "headless" and therefore useless. Therefore, this burden is rotated among the nodes. Specifically, each node decides independent of other nodes whether it becomes a clusterhead, and therefore there is no signaling traffic related to clusterhead election (although signaling traffic is needed for subsequent association of nodes to some clusterhead). This decision takes into account when the node served as clusterhead the last time, such that a node that has not been a clusterhead for a long time is more likely to elect itself than a node serving just recently [346]. The protocol is round based, that is, all nodes make their decisions whether to become a clusterhead at the same time and the nonclusterhead nodes have to associate to a clusterhead subsequently. The nonclusterheads choose their clusterhead based on received signal strengths. The network partitioning into clusters is time variable and the protocol assumes global time synchronization.

After the clusters have been formed, each clusterhead picks a random CDMA code for its cluster, which it broadcasts and which its member nodes have to use subsequently. This avoids a situation where a border node belonging to clusterhead A distorts transmissions directed to clusterhead B, shown in Figure 5.10.

A critical network parameter is the percentage of nodes that are clusterheads. If there are only a few clusterheads, the expected distance between a member node and its clusterhead becomes longer and therefore the member has to spend more energy to reach its clusterhead while maintaining a given BER target. On the other hand, if there are many clusterheads, there will be more energy-expensive transmissions from clusterheads to the sink and less aggregation. Therefore, there exists an optimum percentage of clusterheads, which for the scenario investigated in [344, 346] is $\approx$5 %. If this optimum is chosen, LEACH can achieve a seven to eight times lower overall energy dissipation compared to the case where each node transmits its data directly to the sink, and between four and eight times lower energy than in a scenario where packets are relayed in a multihop fashion. In addition, since LEACH distributes the clusterhead role fairly to all nodes, they tend to die at about the same time.

The protocol is organized in **rounds** and each round is subdivided into a setup phase and a steady-state phase (Figure 5.11). The **setup phase** starts with the self-election of nodes to clusterheads. In the following **advertisement phase**, the clusterheads inform their neighborhood with an advertisement packet. The clusterheads contend for the medium using a CSMA protocol with no further provision against the hidden-terminal problem. The nonclusterhead nodes pick the advertisement packet with the strongest received signal strength. In the following cluster-setup phase, the members inform their clusterhead ("join"), again using a CSMA protocol. After the cluster setup-phase, the clusterhead knows the number of members and their identifiers. It constructs a

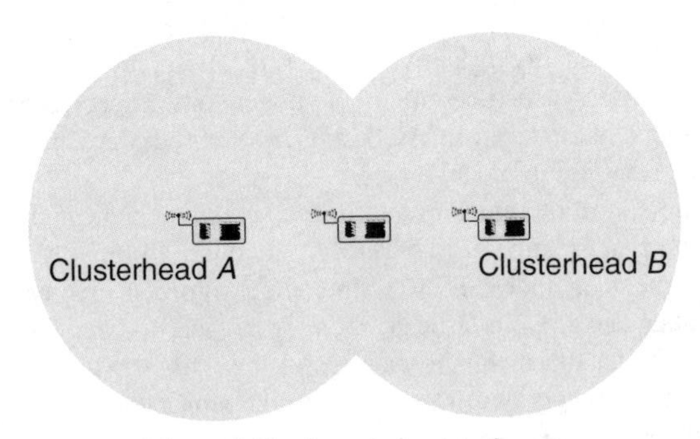

Figure 5.10 Intercluster interference

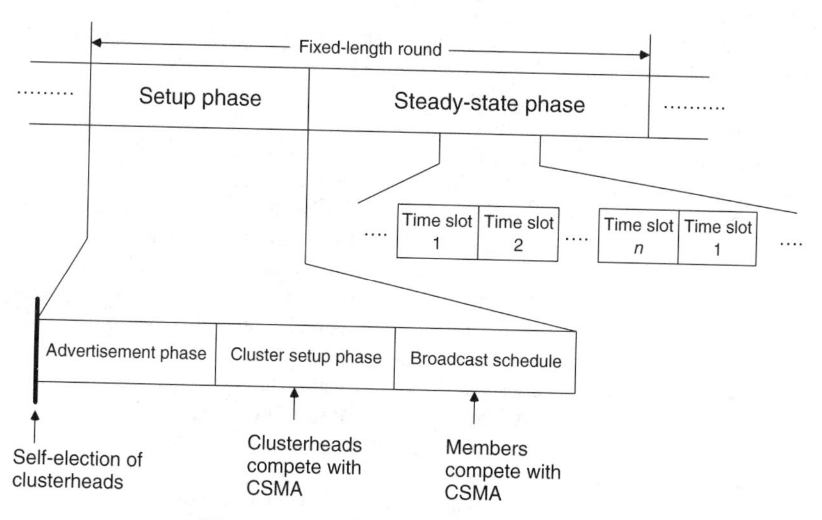

Figure 5.11 Organization of LEACH rounds

TDMA schedule, picks a CDMA code randomly, and broadcasts this information in the broadcast schedule subphase. After this, the TDMA steady-state phase begins.

Because of collisions of advertisement or join packets, the protocol cannot guarantee that each nonclusterhead node belongs to a cluster. However, it can guarantee that nodes belong to at most one cluster.

The clusterhead is switched on during the whole round and the member nodes have to be switched on during the setup phase and occasionally in the steady-state phase, according to their position in the cluster's TDMA schedule.

With the protocol described so far, LEACH would not be able to cover large geographical areas of some square miles or more, because a clusterhead two miles away from the sink likely does not have enough energy to reach the sink at all, not to mention achieving a low BER. If it can be arranged that a clusterhead can use other clusterheads for forwarding, this limitation can be mitigated.

5.4.2 SMACS

The Self-Organizing Medium Access Control for Sensor Networks (SMACS) protocol described by Sohrabi et al. [778], Sohrabi and Pottie [780] is part of a wireless sensor network protocol suite that addresses MAC, neighbor discovery, attachment of mobile nodes, a multihop routing protocol, and a local routing protocol for cooperative signal processing purposes.

SMACS essentially combines neighborhood discovery and assignment of TDMA schedules to nodes. SMACS is based on the following assumptions:

- The available spectrum is subdivided into many channels and each node can tune its transceiver to an arbitrary one; alternatively, it is assumed that many CDMA codes are available.
- Most of the nodes in the sensor network are stationary and such an assignment is valid for fairly long times.
- Each node divides its time locally into fixed-length superframes (of duration T_{frame} seconds), which do not necessarily have the same phase as the neighbor's superframes. However, all nodes have the same superframe length and this requires time synchronization. Superframes are also

subdivided into time slots but this is only loose since transmissions are not confined to occur *only within* a single time slot.

The goal of SMACS is to detect neighboring nodes and to set up exclusive **links** or **channels** to these. A link is directional, that is on a given link all packets are transmitted in one direction. Furthermore, a link occupies a TDMA slot in either endpoint. When two nodes want bidirectional operation, two such links are needed; from the perspective of one node, there is a **receive slot** and a **transmit slot** to the other node. The assignment of links shall be such that no collisions occur at receivers. To achieve this, SMACS takes care that for a single node the time slots of different links do not overlap (using a simple greedy algorithm) and furthermore for each link randomly one out of a large number of frequency channels/CDMA codes is picked and allocated to the link. It is not required that a node and its neighbors transmit at entirely different times. In this case, however, they must transmit to different receivers and have to use different frequencies/codes. After link setup, the nodes wake up periodically (once per superframe) in the respective receive time slots with the receiver tuned to the corresponding frequency or with the correct CDMA code at hand; the transmit time slots are only used when required.

By using a local scheme instead of a global assignment, the task of transmitting the neighborhood information to a central node and the computation results back is avoided.

Regarding the neighbor discovery and link setup, we consider four different cases. Suppose that nodes x and y want to set up a link and x is switched on first. First, we assume that neither x nor y has any neighbor so far, as illustrated by Figure 5.12. Node x listens on a fixed frequency band for a random amount of time. If nothing is received during this time, node x sends an invitation message, more specifically a `TYPE1`(x, unattached) message, indicating its own node identification and the number of attached neighbors, which so far is zero. When any neighbor z of node x receives this message, it waits for a random but bounded amount of time and answers with a `TYPE2`(x, z, n) message, indicating its own address, x's address and its number of neighbors n. Now, suppose that the so-far unconnected node y answers first – with `TYPE2`(x, y, unattached) – and x receives this message properly. Since y sent the first answer, x invites y to construct a link by sending a

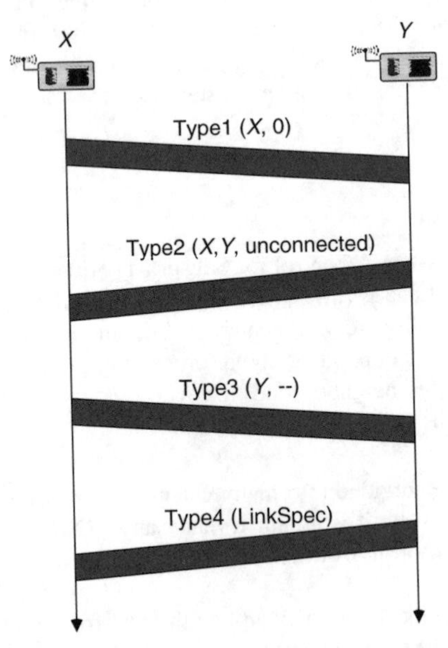

Figure 5.12 SMACS: link setup for two lonesome nodes

TYPE3(y, –) message, carrying the identification of the "winning" node y and no further parameters. This message is sent when the contention period for the TYPE2(·, ·, ·) answer message ends. Now, node y knows that (i) it has been selected, and (ii) it can pick any time slot it wants since neither x nor y has any link allocated so far. Node y answers to node x with a link specification, that is, two time slot specifications and a frequency/code, using a TYPE2(x, y, LinkSpec) message. The time slot specifications have a common time base since node y adopts x's superframe phase upon receiving the TYPE3(y, –) message. By this rule, neighboring nodes that discover each other first, have a common phase (and a common period).

Any other node z loosing against y goes back into sleep mode and tries again at some later time. The nodes repeat their invitations periodically using TYPE1(·, ·) messages.

The second case is where node x already has some neighbors but the winning node y has none so far. Therefore, x sends a TYPE1(x, attached) message and y manages to answer first with its TYPE2(x, y, unattached) message. After this, node x knows that it can schedule the connection to y freely, since y has no obligations so far. Node x picks two convenient time slots and a frequency and sends a TYPE3(y, LinkSpec) message to y. Again, since y has no neighbors so far, y adopts the superframe phase of x. Finally, node y answers with TYPE2(x, y, –) message, carrying an empty link specification (meaning that x's link specification is adopted).

In the third case, node x does not have any neighbor yet, but y has. Therefore, y answers to x's TYPE1(x, unattached) with a TYPE2(x, y, attached) message. Node x proceeds with sending a TYPE3(y, –) message without link specification to y, and it is y's turn to pick the time slots and frequency. Accordingly, y sends back a TYPE2(x, y, LinkSpec) to x.

In the final case, both x and y are already attached to other nodes and their superframes are typically not aligned. Accordingly, x sends a TYPE1(x, attached) message and y answers with a TYPE2(x, y, attached) message. Node x answers with a TYPE3(y, Schedule) message, which contains its entire schedule as well as timing information allowing y to determine the phase shift between x and y's superframes. After receiving this information, node y determines time slots that are free in both schedules, and which are not necessarily aligned with any time slot boundaries in either schedule.

This protocol allows to set up static connections between stationary nodes. Since the neighborhood discovery process is repeated from time to time, the protocol can adapt to changes in topology. In reference [778], an extension ("eavesdrop and register" algorithm) is described that allows a mobile node to set up, maintain, and tear down connections to stationary nodes as it moves through the network.

A critical issue with this protocol is the choice of the superframe length. It should be large enough to accommodate the highest node degree in the network, which is a random variable for random deployments. If the superframe length is too short, some of a node's neighbors may simply not be visible to it. A second problem occurs in a densely populated sensor network with low traffic load, where schedules are highly populated and nodes wake up quite often just to notice that there is no packet destined to them. The number of wakeup slots depends on the node density for this protocol.

5.4.3 Traffic-adaptive medium access protocol (TRAMA)

The Traffic-Adaptive Medium Access (TRAMA) protocol presented by Rajendran et al. [672] creates schedules allowing nodes to access a single channel in a collision-free manner. The schedules are constructed in a distributed manner and on an on-demand basis. The protocol assumes that all nodes are time synchronized and divides time into **random access periods** and **scheduled-access periods**. A random access period followed by a scheduled-access period is called a **cycle**. The nodes broadcast their neighborhood information and, by capturing the respective packets from their neighbors, can learn about their two-hop neighborhood. Furthermore, they broadcast their schedule information, that is, they periodically provide their neighbors with an updated list of receivers for the packets currently in a nodes queue. On the basis of this information, the nodes execute a distributed scheduling algorithm to determine for each time slot of the scheduled-access period the transmitting and receiving nodes and the nodes that can go into sleep mode.

The protocol itself consists of three different components: the **neighborhood protocol**, the **schedule exchange protocol** and the **adaptive election algorithm**. The **neighborhood protocol** is executed solely in the random access phase, which is subdivided into small time slots. A node picks randomly a number of time slots and transmits small control packets in these without doing any carrier sensing. These packets indicate the node's identification and contain incremental neighborhood information, that is only those neighbor identifications are included that belong to new neighbors or neighbors that were missing during the last cycle. When a node does not transmit, it listens to pick up its neighbors' control packets. The length of the random access phase should be chosen such that a node receives its neighbors packets with sufficiently high probability to ensure consistent topology information. It depends thus on the node degree. All nodes' transceivers must be active during the random access period.

By the **schedule exchange protocol**, a node transmits its current transmission schedule (indicating in which time slots it transmits to which neighbor) and also picks up its neighbors' schedules. This information is used to actually allocate slots to transmitters and receivers. How does a node know which slots it can use? All nodes possess a global hash function h, and a node with identification x computes for time slot occurring at t the following priority value p:

$$p = h(x \oplus t)$$

where $x \oplus t$ is the concatenation of x's node identification with the current time t. To compute its schedule, a node looks ahead for a certain number of time slots, called its **schedule interval** (say: 100 slots) and for each of these slots computes its own priority and the priority of all its two-hop neighbors. For higher node densities, this incurs significant computation costs. The slots for which x has the highest priority value can be used by x to transmit its packets. These are called *winning slots*; for the sake of example, let us say these are slots 17, 34, 90, and 94. By looking at its packet queue, x can determine whether it needs all of these slots or can leave some of them to other nodes. Node x assigns to each of its winning slots a receiving node or a set of receivers, and sends this assignment as its schedule packet. The last of the future winning slots (slot 94 in our example) is always used for broadcasting x's next schedule, that is the whole schedule computation has to be repeated immediately before slot 94, spanning again over a full schedule interval. By using the last winning slot, the schedule can be transmitted without risk of collision.

The neighbors of x should wake up at slot 94 to receive x's next schedule (they should also have woken up to receive the current schedule!) and to determine when they have to leave sleep mode to receive a packet from x. In turn, x should also wake up when its neighbors have announced to transmit their next schedule.

With what we have described so far, node x can determine its winning slots and thus its transmit opportunities. The other question is: when must x prepare for receptions and when can x go into sleep mode during the scheduled-access phase? Fix one specific slot. There are two easy cases: Suppose that a one-hop neighbor y of x has the highest priority in x's two-hop neighborhood and that y has announced a packet for this slot. Either x is the receiver or x can go to sleep. A more complicated situation is depicted in Figure 5.13. Here, node D has the highest priority in B's two-hop neighborhood, but, on the other hand node, A has highest priority in *its* two-hop neighborhood. The **adaptive election algorithm** of TRAMA provides approaches for resolving this situation and also for allowing nodes to reuse their neighbors' unused winning slots.

Rajendran et al. [672] compare the performance of TRAMA with S-MAC (having 10 and 50 % duty cycle, respectively), the IEEE 802.11 protocol, a CSMA protocol according to reference [422], and NAMA [54], a precursor of TRAMA. The investigated performance measures are average packet delivery ratio, the achievable percentage of sleep time, the average sleep interval (which should be long since switching on and off transceivers costs energy), and the average

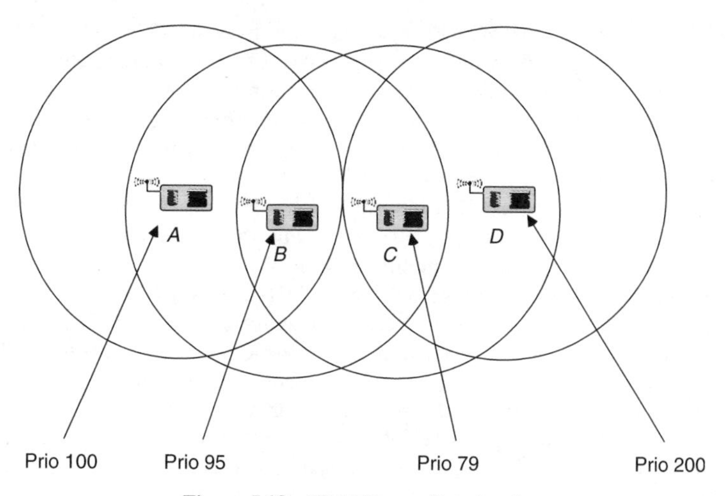

Figure 5.13 TRAMA: conflict situation

queuing delay of a packet waiting for transmission. Two scenarios are simulated: a single-hop scenario and a multihop scenario with one sink node and the sensors transmitting periodically to the sink. The underlying physical layer resembles the RF Monolithics TR1000 transceiver [690] (see also Section 2.1.4). As opposed to S-MAC, the energy savings of TRAMA depend on the load situation, while in S-MAC it depends on the duty cycle. The investigations confirmed also a well-known property of TDMA protocols stating that these have higher delays but also higher maximum throughput than contention-based protocols.

The TRAMA protocol needs significant computation and memory in dense sensor networks since the two-hop neighborhood of a node tends to be large in this case. Therefore, TRAMA is a feasible solution only if the sensor nodes have sufficient resources.

5.4.4 Further solutions

So far, we have mainly concentrated on protocols that assume almost no infrastructure, except some sink nodes. If we can assume the presence of infrastructure nodes, that is, nodes that are interconnected and not energy constrained, we can shift burdens to these nodes. For example, Shih et al. [762] assume an infrastructure setting (base stations close to the sensors) with tight latency requirements. As for the MAC, they investigate a combination of FDMA and TDMA (similar to GSM [848]) and derive a formula for the optimum number of FDMA channels for best energy efficiency under delay constraints.

All the schemes discussed so far need a periodically recurring neighbor discovery/network setup phase to adapt their schedules to changing network topologies. It would be very interesting to have topology-invariant schemes, which do not need these phases; one example scheme is proposed in reference [160] and another can be found in [161]. However, these schemes still need the knowledge of global network parameters (number of nodes, maximum node degree) and contain no provisions to let nodes sleep. Another topology-independent scheme for schedule formation is described in reference [386].

5.5 The IEEE 802.15.4 MAC protocol

The Institute of Electrical and Electronics Engineers (IEEE) finalized the IEEE 802.15.4 standard in October 2003 ([468]; see also [929], [317], and [114]). The standard covers the physical layer and the MAC layer of a low-rate Wireless Personal Area Network (WPAN). Sometimes, people confuse IEEE 802.15.4 with ZigBee[5], an emerging standard from the ZigBee alliance. ZigBee uses the services offered by IEEE 802.15.4 and adds network construction (star networks, peer-to-peer/mesh networks, cluster-tree networks), security, application services, and more.

The targeted applications for IEEE 802.15.4 are in the area of wireless sensor networks, home automation, home networking, connecting devices to a PC, home security, and so on. Most of these applications require only low-to-medium bitrates (up to some few hundreds of kbps), moderate average delays without too stringent delay guarantees, and for certain nodes it is highly desirable to reduce the energy consumption to a minimum. The physical layer offers bitrates of 20 kbps (a single channel in the frequency range 868–868.6 MHz), 40 kbps (ten channels in the range between 905 and 928 MHz) and 250 kbps (16 channels in the 2.4 GHz ISM band between 2.4 and 2.485 GHz with 5-MHz spacing between the center frequencies). There are a total of 27 channels available, but the MAC protocol uses only one of these channels at a time; it is not a multichannel protocol. More details about the physical layer can be found in Section 2.1.4.

The MAC protocol combines both schedule-based as well as contention-based schemes. The protocol is asymmetric in that different types of nodes with different roles are used, which is described next.

5.5.1 Network architecture and types/roles of nodes

The standard distinguishes on the MAC layer two types of nodes:

- A Full Function Device (FFD) can operate in three different roles: it can be a **PAN coordinator** (PAN = Personal Area Network), a simple **coordinator** or a **device**.
- A Reduced Function Device (RFD) can operate only as a device.

A device must be associated to a coordinator node (which must be a FFD) and communicates only with this, this way forming a **star network**. Coordinators can operate in a peer-to-peer fashion and multiple coordinators can form a Personal Area Network (PAN). The PAN is identified by a 16-bit **PAN Identifier** and one of its coordinators is designated as a PAN coordinator.

A coordinator handles among others the following tasks:

- It manages a list of associated devices. Devices are required to explicitly associate and disassociate with a coordinator using certain signaling packets.
- It allocates short addresses to its devices. All IEEE 802.15.4 nodes have a 64-bit device address. When a device associates with a coordinator, it may request assignment of a 16-bit short address to be used subsequently in all communications between device and coordinator. The assigned address is indicated in the association response packet issued by the coordinator.
- In the beaconed mode of IEEE 802.15.4, it transmits regularly **frame beacon** packets announcing the PAN identifier, a list of outstanding frames, and other parameters. Furthermore, the coordinator can accept and process requests to reserve fixed time slots to nodes and the allocations are indicated in the beacon.
- It exchanges data packets with devices and with peer coordinators.

In the remainder of this section, we focus on the data exchange between coordinator and devices in a star network; a possible protocol for data exchange between coordinators is described in Section 5.2.3. We start with the beaconed mode of IEEE 802.15.4.

[5] see `http://www.zigbee.org/`; a brief slide set on ZigBee entitled "ZigBee Overview" can be found under `http://www.zigbee.org/en/resources`.

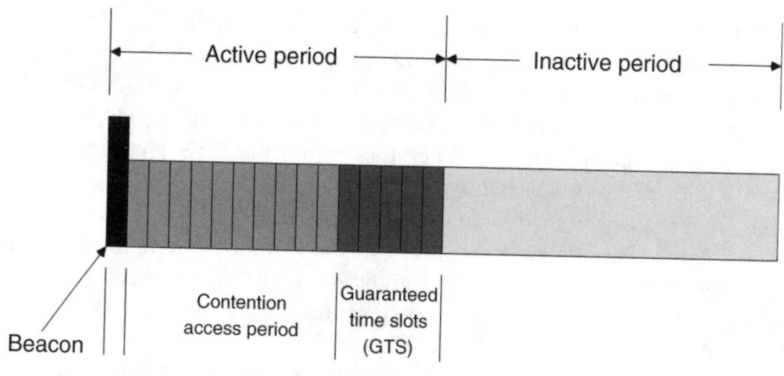

Figure 5.14 Superframe structure of IEEE 802.15.4

5.5.2 Superframe structure

The coordinator of a star network operating in the **beaconed mode** organizes channel access and data transmission with the help of a superframe structure displayed in Figure 5.14.

All superframes have the same length. The coordinator starts each superframe by sending a frame beacon packet. The frame beacon includes a **superframe specification** describing the length of the various components of the following superframe:

- The superframe is subdivided into an **active period** and an **inactive period**. During the inactive period, all nodes including the coordinator can switch off their transceivers and go into sleep state. The nodes have to wake up immediately before the inactive period ends to receive the next beacon. The inactive period may be void.
- The active period is subdivided into 16 time slots. The first time slot is occupied by the beacon frame and the remaining time slots are partitioned into a **Contention Access Period (CAP)** followed by a number (maximal seven) of contiguous **Guaranteed Time Slots (GTSs)**.

The length of the active and inactive period as well as the length of a single time slot and the usage of GTS slots are configurable.

The coordinator is active during the entire active period. The associated devices are active in the GTS phase only in time slots allocated to them; in all other GTS slots they can enter sleep mode. In the CAP, a device can shut down its transceiver if it has neither any own data to transmit nor any data to fetch from the coordinator.

It can be noted already from this description that coordinators do much more work than devices and the protocol is inherently asymmetric. The protocol is optimized for cases where energy-constrained sensors are to be attached to energy-unconstrained nodes.

5.5.3 GTS management

The coordinator allocates GTS to devices only when the latter send appropriate request packets during the CAP. One flag in the request indicates whether the requested time slot is a **transmit slot** or a **receive slot**. In a transmit slot, the device transmits packets to the coordinator and in a receive slot the data flows in the reverse direction. Another field in the request specifies the desired number of contiguous time slots in the GTS phase.

The coordinator answers the request packet in two steps: An immediate acknowledgment packet confirms that the coordinator has received the request packet properly but contains no information about success or failure of the request.

After receiving the acknowledgment packet, the device is required to track the coordinator's beacons for some specified time (called *aGTSDescPersistenceTime*). When the coordinator has sufficient resources to allocate a GTS to the node, it inserts an appropriate **GTS descriptor** into one of the next beacon frames. This GTS descriptor specifies the short address of the requesting node and the number and position of the time slots within the GTS phase of the superframe. A device can use its allocated slots each time they are announced by the coordinator in the GTS descriptor. If the coordinator has insufficient resources, it generates a GTS descriptor for (invalid) time slot zero, indicating the available resources in the descriptors length field. Upon receiving such a descriptor, the device may consider renegotiation. If the device receives no GTS descriptor within *aGTSDescPersistenceTime* time after sending the request, it concludes that the allocation request has failed.

A GTS is allocated to a device on a regular basis until it is explicitly deallocated. The deallocation can be requested by the device by means of a special control frame. After sending this frame, the device shall not use the allocated slots any further. The coordinator can also trigger deallocation based on certain criteria. Specifically, the coordinator monitors the usage of the time slot: If the slot is not used at least once within a certain number of superframes, the slot is deallocated. The coordinator signals deallocation to the device by generating a GTS descriptor with start slot zero.

5.5.4 Data transfer procedures

Let us first assume that a device wants to transmit a data packet to the coordinator. If the device has an allocated transmit GTS, it wakes up just before the time slot starts and sends its packet immediately without running any carrier-sense or other collision-avoiding operations. However, the device can do so only when the full transaction consisting of the data packet and an immediate acknowledgment sent by the coordinator as well as appropriate InterFrame Spaces (IFSs) fit into the allocated time slots. If this is not the case or when the device does not have any allocated slots, it sends its data packet during the CAP using a slotted CSMA protocol, described below. The coordinator sends an immediate acknowledgment for the data packet.

The other case is a data transfer from the coordinator to a device. If the device has allocated a receive GTS and when the packet/acknowledgment/IFS cycle fits into these, the coordinator simply transmits the packet in the allocated time slot without further coordination. The device has to acknowledge the data packet.

The more interesting case is when the coordinator is not able to use a receive GTS. The handshake between device and coordinator is sketched in Figure 5.15. The coordinator announces a buffered packet to a device by including the devices address into the **pending address field** of the beacon frame. In fact, the device's address is included as long as the device has not retrieved the packet or a certain timer has expired. When the device finds its address in the pending address field, it sends a special **data request** packet during the CAP. The coordinator answers this packet with an acknowledgment packet and continues with sending the data packet. The device knows upon receiving the acknowledgment packet that it shall leave its transceiver on and prepares for the incoming data packet, which in turn is acknowledged. Otherwise, the device tries again to send the data request packet during one of the following superframes and optionally switches off its transceiver until the next beacon.

5.5.5 Slotted CSMA-CA protocol

When nodes have to send data or management/control packets during the CAP, they use a slotted CSMA protocol. The protocol contains no provisions against hidden-terminal situations, for example

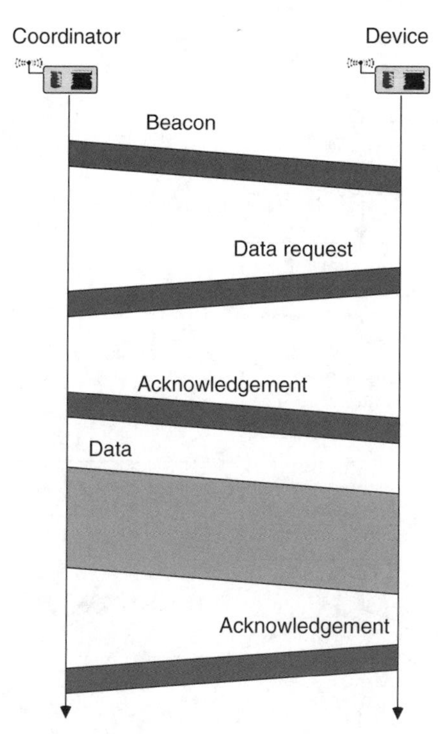

Figure 5.15 Handshake between coordinator and device when the device retrieves a packet [468, Fig. 8]

there is no RTS/CTS handshake. To reduce the probability of collisions, the protocol uses random delays; it is thus a CSMA-CA protocol (CSMA with Collision Avoidance). Using such random delays is also part of the protocols described in Section 5.3.1. We describe the protocol operation in some more detail; please refer to Figure 5.16 also.

The time slots making up the CAP are subdivided into smaller time slots, called **backoff periods**. One backoff period has a length corresponding to 20 channel symbol times and the slots considered by the slotted CSMA-CA protocol are just these backoff periods.

The device maintains three variables *NB*, *CW*, and *BE*. The variable *NB* counts the number of backoffs, *CW* indicates the size of the current congestion window, and *BE* is the current backoff exponent. Upon arrival of a new packet to transmit, these variables are initialized with $NB = 0$, $CW = 2$, and $BE = macMinBE$ (with *macMinBE* being a protocol parameter), respectively. The device awaits the next backoff period boundary and draws an integer random number r from the interval $[0, 2^{BE} - 1]$. The device waits for r backoff periods and performs a carrier-sense operation (denoted as Clear Channel Assessment (CCA) in the standard). If the medium is idle, the device decrements *CW*, waits for the next backoff period boundary, and senses the channel again. If the channel is still idle, the device assumes that it has won contention and starts transmission of its data packet. If either of the CCA operations shows a busy medium, the number of backoffs *NB* and the backoff exponent *BE* are incremented and *CW* is set back to $CW = 2$. If *NB* exceeds a threshold, the device drops the frame and declares a failure. Otherwise, the device again draws an integer r from $[0, 2^{BE} - 1]$ and waits for the indicated number of backoff slots. All subsequent steps are repeated.

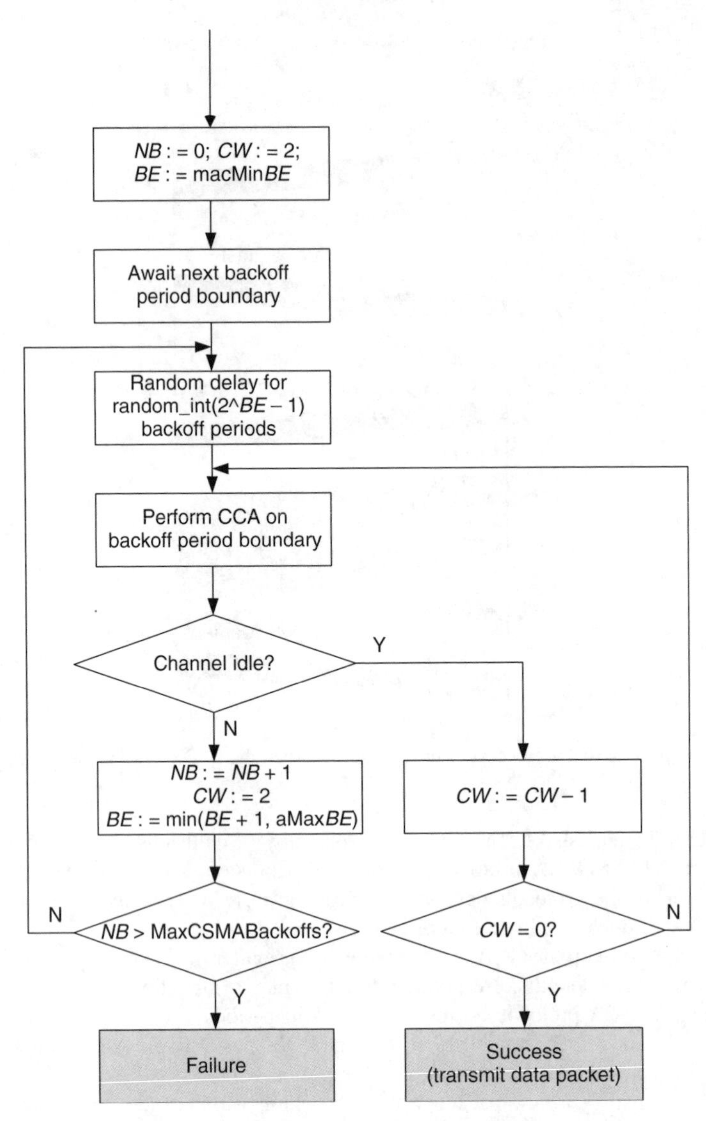

Figure 5.16 Schematic of the slotted CSMA-CA algorithm (simplified version of [468, Fig. 61])

5.5.6 Nonbeaconed mode

The IEEE 802.15.4 protocol offers a **nonbeaconed mode** besides the beaconed mode. Some important differences between these modes are the following:

- In the nonbeaconed mode, the coordinator does *not* send beacon frames nor is there any GTS mechanism. The lack of beacon packets takes away a good opportunity for devices to acquire time synchronization with the coordinator.
- All packets from devices are transmitted using an unslotted (because of the lack of time synchronization) CSMA-CA protocol. As opposed to the slotted CSMA-CA protocol, there is no

synchronization to backoff period boundaries and, in addition, the device performs only a single CCA operation. If this indicates an idle channel, the device infers success.
- Coordinators must be switched on constantly but devices can follow their own sleep schedule. Devices wake up for two reasons: (i) to send a data/control packet to the coordinators, or (ii) to fetch a packet destined to itself from the coordinator by using the data request/acknowledgment/data/acknowledgment handshake (fetch cycle) discussed above. The data request packet is sent through the unslotted CSMA-CA mechanism and the following acknowledgment is sent without any further ado. When the coordinator has a data packet for the device, it transmits it using the unslotted CSMA-CA access method and the device sends an immediate acknowledgment for the data. Therefore, the device must stay awake for a certain time after sending the data request packet. The rate by which the device initiates the fetch cycle is application dependent.

5.5.7 Further reading

Performance evaluations of certain aspects of IEEE 802.15.4 are presented in [521] and [880]. Lu et al. [521] focus on the case of beacon-enabled networks and investigate some of the throughput/energy/delay trade-offs. Some attention is paid to the aspect of synchronizing devices to the beacons, for example, to initiate fetch cycles. In the tracking mode, the device scans the channel for the first beacon, learns the timing from this, and tries to wake up immediately before the next beacon, and so on. Accordingly, nodes have a precise idea when the next beacon will arrive and when they can learn about packets destined to them. In the nontracking mode, a node simply wakes up at some time, seeks the next beacon, and goes back into sleep mode for longer time when no packet is stored at the coordinator. Willig [880] investigates theoretical throughput bounds for the two CSMA-CA variants. The unslotted CSMA-CA variant has a higher throughput because of its lower overhead (no waiting for backoff period boundaries, only one carrier-sense operation). One key parameter for the achievable throughput is the transceiver turnover times since these directly impact the collision rates. In the unslotted version, the number of collisions can be decreased when the turnover times are decreased. This is, however, not true for the slotted version because of the synchronization of devices to the coordinator: when two backlogged devices choose the same number r of backoff periods as their random delay, they will sense the channel at the same time, see the same idle channel, and start to transmit at the same time.

5.6 How about IEEE 802.11 and bluetooth?

An obvious question is the following: Given that there are already a number of proven wireless MAC protocols and wireless products out there, why not simply use them? Specifically: There are two popular and commercially available systems, Bluetooth and IEEE 802.11. What is "wrong" about them?

The Bluetooth system is designed as a Wireless Personal Area Network (WPAN) with one major application, the connection of devices to a personal computer [318]. It already has been used as a means for prototyping wireless sensor network applications [72]. The PHY is based on a FHSS scheme having a hopping frequency of 1.6 kHz and an appropriate allocation of hopping sequences. The nodes are organized into **piconets** with one master and up to seven active slave nodes. The master chooses the hopping sequence, which the slaves have to follow. Furthermore, there can be several passive slave nodes in a piconet. The master polls the active slaves continuously. Two major drawbacks of Bluetooth are the need to constantly have a master node, spending much energy on polling his slaves, and the rather limited number of active slaves per piconet.[6] This is not compatible with the case of dense wireless sensor networks where a huge number of master nodes would be

[6] A newer version of Bluetooth remedies this limitation.

needed. An active slave must always be switched on since it cannot predict when it will be polled by the master. A passive slave has to apply at the master to become an active slave. This fails if there are seven active nodes already. Furthermore, it is required that each node is able to take the role of masters or slaves and thus bears considerable complexity. Also, the fast frequency hopping operations require tight synchronization between the nodes in a piconet.

In the IEEE 802.11 family of protocols, several physical layers are specified sharing a single MAC protocol [284, 815], the DCF, and on top of it the Point Coordination Function (PCF). Without further provisions, IEEE 802.11 requires any node x to constantly be in listen mode since another node y may attempt to transmit a frame to x at any time. Secondly, nodes are required to overhear RTS and CTS packets to adjust their NAV timers properly. IEEE 802.11 has some power-saving functionalities [691], but, in general, the system is targeted towards high bitrates and the available transceivers require orders of magnitude more energy than acceptable in low-bitrate sensor network applications. Furthermore, IEEE 802.11 is a single-hop protocol for both the infrastructure and ad hoc network scenarios and, in general, is targeted at letting a number of independent and competing users share a common channel in a fair manner. These goals do not match the goals of wireless sensor networks.

5.7 Further reading

MAC protocols is a popular issue in the sensor networks literature and many protocols are missing. The following list of references provides some further starting points.

- There are some proposals to add power-control elements to MAC protocols. A recurring theme here is the augmentation of protocols using RTS/CTS handshakes with methods for adapting the transmit power to the minimum level necessary to reach the intended neighbor with a given BER target or packet-loss probability; see, for example, [895] or the BASIC scheme discussed in [389]. In BASIC, the RTS and CTS packets are transmitted at highest power, the data and acknowledgment packets at just the right power, inferred from the signal strengths of either RTS or CTS. The same authors Jung and Vaidya [389] propose a variation of BASIC in which the transmitter increases transmit power periodically *within* the data packet to keep contenders away. In the PCMA protocol described by Monks et al. [572], the transmitter issues a RTS packet and the receiver measures the signal strength of this packet. Using this information, the receiver can derive bounds on acceptable interference from hidden nodes and this bound is transmitted along with the CTS answer. Any neighbor hearing the CTS is allowed to transmit as long as the interference it creates to the receiver is below the given bound. Further references are [446], [573], [48], and [577].
- Chlamtac et al. [163] propose different access protocols for RF ID applications. One of these protocols is based on TDMA, a second one is a random access protocol, and a third one is a directory protocol.
- The DE-MAC protocol presented by Kalidindi et al. [393] is a TDMA-based scheme, which aims to regulate slot assignment such that energy-poor nodes can go into sleep mode more often.
- Tay et al. [813] present a nonpersistent CSMA variant with carefully chosen backoff times specifically targeted for achieving small delays for the first few packets after a sensor network starts to report data about a triggering event.
- The capacity of MAC protocols based on the RTS/CTS virtual carrier-sensing approach in ad hoc networks is discussed in [43].
- One reference focusing on quality-of-service aspects in wireless multihop networks instead of energy is [398].

Table 5.1 Summary of important WSN MAC protocols

Protocol	References	Flat/ clustered	# of required channels	Idle listening avoidance	Overhearing avoidance	Collision avoidance	Overhead
LEACH	[346]	Rotating clusters	1	By TDMA	By TDMA	By TDMA	Cluster election/ formation
STEM	[742]	Both	2	Periodic sleep	STEM-B	Depends on MAC	Depends on MAC, wakeup beacons
S-MAC	[914, 915]	Flat	1	Periodic sleep	Through NAV	RTS/CTS	RTS/CTS, SYNCH, virtual cluster init.
Mediation device	[115, Chap. 4]	Flat	1	Periodic sleep	Implicit	No	Periodic mediator service, query beacons, RTS/CTS
Wakeup radio	[667, 931]	Flat	≥ 2	Wakeup signal	Wakeup signal	Multichannel CSMA	Extra wakeup radio
CSMA protocols	[888]	Flat	1	–	Sleep during backoff	RTS/CTS	RTS/CTS
PAMAS	[668]	Flat	2	–	Yes	RTS/CTS, busy tone	Signaling channel
SMACS	[778, 780]	Flat	Many	By TDMA	By TDMA	By TDMA	Neighborhood discovery, channel setup
TRAMA	[672]	Flat	1	By scheduling	By scheduling	By scheduling	Neighbor protocol, schedule transmission

5.8 Conclusion

Several different MAC protocols for wireless sensor networks have been discussed in this chapter; their most important characteristics are summarized in Table 5.1. All of them are designed with the goal to conserve energy; other goals like small delays or high throughput are often traded off for energy conservation. There is no generic "best" MAC protocol; the proper choice depends on the application, the expected load patterns, the expected deployment (sparse versus dense sensor networks), and the specifics of the underlying hardware's energy-consumption behavior, for example, the relative costs of transmitting, receiving, switching between modes, wakeup times, and wakeup energy from sleep mode as well as the specific computation costs for executing the MAC protocol. We summarize some of the more important features of the discussed protocols in Table 5.1.

6

Link-layer protocols

Objectives of this Chapter

The most important tasks of the link layer are the formation and maintenance of direct communication associations ("links") between neighboring sensor nodes and the reliable and efficient transfer of information across these links. The reliability has to be achieved despite time-variable error conditions on the wireless link, and many mechanisms with different performance and energy-consumption characteristics have been devised. We discuss some of these mechanisms and point to a simple relationship between reliability, energy, and error rates: Increased error rates or increased reliability targets increase the energy consumption. Therefore, the reliability goals should be carefully stated according to application needs and expected error conditions.

Chapter Outline

The Data Link Layer (DLL) – or link layer for short – has the task of ensuring a **reliable** communications link between neighboring nodes, which in the case of wireless (sensor) networks means between nodes in radio range. Therefore, **error control** is one of its most important tasks, although there are others too. Error-control techniques are designed with the goal to achieve a certain level of reliability in the transmission of packets despite errors on the transmission channel. In wireless sensor networks, we have the additional requirement to achieve this reliability level with a minimum amount of energy. The most important control knobs and correction mechanisms that error-control schemes typically use are redundancy, retransmissions, and choice of transmit parameters like packet sizes and transmit powers. These control knobs are discussed in this chapter. One key aspect of these mechanisms is that they do not work under all channel error patterns equally

Protocols and Architectures for Wireless Sensor Networks H. Karl and A. Willig

well. A channel governed by a simple BSC model (independent bit errors with a constant error probability p) is quite different from channels exhibiting "bursty" errors and the performance as well as the energy expenditure depend on the channel type.

A trivial but nonetheless key lesson of this chapter – illustrated by a simple example – is that, in general, reliability costs energy, and increased reliability costs (sometimes exponentially) more energy. Therefore, reliability goals should be set carefully. This chapter has a single-hop perspective; the issue of reliable transmission in the end-to-end case is treated in Chapter 13.

However, the link layer has more tasks than reliable transmission. This chapter gives an overview of the different functions of the link layer, focusing on aspects important for wireless sensor networks. General treatments can be found in standard textbooks, for example, references [68, 458, 808].

6.1 Fundamentals: tasks and requirements

The DLL sits on top of the packet transmission and reception service offered by the MAC layer and offers its services to the network layer and other higher layers. Specifically, the network layer can use the link-layer services for packet delivery and to aid in routing and topology-control operations.

One of the most important tasks of the link layer is to create a reliable communication link for packet transmission between neighboring nodes, that is, nodes in mutual radio range. This can be broken down into the following aspects (a more refined breakdown is discussed in reference [930]):

Framing User data is fragmented and formatted into **packets** or **frames**, which include the user data and protocol-related information for the link layer (and the underlying MAC layer). The format and size of packets can have significant impact on performance metrics like throughput and energy consumption. The issues involved in framing are discussed in Section 6.3.

Error control This function acknowledges the fact that all transmission media and, particularly, wireless media introduce distortions into transmitted waveforms, which may render transmitted packets useless. With error-control mechanisms, the effect of errors shall be compensated for. The efficiency and energy consumption of the different error-control mechanisms depends on the error patterns on the link. Error control is discussed in Section 6.2.

Flow control The receiver of a (series of) packets may be temporarily not willing to accept packets, for example, because of lack of buffer space or processing capacity. Flow-control mechanisms introduce some signaling to let the transmitter slow down transmission. Since many sensor node designs use only very low bitrates, it is reasonable to assume that flow control is not an issue in sensor networks. This holds especially true when the destinations are more capable sink nodes. Accordingly, flow control has not been explicitly investigated in the context of wireless sensor networks, and it seems that the existing mechanisms – some of them integrated in error-control protocols (sliding window mechanism) – are sufficient. Therefore, we refer the reader to the standard networking literature.

Link management This mechanism involves discovery, setup, maintenance, and teardown of links to neighbors. An important part of the link-maintenance process is the estimation of the link quality, which can be used by higher-layer protocols for routing decisions or topology-control purposes. This link-quality estimation is discussed in Section 6.4.2. Some procedures for link setup and teardown have already been discussed in the context of MAC protocols, for example, in the SMACS protocol (Chapter 5).

Sometimes, the link layer has additional tasks, for example, the implementation of certain security mechanisms. Security issues are, in general, out of the scope of this book; a short glimpse on this topic can be found in Section 14.2.

6.2 Error control

In this section, we discuss several error-control methods, that is, methods that deal with transmission errors on wireless links to provide a certain service. The data transport service provided by a link layer can be characterized in terms of the following attributes:

Error-free The information that the receiving node's link layer delivers to its user should contain no errors, that is the transmitted bits are reproduced exactly.

In-sequence If the user of the transmitter's DLL hands over two pieces of information A and B in this sequence, the receiver's DLL must never pass B before A to its user; all other outcomes – A before B, only A, only B, or no packet at all – are allowed.

Duplicate-free The receiver's DLL user should get the same piece of information at most once.

Loss-free The receiver's DLL user should get any piece of information at least once.

Additionally, there can be delay constraints and energy constraints, meaning that either the delay a packet experiences in the DLL and lower layers or the energy spent by the DLL and lower layers should be bounded by an explicitly given value.

The most important error-control techniques are Forward Error Correction (FEC) and Automatic Repeat Request (ARQ) and combinations thereof; these are discussed in Sections 6.2.2 and 6.2.3. ARQ protocols address all the desired service attributes (error-free, in-sequence, duplicate-, and loss-free), while FEC methods are focused primarily on achieving error-free transmission.

There exists a rich literature on error control; some standard references are [74, 178, 322, 504, 511, 551, 661].

6.2.1 Causes and characteristics of transmission errors

It is a well-known fact that transmission on wireless channels is much more error prone than on wired channels. Physical phenomena like reflection, diffraction, and scattering of waveforms, partially in conjunction with moving nodes or movements in the environment, lead to **fast fading** and **intersymbol interference**. Path loss, attenuation, and the presence of obstacles lead to **slow fading**. In addition, there is noise and interference from other nodes/other systems working in overlapping or neighboring frequency bands. All these impairments are discussed in Chapter 4 and in references [124, 592, 682].

The distortion of waveforms translates into bit errors and packet losses. We explain this distinction with the help of an example. In Figure 6.1, we show the format of the Physical-layer Protocol Data Unit (PPDU) of the IEEE 802.11 WLAN standard with DSSS physical layer [467]. The PPDU is subdivided into a preamble, a PHY header, and the MAC-layer Protocol Data Unit (MPDU)

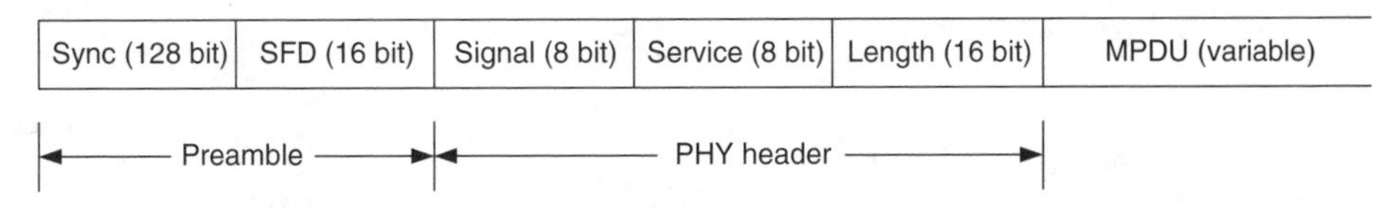

Figure 6.1 Format of an IEEE 802.11/802.11b physical layer frame

data part. The latter carries the MAC packet. The preamble is a constant bit pattern and useful for equalization purposes and to allow the receiver to acquire bit and frame synchronization. The PHY header describes among others the length and the modulation scheme used in the data part; in addition, the header is protected by its own checksum field. The end of the PHY header and the beginning of the actual MPDU is indicated by a fixed SFD. **Packet losses** occur if (i) the receiver fails to acquire bit/frame synchronization, (ii) the SFD is wrong, or (iii) bit errors in the remaining PHY header lead to an incorrect header checksum.[1] The result of a packet loss is that subsequent receiver stages like a FEC decoder or a MAC protocol entity do not see *any data at all*. When synchronization and PHY header have been acquired successfully, the bits making up the MPDU can be processed further by FEC or MAC. If any of these bits is not the same as its transmitted counterpart, we have a **bit error**. ARQ protocols provide checksums to detect bit errors and often drop the entire packet.[2]

In this particular framing scheme, applying FEC [504, 551] may correct bit errors in the data/ MPDU part, but the PHY header is not covered and therefore packet losses cannot be prevented. Measurements with an IEEE 802.11-compliant radio transceiver taken in an industrial Non Line Of Sight (NLOS) environment [882] have shown that indeed both types of errors occur, at sometimes impressive (or disturbing) rates. The frame structure shown in Figure 6.1 is quite common and packet losses may therefore occur in other systems as well.

The bit error and packet-loss statistics depend on a multitude of factors, including frequency, modulation scheme, distance, propagation environment (number of paths, materials), and the presence of interferers. Several studies of these statistics [13, 213, 223, 594, 882, 888] show some common properties:

- Both bit errors and packet losses are "bursty", that is they tend to occur in clusters with error-free periods ("runs") between the clusters. The empirical distributions of the cluster and run lengths often have a large coefficient of variation or sometimes even seem to be heavy tailed [436].
- The error behavior even for stationary transmitter and receiver is time varying, and the instantaneous bit-error rates can be sometimes quite high ($10^{-4} \dots 10^{-2}$). The same is true for packet-loss rates, which can reach values well beyond 50 %.

The bursty nature of wireless channel errors is a source of both problems and opportunities as we will see in the following sections.

6.2.2 ARQ techniques

The basic idea of ARQ protocols [322, 511] can be described as follows. The transmitting node's link layer accepts a data packet, creates a link-layer packet by prepending a header and a checksum, and transmits this packet to the receiver. The receiver checks the packet's integrity with the help of the checksum and provides **feedback** to the transmitter regarding the success of packet transmission. On receiving negative feedback, the transmitter performs a **retransmission**.

Key ingredients of ARQ protocols

Let us discuss some of the steps in some more detail:

[1] It is important to note that the meaning of the word "packet loss" is not uniform throughout the literature: many authors subsume both bit errors and synchronization errors as packet losses, since either way the receiver gets no valid packet.

[2] For certain types of user data, bit errors do not matter (much), and the MAC/ARQ checksum does not need to span the whole user data but only the MAC and link-layer header. An example is image transmission, where a few pixels distorted by bit errors are not visible to the human eye.

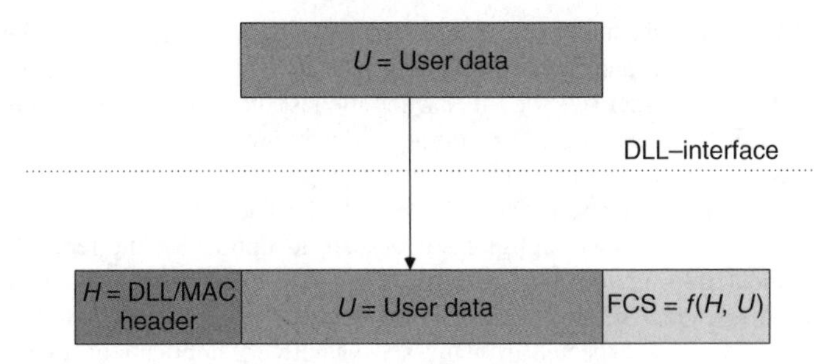

Figure 6.2 Generic packet formatting at the DLL

Packet formatting The DLL at the transmitter accepts user data U from upper layers, typically subject to some restriction in size (Figure 6.2). This restriction can be imposed, for example, by technological parameters or restrictions of the physical layer like carrier or bit synchronization schemes having poor tracking algorithms (see Section 4.2.6. The DLL prepends a **header** H to the packet, which contains control and address information. The address information is needed because wireless media are broadcast media by nature and a packet can theoretically be received by all neighbors in radio range. The address distinguishes the intended receiver and the transmitter. The control information can include sequence numbers, or flags, depending on the specific ARQ protocol.

Checksum The checksum (also often called **frame check sequence**) is appended to a packet after the packet-formatting process. In general, the checksum is a function of both the user data U as well as the header H. A widespread class of checksums are CRC values, often 8-, 16-, or 32-bits wide. The computation of CRC values is comparably easy, since they can be implemented with a linear-feedback shift register [322]. The receiver repeats the checksum calculation for the received header H' and the received user data U'. If the newly computed checksum is the same as the one carried in the packet, the receiver accepts the packet as correct. Typically, the size of user data U and header H is larger than the width of the checksum field and the checksum computation is not one-to-one. It could thus happen that a certain error pattern transforms the user data U into other user data U' such that for both U and U' the same checksum is computed – the packet would be accepted despite the (undetected) presence of errors. Hence, there is always a **residual error rate**, and the residual becomes larger as the checksum field becomes smaller; it depends, furthermore, on the statistical characteristics of both the user bits and the error patterns [797]. Small checksum fields are desirable in sensor networks to keep the number of transmitted bits low.

Feedback generation The **feedback-generation** step provides the transmitter with information about the outcome of the packet transmission. Two often-used mechanisms for obtaining feedback are timers (at the transmitter) and acknowledgment packets. Acknowledgements can be positive or negative. With a **positive acknowledgement**, the receiver confirms that he has received a packet. Positive acknowledgements can be sent for each packet or one acknowledgment packet can carry information about several data packets. The latter approach is clearly more energy efficient as is shown in reference [162] for the Selective Reject ARQ protocol. A **negative acknowledgement** is sent when the receiver detects a reception failure. However, in environments where multiple nodes can hear each other, negative acknowledgments are only feasible if (i) the MAC/DLL header has a separate checksum and is correctly received and the receiver thus knows which node has transmitted the packet, or (ii) the receiver has detected the reception failure by other means, for example, by finding a hole in the received sequence numbers. Hence, the *next* packet triggers detection of reception failure. Since both data and acknowledgment packets can get lost, the transmitter needs to use **timers**. The transmitter sets the timer when the last bit of the data packet has been sent to alert itself when the likely time for receiving the acknowledgment has expired. The timeout value must be large enough to include the processing time at the receiver, the duration of the ack packet, and the propagation delay. In WSNs, the nodes are only a short distance apart and the propagation delay is negligible. It is also an option for the receiver to piggyback acknowledgment information onto data packets going in the opposite direction. However, there must be sufficient traffic to avoid large delays for the acknowledgment information. In light traffic cases, as expected in many sensor network applications, such a scheme will often fail and extra ack packets must be sent.

Retransmissions Upon receiving negative feedback for a packet, the transmitter performs **retransmissions**. A first consequence of this fact is that the transmitter has to buffer the packets. Secondly, the transmitter must decide *when* to retransmit and *what* to retransmit. We will discuss these issues in some more detail below.

Standard ARQ protocols

Three standard ARQ protocols have emerged in the literature. They differ in their buffer requirements and retransmission strategies [68, 322, 505, 743, 808]:

Alternating bit The transmitter buffers one packet, sends it, and sets a timer. The receiver either receives the packet and sends a positive acknowledgment back or nothing is received and the receiver keeps quiet or sends a negative acknowledgment. If the transmitter receives a positive ack, the buffer is freed and the next packet can be transmitted. Otherwise, the transmitter retransmits the packet. The transmitter stamps each new packet with sequence numbers alternating between 0 and 1. Retransmitted packets are literal copies of the original packet and have thus the same sequence number. The sequence numbers allow the receiver to detect duplicates, which result if the positive acknowledgment, not the data packet, is lost. Alternating bit can provide loss-free, duplicate-free and in-sequence delivery of data given that the round-trip time (RTT) can be tightly bounded and the transmitter timeouts are larger than the RTT. Alternating bit [57] is also often referred to as Send-and-Wait.

Goback N Alternating bit is inefficient in case of "long fat pipes", that is, links where multiple packets can be in transit during a round-trip time (links with a large product of bandwidth and delay). Goback N is a protocol that allows the transmitter to have multiple outstanding, that is, unacknowledged frames. The transmitter keeps a buffer for up to N packets, called its **window**. Each packet in the window has its own timer, started upon the packet's transmission. The receiver accepts frames only in sequence and drops frames that are correctly received but do not have the expected sequence number (typically, because some previous frame had been lost). Therefore, the receiver only needs buffer space for a single frame. One common strategy for acknowledgements is to let the receiver always acknowledge the last packet arrived in sequence. If at the transmitter the timer for the oldest frame expires because the corresponding acknowledgement has not been received, this frame and all other frames in the window are retransmitted.

Selective Repeat/Selective Reject Selective Repeat has similarities to Goback N. However, unlike the Goback N protocol, in Selective Repeat the receiver also has N buffers and uses them to buffer frames arriving out of sequence. To achieve in-sequence delivery of data to the user, the receiver keeps out-of-sequence packets in the buffer until the missing packets have arrived.

The receiver can use both positive acknowledgements and negative acknowledgements. On the other side, the transmitter retransmits only those packets for which no acknowledgment has been received within the timeout period.

Send-and-Wait and Selective Repeat have the important property that only erroneous packets are retransmitted while Goback N potentially also retransmits correctly received packets, which is a waste of energy. In practice, often, the number of retransmissions allowed per packet is bounded to avoid spending too much energy in hopeless cases. In this case, a loss-free service cannot be guaranteed. Such a protocol is also said to be **semireliable**.

We illustrate the dependency between the desired reliability and the energy consumption with a simple example.

Example 6.1 (Energy consumption of alternating bit) Let us consider the alternating bit protocol over a BSC with a fixed bit error probability p, a packet of length l bits, and an infinite number of trials. The packet-success probability P_s is thus given by

$$P_s(l) = (1 - p)^l$$

and the packet-error probability is $P_e(l) = 1 - P_s(l)$. The number of trials $i \in \mathbb{N}$ needed to successfully transmit the packet over the link is a geometric random variable X with probability mass function:

$$\Pr[X = i] = P_s(l) \cdot P_e(l)^{i-1}$$

and cumulative distribution function

$$F(k) = \Pr[X \leq k] = \sum_{i=1}^{k} \Pr[X = i] = 1 - P_e(l)^k \qquad k \in \mathbb{N}.$$

If we prescribe ourselves a desired delivery probability $\delta \in (0, 1)$, we can ask for the number k^* of trials needed to deliver the packet with at least probability δ. Any smaller number of transmissions does not reach the reliability target. The value k^* is directly proportional to the energy consumption. The number k^* is given by:

$$k^* = F^{-1}(\delta) = \min\{k \in \mathbb{N} : F(k) \geq \delta\}.$$

Technically, k^* is the δ-quantile of the random variable X. With simple algebra it follows that

$$k^* = \left\lceil \frac{\log(1 - \delta)}{\log P_e(l)} \right\rceil.$$

The number k^* is graphed in Figure 6.3 for packets of length $l = 1023$ bits and two different reliability values $\delta_1 = 0.9$ and $\delta_2 = 0.99$; in the figure, the integer constraint for k^* is neglected.

The figure shows clearly that for relaxed reliability requirements $\delta = 0.9$, and for moderate-to-high bit-error rates (in the range $10^{-5} \ldots 10^{-2}$), we have to spend significantly less energy than for the higher reliability requirement. Below a BER of 10^{-5}, the channel is already good enough to transmit the packet successfully with the first trial and to guarantee any of the desired reliability bounds.

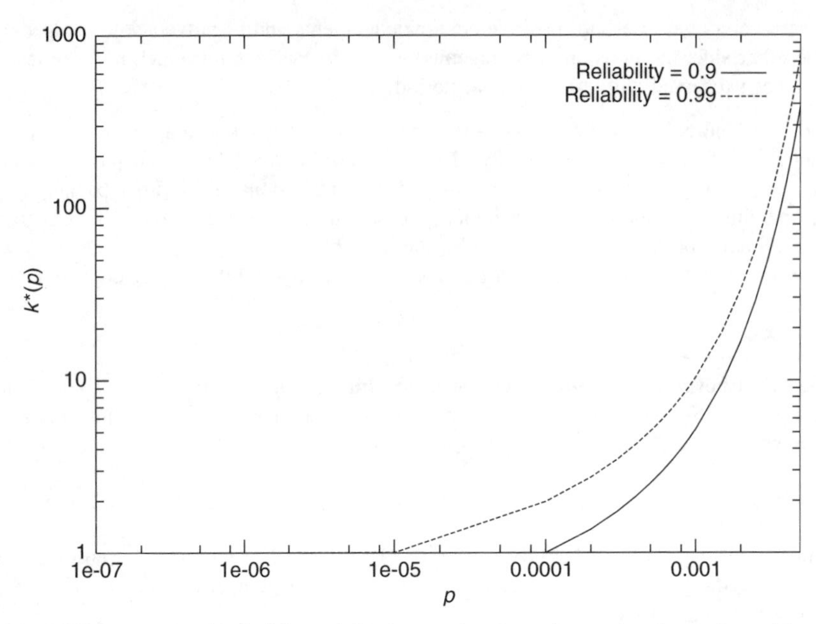

Figure 6.3 Minimum number k^* of trials needed to be sure that the packet reaches the receiver with prescribed probability $\delta \in \{0.9, 0.99\}$ with alternating bit for varying bit-error rate p

How to use acknowledgments?

The acknowledgment packets discussed so far in this chapter are link-layer acknowledgments and their meaning is that the receiver (i) has correctly received the packet, (ii) has sufficient buffer space to process it further, and (iii) actually accepts the packet because it is received in sequence or an out-of-sequence packet is accepted in Selective Reject. On the other hand, several MAC protocols use MAC-layer acknowledgments, for example, the mediation device protocol (Section 5.2.3) or S-MAC (Section 5.2.2). Sending out a MAC-layer ack implies at least condition (i) and for Alternating Bit and Selective Repeat, often (ii) and (iii) are implied too. This depends on the implementation: The receiver should place the packets immediately into link-layer buffers instead of using a MAC layer buffer and doing the transfer later. In the case that all three conditions are met, the MAC layer acks can be used simultaneously as link-layer acks.

If there are no MAC layer acknowledgments or the MAC acks satisfy only condition (i), either extra link-layer acknowledgment packets are required or the receiver can piggyback the acknowledgment information onto outgoing data packets destined for the transmitting node. However, piggybacking is not effective in case of low traffic load or in scenarios where almost all traffic is directed to a few sink nodes.

It is clear that we can save energy by reducing the number of acknowledgment packets, that is, if we let one acknowledgment packet acknowledge multiple data packets. On the downside, such a scheme requires the transmitter and, in case of Selective Repeat also the receiver, to provide a certain number of buffers, possibly creating a problem on memory-constrained sensor nodes.

Two such schemes have been discussed in reference [162] for the Goback N and Selective Repeat protocols. Their work is similar to the AIRMAIL link-layer protocol [33]. We focus on the Selective Repeat case, since it avoids retransmission of already received frames. In the first scheme, named **windowed feedback with selective repeat**, the receiver sends acknowledgments either after receiving a fixed number W of packets (called *window size*), after a timeout, or a duplicate packet reception. The acknowledgment contains a bitmap indicating the receiver's view on the packets' status since sending the last acknowledgment. In the **instantaneous feedback with selective repeat scheme**, the receiver in addition sends a (negative) acknowledgment upon receiving an out-of-order packet. The behavior of the transmitter is simple. It can continue to send as long as the new packets belong to the current window of size W. Upon reaching the end of the window without having received any feedback, the transmitter repeats the newest frame, triggering acknowledgment generation in the receiver. When the transmitter receives the acknowledgment, it can free the buffers for the correctly received packets and the erroneous packets are retransmitted. The window is advanced according to how many of the oldest packets in the window are acknowledged.

To judge the performance of these schemes, consider the following metrics. For a slotted system with fixed-size packets, the energy efficiency is measured as the average fraction of slots carrying acknowledgment frames, this way characterizing only the savings in ack frames. The throughput is evaluated as the average acceptance rate of new packets into the transmitter's window per slot given a saturated source. The delay is measured as the time between arrival of a packet at the transmitter and the time instant when the packet is acknowledged and its buffer is freed. Some of the results are:

- For both fast fading and slow fading scenarios with a mean packet-error rate of 30 %, increasing the window size increases the throughput and reduces the average energy consumption for both schemes, with the windowed feedback scheme giving better energy efficiency at the same throughput levels.
- When considering the relationship between energy consumption and delay, an increase in window size leads, for the instantaneous feedback scheme, to a reduction of energy consumption with only moderately varying delays. For the windowed feedback scheme, an increasing window size leads also to a decreased energy consumption but also to vastly increased delays.

Summarizing, given a fixed energy-consumption target, the scheme of choice depends on the second performance metric: if delay is at premium, the instantaneous feedback scheme is preferable, but if throughput is more important, the windowed feedback scheme is better.

In reference [689], these considerations are extended to allow a central node (base station) to explicitly control the acknowledgment generation behavior of its associated nodes based on traffic types and remaining energy. Here, much protocol complexity is moved to the central node, which should have appropriate memory and computation resources.

When to retransmit?

In ARQ protocols, a transmitter does not only have to decide *that* he has to retransmit but he also has to decide *when* to do so. Does the point in time make a difference? This depends on the channel error characteristics:

- In case of a static BSC channel, any time instant is just as good as any other – we cannot improve the chances of successful transmission by waiting.
- In case of fading channels, the situation is different. If a packet is hit by a deep fade and if the packet length is short as compared to the average fade duration, an immediate retransmission will likely be hit by the same deep fade and is thus a waste of energy. If the protocol is semireliable, the maximum number of trials may be exhausted before the channel turns back into a good state.

Therefore, in case of fading channels, it is wise to **postpone retransmissions**. It has been shown in other contexts [73, 116, 117] that postponing retransmissions and serving packets destined to other nodes meanwhile can significantly increase the throughput on a wireless network since no precious time and bandwidth (and energy!) is spent on useless immediate retransmissions. But how long shall we wait?

The **probing protocol** presented by ZORZI and RAO [942] distinguishes two different "channel modes", the **normal mode** and the **probing mode**. During the normal mode, the transmitting node sends packets according to an ARQ protocol, for example, Goback N or Selective Repeat. If the transmitter receives negative feedback (for example, lack of acknowledgement and timer expiration), it switches into the probing mode. In this mode, the transmitter periodically sends small **probe packets**. These probes are acknowledged by the receiver. Upon receiving such a probe acknowledgment, the transmitter assumes that both the forward channel (transmitter to receiver) and the backward channel (receiver to transmitter) are okay and continues in normal mode.

Two versions of the probing protocol are proposed, corresponding to Goback N and Selective Repeat. In the first one, upon switching back to normal mode, the transmitter simply transmits the failed packets and all subsequent packets again to the receiver. In the selective probing protocol, the receiver has buffers for the packets following a missed/corrupted packet. In the probe packet acknowledgment, the receiver indicates the buffered frames and the transmitter retransmits only the missing ones. Such an approach is useful if, because of long links, there can be several outstanding frames and the error burst was short enough to hit only one of them or in the case that the forward channel is good but the return channel is currently bad. The probing protocols decrease throughput since after the channel switches back to a good state it takes some time before the transmitter notices this and switches back to normal mode. This time is related to the period of the probing packets. The energy efficiency of the probing protocols and a classical Goback N protocol are compared by looking at the average number of trials needed to transmit a data packet. For bad channel conditions (long error bursts) and channels with short round-trip times, the probing protocol manages to keep this number close to one. The average number of probing packets needed, however, increases also with degrading channel conditions/longer error bursts.

The usage of probing packets in the probing protocol is questionable if the data packets are small themselves. In this case, one can directly use the data packets and avoid one extra transmission when the channel switches back into the good state. Such an approach amounts to inserting waiting times of fixed duration before the next trial, as described in reference [73]. When the fade statistics are known a priori or can be estimated with sufficient precision, such an approach can be successful. Another variation is to use linearly or exponentially growing inter-packet spaces between successive probing packets instead of using constant spaces.

6.2.3 FEC techniques

In all FEC mechanisms, the transmitter accepts a stream or a block of user data bits or source bits, adds suitable redundancy, and transmits the result toward the receiver. A conceptual view of where FEC could be placed in a communication system is shown in Figure 6.4.

Depending on the amount and structure of the redundancy, the receiver might be able to correct some bit errors. FEC can be used as an **open loop technique**, which means that there is no feedback from the receiver. Accordingly, the transmitter uses the same coding method all the time. This can be an interesting feature in terms of energy, since feedback is usually provided through acknowledgment packets. These would require the transmitting node to switch its transceiver into receive mode and wait for the acknowledgment; therefore, we incur both the reception costs (for the acknowledgment) and the costs for receiver turnaround. Furthermore, since typically the data packets in wireless sensor networks tend to be small, the acknowledgment packets make up a significant share of the total energy to transmit a packet.

Several coding schemes have been developed since Claude Shannon opened up the field of coding and information theory in 1948 [756]. Two widely used classes are block codes and convolutional codes, which are discussed next. The recently investigated class of **turbo codes** [66, 67, 773] has the

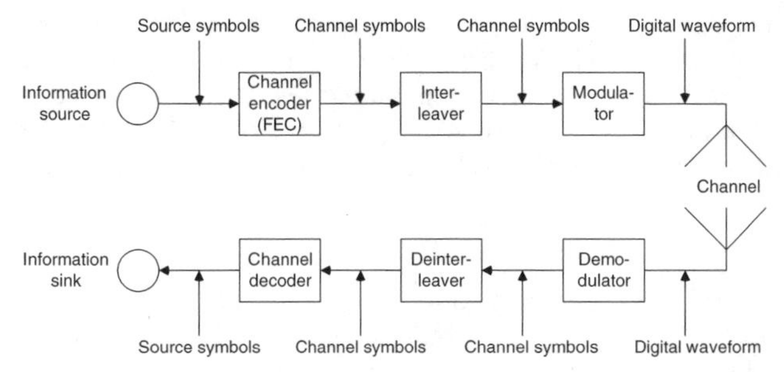

Figure 6.4 Conceptual view of FEC placement in a send/receive chain; channel encode and decode implement FEC

potential to nearly achieve the Shannon capacity of channels, but requires complex implementation and so far has not been considered as a candidate for wireless sensor networks.

Block-coded FEC

A block FEC coder takes a **block** or a **word** of a number k of p-ary **source symbols** and produces a block consisting of n of q-ary **channel symbols**; mostly, we have $p = q = 2$, $n \geq k$, and the symbols correspond to bits. Different source blocks are coded *independent* of each other [504, 772]. The mapping of the 2^k different source words into the 2^n different channel bits is injective and the range of this mapping is the set of **valid channel symbols**. Such a mapping is also called a **code**, and the ratio $\frac{k}{n}$ is called the **code rate** (small code rate equals high redundancy and small useful data rate). The number t of reliably correctable bits in a channel block of length n bits depends on the coding scheme; however, an upper bound is imposed by the **Hamming bound**, which states that a block code with k user bits mapped to n channel bits can correct up to t bit errors only if the relation

$$2^{n-k} \geq \sum_{i=0}^{t} \binom{n}{i}$$

holds. The fact that a triple (n, k, t) satisfies this relation does not imply that a code with this properties really exists.

An important metric for a block code is its **Hamming distance**: the Hamming distance of two valid channel words w_1 and w_2 is defined as the number of bits in which they differ and the Hamming distance $d_{\min}$ of the whole code is defined as the minimum Hamming distance of all pairs of valid channel words. Any code with a Hamming distance $d_{\min}$ can reliably detect up to and including $d_{\min} - 1$ bit errors and can reliably correct up to and including $\frac{d_{\min}-1}{2}$ bit errors. For practical applications, the set of valid codewords should be structured to allow easy coding and decoding. For example, in so-called linear block codes [551, Chap. 3] a vector space structure is imposed on the 2^n codewords and the subset of 2^k valid codewords is a subspace. Decoding can be regarded as finding the orthogonal projection of the received codeword onto the subspace of valid codewords. Popular and widely used examples of block codes are Reed–Solomon (RS) codes and Bose–Chaudhuri–Hocquenghem (BCH) codes.

If block coding FEC is applied as an open-loop technique, a constant overhead is incurred in every packet. On the transmitting node, this overhead consists of transmitting some extra bits and in doing the computations necessary for coding. For binary BCH codes, the coding process uses a linear-feedback shift register [551, Sec. 4.4] and can be assumed to have negligible energy costs [411, 761]. For decoding, several efficient algorithms have been developed, for example the Berlekamp-Massey algorithm [551, Sec. 6.6]. The energy costs of these algorithms depend both on the block length n and the code rate/number of correctable bits. With respect to the block length n, these algorithms show a linear increase in energy costs [294, 721]. According to [721, Eq. 13], the decoding energy depends on the block length n and the number of correctable bits t as

$$E_{dec} = (2nt + 2t^2) \cdot (E_{add} + E_{mult}), \tag{6.1}$$

where E_{add} and E_{mult} are the energy needed to carry out addition and multiplication in the Galois field $GF(2^m)$ with $m = \lfloor \log_2 n + 1 \rfloor$. For Reed–Solomon codes, similar relations hold, since they are a subclass of nonbinary BCH codes. In [478], it has been shown experimentally that the energy costs also depend on the code rate, with increasing code rates (less protection) having less energy costs (Figure 6.5).

If the error conditions on the channel vary over time, the **residual error rate** (rate of uncorrectable errors) of FEC techniques varies too. If the bit-error rate is very small all the time – or at least during good channel periods in a fading channel – a large FEC overhead would not be justified. Conversely, in case of extremely high error rates like, for example, during a deep fade in a Rayleigh fading channel model (see Section 4.2.4) or if the received signal strength of a transmitter node at a receiver node is close to the receive threshold, very low code rates would be needed to ensure error-free transmission. As discussed before, decoding codes with very low code rates is energy consuming. Therefore, it might be more appropriate to await the end of the fade and continue with a moderate code rate in the following good channel period.

Convolutional codes

In convolutional codes [845] also k bits of user data are mapped to n channel symbols; however, the coding of two successive k-bit blocks is not independent. We explain the operation of a convolutional coder briefly (Figure 6.6).

The encoding procedure runs in **steps**. During each step, k bits of user data are shifted into a shift register. This shift register has a total length of $K \cdot k$ bits, where K is called the **constraint**

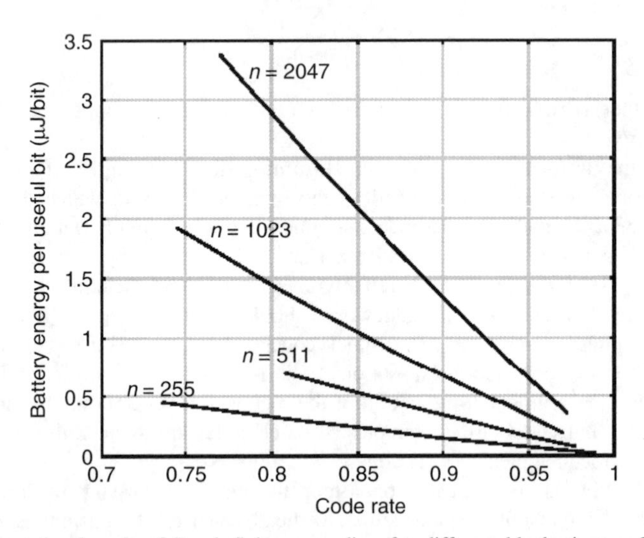

Figure 6.5 Computational costs of Reed–Solomon coding for different block sizes and code rates [478, Fig. 5]. Reproduced with kind permission of Springer Science Business Media

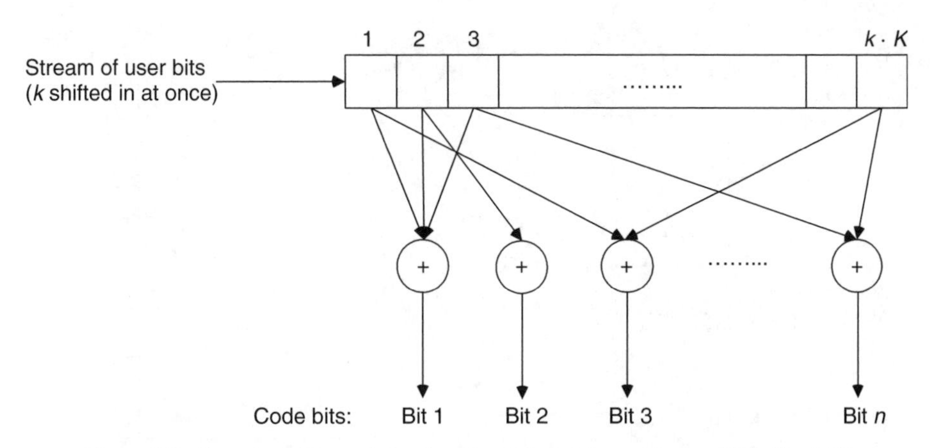

Figure 6.6 Operation of convolutional coding (adapted from reference [772, Fig. 6.2])

length of the code. The k user data bits are shifted into the register on one side while on the other side the "oldest" block of k bits is removed. Furthermore, there are n modulo-2 adders, each of which sums up a specific subset of the $k \cdot K$ registers. The outputs of the n adders are transmitted once per step. This coding scheme has the property that the k bits are present in the shift register for K steps and thus the coding of any k user bits depends not only on these bits themselves but also on the previous $(K - 1) \cdot k$ user bits. Again, the ratio $\frac{k}{n}$ is called the textitcode rate. The constraint length K controls the amount of redundancy contained in the code bits. If we increase K, we can also increase the **coding gain**, that is we can reduce the transmit energy needed to achieve a given BER target or reduce the BER for fixed transmit power [772, Sec. 6.4.5].

Similar to block codes, the encoding procedure of convolutional codes is cheap in terms of energy [762]. For decoding convolutional codes, often the Viterbi algorithm is used [844], [772, Sec. 6.3]. For this algorithm, the memory requirements of the receiver depend exponentially on the constraint length K [772, Sec. 6.3.5]. As shown in reference [761] for both a software and a hardware implementation of the Viterbi algorithm, the energy consumption increases also exponentially with K. However, the hardware implementation is doing so at energy levels being some orders of magnitude below the energy required by the software solution.

Certain block codes and convolutional codes have been compared for their energy efficiency in reference [721]. The chosen energy-efficiency criterion is the product of the packet success probability and the energy costs per packet, which include transceiver and decoding costs. For a BSC with fixed bit-error rate of $p = 10^{-3}$ and with choosing the packet lengths in an optimal way, the block coding schemes achieve better energy efficiency than convolutional coding schemes. With respect to the latter, high rate convolutional codes have been shown to achieve comparably small packet success probabilities, while low rate codes achieve high reliability but at significant energy costs. However, even the favorable medium rate codes with code rates of 3/4 and 5/6 and with the maximum investigated constraint length of $K = 9$ are outperformed by BCH codes in terms of energy efficiency.

Min et al. [561] compare several convolutional codes with respect to their energy efficiency when the goal is to achieve a certain acceptable residual error probability on a fixed wireless link with a path loss of 70 dB and the energy consumption characteristics of MITs μAMPS-1 nodes (Section 2.2.4). These characteristics include the energy consumed by the transceiver for coding and decoding as well as further processing costs. All the chosen codes have a constraint length of $K = 3$ but different code rates: 1/2, 2/3, and 3/4. For very small desired residual error probabilities ($\approx 10^{-9}$ and smaller), the rate-1/2 code is the most energy efficient, while in the range of $\approx 10^{-9} \ldots 10^{-7}$,

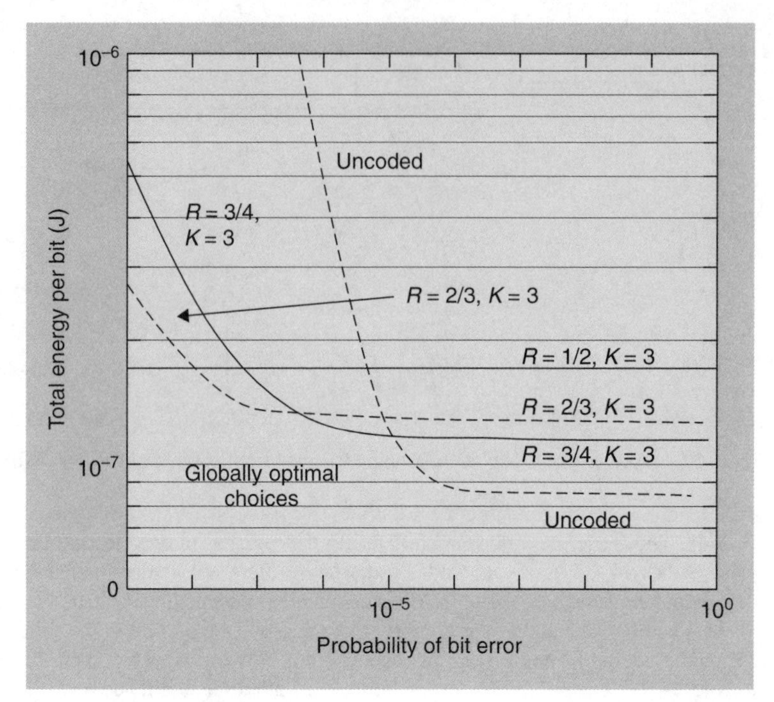

Figure 6.7 Energy consumption during transmission of 10 kB of data over a link with 70-dB path loss and different coding schemes [561, Fig. 9b]. Reproduced by permission of the IEEE

the rate-2/3 code is the best one, and, finally, if the residual error rate requirements are lenient, uncoded modulation is the method of choice; compare Figure 6.7. This shows that by relaxing the reliability requirements codes with higher code rates, and thus less overhead, can be used, saving transmit energy.

Interleaving

Many coding schemes, particularly convolutional schemes, do not achieve optimum performance when they are faced with a medium with bursty error characteristics. A common way to circumvent this (and to make the energy investment into FEC coding pay off more likely) is to use **interleaving**. Here, an **interleaver** in the transmitting node accepts a fixed-length data packet generated by the FEC encoder (compare Figure 6.4), which often consists of multiple coded blocks. The bits in this packet are permuted before transmitting. The **deinterleaver** at the receiving node inverts the permutation before the packet is handed to the FEC decoder. When an error burst hits such an permuted packet, the deinterleaving spreads this concentrated burst over this entire packet length. Hence, error bursts are spread over multiple coding blocks instead of just being concentrated to one or a few blocks. This increases the chance that each block can be successfully decoded.

A typical mode of operation of an interleaver is as follows:

- The interleaver waits for m codewords w_i $(i = 1, \ldots, m)$ of n bits each $(w_i = (w_{i,1}, w_{i,2}, \ldots, w_{i,n}))$.
- These are arranged in an $m \times n$ matrix.

- Example with $n = 6$, $m = 4$; successive codewords are arranged in rows:

$$\begin{array}{cccccc} w_{1,1} & w_{1,2} & w_{1,3} & w_{1,4} & w_{1,5} & w_{1,6} \\ w_{2,1} & w_{2,2} & w_{2,3} & w_{2,4} & w_{2,5} & w_{2,6} \\ w_{3,1} & w_{3,2} & w_{3,3} & w_{3,4} & w_{3,5} & w_{3,6} \\ w_{4,1} & w_{4,2} & w_{4,3} & w_{4,4} & w_{4,5} & w_{4,6} \end{array}$$

- These symbols are transmitted as

$$w_{1,1}\, w_{2,1}\, w_{3,1}\, w_{4,1}\, w_{1,2}\, w_{2,2}\, \ldots\, w_{3,6}\, w_{4,6}.$$

In this example, all error bursts of length $\leq m = 4$ are distributed by the deinterleaving operation "fairly" over all the codewords w_i such that each one has only one bit error. Accordingly, interleaving is most effective if m channel symbols have a duration of at least the mean fade duration. It must be noted that interleaving does not reduce the pre-FEC decoding bit-error rate, but only arranges the errors in a "nicer" way for the FEC decoder. Interleaving has one significant drawback, though: The transmitting node must wait with its transmission until $m \cdot n$ data symbols have been collected; thus significant delay can be introduced by interleaving. The use of interleaving in sensor networks is mentioned in references [411, 648] but is, to our knowledge, not explicitly addressed yet.

Multihop FEC

So far, we have discussed FEC solely in the context of a single hop. Zorzi and Rao [943] consider the multihop case, investigating three different schemes for their energy efficiency with respect to a given constraint on received power at the final destination: (i) Direct FEC-coded transmission from source to destination, (ii) multihop transmission with the intermediate nodes doing FEC decoding and recoding again (and expecting the target receive power, too), and (iii) multihop transmission without letting intermediate nodes do FEC decoding but only have them forward the packets. The relative advantages of the three schemes depend on the distance between source and destination, with the multihop schemes being clearly preferable over longer distances. FEC decoding at the intermediate nodes only pays off over the longest investigated distances between source and destination since accumulation of too many errors is avoided.

6.2.4 Hybrid schemes

From the discussion so far, it is clear that no single fixed error-control strategy will give optimum energy efficiency at all times. We illustrate the involved trade-offs with an example:

Example 6.2 (Energy efficiency of FEC and ARQ) Let us consider a transmitter and a receiver node connected by a wireless channel. The channel is a BSC with a fixed bit-error rate p. As for FEC, we consider the special case of binary BCH codes. For these codes, the following property holds [551, Sec. 4.3]: For all positive integers m and t, there exists a binary BCH code that has a code block length of $n = 2^m - 1$, of which at most $t \cdot m$ are overhead bits and that can reliably correct up to t errors. Additionally, we assume that uncorrectable errors are at least reliably detected. Furthermore, we assume that a simple alternating bit protocol with unlimited number of retransmissions runs between the two nodes. We fix the block length n at $n = 1023$. According to reference [551, Table 4.6], for error-correcting

Table 6.1 Amount of user data k for various numbers of correctable bits t in a block $n = 1023$ bits long for BCH coding [551, Table 4.6]

t	k
0	1023
2	1003
4	983
6	963
8	943
10	923

capabilities of $t = 0, 2, 4, 6, 8, 10$ bits within $n = 1023$ bits, an amount of k-bits user data can be transported as given by Table 6.1.

For all t, the probability that a packet of $n = 1023$ is transmitted successfully (i.e., the number of bit errors is $\leq t$) is given by:

$$P(n, t, p) = \sum_{i=0}^{t} \binom{n}{i} (1-p)^{n-i} p^i. \tag{6.2}$$

With the alternating bit protocol over a BSC the number X of trials needed to transmit the packet successfully is a geometric random variable with expectation

$$E[X] = \frac{1}{P(n, t, p)}.$$

If we assume unit costs to transmit and receive a single bit ($E_t = 1$) and if we furthermore take the decoding costs E_{dec} according to Equation 6.1 into account (with $E_{\text{add}} = 3.3 \cdot 10^{-4}$ and $E_{\text{mult}} = 3.7 \cdot 10^{-2}$ energy units – these numbers are chosen only for illustration purposes), we can express the expected amount of energy spent for any of the k user bits as:

$$E[Y] = \frac{n \cdot E_t + E_{\text{dec}}}{k \cdot P(n, t, p)}.$$

This expected value is visualized for the different values of t in Figure 6.8.

It can be seen that for extremely low bit-error rates FEC coding is more costly than going without any FEC. Beyond $p \approx 10^{-4}$, it is more appropriate to use FEC. However, for poor channels ($p \approx 10^{-2}$ and higher), all schemes would need a large amount of energy, and in such a case, other mechanisms should be employed, for example, choosing more appropriate packet sizes (see Section 6.3) or dropping the packet and hoping that a neighboring sensor node has made the same environmental observation and finds a better channel.

While the exact numbers may in real systems differ from the above illustrative choice, the conclusions are likely to remain valid.

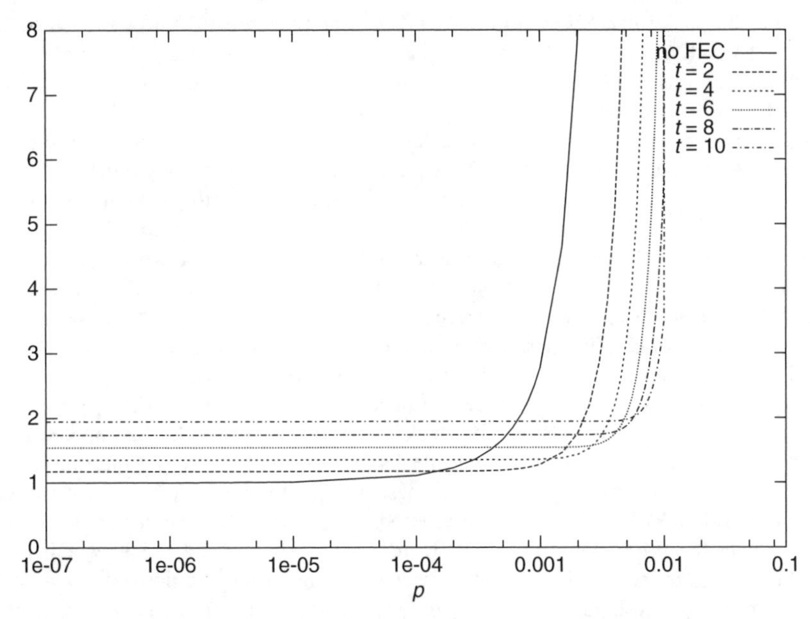

Figure 6.8 Energy spent per successfully transmitted data bit for reliable packet transmission with various BCH codes and without any FEC with $n = 1023$ bits code length and various numbers t of correctable errors

In the example, we have examined a simple scheme where ARQ and FEC are combined; in this scheme a light FEC code is applied to any packet and uncorrectable errors are handled by the ARQ protocol. Without going too much into details one can imagine other schemes, for example:

- Normal packets are transmitted uncoded and only retransmitted packets enjoy FEC coding. This is clearly the most energy-efficient solution for the good periods of a fading channel. In this case, however, it makes sense to keep FEC enabled not only during a retransmission (likely in a bad channel period) but also for some subsequent packets, lest they be destroyed by the same channel fade.
- Instead of retransmitting the data packets, special **parity packets** can be transmitted. Such an approach is discussed in another context in reference [119].
- In **packet combining schemes**, the receiving node tries to take advantage of the information contained in already received erroneous packets, for example, by using packet-combining methods like equal-gain combining or bit-by-bit majority voting [334, 394, 860]. The goal is to reduce the number of retransmissions needed. Such an approach makes sense if only a few bits of a packet are hit by bit errors; on the other hand, it requires buffering and significant signal processing at the receiver. A somewhat similar approach, the intermediate checksum scheme, is discussed in Section 6.3.2.

To the best of our knowledge, so far these techniques have not been considered in the context of wireless sensor networks.

6.2.5 Power control

Another control knob for increasing the reliability of packet transmission over a link is the transmit power, more precisely, the radiated output power of the transmitter. Increasing this power increases the energy per bit E_b/N_0/the SNR and thus decreases the bit-error rate and the need for retransmissions. EBERT and WOLISZ [222] show that for a single-hop ad hoc network scenario there is in fact an optimal transmit power (which is equivalent to a BER target) balancing the radiated energy and the need for retransmissions for a given packet length.

In larger networks with multihop communications, however, things are a bit different. If one node increases its transmit power, it also increases the interference seen by other nodes and thus effectively the bit error rates they have to deal with. To which extent, this is a problem depends on the expected load situation.

NARENDRAN et al. [583] describe a fully distributed scheme in which two nodes sharing a link jointly control the transmit power and the code rate. The transmitter adapts power and code based on measurements of word error rates, interference, and received power taken at the receiver and fed back to the transmitter. The algorithm is interesting, although it is developed and evaluated in a cellular system scenario where all this information is readily available at the base stations. However, the approach is sufficiently generic to be applied in sensor networks as well.

The approach is most easily explained for block coding FEC schemes. It is assumed that the FEC decoder at the receiver does not only deliver blocks of user bits but also the information whether the block contained uncorrectable errors or not. Using this information, the receiver can compute the **word error rate** (WER). The algorithm proceeds in **iterations**. At the end of an iteration, the receiver computes the WER over all received words in this iteration and checks whether an upper or lower threshold is exceeded. Somewhere between these bounds is the **desired word error rate** (DWER). If indeed a threshold is exceeded, the receiver provides the transmitter with the word error rate, the received power, and the observed interference level. The transmitter has a range $[P_{min}, P_{max}]$ available within which it can choose the output power, as well as a set $\{c_1, \ldots, c_N\}$ of coding schemes. For each coding scheme, one can compute a priori the Carrier to Interference Ratio (CIR) needed to obtain DWER and, accordingly, one can compute for each coding scheme c_i the output power level p_i needed to obtain this specific CIR at the receiver. Ultimately, the transmitter picks the power-level/code pair that can reach the goal with minimum overall energy. If the word error rate exceeds the upper threshold, we have a **bad iteration**. Since nodes act independently, oscillations can occur, or, even worse, the power increase of one node x triggers a power increase at its neighbors, in turn triggering a further power increase at x, and so forth. Therefore, a connection drop policy is adopted, by which the link is shut down with increasing probability as more and more bad iterations occur. In order to reduce the amount of feedback traffic from the receiver, it is also possible to send feedback only if the averaged WER, taken over several iterations, exceeds a threshold. NARENDRAN et al. [583] show that such an approach can increase the lifetime of mobile nodes significantly. Clearly, if such an assignment of transmit power settings could be made a priori and if this could also be combined with scheduling, even better power savings would be possible [185].

6.2.6 Further mechanisms to combat errors

In this section, we briefly touch upon further strategies to deal with errors, not necessarily tied to the link layer.

An interesting strategy is **error concealment**. The idea is to not correct all transmission errors but to live with them to some extent and to take other measures to let the influence of errors disappear for the application. This relaxes the reliability requirements and energy consumption, but sometimes at the price of higher computational efforts. One interesting example in the context of wireless sensor networks is discussed by HONARBACHT and KUMMERT [355]. Let us assume that a source sensor observes a continuous and slowly varying signal from the environment. The sink node uses an asynchronous Kalman filter and a Taylor approximation of the sensor signal to predict missing values, at moderate computational costs. HONARBACHT and KUMMERT [355] show that using this filtering technique the number of signal samples needed to reconstruct the signal at the receiver can be reduced significantly; in one example, 5 % of ≈2800 samples are sufficient to let the predicted signal look almost the same as the signal consisting of all samples. If such a technique can be applied, the reliability requirements can be relaxed and thus the energy costs can be decreased. However, error concealment is not primarily a link-layer technique but needs to incorporate application information.

In reference [740] (and in a much more general context also in reference [835]), the variation of modulation schemes and therefore bit rates have been considered for given BER constraints and in the presence of deadlines.

6.2.7 Error control: summary

In this section, we have considered several error-control approaches and their energy-consumption aspects. Error control is carried out to improve the reliability of packet transmission over a link and the fundamental trade-off is that both increased reliability requirements and increased channel error rates demand more and more energy. Therefore, before starting to design error-control schemes, it is of paramount importance to properly assess the required reliability. In wireless sensor networks we have the fortunate situation that often the same or at least correlated information about physical events is present in multiple nodes and thus there is no need for a single node to give its very best to forward a packet with very high reliability.

FEC coding schemes have the potential to achieve a real gain in energy efficiency for a given reliability target and in case of a medium with not too low and not too high bit-error rates. For very low BER, the coding overhead is wasted, and for extremely high BERs, the overhead is wasted too, since practical coding schemes do not have sufficient error-correction capabilities to justify the energy investment. For both convolutional and block coding schemes, the energy needed for the encoding operation is negligible but the decoding energy can be significant and outweigh the energy gains. Block coding schemes tend to be more energy efficient than convolutional coding, but convolutional codes often have better error-correction capabilities. If the encoding and decoding can be done in hardware, FEC schemes can be attractive.

ARQ schemes, on the other hand, can adapt their overhead to the channel conditions. On excellent channels, the only overhead is the acknowledgement frames and their number can be reduced by clever choice of acknowledgement schemes. On the other hand, in case of channel errors, the standard ARQ protocols retransmit whole packets, even when only a few bits were wrong.

There are at least two ways out. First, we have discussed hybrid schemes that combine FEC and ARQ in appropriate ways, for example, by applying a computationally moderately expensive FEC scheme to all packets and let the ARQ protocol correct the remaining errors. Second, one can try to find **adaptive schemes** that actually change their error-control strategy depending on the channel conditions, for example, switch from Goback-N_1 to Goback-N_2, adaptive multicopy ARQ [25] or adaptive error coding [145, 223, 230]. However, it depends on the application, the node hardware, and the expected error patterns whether the extra complexity needed for adaptation (channel state estimation, signaling of adaptation control operations) really pays off in terms of energy or whether a carefully designed "fixed" scheme will work well enough.

Another discussion of the relative advantages of FEC and ARQ can be found in reference [17].

6.3 Framing

In the process of framing, the link-layer constructs a frame that is then transmitted. Some general considerations regarding framing have already been discussed in Section 6.2.2 and the important aspect of addressing is discussed in Chapter 7. Here we discuss the important issue of the **choice of packet size**. We illustrate in the following example why this issue is important.

Example 6.3 (Energy efficiency for different packet sizes) Consider two nodes connected through a BSC with some bit-error probability $p \in (0, 1)$. Each packet has a fixed overhead of o bits (header, trailer) and a variable number u of user data bits. Accordingly, the probability of successful transmission of such a packet is given by:

$$P(o, u, p) = (1 - p)^{o+u}, \tag{6.3}$$

and the probability of a packet error (at least one bit is erroneous) is just $Q(o, u, p) = 1 - P(o, u, p)$. In case of a simple Alternating bit ARQ protocol with an infinite number of retransmissions, the number of trials needed to transmit a packet, X, is a geometric random variable with success probability $P(o, u, p)$ and expectation

$$E[X] = \frac{1}{P(o, u, p)} = \frac{1}{(1 - p)^{o+u}}.$$

The positive acknowledgment packets consist solely of overhead and are also of o bits length. We concentrate on the transmitter. Suppose that the energy spent on transmitting/receiving a packet of size l bits is given by:

$$e_t(l) = e_{t,0} + e_t \cdot l \qquad e_r(l) = e_{r,0} + e_r \cdot l,$$

respectively, where $e_{t,0}$ and $e_{r,0}$ are fixed energy costs spent each time transmitting/receiving a packet, resulting, for example, from transceiver switching, warm-ups, and so forth. Parameters e_t and e_r are the energy spent on transmitting/receiving a single bit. Other sources of energy consumption are not considered. The overall average energy spent to successfully transmit a packet is given by the amount of energy spent per packet multiplied by the average number of trials needed until success. The transmitter spends the following energy per packet:

$$e_t(o + u) + e_r(o),$$

where the first term corresponds to transmitting the data packet and the second term corresponds to a trial to receive an acknowledgment (idle and receive mode are assumed to have the same costs). The expected overall energy spent per packet is thus

$$\frac{e_t(o + u) + e_r(o)}{(1 - p)^{o+u}}$$

and the energy spent per user bit is therefore

$$h(o, u, p) = \frac{e_t(o + u) + e_r(o)}{u \cdot (1 - p)^{o+u}}$$

energy units. This function is plotted in Figure 6.9 for varying bit-error rate p and fixed values for $e_{r,0} = e_{t,0} = 100$, $e_r = e_t = 1$ and $o = 100$ for two different user data sizes ($u = 100$ bits and $u = 500$ bits) and varying bit-error rate p.

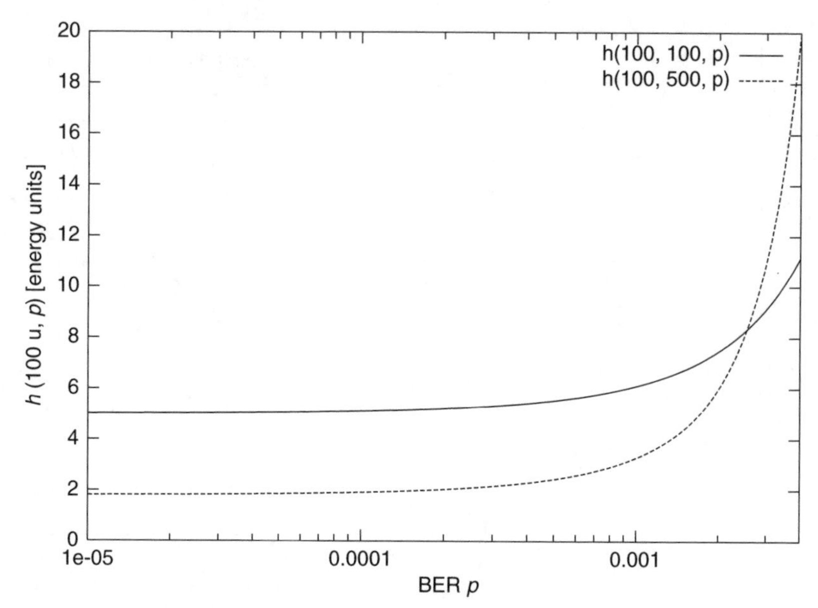

Figure 6.9 Energy per useful bit (in energy units) for user data sizes $u = 100$ and $u = 500$ bits for varying bit error rate p

Two points are remarkable: the relative differences can be significant and it depends on the bit error rate p and on the parameters $e_{r,0}$, $e_{t,0}$, e_r, and e_t which choice of u is more energy efficient for any given p. In general, a lot of energy can be saved if the packet sizes are chosen properly. Experimental results using a WaveLAN radio confirm this [479].

Clearly, a large framing overhead favors large packet sizes to achieve a reasonable energy efficiency per user bit. On the other hand, larger packets are more susceptible to bit errors if no mechanisms like FEC are applied. Therefore, it can be expected that for a given instantaneous bit-error rate p there is an **optimal packet size**, which is easy to obtain analytically (solve $\frac{\partial h(o,u,p)}{\partial u} = 0$ and check for minimality). For the setting given in the example, we show in Figure 6.10 the energy per user bit for varying size of user data u and a fixed bit error rate of $p = 0.001$. This figure has a minimum at $u \approx 463$ bits. However, it is remarkable that if we choose smaller values of u, the energy consumption rises steeply and thus the transmitter should carefully control its packet size. Similar results have been obtained by Sankarasubramaniam et al. [721], who propose to use a fixed, but nearly optimal packet size, arguing that the adaptation to varying channel conditions itself can be a costly process and requires some resource management. This packet size is derived from the technological parameters and an estimate of the bit error probability p.

It must be noted that the assumption of a BSC is rather simplistic and typically far from reality. If the channel errors are dominated by fading, then likely no packet will go through the channel during bad channel states no matter what its size is; in good channel states, larger packets achieve better throughput and energy efficiency. Siew and Goodman [766] derive the optimal packet size with respect to throughput for a Rayleigh fading channel alternating between deep fades, where packets are assumed to be erroneous, and good channel periods, where packets go through unharmed. In order for a packet to be transmitted correctly, it must lie entirely within a good period. The longer a packet, the more likely it will be hit by (parts of) a fade. Again, there is some optimal packet size.

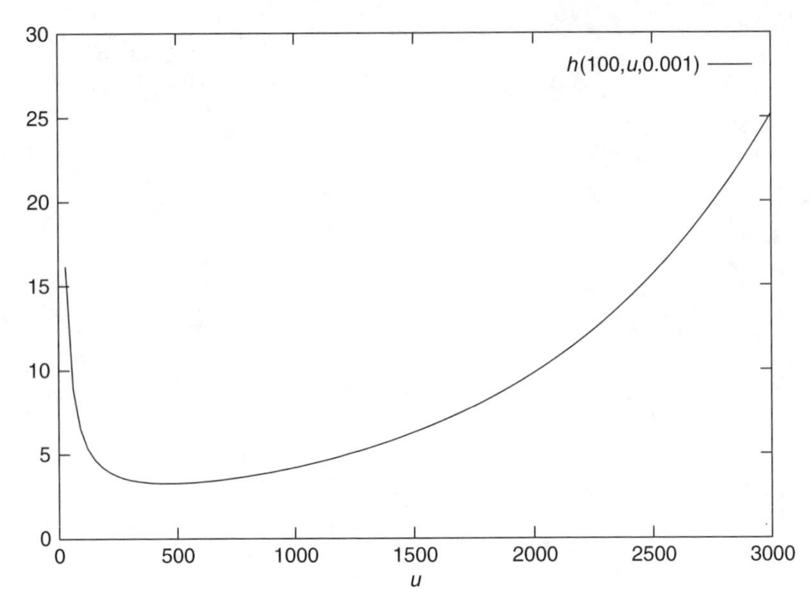

Figure 6.10 Energy per useful bit for fixed bit error rate p and varying user data size u

6.3.1 Adaptive schemes

Wireless channels do not have stationary error conditions but fluctuate both over short timescales (fast fading) and longer timescales (slow fading). For such a time-variable environment, it seems appropriate to let the transmitter and receiver nodes continuously **estimate channel conditions** and to adapt the packet size accordingly.

However, this idea has one drawback. How shall an application or an upper-layer protocol running on a sensor node cope with varying packet sizes? If the link layer signals the variations to the application/upper layers, the latter have to take care of fragmenting their data appropriately for transmission, introducing extra application logic. On the other hand, if the different packet sizes shall be transparent to the application, this **fragmentation and reassembly** work has to be carried out at the link layer, again adding extra complexity. A solution where fragmentation and reassembly is carried out in the link layer is described in references [478, 479]. Their architecture also includes packet classification and FEC coding at appropriate places, which allows to treat packets from different applications differently. Ideally, the MAC layer supports transmission of fragment series or packet trains, like, for example, the S-MAC protocol (Section 5.2.2).

How to estimate the instantaneous channel conditions, specifically the instantaneous bit-error rate p? A standard scheme used, for example, by LETTIERI and SRIVASTAVA [479], lets the receiver collect channel quality information and transmit this back to the transmitter with the acknowledgment frames, allowing the latter to adjust its frame size. The receiver might use different metrics:

- The success information obtained by a pure ARQ scheme (for example number of observed retransmissions per packet) can be used to estimate the instantaneous bit-error rate.
- If FEC is used, the FEC decoder might provide information about the number of corrected errors, which the receiver feeds back to the transmitter [479].
- If the transceiver delivers additional receive information like RSSI, this can also be taken into account.

The first method has the advantage of making the fewest assumptions about the capabilities of the underlying transceiver. We discuss it in more detail.

Under the assumption that the channel can be modeled as a BSC for a certain **observation period** and that the packet length has not changed over this period, either the receiver or the transmitter might estimate the current packet success rate $P(o, u, p)$, for example, by counting the overall number of trials T spent to transmit a number M of packets (if the receiver collects this information it must return it to the transmitter). The number T/M is taken as expectation of a geometric random variable with success parameter $P(o, u, p)$ and this value is used when solving Equation 6.3 for p. Similar approaches are described in references [479] and [332] where packet sizes are picked from a set of a few given packet sizes. MODIANO [570] proposes two different approaches. In the first one, a table is precomputed, relating the number of retransmissions R requested for M packets transmitted with a given packet size k to a new optimal packet size chosen from a set of fixed ones. In the second approach, a Maximum Likelihood Estimation (MLE) method is devised, working as follows.

Example 6.4 (MLE Method for adapting packet sizes) For a BSC with bit-error probability p and packets consisting of o overhead bits and u user bits, the packet error probability is given by:

$$Q(o, u, p) = 1 - (1 - p)^{o+u}$$

and the probability that R out of the M last packets need retransmissions is given by:

$$\Pr\left[R \mid p\right] = \binom{M}{R} \cdot Q(o, u, p)^R \cdot (1 - Q(o, u, p))^{M-R}$$

Assuming that o and u are known and R and M have been observed, the value of p most likely to have caused this situation is needed; it is the value p that maximizes $\Pr\left[R \mid p\right]$. It can be determined by taking the derivative $\mathrm{d}/\mathrm{d}p \Pr\left[R \mid p\right]$ and solving $\mathrm{d}/\mathrm{d}p \Pr\left[R \mid p\right] = 0$ for p. The result is [570]

$$\widehat{p} = 1 - \left(\frac{M - R}{M}\right)^{\frac{1}{u}}$$

and it is then straightforward to find the packet size u' that gives the best energy efficiency.

A problem with this estimator is the occurrence of very high or very low error rates: in case of very high error rates we often have $M = R$ and the estimator gives $\widehat{p} = 1$; conversely, for extremely low error rates we would find $\widehat{p} = 0$, leading to infinite packet lengths, respectively. Therefore, reasonable minimum and maximum packet lengths should be chosen as boundary values.

A critical issue is the choice of the **observation period length**. If the period is too short, there might not be enough data to obtain an accurate estimate of the bit error probability p. On the other hand, if the period is too long, the algorithm can take too much time to adapt to changing channel conditions and suboptimal packet sizes are in use for too long.

All methods to estimate the instantaneous channel conditions use this as a short-term prediction of channel quality. If a node wants to send a new packet after a long period of silence, all available channel information is old and should be considered useless.

6.3.2 Intermediate checksum schemes

Lettieri and Srivastava [479] and Willig [879] discuss a scheme that allows to use long packets without requiring a fragmentation/reassembly scheme. This approach tries to take advantage of cases where only a few bits in a packet are erroneous and it rescues most of the correct bits. Retransmissions are restricted only to those parts of a packet where errors actually occurred. The intermediate checksum scheme can be integrated with the standard ARQ protocols.

The idea is as follows [879], compare Figure 6.11. If there are u bits of user data, protocols with conventional header/data/trailer framing schemes (Figure 6.11(a)) put a header of o bits in front of the user data and a trailer of h bits behind the user data. The header typically carries source and destination address, frame length information and further control information whereas the trailer consists of the frame's checksum. Thus, the overall frame has size $o + s + h$. In case of a retransmission, all of these $o + s + h$ bits would have to be transmitted again. In contrast, in the **intermediate checksum scheme** (Figure 6.11(b)), the u user data bits are partitioned into a number L of **chunks**, with each chunk having a raw size of c bits to which a checksum of h' bits is appended. The last chunk might have less than c bits. A frame is created by appending all the chunks to a frame header of size $o' \geq o$ bits, and the overall frame has size $o' + L \cdot (c + h')$ bits. The frame header in the intermediate checksum scheme additionally includes a separate header checksum and information about the chunk size/number of chunks.

The receiver behaves as follows. If it detects an error in the frame header, the whole frame is discarded and the transmitter has to retransmit the last frame entirely. If the header is correct, the receiver checks each chunk separately and buffers the correct chunks. If all chunks are correct, the receiver delivers the frame to its upper layers and sends a **final acknowledgment**. If some chunks are incorrect, the faulty chunks are indicated to the transmitter with an **incomplete acknowledgment**. The transmitter retransmits *only the faulty chunks*. For example, if the first frame has $L = 8$ chunks and the receiver receives five out of eight, it requests the missing three chunks. This has the beneficial effect that the retransmission frame is much smaller, consumes less energy, produces less interference, is less likely hit by errors, and reaches the receiver with smaller delay. There are also disadvantages: The intermediate checksums impose a higher overhead, which may void any gains in goodput and energy reduction for small bit error rates. However, again the question for the optimal chunk size must be raised. Willig [879] shows that for a BSC with moderate-to-high bit error rates ($\geq 10^{-4}$), the intermediate checksum scheme can achieve higher throughput and needs less frames than the normal framing scheme even if the frame sizes are chosen optimally. The chunk sizes are adapted according to a simple scheme where the transmitter counts the chunks requested for retransmissions and the overall number of chunks transmitted and uses this data for estimating the instantaneous bit-error rate.

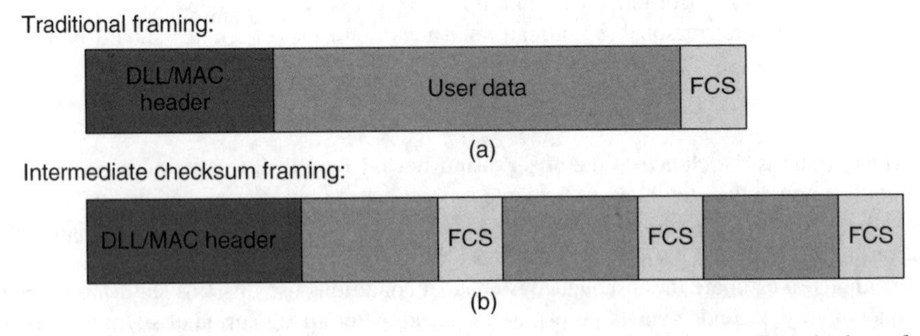

Figure 6.11 "Traditional" framing with header/data/checksum compared with intermediate checksum framing

A disadvantage of such a scheme is that CRC computation is not so easy anymore. In the traditional scheme, the CRC can be obtained by shifting the header and data through a linear-feedback shift register, whose content is simply appended. The intermediate checksum scheme needs more elaborate control of the CRC computation process [479].

6.3.3 Combining packet-size optimization and FEC

Packet-size optimization can also be combined with FEC [332, 721]. SANKARASUBRAMANIAM et al. [721] investigate BCH codes and convolutional codes regarding their energy efficiency. For a BSC-type channel and for a moderate bit-error rate $p = 0.001$, the optimum packet length (user data plus overhead) with respect to energy consumption can be significantly increased when BCH codes are used as compared to the case without error correction. With being able to correct six bits, the optimal packet size would be 2047 bits, and according to the relationship between block size $n = 2047$, user data size k, and error-correction capability $t = 6$ described in Section 6.2.4, a packet can contain at least $k \geq 2047 - 6 \cdot 11 = 1981$ bits of user data and such a frame is successfully delivered with probability $P(2047, 6, 0.001) \approx 0.995$ (compare Equation 6.2). HARA et al. [332] code packets of different sizes with a Reed–Solomon code and the throughput for an alternating bit protocol is evaluated. For very good signal-to-noise ratios (low bit-error rates), the uncoded protocol is better than the coded one; for higher SNRs the coded version is better.

In reference [478], packet-size optimization, an error-control scheme combining FEC (Reed–Solomon codes and rate-1/2 convolutional codes) and Selective Repeat ARQ have been applied to Rayleigh fading channels. The channel is modeled as a two-state Markov chain with states "good" and "bad", with the channel behaving according to a BSC model in either state. In the bad state, the bit-error rate is assumed to be 0.5 while the BER of the good state varies. The transition probabilities between the states are derived from a physical model taking only simple physical parameters into account [858]. The duration of channel fades/bad states and good states depends essentially on the speed of a mobile station (via its Doppler frequency). For example, if the mobile is fast, then many channel fades can occur during a single packet, letting the channel almost look like a BSC; for a very slow mobile, on the other hand, a single fade can span several packets. The effects of packet size variation, FEC, and ARQ have been investigated for three different application data types: (i) for simple datagram data generated from a saturated data source, (ii) for periodic speech data with an additional delay constraint (packets not successfully delivered until their deadline are dropped), and (iii) TCP traffic. The bit-error rate in the good state was either 10^{-2} or 10^{-8}, corresponding to two different channels named *good* and "bad". Some interesting points are:

- For the datagram traffic and small packet sizes (50 bytes), the results are shown in Figure 6.12. Here, a combined FEC and ARQ scheme shows, for both the good and the bad channel, almost the same values on energy per useful bit and expected packet delay. The scheme with ARQ-only offers small delays and small energy consumption for the good channel and large delays as well as large energy consumption are encountered in case of a bad channel. For a packet size of 1500 bytes, ARQ plus FEC offers better delays at higher energy costs for the good channel while for the bad channel both schemes offer the same average delay, with FEC plus ARQ requiring significantly more energy.
- For the speech source, it is shown that for low bit-error rates during the good channel period, the ARQ-only scheme requires lesser energy than FEC plus ARQ and both schemes drop no packets because of deadline violations. For higher bit error rates, the ARQ-only scheme requires much more energy and at the same time drops much more packets than the ARQ plus FEC scheme.

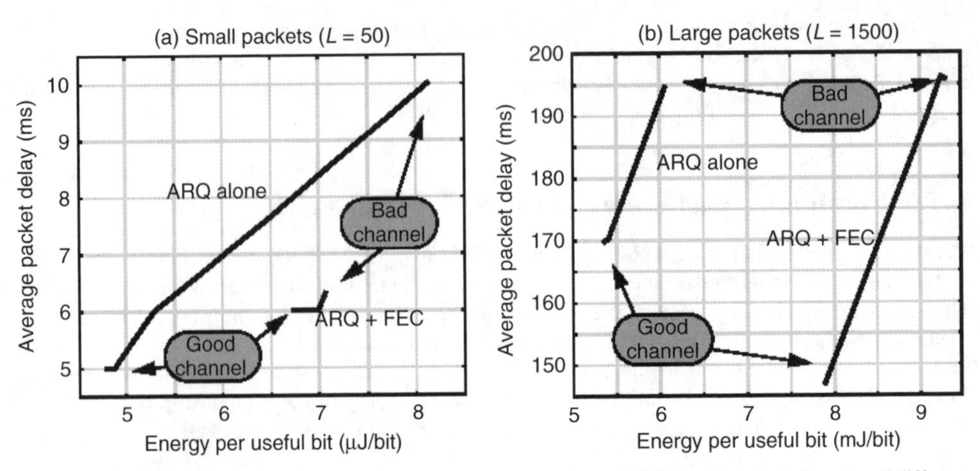

Figure 6.12 Expected delay and energy consumption for datagrams of different sizes L (in bytes) and different good state error rates [478]. Reproduced with kind permission of Springer Science Business Media

6.3.4 Treatment of frame headers

Virtually all frame formats have some **frame header** that contains control information like sequence numbers, addresses, packet-length information, and flow-control information. If such a frame contains only a single checksum covering both header and data part (compare Figure 6.11(a)), upon occurrence of errors, it cannot be said whether the header information or the user data is bogus. It is recommended [478] [879] to either protect the header with its own checksum or even apply FEC coding to it separately in order to recover at least partial information from the user data in case of a correct header. The use of such a feature in intermediate checksum schemes is obvious. In addition, several user data types can tolerate some bit errors, and in wireless sensor networks, several noisy sensor measurements of the same phenomenon can be aggregated to provide higher-quality data.

Another consideration in defining and treating headers is to reduce the header size as much as possible. A good point of attack is address fields. If locally unique addresses are sufficient, then the address field can be significantly smaller than if network-wide or globally unique addresses are required. This is explored in some more detail in Chapter 7.

6.3.5 Framing: summary

The two aspects to framing discussed in this chapter (choice of packet size, intermediate checksum schemes) have a common theme. They try to reduce the amount of information that must be retransmitted in case of errors without imposing too much additional overhead. Indeed, significant energy savings are possible but the additional complexities like obtaining the necessary feedback for adaptive schemes or the necessary fragmentation and reassembly processes impose energy as well as memory costs. The pros and cons of such schemes depend on the application and must be carefully investigated.

6.4 Link management

The upper layers, specifically the routing protocol, need to know about the **available neighbors** and also about the **link quality** of these neighbors. This quality information can be used to make sensible routing decisions by avoiding bad links with a high chance of packet loss. It is important to realize the following:

- The quality of a link is not binary, that is there are more link qualities than just "good" and "bad". One way to characterize the link quality is the probability of loosing a packet over this link.
- The quality of a link is time variable, for example, because of mobility or when some obstacle has moved between the two nodes.
- The quality has to be *estimated*, either actively by sending probe packets and evaluating the responses or passively by overhearing and judging the neighbors' transmissions. Both approaches incur energy costs, which in some cases are already expended by the underlying MAC protocol as part of the neighborhood discovery, for example, in TRAMA (Section 5.4.3) or in S-MAC (Section 5.2.2).

The neighboring nodes and their associated link qualities are often stored in a **neighborhood table** [890, 930], which can be accessed by upper layers. In the case of very dense sensor networks of cheap and memory-constrained nodes, it might happen that there is not enough memory available to store all the possible neighbors. In such a case, it is desirable to select the neighbors with the best link qualities. How can this be done with constrained table space? This problem is discussed in reference [890] and will not be covered here. In the remaining section, we look more closely at the notion of link quality as well as requirements and approaches for its estimation.

The process of neighborhood discovery itself is often an integral part of MAC protocols (for example TRAMA and S-MAC, see Chapter 5) or address allocation protocols (for example [739], see Chapter 7). It must be repeated from time to time to accommodate changing topologies. Neighborhood discovery is not covered in more detail here; a reference dealing with this is Borbash and McGlynn [89].

6.4.1 Link-quality characteristics

Woo et al. [890] and Ganesan et al. [277] express the link quality in terms of packet loss rates. Specifically, Ganesan et al. [277] present measurements of the link quality in a 13×13 grid of 169 motes placed on an open parking place, spaced two feet apart. In this experiment, one node at a time transmits packets and all the others try to receive them. The most important findings of this experiment are as follows:

- For a given transmit power, there is no deterministic relationship between distance and link quality; nodes at the same distance from the transmitter can experience **widely varying packet loss rates**. In extreme cases, nearby neighbors cannot hear a node's packets but far away nodes (occasionally) can.
- The region around a node having a certain packet loss rate does not have the shape of a circle, but is irregularly shaped. This is illustrated in Figure 6.13, showing a contour plot of reception probability when one central node transmits packets. The lines are isolines, that is the points on a line have the same reception probability.
- There is a significant degree of **asymmetric links**. In an asymmetric link, packets sent from node A to node B are received by B with few losses but conversely A receives B's packets with much higher loss probability. The fraction of asymmetric links grows with the distance, taking values between 5 and 15 % of all links.
- The packet loss rate is time variable even when the neighbors in question are stationary. Although the mean loss rate for a given distance over time is more or less fixed, there can be significant short-term variations.

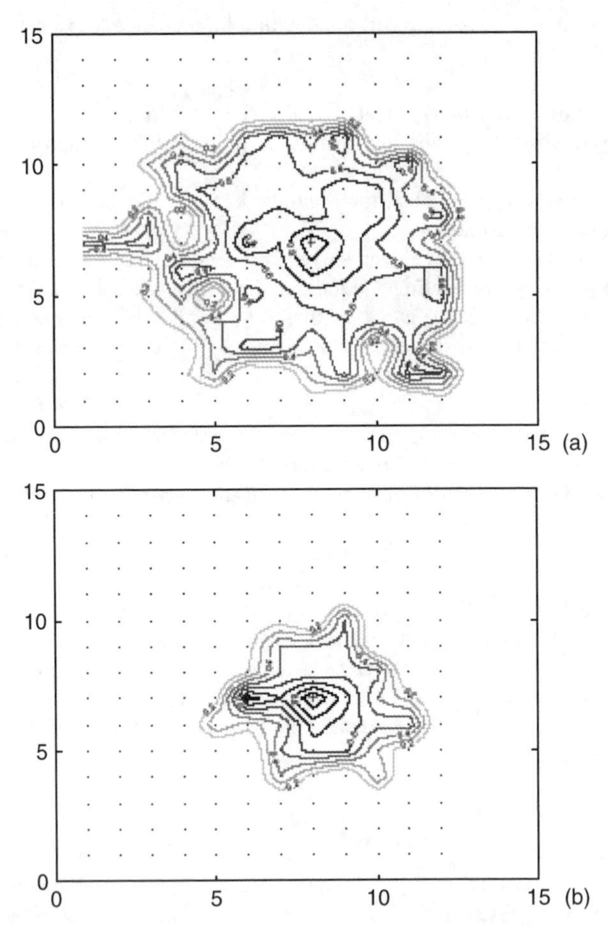

Figure 6.13 Contour plot of packet reception probability for packets generated by a central node and for two different power settings [277, Fig. 4]. Reproduced by permission of Deepak Ganesan, Bhaskar Krishnamachari, Alec Woo, David Culler, Deborah Estrin, and Stephen Wicker

Woo et al. [890] present results of a measurement in a linear network. A number of nodes are arranged in a line with a spacing of two feet. Each node transmits 200 packets and all other nodes try to capture the packets, but only one node transmits at a time. This way there can be many different measurements of the same distance between transmitter and receivers. The quality-vs-distance relationship (with quality measured as packet reception rate) shows three different **regions**, compare Figure 6.14:

- In the **effective region**, receiving nodes have a distance of at most 10 ft to the transmitter and consistently more than 90 % of the packets are received by nodes in this region.
- The **poor region** starts at a distance of 40 ft between transmitter and receivers and the nodes consistently have loss rates well beyond 90 %.
- In the **transitional region** in between, the variance of the experienced loss rates for nodes at the same distance is significant.

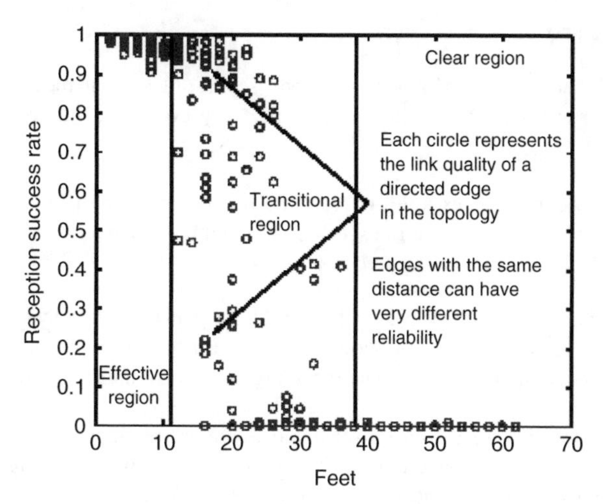

Figure 6.14 Packet reception rates for receivers at varying distance [890, Fig. 1a]. Reproduced by permission of ACM

All these findings occur with different transmit powers, although at different scales. To summarize:

> Link quality should be understood
> in a statistical and time-varying sense.

6.4.2 Link-quality estimation

If a node wants to estimate the quality of a link toward a neighboring node, it has to do so by receiving packets from the neighbor and judging their quality or computing loss rates. Because of the large variability of loss rates for the same distance, it would not be sufficient to derive the packet loss information from the known distance to the next node. The estimation of link qualities in wireless and mobile networks has been considered, for example, in references [417, 889, 890].

There are several desirable properties for an **estimator** [889, 890]:

Precision It should collect enough results and give statistically meaningful results.

Agility It should detect significantly changing link conditions quickly. These can, for example, result from node movements.

Stability The estimation should be immune to short/transient fluctuations in the link quality. This, in general, requires averaging over multiple samples/events to smooth out the transient.

Efficiency It should avoid too much listening for other nodes' transmissions since this can cost precious energy. Furthermore, the computational complexity and the amount of memory needed to keep the link statistics should stay within the bounds permitted by a wireless sensor node. This can be an obstacle for implementation of more advanced filtering algorithms like Wiener or Kalman filters [414].

Clearly, these goals are conflicting. For example, one needs to find good trade-offs between stability and agility [417].

There are passive and active estimators:

Active estimator In an **active estimator**, the node sends out special measurement packets and collects responses from its neighbors. By repeatedly doing so, the necessary loss statistics can be obtained.

Passive estimator In a **passive estimator**, the node overhears the transmissions of its neighbors and estimates the loss rates from observing the neighbor's sequence numbers; packet losses are detected from gaps in the received numbers.

Passive estimation is especially feasible if the neighbors generate sufficient traffic within a certain amount of time, for example, if the nodes generate traffic with a minimum message rate. If transmissions cost much more energy than receiving or idling, then passive estimation may be preferable.

The setting for a passive estimation is illustrated in Figure 6.15. Input events to the estimator are packet arrivals; packet losses have to be inferred indirectly from these using sequence numbers. The estimator should observe several events before producing an estimate to obtain statistically meaningful results.

Suppose that the estimator shall at time $t + T$ produce an estimate of the packet loss rate in the interval $(t, t + T]$, where T is the observation period (for example, $T = 30$ s [889]). Suppose furthermore that somehow the estimator has determined that the last sequence number it could have seen immediately before time t is number seven. Upon arrival of packet number 10 at time τ_0, it knows that two packets are missing in the time interval $(t, \tau_0]$. Upon arrival of packet 15 at time τ_1, it knows that so far only 3 out of 8 packets in the interval $(t, \tau_1]$ have been received. But which value shall be produced at time $t + T$? To produce a reliable estimate, we need the number of packets lost in the last gap. If the packets are generated periodically, this issue seems easy to resolve. But even then, the MAC layer and the application introduce some random jitter, and one can give only a reasonable guess of the gap size.

Assuming that these numbers are available, they can be used in different ways within estimators. Woo and Culler [889] investigate several passive estimators, including exponentially weighted moving average (EWMA), flip-flop estimators that switch between a stable and an agile EWMA estimator if these two deliver vastly different results, pure moving average, time-weighted moving average, and window mean with EWMA (abbreviated WMEWMA). The latter has been found to give the best compromise between stability and agility and works as follows: The estimator produces predictions $\widehat{P_n}$ only at times $t_n = t + n \cdot T$; denote by $\widehat{P_1}, \widehat{P_2}, \widehat{P_3}, \ldots$ the sequence of estimates at these times. The estimator has two tuning parameters, $\alpha \in (0, 1)$ and the observation period $T \in \mathbb{N}$, expressed in durations of fixed-size packets. At an update instant t_n, let r_n be the number of received packets in $(t_{n-1}, t_n]$ and f_n be the number of packets identified as lost during

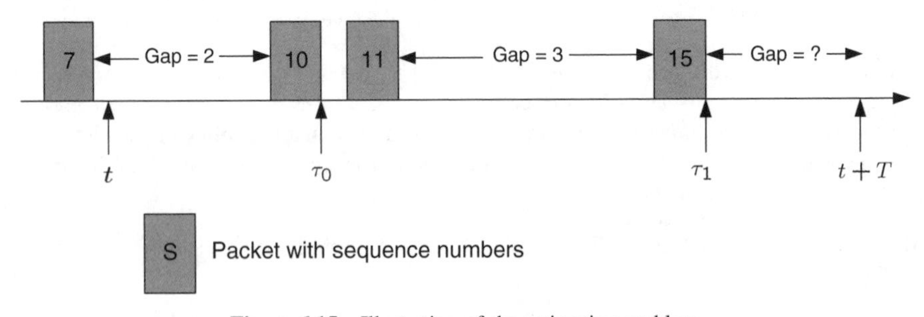

Figure 6.15 Illustration of the estimation problem

$(t_{n-1}, t_n]$. Then,

$$\mu_n = \frac{r_n}{r_n + f_n}$$

$$\widehat{P}_n = \alpha \cdot \widehat{P}_{n-1} + (1 - \alpha) \cdot \mu_n$$

The history needed by this estimator is summarized in $\widehat{P}_{n-1}$, and except the storage for $\widehat{P}_{n-1}$, f_n, and r_n, no further memory is needed.

6.5 Summary

The design of the link layer should strive for energy-efficiency and therefore depends to a good degree on the energy consumption characteristics of the underlying physical layer and on the expected load characteristics. There are several control knobs that can be used to save energy, for example, FEC or packet-size optimization. If the sensor network application is characterized by mostly periodic data transfer, the different adaptation mechanisms are attractive since energy is drained constantly and we can decrease the drain rate. On the other hand, if the network waits for the occurrence of rare events before it starts any transmission, it may make more sense to be not too clever and start with robust settings (FEC, small packets) from the very beginning.

It is an interesting and important task to find suitable models for the energy expenditure of the link layer, possibly considered jointly with the MAC layer and physical layer. Research in this direction has started, for example, in references [932] and [746], but much more work appears to be necessary.

7

Naming and addressing

Objectives of this Chapter

Naming and addressing schemes are used to denote and to find things. In networking, names and addresses often refer to individual nodes as well as to data items stored in them.

Addresses/names are always tied to a representation, which has a certain length when considered as a string of bits. As opposed to other types of networks, representation size is a critical issue in wireless sensor networks, since addresses are present in almost any packet. However, coordination among nodes is needed to assign reasonably short addresses.

A second key aspect is content-based addressing, where not nodes or network interfaces but data is addressed. Content-based addressing can be integrated with data-centric routing and is also a key enabler of in-network processing.

Chapter Outline

Naming and addressing are two fundamental issues in networking. We can say very roughly that **names** are used to denote things (for example, nodes, data, transactions) whereas **addresses** supply the information needed to *find* these things; they help, for example, with routing in a multihop network. This distinction is not sharp; sometimes addresses are used to denote things too – an IP address contains information to both find a node (the network part of an address) and to identify a node – more precisely: a network interface within a node – within a single subnetwork (the host part).

In traditional networks like the Internet or ad hoc networks, frequently independent **nodes** or **stations** as well as the data hosted by these are named and addressed. This is adequate for the

Protocols and Architectures for Wireless Sensor Networks H. Karl and A. Willig

intended use of these networks: They connect many users and let them exchange data or access servers. The range of possible user data types is enormous and the network can support these tasks best by making the weakest assumption about the data – all data is just a pile of bits to be moved from one node to another. In wireless sensor networks, the nodes are not independent but collaborate to solve a given task and to provide the user with an interface to the external world. Therefore, it might be appropriate to shift the view from naming nodes toward naming aspects of the physical world or **naming data**.

The issue of naming and addressing is often tightly integrated with those parts of a protocol stack using them, for example, routing or address resolution protocols. These protocols are *not* the subject of this chapter but treated in subsequent chapters. Here we focus on aspects like address allocation, address representation, and proper use of different addressing/naming schemes in wireless sensor networks.

7.1 Fundamentals

7.1.1 Use of addresses and names in (sensor) networks

In most computer and sensor networks, the following types of names, addresses, and identifiers can be found [458, 718]:

Unique node identifier A **unique node identifier (UID)** is a persistent data item unique for every node. An example of a UID might be a combination of a vendor name, a product name, and a serial number, assigned at manufacturing time. Such a UID may or may not have any function in the protocol stack.

MAC address A **MAC address** is used to distinguish between one-hop neighbors of a node. This is particularly important in wireless sensor networks using contention-based MAC protocols, since by including a MAC addresses into unicast MAC packets a node can determine which packets are *not* destined to it and go into sleep mode while such a packet is in transit. This **overhearing avoidance** is an important method of conserving energy at the MAC layer (Chapter 5).

Network address A **network address** is used to find and denote a node over multiple hops and therefore network addresses are often connected to routing.

Network identifiers In geographically overlapping wireless (sensor) networks of the same type and working in the same frequency band, it is also important to distinguish the networks by means of **network identifiers**. An example is given in reference [45] where medical body area sensor networks for clinical patients in the same room have to be distinguished to prevent confusion of sensor data belonging to different patients.

Resource identifiers A **name** or **resource identifier** is represented in user-understandable terms or in a way that "means something" to the user. For example, upon reading the name `www.xemacs.org`, an experienced user knows that (i) the thing the name refers to is likely a web server and (ii) the user can find information about a great text editor. In contrast, upon looking at the IP address `199.184.165.136`, hardly any user draws either conclusion. Names can refer to nodes, groups of nodes, data items, or similar abstractions.

A single node can have many names and addresses. For example, the WWW server `www.xemacs.org` has the name `www.xemacs.org`, it has the IP address `199.184.165.136` and, assuming that the server is attached to an Ethernet, it has a 48-bit IEEE MAC address. The mapping between user-friendly names like `www.xemacs.org` and the addresses relevant for network operation is carried out by **binding services**. This mapping is also often referred to as **name resolution**. In our example, the domain name service (DNS) provides the mapping from the name to the IP address while the address resolution protocol (ARP) maps the IP address to a MAC address [790].

7.1.2 Address management tasks

We summarize the fundamental tasks of address management, which are independent of the type of addresses:

Address allocation In general, this denotes the assignment of an address to an entity from an address pool.

Address deallocation In on-demand addressing schemes, the address space often has a small-to-moderate size. The node population in sensor networks is intrinsically dynamic, with nodes dying or moving away and new nodes being added to the network. If the addresses of the leaving nodes were not put back into the address pool for **reuse**, the address pool would be exhausted eventually and no addresses could be allocated to new nodes. Address deallocation can be either graceful or abrupt. In **graceful deallocation**, a node explicitly sends out control packets to give up its address. In **abrupt deallocation**, the node disappears or crashes and consequently does not send appropriate control packets, leaving the responsibility to detect and deallocate the node's address to the network. When very large address spaces are used, like for example the IEEE 802.3 MAC addresses of 48-bits length, address deallocation is not an issue.

Address representation A format for representing addresses needs to be negotiated and implemented.

Conflict detection/resolution Address conflicts can occur in networks with distributed assignment of on-demand addresses or in case of **mergers** of so-far distinct networks. If conflicts cannot be tolerated, they must be resolved.

Binding If several addressing layers are used, a mapping between the different layers has to be provided. For example, in IP networks, an IP address has to be mapped to a MAC address using the ARP protocol.

Any address management scheme for sensor and ad hoc networks is occasionally faced with **network partitions** and **network merge** events. Consider for example the network shown in Figure 7.1. If the critical node runs out of energy, the network is split into two partitions that have no connectivity anymore. The critical issue here is address deallocation. Both subnetworks should detect that several nodes cannot be reached anymore and their addresses should be reclaimed. On the other hand, address deallocation and reallocation of reclaimed addresses should not happen too quickly. If for some reason the network remerges, the same address allocation as before the partition event is in place and no address conflicts need to be resolved [936].

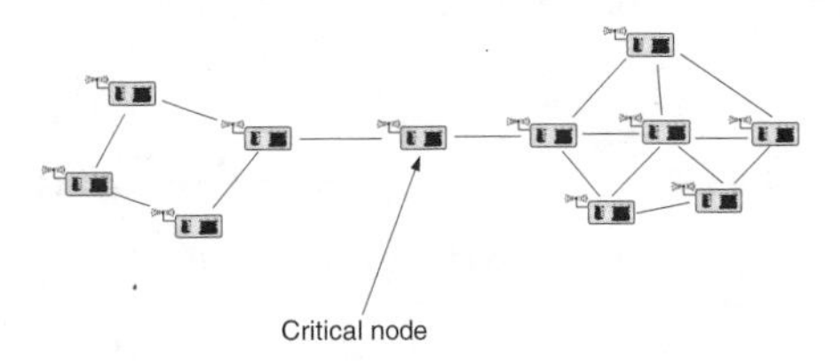

Figure 7.1 Example for network partition

7.1.3 Uniqueness of addresses

We can distinguish the following **uniqueness requirements** for network names and addresses.

Globally unique A **globally unique address** or identifier is supposed to occur at most once all over the world. An example is the 48-bit IEEE MAC addresses used in Ethernet and Token Ring networks.[1] The binary representation of such addresses must be sufficiently large to accommodate all devices worldwide.

Networkwide unique A **networkwide unique address** is supposed to be unique within a given network, but the same address can be used in different networks. By having different networks A and B, we mean that there is no pair of nodes $a \in A$ and $b \in B$ that can communicate.

Locally unique A **locally unique address** might occur several times in the same network, but it should be unique within a suitably defined **neighborhood**. To illustrate this:

- For MAC addresses it is reasonable to require that they are unique only within a two-hop neighborhood. The problem underlying this requirement is displayed in Figure 7.2: node C is a one-hop neighbor of B and a two-hop neighbor of A. If A and C have the same MAC addresses, B would not be able to infer the transmitter of an incoming packet, nor would B be able to direct its packets to a unique intended receiver.
- Another example is given by a sensor network with different sensor types like temperature, humidity, and light sensors. We might require that no two temperature sensors have the same address but a temperature and a humidity sensor may. In this case, the neighborhood is constituted by the sensors of the same type.

7.1.4 Address allocation and assignment

The address assignment can happen **a priori** (e.g. during the manufacturing process or before network deployment) or **on demand**, by using an address assignment protocol. Such an on-demand address assignment protocol can be either **centralized** or **distributed**. In a centralized solution, there is one single authority/node taking care of (parts of) the address pool, whereas in distributed solutions, there is no such exposed node. Instead, potentially all nodes play the same role in address assignment. Address release/deallocation plays an important role when networkwide or locally unique addresses are assigned on demand.

In distributed address assignment, it might not always be possible to guarantee networkwide uniqueness at all times. One can either decide to simply live with some few address conflicts or to detect and **resolve** them. For the latter case, Vaidya [836] introduces the distinction between strong and weak Duplicate Address Detection (DAD):

- In **strong DAD**, it is required that if address x is already assigned to node A at time t_0 and subsequently assigned to node B at time t_1, then this duplicate assignment must be detected latest at time $t_1 + T$ where T is some fixed time bound.

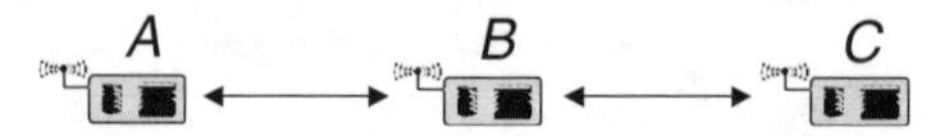

Figure 7.2 Example for network partition

[1] Uniqueness is jeopardized or destroyed by the possibility to reprogram the MAC address of an Ethernet card and by the fact that some manufacturers assign the same address several times [591].

- In **weak DAD**, duplicate addresses are tolerated as long as they do not distort ongoing sessions. For example, if two networks A and B merge and one address x is assigned in both networks, no action should be taken as long as still all packets from nodes of the former network A destined to x reach the node in A with address x and not the node with the same address in the other network.

An example for a centralized address assignment scheme is the DHCP protocol [211] known from the Internet world. However, there are some problems with centralized schemes:

- Centralized solutions do not scale well to sensor networks. The sheer number of nodes creates significant traffic, which is directed to one or a few address servers and the area around these servers becomes a hot spot. This can be circumvented to some extent by employing cluster-based techniques, where disjoint shares of the address space are allocated to clusterheads, which in turn allocate these addresses to their cluster members.
- If the network is partitioned before a new node enters, the central address server might not be reachable.
- The DHCP protocol requires nodes to renew their addresses periodically to detect abrupt deallocations.

7.1.5 Addressing overhead

One of the most important aspects of addresses is the number of bits needed for their representation or their *overhead*. This overhead and consequently the energy needed to transmit addressing information is related to two factors: (i) The frequency with which addresses are used and (ii) the size of their representation. Consider MAC addresses as an example. There are some MAC protocols like TRAMA (Section 5.4.3) or SMACS (Section 5.4.2) that set up dedicated links between two neighboring nodes by assigning conflict-free time slots or frequencies. If such a link is used for a data packet, there is no need to carry address information in these packets since source and destination nodes are implicitly given.

In contrast, in contention-based MAC protocols, at a given time any node might transmit to any other node and addressing information is thus vital to identify source and destination and to achieve overhearing avoidance. Accordingly, the fewer bits spent per address, the better. Let us look into the trade-offs involved here:

- Say we choose a priori assigned, globally unique addresses like in IEEE 802.3/Ethernet. Here, 48 bits are used to accommodate the current and anticipated number of devices. Given that in Ethernet networks often comparably large frames of several hundreds of bytes are used, the six-byte addresses are a negligible overhead, and the a priori assignment eliminates the need for an address assignment protocol. On the other hand, in wireless sensor networks, there will be many small data packets and a single 48-bit address can be larger than the data!
- A networkwide unique address must have a sufficient number of bits to accommodate all the nodes in the network. In a sensor network with 10,000 nodes, an address of 14 bits suffices. However, in order to minimize the number of address bits, the size of the network must be known in advance. Some safety margin in address field width is important if multiple deployment phases can occur, that is, if further nodes are deployed long after the network first became operational, for example, to replace nodes with depleted energy reserves.
- A locally unique address must be unique within a certain neighborhood, which is typically much smaller than the entire network. For example, MAC addresses should be unique within a two-hop neighborhood, which may consist of some dozens of nodes, depending on the node density. Accordingly, the addresses can use fewer bits than would be needed for networkwide unique

addresses. On the other hand, since the exact topology is rarely known in advance, an address assignment protocol is needed. Having fewer address bits is important if the number of data bits in a packet is small too. A scheme for choosing small address representations is discussed in Section 7.4.

An important trade-off found here is that the use of shorter local MAC addresses can save significant energy in case of small data packets, but require overhead in terms of address assignment/negotiation protocols. In sensor networks with mostly stationary nodes, such a protocol needs to run at the beginning and occasionally later on to handle new and deleted nodes. In this case, the gains from saving address bits in every data packet can outweigh the costs of the negotiation protocol. On the other hand, in highly mobile sensor or ad hoc networks, the negotiation protocol would need to run too often to result in energy savings.

7.2 Address and name management in wireless sensor networks

We have seen that MAC addresses are indispensable if the MAC protocol shall employ overhearing avoidance and go into sleep mode as often as possible. However, do MAC addresses need to be globally or networkwide unique? No, since the scope of a MAC protocol is communication between neighboring nodes and it is sufficient that addresses are locally unique within a two-hop neighborhood (see Section 7.1.3). This requirement ensures that no two neighbors of a selected node have the same MAC address. As discussed above, locally unique addresses potentially are short but need an address assignment protocol. These issues are treated in Sections 7.3 and 7.4.[2]

How about higher-layer addresses, specifically network layer addresses, which for traditional routing protocols must be globally or networkwide unique? We will discuss briefly that fulfilling this requirement is a formidable task. We will argue also that this requirement is not really necessary in wireless sensor networks since after all the whole network is *not* a collection of individual nodes belonging to individual users but the nodes *collaborate* to process signals and events from the physical environment. The key argument is that users ultimately are interested in the *data* and not in the individual or groups of nodes delivering them. Taking this a step further, the data can also influence the operation of protocols, which is the essence of data-centric networking. Data-centric or content-based addressing schemes are thus important and will be discussed in Section 7.5.

7.3 Assignment of MAC addresses

In this section, we discuss assignment methods for MAC addresses. As already discussed in Section 7.1.5, the assignment of globally unique MAC addresses is undesirable in sensor networks with mostly small packets.

An a priori assignment of networkwide unique addresses is feasible only if it can be done with reasonable effort. But there is still the problem that the overhead to represent addresses can be considerable although not as large as in globally unique addresses. For example, up to 16,384 nodes can be addressed with 14 bits and this number is much friendlier than 48 bits used for globally unique IEEE addresses.

[2] On this level, wireless sensor networks leverage two important differences to MANETs. In MANETs, the assumed mobility is much higher than in sensor networks. Furthermore, typical packets are significantly longer, rendering efforts to shorten address fields almost meaningless.

Therefore, we concentrate on dynamic and distributed assignment of networkwide and local addresses. The protocols discussed in this section differ in the amount and scope of collaboration with other nodes.

7.3.1 Distributed assignment of networkwide addresses

Let us start with a very simple approach: A node randomly picks an address from a given address range and hopes that this address is unique. For ease of exposition, we assume that this address range is given by the integers between 0 and $2^m - 1$ and an address can thus be represented with m bits. The address space has a size of $n = 2^m$ addresses.

A node chooses its address without any prior information, in which case it is best to use a uniform distribution on the address range since this has maximum entropy. However, this approach is not without problems, as is shown in the following example.

Example 7.1 (Random address assignment) Suppose that we have k nodes and each of these nodes picks uniformly and independently a random address from 0 to $2^m - 1$. What is the probability that these nodes choose a conflict-free assignment? A similar problem is known as the "birthday problem"[3] [255, Chap. II] and can be answered by simple combinatorial arguments. For $k = 1$ this probability is one. For $k = 2$, the second node picks with probability $\frac{n-1}{n}$ an address different from the first node's choice. For $k = 3$, the third node picks with probability $\frac{(n-1)\cdot(n-2)}{n^2}$ an address different from the first two and so forth. Hence, we have the probability $P(n, k)$ to find a conflict-free assignment

$$P(n,k) = 1 \cdot \frac{n-1}{n} \cdot \ldots \cdot \frac{n-k+1}{n} = \frac{1}{n^k} \cdot \frac{n!}{(n-k)!} = \frac{k!}{n^k} \cdot \binom{n}{k},$$

which, by Stirlings approximation ($n! \approx \sqrt{2\pi} \cdot n^{n+1/2} \cdot e^{-n}$ [255, Chap. II]), is approximately given by:

$$P(n,k) \approx e^{-k} \cdot \left(\frac{n}{n-k}\right)^{(n-k)+1/2}.$$

For an address field of $m = 14$ bits size, corresponding to $n = 2^{14} = 16384$ distinct addresses, we show in Figure 7.3 the probability $P(n, k)$ for different values of k. Already, for quite small values of k, the probability of conflicts becomes close to one. For example: for $k = 275$ the conflict probability is already larger than 90 % but only ≈1.7 % of the address space is used!

Therefore, this method of random assignment quickly leads to address conflicts. To preserve networkwide uniqueness, either a conflict- resolution protocol is needed or more clever assignment schemes should be chosen.

Can we do better? A node can try to obtain information about already-allocated addresses by overhearing packets in its vicinity [234] and avoiding these addresses. In many sensor network applications, where nodes transmit their sensor data to a local coordinator aggregating and processing

[3] What is the probability that in a room with n people, no two of them have the same birthday?

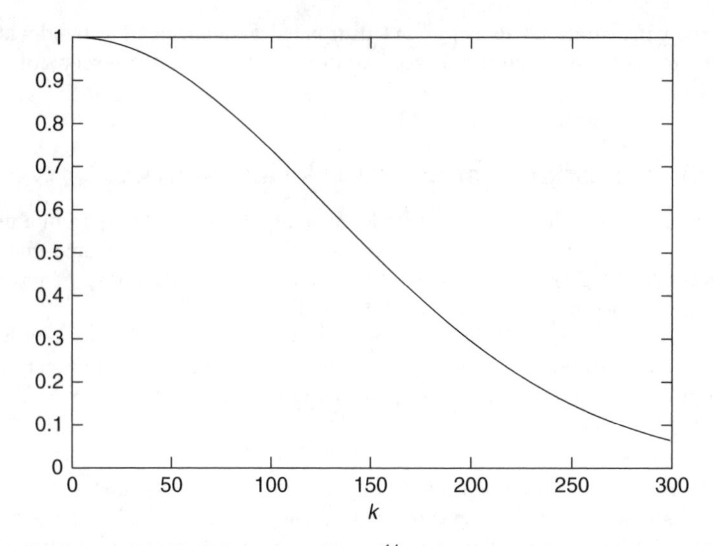

Figure 7.3 "Birthday probability" that k out of $n = 2^{14}$ station pick random addresses without conflicts

the data, overhearing can avoid many conflicts with other local nodes transmitting to the same coordinator.

With random address assignment we are faced with address collisions with high probability. How do we deal with them? The first solution is to simply accept them and do nothing. Other techniques have been investigated in the context of IP address assignment in MANETs:

- PERKINS et al. [636] present an address autoconfiguration protocol suitable for MANETs. A node starts by randomly selecting a temporary address and a proposed fixed address and sends out an **address request** control packet, carrying the chosen temporary and fixed addresses. The temporary address is allocated from a dedicated address pool, being disjoint from the pool of true node addresses. The underlying routing protocol tries to find a path to a node having the same fixed address. If there exists such a node (and a path to it), an **address reply** packet is generated and sent toward the temporary address. Upon receiving this reply, the node knows that the chosen fixed address is allocated and tries another address. If no address reply is received within a certain time, the node repeats the address request packet a configurable number of times to compensate for possibly lost address reply packets. If still no address reply is received after all trials are exhausted, the node accepts the chosen IP address. It is proved by VAIDYA [836] that this protocol breaks down if the delays cannot be bounded, for example, after network partitions. If in sensor networks this scheme is applied to MAC addresses instead of network addresses, then other nodes do not have any routing information, and the address request/reply packets must be flooded into the network. Further problems of this approach are discussed in reference [591].
- NESARGI and PRAKASH [591] regard the address assignment problem as a distributed agreement problem, a well-known problem in distributed systems [526]. A requesting node (the **requester**) contacts a neighboring node already having an address, the **initiator**. The initiator keeps a table of all known address assignments and picks an unused address. The initiator then disseminates the proposed new address to all nodes in the network and collects the answers. All nodes put the proposed address into a list of candidate addresses. If a node finds the address either in the candidate list or in its local list of known assignments, it answers with a reject packet, otherwise it answers with an accept packet. If all known nodes have answered with an accept packet, the initiator assigns the address to the requester and informs all other nodes in the network that the

assignment now is permanent. Otherwise, the initiator picks another address and tries again. This approach is similar to a two-phase commit protocol and clearly produces too much overhead in terms of transmitted packets and buffer space requirements to be feasible in wireless sensor networks.

- A hierarchical address autoconfiguration algorithm for IPv6 addresses intended for MANETs is described in reference [871]. Some nodes in the network become leader nodes and choose a subnet ID randomly. A DAD is executed between leader nodes to guarantee uniqueness of subnet IDs. Other nodes create their addresses from the subnet ID of their leader and a local address (for example, based on the nodes MAC address). In reference [828], a leader is elected for each network partition, assigning addresses to newly arriving nodes. Mergers of networks are detected by introducing separate network identifiers.

The observation that the networkwide uniqueness requirement translates into a distributed consensus problem [591] gives some insight into lower bounds on the complexity and communication overhead involved in this assignment problem. The price in terms of communication overhead is to be paid upon an address assignment trial, for example, when the network must be flooded because the requesting node has no routable address. Alternatively, if a proactive routing protocol is used, the nodes possess tables of used addresses that can be consulted quickly or can infer the presence of duplicate nodes because of receiving bogus routing messages carrying the node's source address [870]. However, on-demand routing protocols are more popular in sensor networks than are proactive protocols (see Chapter 11) because of their overhead.

7.4 Distributed assignment of locally unique addresses

SCHURGERS et al. [734, 739] discuss a protocol that assigns locally unique MAC addresses to nodes, utilizing a localized protocol in which a node communicates only with immediate neighbors. By restricting the uniqueness requirement to a small local neighborhood, fewer bits are needed for address representation than for networkwide or globally unique addresses. The energy reduction due to the saved address bits can be significant if the size of the data payload has the same order of magnitude as the address size.

By using locally unique addresses, we can *reuse* the same address several times in the overall network. In the scheme described in references [734, 739], the second key ingredient besides address reuse is *greediness*, in the sense that a node allocates the numerically lowest nonallocated address. By this approach, lower addresses tend to be used more often than higher addresses and the relative frequencies of how often each address is used in the network becomes uneven, with most of the probability mass at lower addresses. This opportunity is taken by not transmitting addresses directly but by *encoding* them according to the lossless Huffman coding procedure and transmitting the codewords. The mapping from addresses to codes is called the ***codebook*** and must be known a priori to nodes. In Huffman coding [180, Chap. 5], short codes are assigned to frequent addresses while longer codes are assigned to less frequent ones. According to a well-known theorem, Huffman coding is optimal, that is, gives the shortest expected codeword length of all codes [180, Theorem 5.8.1] and is within one bit close to the entropy of the underlying distribution. The relative frequencies of the addresses are a function of the node deployment and have to be known in advance to compute the optimal codebook. Each node needs access to the codebook.

7.4.1 Address assignment algorithm

Before describing an assignment protocol, the exact requirements have to be clarified [739]. The assignment algorithm must take the existence of asymmetric links into account (Section 6.4.1),

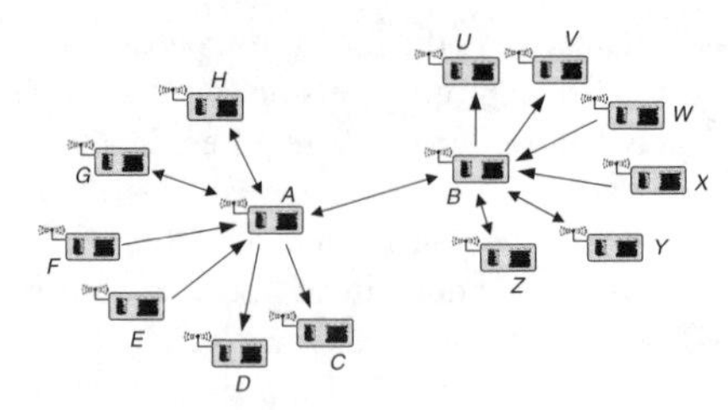

Figure 7.4 Example situation for address assignment with bidirectional, inbound, and outbound neighbors [739]

which are links where a node A may hear some other node B but not vice versa. It is assumed that each node has already discovered its bidirectional and **inbound neighbors** from a previous run of a neighborhood discovery protocol. An inbound neighbor of a node A is a neighbor whose transmissions A can hear but (apparently) not vice versa. Similarly, an **outbound neighbor** of A is a node that receives A's transmissions but not vice versa.

Consider the situation shown in Figure 7.4. We want to assign addresses to nodes A and B. The requirements are:

- Node A and B are assigned different addresses.
- The address of node A is different from those of nodes W, X, Y, and Z since all these nodes can direct packets to B.
- By symmetry, B's address must be different from the addresses of E, F, G, and H.
- Do B's and C's (or D's) addresses have to be different? If they are the same, all packets from node A destined to node B would also be received by C. If, however, the link between A and C is perfectly unidirectional and if the additional constraint is introduced that any node accepts packets only from bidirectional neighbors, there is no need to let C choose another address than B (how shall C tell A about the conflict anyway?). By the same argument, both A and B can also have the same address as W and X, since B would not accept packets from any of these. It is not even required that W and X have different addresses as far as node B is concerned.

The requirements can be summarized as follows [739]. Under the assumption that nodes communicate only with bidirectional neighbors, then for any node A all its bidirectional neighbors must have distinct addresses. Furthermore, the address of any inbound neighbor must be different from the address of all bidirectional neighbors.

The classification of links into bidirectional, inbound, and outbound links is an immediate result of the neighborhood discovery protocol. In order to accommodate the time-varying and stochastic nature of wireless links (Section 6.4.1), this neighborhood discovery should be repeated from time to time.

We briefly describe the distributed address assignment algorithm described by Schurgers et al. [739]. Many details are omitted, for example, those concerning compensation for packet losses or the maintenance of a soft state for stored neighbor addresses to detect leaving nodes [739]. After a node A has performed neighborhood discovery, it starts by broadcasting a HELLO message. The bidirectional and outbound neighbors of A reply to this message by sending INFO messages.[4] With reference to Figure 7.4, let us assume that node B replies with an INFO packet. This packet contains the following information:

- the unique node identifier of B, which can also be a networkwide or globally unique and routable address;

[4] By having the neighbors delay their INFO messages randomly, the amount of collisions is reduced.

- B's MAC address;
- and the MAC addresses of all bidirectional neighbors of B.

Only neighbors that already have a MAC address send INFO messages. By assumption, node A will only hear the INFO messages of its bidirectional neighbors and therefore knows their identity. To see why it is useful that B includes its neighbors as well, remember that A must choose its MAC address distinct from nodes Y and Z.

The time that node A listens for INFO messages is bounded. After this time has elapsed, node A knows the entire one-hop and two-hop neighborhood. Now, two important cases can occur:

- If all of A's one-hop neighbors have different addresses, there is no conflict and A can choose an address such that this address is unique within A's two-hop neighborhood.
- If there is a conflict between A's one-hop neighbors, node A issues a CONFLICT message, indicating the conflicting address. Reception of this message by the conflicting nodes triggers a new address selection round.

7.4.2 Address selection and representation

When node A has successfully executed the assignment algorithm, it can pick an own address. Instead of picking a random address, A selects the lowest possible nonconflicting address. In a sense, address selection is **greedy**. By this approach, lower addresses are preferred and occur more often in the network. Accordingly, lower addresses have a higher relative frequency and we have a *nonuniform address distribution*. SCHURGERS et al. [734] investigate the address distribution under the greedy allocation scheme for the case of uniform node distribution with different node densities.[5] The node density is defined in [734] as the expected number of neighbors of a node. It turns out that the address distribution converges to the uniform distribution as the density increases. On the other hand, for lower densities, the distribution has most of its weight at lower addresses. This is illustrated in Figure 7.5.

Such skewed, nonuniform address assignments are useful because of the following reason. The minimum average number of bits needed to describe the outcome of a random experiment with finite and discrete range is given by the entropy $H(X)$ of the underlying random variable X [27, 180]. The uniform distribution over a given range has maximum entropy, and thus needs, on average, most bits as compared to other more-skewed distributions. An optimum coding scheme for discrete random variables is the Huffman coding scheme, requiring that an average number of bits be within one bit close to the entropy $H(X)$.

If we can determine the relative frequencies/address distribution of MAC addresses under the greedy allocation scheme, the Huffman algorithm gives an optimum codebook for allocating *variable-length codewords* to MAC addresses. This codebook is disseminated to the sensor nodes in advance. The node includes the codewords instead of the true addresses into the packets, on average saving a number of bits.[6] To implement this, the receiver has to spend some extra effort to parse the variable-length addresses and find the remaining packet fields. Fortunately, Huffman codes are prefix-free codes that can be decoded easily.

After the codebook has been constructed, a slight modification is possible, illustrated in Figure 7.6. Suppose that a node allocates an address and that the codeword for the lowest possible address has a width of m bits. The codebook may contain several codewords (say, M) of the same width, and instead of simply choosing the smallest address, one of the M codewords of width

[5] Technically, such a setup corresponds to a Poisson point process, which are briefly explained in Section 13.2.3.

[6] What if the actual address distribution differs from the assumed distribution? The expected loss in coding performance is given by the **relative entropy** or **Kullback–Leibler distance** between the assumed and the actual distributions [180, Chap. 2].

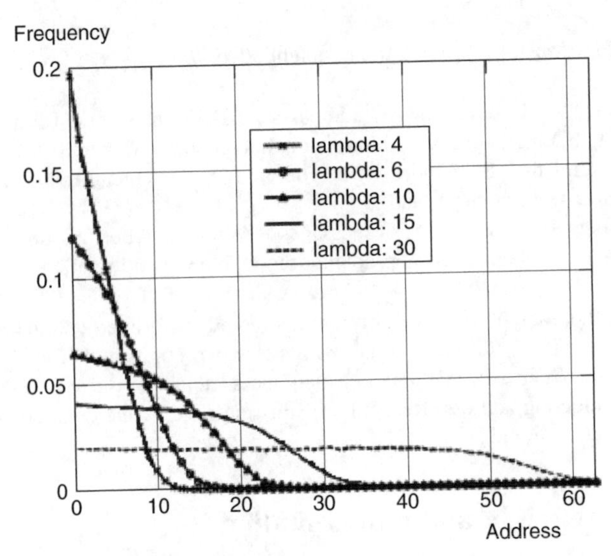

Figure 7.5 Relative frequencies of MAC addresses under the greedy algorithm for different node densities λ [734]. Reproduced by permission of ACM

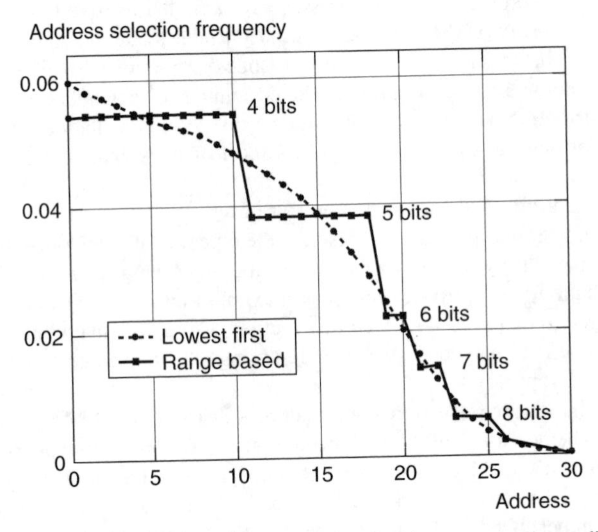

Figure 7.6 Lowest-first versus range-based allocation of MAC addresses under the modified greedy algorithm [739]. Reproduced by permission of IEEE

m is picked at random. The corresponding relative address frequencies are shown in Figure 7.6. The range-based choice reduces the address collision probability and thus decreases the number of CONFLICT messages [734].

A performance evaluation study of this scheme is presented in reference [739]. The assumed node hardware contains an RF Monolithics transceiver, running at 2.4 kb/s and a StrongARM processor at 150 MHz. $N = 500$ nodes are uniformly placed over a rectangular area with a node density of $\lambda = 10$ average neighbors per node. The extra overhead for decoding variable-length

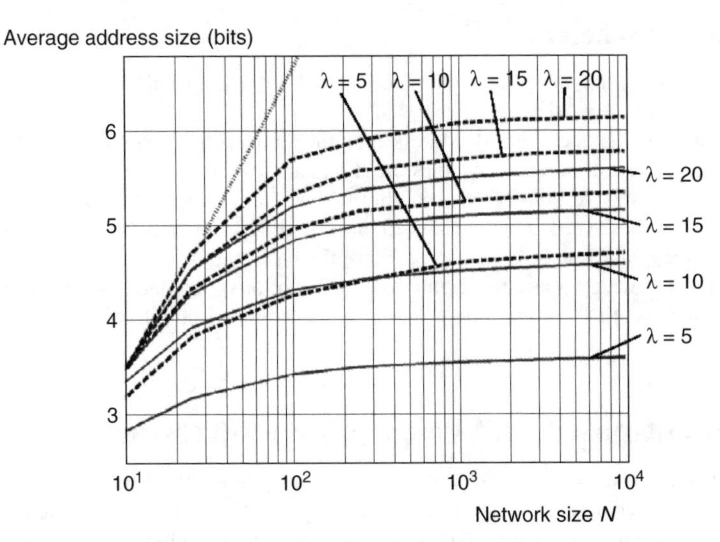

Figure 7.7 Average address size versus number of nodes N in the network for different node densities λ (solid lines represent variable-length scheme, dashed lines fixed-length scheme) [739]. Reproduced by permission of IEEE

addresses is assumed to require 50 processor instructions on average. Energy-wise, transmitting a single bit corresponds to about 120 instructions. The variable-length scheme is compared against two other schemes. The first one uses preassigned networkwide unique addresses with an address representation having fixed and minimum length. The second one ("fixed-length scheme") uses the same address allocation protocol as the variable-length scheme but represents addresses by a fixed number of bits, namely, the minimum number of bits needed to represent the highest address found in the network. The study revealed the following results (Figure 7.7):

- For the same network size N, the variable-length coding scheme (solid lines) has the lowest average address size, typically around one bit better than the fixed-length scheme (dashed lines). For comparison, the light gray curve indicates the number of bits needed for networkwide unique addresses.
- The average address size for the variable-length and fixed-length schemes tends to converge to a fixed value as the network size N increases; therefore, the average address size is dominated by the node density and is almost independent of the actual number of nodes. Only for fewer nodes, edge effects play a role. This is not true for networkwide unique addresses, whose representation size grows logarithmically with the number of nodes.
- The energy saved by having fewer addressing bits with the variable-length/fixed-length scheme pays off already after a few dozens of data packets and amortizes the extra overhead introduced by the address assignment protocol (HELLO, INFO, CONFLICT packets). The point of reference for this statement is a version where addresses are assigned a priori and a 14-bit address field is used. The exact point where the variable-length/fixed-length scheme amortizes depends on the node density.

The average address size given by the codeword-based scheme depends on how well the actual (local) node density fits the density for which the codebook was computed. A general advantage of variable-length addresses over fixed-length ones is the increased flexibility in the size or density of the network, as variable-length codewords can become arbitrarily long, whereas for fixed-length codewords the number of distinguishable one- or two-hop neighbors is naturally bounded.

7.4.3 Further schemes

A scheme that is similar in spirit to the codeword-based scheme but assigns addresses to *links* instead of nodes is presented in reference [454].

We have argued that centralized address assignment is bad. A compromise between purely centralized and purely distributed schemes are assignment schemes that use a clustered approach where a clusterhead is responsible for conflict-free address assignment to its cluster members. Such a scheme is described in reference [76] in the context of a cellular wireless LAN. Here, address assignment takes place in two steps. First, the base stations are assigned addresses that are locally unique (i.e. with respect to neighboring base stations). Second, mobile nodes apply for a fixed-length address at their current base station. This address is unique with respect to the current base station.

7.5 Content-based and geographic addressing

Traditional fixed and ad hoc networks offer services and protocols that allow a number of independent users to exchange data among each other and with the remaining world. On the other hand, in wireless sensor networks, the nodes interact with the physical environment, and they *collaborate*, that is they are not independent of each other. A user of a wireless sensor network ultimately wants to know something about the physical environment the network interacts with, but he typically does not care about the individual sensor nodes. As an example, a user wants to ask "Give me the mean temperature in room C-1.3 in the HPI building" instead of "Obtain the temperature values of sensor nodes 13, 47, 2225, 14592, and 14593 and give me the mean value". It is preferable to allow users to *name the data they are interested in and not the (set of) nodes producing the data.*

In traditional IP-based networks, this requirement corresponds to introducing a naming system *on top of* IP addresses and to introducing appropriate binding services like DNS or other directory services, providing a mapping from names (meaningful to the user) to IP addresses (meaningful for the routing protocol). In sensor networks, however, these levels of indirection can be eliminated and the user-specified attributes can be directly used to find (groups of) nodes. This idea is often referred to as **data-centric addressing** (Section 3.3.4). Such an approach to make the application data meaningful to the operation of network (especially routing) protocols is also a key enabler for in-network processing techniques.

Geographic addressing can be regarded as a special case of content-based addressing. Here, some of the user-specifiable attributes refer to spatial coordinates. Geographic addressing assumes that each node knows its own location with respect to some agreed-upon coordinate system. Thus, locationing techniques (Chapter 9) are essential for working with geographic addresses. On the other hand, geographic addresses can help with routing (Section 11.5). For example, in the directed diffusion protocol [378], location information can help to make the flooding/interest propagation step *directional* and to reduce the number of interest packets significantly.

Both content-based and geographic addressing are no replacements for MAC protocols but can be used, for example, on the network layer to help with routing decisions.

7.5.1 Content-based addressing

Several content- or attribute-based naming systems have appeared in the literature; some examples are references [10, 342]. Before turning to a more general discussion, we present an example naming scheme, developed in the context of wireless sensor networks.

Example 7.2 (A low-level-naming mechanism [342]) In this approach, content-based addressing is integrated with **directed diffusion** routing. In a nutshell, in directed diffusion a sink node issues an **interest message**, specifying a set of attributes to describe the desired data. This message is disseminated into the network. The nodes that can produce sensor data matching the interest are called **source nodes**. A data packet generated by a source node travels through intermediate nodes to the sink. An intermediate node stores the interest along with a (set of) possible upstream neighbors in the **interest cache**. Upon receiving a data packet, the intermediate node searches its cache for an interest matching the data and forwards the data packet to the associated upstream neighbor. Directed diffusion is described in more detail in Chapter 12.

Both the interests and the data packets are represented as sets of Attribute Value Operation (AVO) tuples. The set of attributes is predefined and each attribute possesses a unique, well-known key as well as an understanding of the data type for the corresponding value.[7] The different operators and their meaning are shown in Table 7.1. The IS operator specifies that the corresponding attribute actually has the indicated value and is typically generated by the data source; IS is also called an **actual operator**. All the other operators are called **formal operators**, and they are used to specify the interests against which the actual values generated by the source are matched.

The intermediate nodes use an operation called **one-way match** (shown in Listing 7.1 in a pseudocode notation) to decide which interest a received data packet matches. Basically, for a given interest, the intermediate node goes through all its formal attributes and checks whether (i) the data packet possesses a corresponding attribute at all and (ii) whether the actual value carried in the data packet's attribute matches the formal operator in the interest.

Let us look at an example. Suppose that the user at the sink node wants to know when the temperature in a certain geographic area exceeds a certain threshold. He combines several attributes to an interest message shown in Listing 7.2. In this example, the attribute `type` specifies the kind of sensors to which the interest is directed (here: `temperature` sensors). The next attribute `threshold-from-below` specifies that the sink is interested in cases where a threshold of 20 °C is crossed from below. The area under observation is a square between (0,0) and (20,20) (expressed in meters). If the temperature exceeds this

Table 7.1 Set of operators [342]

Operator name	Meaning
EQ	Matches if actual value is equal to `value`
NE	Matches if actual value is not equal to `value`
LT	Matches if actual value is smaller than `value`
GT	Matches if actual value is greater than `value`
LE	Matches if actual value is smaller or equal to `value`
GE	Matches if actual value is larger or equal to `value`
EQ_ANY	Matches anything, `value` is meaningless
IS	Specifies a literal attribute

[7] For sensor networks, the representation of all AVO tuple components should be as short as possible to reduce the number of bits needed.

threshold, a matching sensor shall report this event every 0.05 s for a duration of 10 s. The final attribute `class` expresses that the present AVO tuple is an interest and not a data message.

Suppose we have different types of sensors, including temperature sensors. Each sensor possesses a set of AVO tuples describing itself; a self description of a temperature sensor could look like the one shown in Listing 7.3. If the interest message shown in Listing 7.2 reaches the temperature sensor, the one-way check reveals that the self-description of the sensor matches the attributes in the interest. Accordingly, the sensor (more precisely: *all* sensors matching this interest and actually willing to serve interests at all) start to observe their environment and generate data messages when the event of interest actually occurs. An example of such a data message is shown in Listing 7.4. As explained above, nonmatching sensors store the interest in their interest cache.

It must be noted that the set of attributes used in the listings are really just examples; in general, the attributes and their values depend on the application.

Listing 7.1: One-way matching Algorithm [342]

```
parameters: attribute sets A and B
   // A corresponds to the interest, B to the data message

foreach attribute a in A where a.op is formal {
  matched = false
  foreach attribute b in B where
          a.key == b.key and b.op is actual {
    if b.val satisfies condition
       expressed by a.key and a.val then {
      matched = true
    }
  }
  if (not matched) then {
    return false
  }
}
return true;    // matching successful!
```

Listing 7.2: Example interest message

```
<type,temperature,EQ>
<threshold-from-below,20,IS>
<x-coordinate,20,LE>
<x-coordinate,0,GE>
<y-coordinate,20,LE>
<y-coordinate,0,GE>
<interval,0.05,IS>
<duration,10,IS>
<class,interest,IS>
```

Listing 7.3: Example temperature sensor

```
<type,temperature,IS>
<x-coordinate,10,IS>
<y-coordinate,10,IS>
```

Listing 7.4: Example data message

```
<type,temperature,IS>
<x-coordinate,10,IS>
<y-coordinate,10,IS>
<temperature,20.01,IS>
<class,data,IS>
```

One important characteristic of any naming system is its **expressiveness**. This notion is difficult to define; it includes both a measure of the number of things that can actually be named by *the naming system* as well as the effort needed to represent/write down the name.[8] A second important characteristic is the *computational effort* needed to check whether an entity (in the preceding example: the data packet with its actual attributes) matches a name (here: an interest). With respect to the example, one can see that (i) the attributes are simple to evaluate and (ii) the one-way match procedure allows only one mode of attribute combination – all of the candidate attributes must *match simultaneously*, corresponding to a logical AND between them. By restricting attribute combination to AND, at most as many attribute checks are needed as there are formal attributes. Therefore, the computational effort is linear in the number of formal attributes. Allowing general Boolean functions over the single attribute matches would potentially require significantly more computation time.

The naming system of HEIDEMANN et al. [342] has been implemented (together with directed diffusion), and two specific application examples are discussed: in-network aggregation and a possibility to specify **nested queries**. Nested queries are one specific way to detect composite events. In a composite event, the occurrence of one event triggers interest in another event. The key point in these experiments is that by using a content-based naming system and by deploying so-called **filters** into the network, which can process the named *data*, a good amount of computation can be moved into the network, close to the event sources. Local aggregation and processing can save significant bandwidth.

An example where the approach to name data is investigated for possible energy savings is given in [453]. The SPIN protocol introduces so-called **metadata**, that is, a description of the data a node can produce. The metadata is assumed to be shorter than the actual data. A source node producing data disseminates an advertisement containing the metadata through the network. Other nodes interested in the advertised data send back a request packet. Upon receiving a request, the source node sends the full data packet.

[8] Example: the programming language Smalltalk has the same computational capabilities as a Turing machine, but Smalltalk programs are easier to write and read for humans. So a Smalltalk program developed within some time T typically has much more functionality than a Turing machine whose development took the same time T. Therefore, we would rate Smalltalk as the more expressive language.

Clearly, there are much more sophisticated naming schemes and accompanying resolution/binding protocols allowing much more sophistication. CARZANIGA and WOLF [123] integrate content-based addressing with publish/subscribe middleware for IP-based wireless and ad hoc networks without specifying a particular naming scheme. In reference [426], a sensor network is envisioned as keeping one single shared XML document and each node is responsible for producing some of its parts. Accordingly, data items correspond to XML elements and are addressed as such.

Naming systems and resolution protocols also play a significant role in service discovery (see, for example, the intentional naming system [10]), peer-to-peer file-sharing networks (for example, content-addressable networks [686]), and the X.500 directory system. However, these systems are not directly applicable to sensor networks because of their different architecture and requirement sets. Energy efficiency played no role in the design of these systems.

7.5.2 Geographic addressing

It is often convenient for users to express their queries to a sensor network in terms of not only the *type* and modality of data they want to receive but also the *region* or *location* from where the data should originate. By the same token, as for content-based addresses, users do not want to separately specify each node belonging to the region of interest but they prefer to **specify a region** and let the network figure out which sensors are appropriate. Furthermore, if the location of a sensor node is known, **geographic routing** schemes can be applied, which are discussed in Section 11.5. There are many different ways to specify a region, for example [589]:

- Specify a single point.
- Specify a circle or a sphere by giving center point and radius.
- Specify a rectangle or a parallelepiped by giving two or three corner points.
- Specify a polygon (two-dimensional) or a polytope (three-dimensional) by giving a list of points.

Geographic addressing [589] requires that a sensor can check whether its position lies within a given area. This test can be complex if general polygons/polytopes are used to specify the region of interest. Such a point-in-polygon test is a standard task in computational geometry. The use of more complex shapes also has the disadvantage that more points and thus more data bits are needed to specify them.

7.6 Summary

In any network including sensor networks, there are different levels of addresses and names, for example MAC addresses and network-layer addresses. MAC addresses are used to distinguish between immediate neighbors and network-layer addresses are used to identify (groups of) nodes in a multihop network. A prime concern regarding naming and addressing in sensor networks is the overhead and energy consumption incurred with these schemes. Energy can, for example, be wasted by having inefficient address representations, by running expensive address assignment and deallocation protocols, or by requiring several binding/address resolution protocols.

At the lowest level are MAC addresses, which in contention-based MAC protocols are indispensable to realize energy savings from overhearing avoidance. If required by the MAC protocol, they need to be present in all data packets and can induce significant overhead, especially if the user data is small. As opposed to schedule-based MAC protocols, MAC addresses are always needed in contention-based MAC protocols since a transmitted packet can potentially have many receivers. A general trade-off exists here between stricter uniqueness requirements and the size of the address. On the other hand, locally unique addresses (which are sufficient for the MAC layer) require an address assignment protocol. However, if most of the sensor nodes are stationary, the savings achieved by (efficient representations of) locally unique addresses pay off quickly. Running address assignment protocols with stricter than local uniqueness requirements (say, networkwide uniqueness) quickly becomes impractical in wireless sensor networks since a distributed consensus problem has to be solved, which inevitably has substantial overhead. Furthermore, the address representation size of locally unique addresses depends only on the network density but not on the absolute number of nodes in the network. This is not the case for networkwide unique addresses.

Networkwide or globally unique addresses are needed by traditional routing protocols to denote and find individual nodes. In wireless sensor networks, however, content-based addressing provides an attractive alternative. A key to their usefulness is the integration of content-based addresses with routing and their ability to enable in-network processing.

8

Time synchronization

Objectives of this Chapter

Time is an important aspect for many applications and protocols found in wireless sensor networks. Nodes can measure time using local clocks, driven by oscillators. Because of random phase shifts and drift rates of oscillators, the local time reading of nodes would start to differ – they loose synchronization – without correction.

The time synchronization problem is a standard problem in distributed systems. In wireless sensor networks, new constraints have to be considered, for example, the energy consumption of the algorithms, the possibly large number of nodes to be synchronized, and the varying precision requirements.

This chapter gives an introduction to the time synchronization problem in general and discusses the specifics of wireless sensor networks. Following this, some of the protocols proposed for sensor networks are discussed in more detail.

Chapter Outline

8.1 Introduction to the time synchronization problem

In this section, we explain why time synchronization is needed and what the exact problems are, followed by a list of features that different time synchronization algorithms might have. We also discuss the particular challenges and constraints for time synchronization algorithms in wireless sensor networks [238, 239].

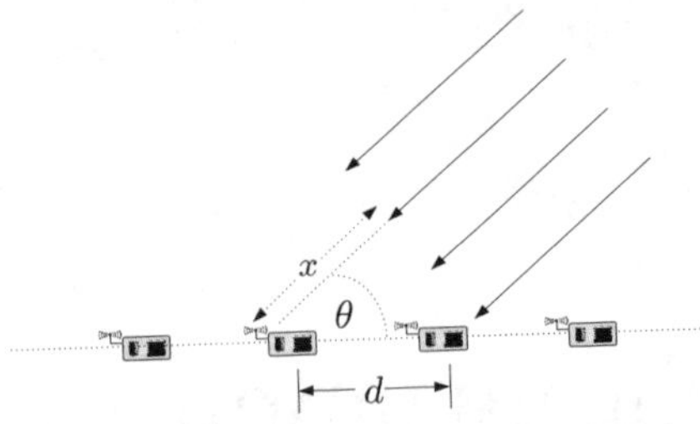

Figure 8.1 Determination of angle of arrival of a distant sound event by an array of acoustic sensors

8.1.1 The need for time synchronization in wireless sensor networks

Time plays an important role in the operation of distributed systems in general and in wireless sensor networks in particular, since these are supposed to observe and interact with physical phenomena.

A simple example shall illustrate the need for accurate timing information (Figure 8.1). An acoustic wavefront generated by a sound source a large distance away impinges onto an array of acoustic sensors and the angle of arrival is to be estimated. Each of the sensors knows its own position exactly and records the time of arrival of the sound event. In the specific setup shown in the figure, the angle θ can be determined when the lengths d and x are known, using the trigonometric relationship $x = d \cdot \sin\theta$, and accordingly $\theta = \arcsin \frac{x}{d}$.[1] The sensor distance d can be derived from the known position of the sensors and the distance x can be derived from the time difference Δ_t between the sensor readings and the known speed of sound $c \approx 330$ m/s, using $x = c \cdot \Delta_t$. Assuming $d = 1$ m and $\Delta_t = 0.001$ s gives $\theta \approx 0.336$ (in radians). If the clocks of the sensors are only within 500 μs accurate, the true time difference can be in the range between 500 and 1500 μs, and thus the estimates for θ can vary between $\theta \approx 0.166$ and $\theta \approx 0.518$. Therefore, a seemingly small error in time synchronization can lead to significantly biased estimates.

There are at least two ways to get a more reliable estimate. The first one (and the one focused on in this chapter) is to keep the sensors clocks as tightly synchronized as possible, using dedicated **time synchronization algorithms**. The second one is to combine the readings of multiple sensors and to "average out" the estimation errors. There are many other applications requiring accurate time synchronization, for example, beamforming [237, 856, 907].

However, not only WSN applications but also many of the networking protocols used in sensor networks need accurate time. Prime examples are MAC protocols based on TDMA or MAC protocols with coordinated wakeup, like the one used in the IEEE 802.15.4 WPAN standard [468]. Sensor nodes running a TDMA protocol need to agree on boundaries of time slots; otherwise their transmissions would overlap and collide.

It is important to note that the time needed in sensor networks should adhere to **physical time**, that is two sensor nodes should have the same idea about the duration of 1 s and additionally a sensor node's second should come as close as possible to 1 s of **real time** or **coordinated universal time (UTC)**.[2] The physical time has to be distinguished from the concept of **logical time** that allows

[1] To keep the example simple, we assume that further sensors provide the information necessary to distinguish between the angles θ and $2\pi - \theta$.

[2] The duration of a second is precisely defined in terms of the number of transitions between the two hyperfine levels of the ground state of Caesium-133 [643] and this information is provided by a number of atomic clocks spread around the world. The operators of these clocks work together to provide the international UTC time scale. The UTC time does not use the atomic times directly but occasionally seconds are inserted or deleted to keep the UTC time in synch with astronomical timescales as they are used for navigation purposes. See the website of your favorite standardization institute, for example, `www.ptb.de` or `www.nist.gov`. Many countries have (sometimes several) local times that are based on UTC; for example, the central European daylight saving time CEDST is just UTC plus 2 h. These local times are transmitted as reference times.

to determine the ordering of events in a distributed system [464, 526] but does not necessarily show any correspondence to real time.

Many time synchronization algorithms have appeared over time; overviews and surveys of algorithms developed for classical distributed systems can be found in references [24, 179, 677]. We restrict the discussion to algorithms relevant for wireless sensor networks, that is, algorithms that care about their energy consumption and which can run in large-scale networks.

8.1.2 Node clocks and the problem of accuracy

Almost all clock devices of sensor nodes and computers share the same common structure [24, 179]: The node possesses an **oscillator** of a specified frequency and a **counter register**, which is incremented in hardware after a certain number of oscillator pulses. The node's software has only access to the value of this register and the time between two increments determines the achievable **time resolution**: events happening between two increments cannot be distinguished from their timestamps.

The value of the **hardware clock** of node i at real time t can be represented as $H_i(t)$. It can be understood as an abstraction of the counter register providing an ever-increasing time value. A common approach to compute a local **software clock** $L_i(t)$ from this value is to apply an affine transformation to the hardware clock:[3]

$$L_i(t) := \theta_i \cdot H_i(t) + \phi_i.$$

ϕ_i is called **phase shift** and θ_i is called **drift rate**. Given that it is often neither possible nor desirable to influence the oscillator or the counter register, one can change the coefficients θ_i and ϕ_i to do **clock adjustment**. Now we can define the notion of **precision** of the clocks within a network, where we distinguish two cases:

External synchronization The nodes $1, 2, \ldots, n$ are said to be **accurate** at time t within a bound δ if

$$|L_i(t) - t| < \delta$$

holds for all nodes $i \in \{1, 2, \ldots, n\}$.

Internal synchronization The nodes $1, 2, \ldots, n$ are said to **agree** on the time with a bound of δ if

$$|L_i(t) - L_j(t)| < \delta$$

holds for all $i, j \in \{1, 2, \ldots, n\}$.

To achieve external synchronization, a reliable source of real time/UTC time must be available, for example, a GPS receiver [401]. Clearly, if nodes $1, 2, \ldots, n$ are externally synchronized with bound δ, they are also internally synchronized with bound 2δ.

There are three problems:

- Nodes are switched on at different and essentially random times, and therefore, without correction, their initial **phases** ϕ_i are random too.

[3] We adopt the convention to write real times with small letters (for example, t, t') and values of local clocks with capital letters. For simplicity, we do not consider overruns of the counter register.

- Oscillators often have a priori a slight random deviation from their nominal frequency, called **drift** and sometimes **clock skew**. This can be due to impure crystals but oscillators also depend on several environmental conditions like pressure, temperature, and so on, which in a deployed sensor network might well differ from laboratory specifications. The clock drift is often expressed in **parts per million (ppm)** and gives the number of additional or missing oscillations a clock makes in the amount of time needed for one million oscillations at the nominal rate. In general, cheaper oscillators – like those used in designs for cheap sensor nodes – have larger drifts with higher probability. In several publications from the field of wireless sensor networks, clock drifts in the range between 1 and 100 ppm are assumed [236, 839]. For Berkeley motes, the datasheets specify a maximum drift rate of 40 ppm. A deviation of 1 ppm amounts to 1 s error every $\approx$11.6 days, a deviation of 100 ppm to 1 s every $\approx$2.78 h.
- The oscillator frequency is time variable. There are short-term variations – caused by temperature changes, variations of electric supply voltage, air pressure, and so on – as well as long-term variations due to oscillator aging, see VIG [843] and `http://www.ieee-uffc.org/fcmain.asp?view=review` for detailed explanations. It is often safe to assume that the oscillator frequency is reasonably stable over times in the range of minutes to tens of minutes. On the other hand, this also implies that time synchronization algorithms should resynchronize once every few minutes to keep track of changing frequencies.

This implies that even if two nodes have the same type of oscillator and are started at the same time with identical logical clocks, the difference $|L_i(t) - L_j(t)|$ can become arbitrarily large as t increases. Therefore, a time synchronization protocol is needed.

How often must such a time synchronization protocol run? Suppose that a node adjusts only its **phase shift** ϕ_i as a result of a synchronization round. If the node's oscillator drift rate is constant and known to be x ppm and if the desired accuracy is δ s, then after at most $\frac{\delta}{x \cdot 10^{-6}}$ s the accuracy constraint is violated. For $x = 20$ ppm and an accuracy of 1 ms, a **resynchronization** is needed every 50 s. More advanced schemes try to estimate and correct not only the phase shift ϕ_i but also the current drift θ_i, hopefully prolonging the periods before resynchronization is required. Again, a one-time synchronization is not useful, since the drift rate is time varying. It is, however, often possible to bound the maximum drift rate, that is there is a $\rho_i > 0$ such that

$$\frac{1}{1+\rho_i} \leq \frac{\mathrm{d}}{\mathrm{d}t} H_i(t) \leq 1 + \rho_i \tag{8.1}$$

and this can be used to find a conservative resynchronization frequency.

8.1.3 Properties and structure of time synchronization algorithms

Time synchronization protocols can be classified according to certain criteria:

Physical time versus logical time In wireless sensor networks, applications and protocols mostly require physical time.

External versus internal synchronization Algorithms may or may not require time synchronization with external timescales like UTC.

Global versus local algorithms A global algorithm attempts to keep *all* nodes of a sensor network (partition) synchronized. The scope of local algorithms is often restricted to some geographical neighborhood of an interesting event. In global algorithms, nodes are therefore required to keep synchronized with not only single-hop neighbors but also with distant nodes (multihop). Clearly, an algorithm giving global synchronization also gives local synchronization.

Absolute versus relative time Many applications like the simple example presented in Section 8.1.1 need only accurate *time differences* and it would be sufficient to estimate the drift instead of phase offset. However, absolute synchronization is the more general case as it includes relative synchronization as a special case.

Hardware- versus software-based algorithms Some algorithms require dedicated hardware like GPS receivers or dedicated communication equipment while software-based algorithms use plain message passing, using the same channels as for normal data packets.

A priori versus a posteriori synchronization In a priori algorithms, the time synchronization protocol runs all the time, even when there is no external event to observe. In a posteriori synchronization (also called **post-facto synchronization** [235]), the synchronization process is triggered by an external event.

Deterministic versus stochastic precision bounds Some algorithms can (under certain conditions) guarantee absolute upper bounds on the synchronization error between nodes or with respect to external time. Other algorithms can only give stochastic bounds in the sense that the synchronization error is with some probability smaller than a prescribed bound.

Local clock update discipline How shall a node update its local clock parameters ϕ_i and θ_i? An often-found requirement is that backward jumps in time should be avoided, that is for $t < t'$ it shall not happen that $L_i(t) > L_i(t')$ after an adjustment.[4] An additional requirement might be to avoid sudden jumps, that is the difference $L_i(t') - L_i(t)$ for times t immediately before and t' immediately after readjustment should be small.

The most important **performance metrics** of time synchronization algorithms are the following:

Precision For deterministic algorithms, the maximum synchronization error between a node and real time or between two nodes is interesting; for stochastic algorithms, the mean error, the error variance, and certain quantiles are relevant.

Energy costs The energy costs of a time synchronization protocol depend on several factors: the number of packets exchanged in one round of the algorithm, the amount of computation needed to process the packets, and the required resynchronization frequency.

Memory requirements To estimate drift rates, a history of previous time synchronization packets is needed. In general, a longer history allows for more accurate estimates at the cost of increased memory consumption; compare Section 8.2.2.

Fault tolerance How well can the algorithm cope with failing nodes, with error-prone and time-variable communication links, or even with network partitions? Can the algorithm handle mobility?

It is useful to decompose time synchronization protocols for wireless sensor networks into four conceptual building blocks, the first three of which are already identified in reference [24]:

- The **resynchronization event detection block** identifies the points in time where resynchronization is triggered. In most protocols, resynchronizations are triggered periodically with a period depending on the maximum drift rate. A single resynchronization process is called a **round**. If rounds can overlap in time, sequence numbers are needed to distinguish them and to let a node ignore all but the newest resynchronization rounds.

[4] Users of `make` tools tend to have strong opinions on this...

- The **remote clock estimation block** acquires clock values from remote nodes/remote clocks. There are two common variants. First, in the **time transmission technique**, a node i sends its local clock $L_i(t)$ at time t to a neighboring node j, which receives it at local time $L_j(t')$ (with $t' > t$). Basically, node j assumes $t \approx t'$ and uses $L_i(t)$ as estimation for the time $L_i(t')$. This estimation can be made more precise by removing known factors from the difference $t' - t$, for example the time that i packet occupies the channel and the propagation delay. Second, in the **remote clock reading** technique, a node j sends a request message to another node i, which answers with a response packet. Node j estimates i's clock from the round-trip time of the message and the known packet transmission times. Finally, node j may inform node i about the outcome. This technique is discussed in more detail below.
- The **clock correction block** computes adjustments of the local clock based on the results of the remote clock estimation block.
- The **synchronization mesh setup block** determines which nodes synchronize with each other in a multihop network. In fully connected networks, this block is trivial.

8.1.4 Time synchronization in wireless sensor networks

In wireless sensor networks, there are some specifics that influence the requirements and design of time synchronization algorithms:

- An algorithm must scale to large multihop networks of unreliable and severely energy-constrained nodes. The scalability requirement refers to both the number of nodes as well as to the average node degree/node density.
- The precision requirements can be quite diverse, ranging from microseconds to seconds.
- The use of extra hardware only for time synchronization purposes is mostly ruled out because of the extra cost and energy penalties incurred by dedicated circuitry.
- The degree of mobility is low. An important consequence is that a node can reach its neighbors at any time, whereas in networks with high degree of mobility, intermittent connectivity and sporadic communication dominates (there are some publications explicitly targeting MANETs [84, 695]).
- There are mostly no fixed upper bounds for packet delivery delay, owing to the MAC protocol, packet errors, and retransmissions.
- The propagation delay between neighboring nodes is negligible. A distance of 30 m needs 10^{-7} s for speed of light $c \approx 300.000.000$ m/s.
- Manual configuration of single nodes is not an option. Some protocols require this, for example, the network time protocol (NTP) [553, 554, 556], where each node must be configured with a list of time servers.
- It will turn out that the accuracy of time synchronization algorithms critically depends on the delay between the reception of the last bit of a packet and the time when it is timestamped. Optimally, timestamping is done in lowest layers, for example, the MAC layer. This feature is much easier to implement in sensor nodes with open firmware and software than it would be using commodity hardware like commercial IEEE 802.11 network cards.

Many of the traditional time synchronization protocols try to keep the nodes synchronized all the time, which is reasonable when there are no energy constraints and the topology is sufficiently stable. Accordingly, energy must be spent all the time for running time synchronization protocols. For several sensor network applications this is unnecessary, for example, when the main task of a network is to monitor the environment for rare events like forest fires. With **post-facto synchronization** [235] (or a posteriori synchronization), a time synchronization on demand can be achieved. Here, nodes are unsynchronized most of the time. When an interesting external event is observed at time t_0, a node i stores its local timestamp $L_i(t_0)$ and triggers the synchronization protocol, which for example, provides global synchronization with UTC time. After the protocol has finished at some later time t_1, node i has learned about its relative offset Δ to UTC time, that is, $t_1 = L_i(t_1) + \Delta$. Node i can use this information to relate the past event at t_0 also to UTC time. After node i has delivered the information about the event, it can go into sleep mode again, dropping synchronization. In a nutshell, post-facto synchronization is synchronization on demand and for a short time, to report about an important event.

Before discussing some of the proposals for time synchronization protocols suitable for sensor networks, let us briefly discuss some of the "obvious" solutions and why they do not fit.

- *Equip each node with a GPS receiver*: GPS receivers still cost some few dollars, need a separate antenna, need energy continuously to keep in synch (acquiring initial synchronization takes minutes!), and have form factors not compatible with the idea of very small sensor nodes [401]. Furthermore, to be useful, a GPS receiver needs a line of sight to at least four of the GPS satellites, which is not always achievable in hilly terrains, forests, or in indoor applications. One application of GPS in a sensor network for wildlife tracking is reported by JUANG et al. [388].
- *Equip each node with some receiver for UTC signals*:[5] the same considerations apply as for a GPS receiver.
- *Let some nodes at the edge of the sensor network send strong timing signals*: Such a solution can be used indoors and in flat terrain but requires a separate frequency and thus a separate transceiver on each node to let the time server not distort ongoing transmissions.

In the following sections, we present several proposals for time synchronization protocols in wireless sensor networks.

8.2 Protocols based on sender/receiver synchronization

In this kind of protocols, one node, called the *receiver*, exchanges data packets with another node, called the *sender*, to let the receiver synchronize to the sender's clock. One of the classic protocols for this is the network time protocol (NTP), widely used in the Internet [553, 554, 556]. In general, sender/receiver based protocols require bidirectional links between neighboring nodes.

8.2.1 Lightweight time synchronization protocol (LTS)

The lightweight time synchronization (LTS) protocol presented by VAN GREUNEN and RABAEY [839] attempts to synchronize the clocks of a sensor network to the clocks held by certain reference nodes, which, for example, may have GPS receivers. The protocol has control knobs that allow to trade off energy expenditure and achievable accuracy, and it gives stochastic precision bounds under certain assumptions about the underlying hardware and operating system. LTS makes no restrictions with respect to the local clock update discipline and it does not try to estimate actual drift rates.

LTS subdivides time synchronization into two building blocks:

- A pair-wise synchronization protocol synchronizes two neighboring nodes.
- To keep all nodes or the set of interesting nodes synchronized to a common reference, a **spanning tree** from the reference node to all nodes is constructed. If the single-hop synchronization errors are independent and identically distributed and have mean zero, the leaf nodes of the tree also

[5] For example, in Germany, a DCF77 receiver can be used for this.

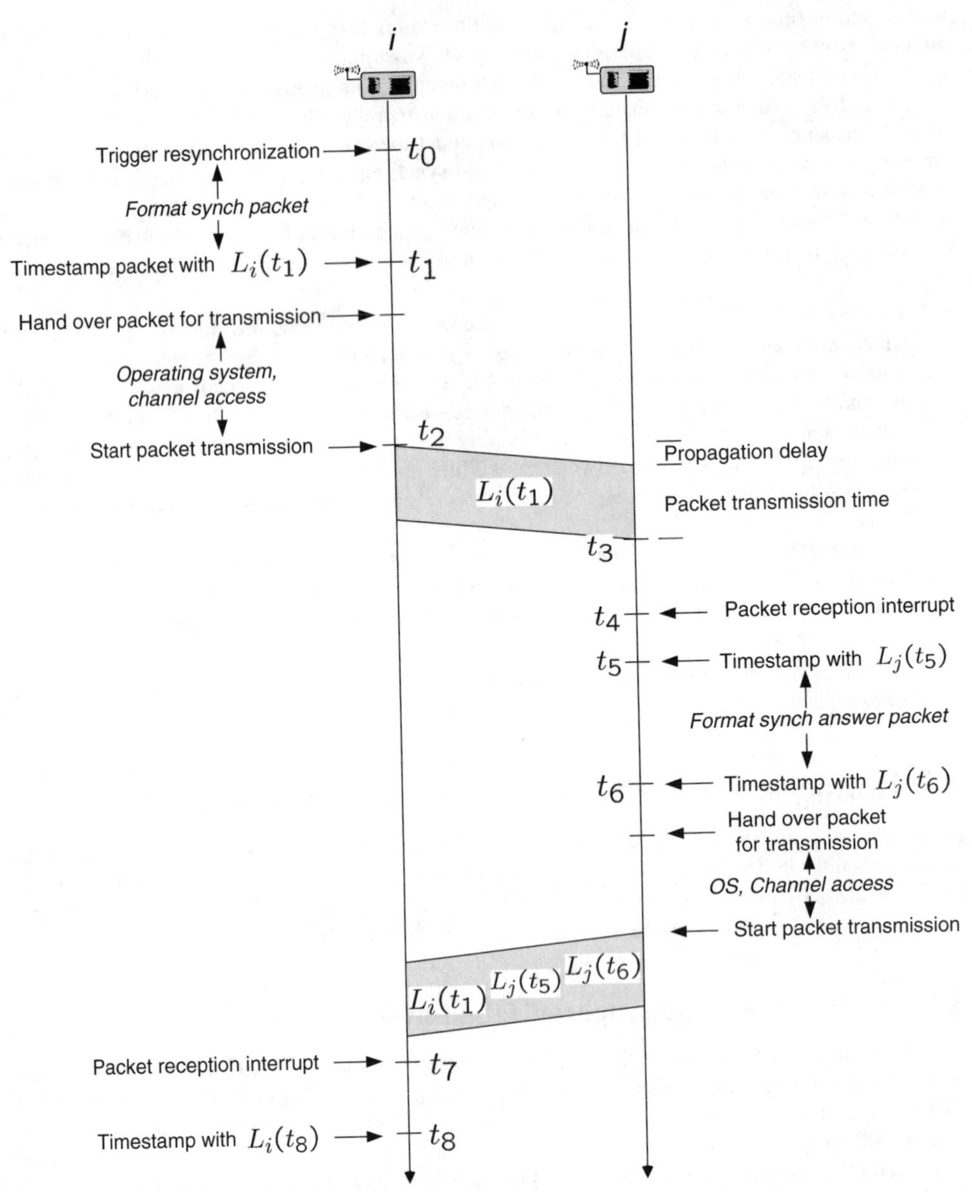

Figure 8.2 Partial sketch of operation in sender/receiver synchronization

have an expected synchronization error of zero but the variance is the sum of the variances along the path from the reference node to the leaf node. Therefore, this variance can be minimized by finding a minimum-height spanning tree.

Pair-wise synchronization

We first explain the pair-wise synchronization protocol (Figure 8.2). The protocol uses a remote clock reading technique. Suppose a node i wants to synchronize its clock to that of a node j.

After the resynchronization is triggered at node i, a synchronization request packet is formatted and timestamped at time t_1 with time $L_i(t_1)$. Node i hands the packet over to the operating system and the protocol stack, where it stays for some time. The medium access delay (Section 5.1.1) can be highly variable and make up for a significant fraction of this time. Often, this delay is a random variable. When node i is sending the first bit at time t_2, node j receives the last bit of the packet at time $t_3 = t_2 + \tau + t_p$, where τ is the propagation delay and t_p is the packet transmission time (packet length divided by bitrate). Some time later (interrupt latency), at time t_4, the packet arrival is signaled to node j's operating system or application through an interrupt and it is timestamped at time t_5 with $L_j(t_5)$. At t_6, node j has formatted its answer packet, timestamps it with $L_j(t_6)$, and hands it over to its operating system and networking stack. This packet includes also the previous timestamps $L_j(t_5)$ and $L_i(t_1)$. Node i receives the packet reception interrupt at time t_7 (which is t_6 plus operating system/networking overhead, medium access delay, propagation delay, packet transmission time, and interrupt latency) and timestamps it at time t_8 with $L_i(t_8)$.[6]

Let us now analyze how node i infers its clock correction. More precisely, node i wants to estimate $O = \Delta(t_1) := L_i(t_1) - L_j(t_1)$. To do this, we make the assumption that there is no drift between the clocks in the time between t_1 and t_8, that is $O = \Delta(t^*)$ for all $t^* \in [t_1, t_8]$, and in fact node i estimates O by estimating $\Delta(t_5)$. From the figure, the timestamp $L_j(t_5)$, which node i's gets back, is generated at some unknown time between t_1 and t_8. However, we can reduce this uncertainty by the following observations:

- There is one propagation delay τ plus one packet transmission time t_p between t_1 and t_5.
- There is another time $\tau + t_p$ between t_5 and t_8 for the response packet. For stationary nodes, we can safely assume that propagation delays are the same in both directions.
- The time between t_5 and t_6 is also known to node i from the difference $L_j(t_6) - L_j(t_5)$.

Therefore, the uncertainty about t_5 can be reduced to the interval $I = [L_i(t_1) + \tau + t_p, L_i(t_8) - \tau - t_p - (L_j(t_6) - L_j(t_5))]$. If we assume that the times spent in the operating system and networking stack, the interrupt latencies as well as the medium access delay are the same in both directions, then node i would conclude that j has generated its timestamp $L_j(t_5)$ at time (from i's point of view)

$$L_i(t_5) = \frac{L_i(t_1) + \tau + t_p + L_i(t_8) - \tau - t_p - (L_j(t_6) - L_j(t_5))}{2}.$$

Therefore,

$$O = \Delta(t_5) = L_i(t_5) - L_j(t_5) = \frac{L_i(t_8) + L_i(t_1) - L_j(t_6) - L_j(t_5)}{2}.$$

Now node i can adjust its local clock by adding the offset O to it. This way, node i synchronizes to j's local time, at the cost of two packets. When the goal is additionally to let node j learn about O, a third packet, sent from i to j and including O, is needed. In this case, the whole synchronization needs three packets.

The maximum synchronization error of this scheme is $|I|/2$ if the times τ and t_p are known with high precision. The actual synchronization error can essentially be attributed to different interrupt latencies at i and j, to different times between getting a receive interrupt and timestamping the packet, and to different channel access times. These uncertainties can be reduced significantly if the requesting node can timestamp its packet as lately as possible, best immediately before transmitting or right after obtaining medium access [272].

[6] Another account of the different variable delays in this process can be found in reference [430].

Several authors including Elson et al. [236] propose to let the receiver timestamp packets as early as possible, for example, in the interrupt routine called upon *packet arrival* from the transceiver. For the case of Berkeley motes, Elson et al. [236] show that for several receivers tasked with timestamping the same packet in their interrupt routines, the pair-wise differences in the actual timestamp generation times are normally distributed with zero mean and a standard deviation of $\sigma = 11.1$ µs (compare Figure 8.6). van Greunen and Rabaey [839] make the additional assumption that the differences in channel access times obey the same distribution. Depending on the degree of correlation between access time difference and timestamping difference, the variance of the sum of these variables is at most four times the variance σ^2 of either component. Accordingly, van Greunen and Rabaey [839] characterize this sum (i.e. the overall error in the above estimation of O) as a normal random variable with variance $4\sigma^2$ and standard deviation 2σ. It is well known for the normal distribution that 99% of all outcomes have a difference of at most 2.3 times the standard deviation from the mean, and therefore under these assumptions, the maximum error after adjusting i's clock to j's is with 99% probability smaller than $2.3 \cdot 2 \cdot \sigma$ µs.

Networkwide synchronization

Given the ability to carry out pair-wise synchronizations, LTS next solves the task to synchronize all nodes of a (connected) sensor network with a reference node. If a specific node i has a distance of h_i hops to the reference node, and if the synchronization error is normally distributed with parameters $\mu = 0$ and $\sigma' = 2\sigma$ at each hop, and if furthermore the hops are independent, the synchronization error of i is also normally distributed with variance $\sigma_i^2 = 4h_i\sigma^2$.[7] On the basis of this observation, LTS aims to construct a **spanning tree** of minimum height and only node pairs along the edges of the tree are synchronized. If the synchronization process along the spanning tree takes a lot of time, the drift between the clocks will introduce additional errors.

Two different variants are proposed, a centralized and a distributed one.

Centralized multihop LTS

The reference node – for example, a node with a GPS receiver or another high-quality time reference – constructs a spanning tree $\mathcal{T}$ and starts synchronization: First the reference node synchronizes with its children in $\mathcal{T}$, then the children with their children, and so forth. Hence, each node must know its children. There are several algorithms available for distributed construction of a spanning tree [31, 526]; for LTS, two specific ones are discussed in reference [839], namely the distributed depth-first search (DDFS) and the Echo algorithms.

The reference node also has to take care of frequent resynchronization to compensate for drift. It is assumed that the reference node knows four parameters: the maximum height h of the spanning tree, the maximum drift ρ such that Equation 8.1 is satisfied for all nodes in the network, the single-hop standard deviation $2 \cdot \sigma$ (discussed above), and the desired accuracy δ. The goal is to always have a synchronization error of leaf nodes smaller than δ with 99% probability.[8] Immediately after resynchronization, a leaf node's accuracy is smaller than $h \cdot 2.3 \cdot 2 \cdot \sigma$ and it is allowed to grow at most to level δ. With maximum drift rate ρ, this growth takes $\frac{\delta - 2 \cdot 2.3 \cdot h \cdot \sigma}{\rho}$ time. The actual choice of the synchronization period should be somewhat smaller to account for the drift occurring during a single resynchronization, possibly harming the initial accuracy $2.3 \cdot 2 \cdot h \cdot \sigma$.

A critical issue is the communication costs. A single pair-wise synchronization costs three packets, and synchronizing a network of n nodes therefore costs on the order of $3n$ packets, not taking channel errors or collisions into consideration. Additionally, significant energy is needed to construct

[7] Please note that *any single* node i is with 99% probability synchronized within an error bound of $2.3 \cdot 2 \cdot \sqrt{h_i} \cdot \sigma$. However, this does *not* imply that all nodes simultaneously are synchronized with 99% within the same bounds. This can be easily seen from the Bonferroni inequations.

[8] We will leave out the 99% qualification henceforth.

the spanning tree, and it is proposed to repeat this construction upon each synchronization round to achieve some fault tolerance.

For reasons of fault tolerance, it is also beneficial to have multiple reference nodes: If one of them fails or if the network becomes partitioned, another one can take over. A leader election protocol is useful to support dynamic reference nodes.

Distributed multihop LTS

The second variant is the **distributed multihop LTS** protocol. No spanning tree is constructed, but each node knows the identities of a number of reference nodes along with suitable routes to them. It is the responsibility of the nodes to initiate resynchronization periodically.

Consider the situation shown in Figure 8.3 and assume that node 1 wants to synchronize with the reference node R. Node 1 issues a synchronization request toward R, which results in a sequence of pair-wise synchronizations: node 4 synchronizes with node R, node 3 synchronizes with node 4, and so forth until node 1 is reached. Two things are noteworthy:

- As a by-product, nodes 2, 3, and 4 also achieve synchronization with node R.
- Given the same accuracy requirement δ and the same drift rate ρ for all nodes, the resynchronization frequency for a node i that is h_i hops away from the reference node is given by $\frac{\delta - 4\cdot 2.3\cdot h_i\cdot\sigma}{\rho}$. Therefore, in the figure, nodes 1 and 6 have the shortest resynchronization period. If these two nodes always request resynchronization with node R, the intermediate nodes 2, 3, 4, and 5 never have to request resynchronization by themselves.

A node should choose the closest reference node to minimize its synchronization error. This way, a minimum weight tree is not constructed explicitly, but it is the responsibility of the routing protocol to find good paths. For certain network setups and routing protocols, the issue of routing/synchronization cycles may arise, which have to be avoided. From the previous example, it is also beneficial if a node can take advantage of ongoing synchronizations. Consider for example node 5 in Figure 8.3. Instead of synchronizing node 5 independently with the reference node R (which would cause nodes 3 and 4 to handle two synchronization requests simultaneously), node 5 can ask all nodes in its neighborhood about ongoing synchronization requests. If there is any, node 5 can wait for some time and then attempt to synchronize with the responder.

Another optimization of LTS is also explained in Figure 8.3, using dashed lines. Suppose again that node 5 wants to synchronize. As explained above, one option would be to let node 5 join an ongoing synchronization request at node 3. On the other hand, it might be necessary to keep nodes 7 and 8 also synchronized with R. To achieve this, node 5 can issue its request through nodes 7 and 8 to R and synchronize the intermediate hops as a by-product. This is called **path diversification**.

The properties of the LTS variants were investigated by simulating 500 randomly distributed nodes in a 120 m (120 m rectangle, each node having a transmit range of 10 m. The single reference

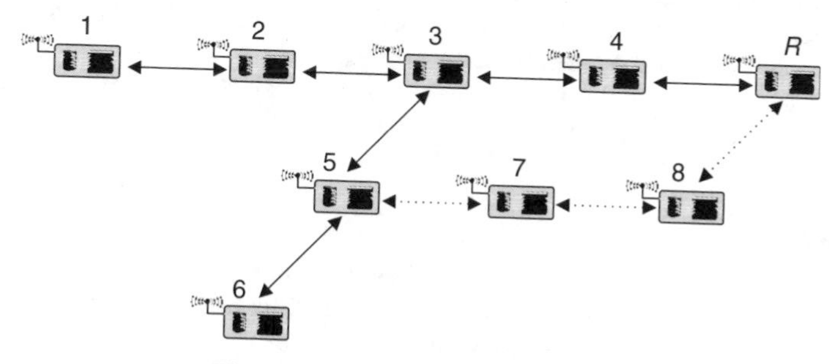

Figure 8.3 Distributed multihop LTS

node is placed at the center. It is shown that the distributed multihop LTS is more costly in terms of exchanged packets when *all* nodes of a network have to be synchronized (between 40 and 100 % overhead to the central algorithm), even when optimizations like path diversification or joining ongoing synchronizations are employed (reducing overhead to 15 to 60 %). However, if only a fraction of nodes has to be synchronized, the distributed algorithms can restrict its overhead to the interesting set and conserve all the other nodes' energy whereas the centralized algorithms always include all nodes and thus have fixed costs that occur whether or not time synchronization is currently requested.

The distributed multihop LTS has the advantage that it is capable of post-facto synchronization. Node i can decide by itself when to synchronize with neighbors closer to the reference node.

8.2.2 How to increase accuracy and estimate drift

We take the opportunity and use the pair-wise synchronization protocol of LTS to explain how the synchronization error between nodes can be decreased and how the drift of node x's clock with respect to a reference node R's clock can be estimated. Both can be transformed into standard estimation problems [414].

Increasing accuracy

Assuming that the drift between x and R is negligible for a certain time and node x wants to estimate the phase offset to R's clock. Node x can increase the accuracy of its estimation by repeating the packet exchange, obtaining multiple estimates $O(t_0), O(t_1), \ldots, O(t_{n-1})$. Let A be the initial phase offset at time t_0 and let us assume that the synchronization errors observed at the different times $t_0, \ldots, t_{n-1}$ are independent. One can therefore model each $O(t_k)$ as:

$$O(t_k) = A + w(t_k)$$

where the $w(t_k)$ are sampled from a white Gaussian noise process with zero mean and standard deviation $\sigma_x = 2 \cdot \sigma$, that is all $w(t_k)$ are independent. Therefore,

$$\widehat{A} = \frac{1}{n} \sum_{k=0}^{n-1} O(t_k)$$

is an unbiased minimum-variance estimator for A, and its variance is $\frac{\sigma_x^2}{n}$, which is linearly decreasing in n. Therefore, the estimator variance scales with n as $O(n^{-1})$.

If the node responding to a request packet is known to send its answer as quickly as possible, another approach making fewer assumptions about the observations and the noise can be used. Referring to Figure 8.2, we can observe the following: When node i makes the k-th observation O_k, it measures the time difference between times $t_{1,k}$ and $t_{8,k}$, that is the times between timestamping the request and the response packets. Clearly, the observation k with the *minimum* difference $t_{8,k} - t_{1,k}$ has the smallest uncertainty and is the most precise one [184].

Drift estimation

Clearly, it is impossible to estimate the drift from just one resynchronization. Therefore, let $O(t_0), O(t_1), \ldots, O(t_{n-1})$ be the estimated offsets obtained by node x for the resynchronizations carried out at times $t_0, t_1, \ldots, t_{n-1}$. We assume that node x's drift ρ_x is constant through the interval $[t_0, t_{n-1}]$.

Let us first consider the case where the pair-wise synchronization protocol of Section 8.2.1 runs repeatedly but node x does not readjust its clock after making an observation. In this case, we can write $O(t_i)$ as follows:

$$O(t_i) = A + \rho_x(t_i - t') + w(t_i)$$

where A is the exact offset between the two clocks at a reference time t' and $w(t_0), w(t_1), \ldots, w(t_{n-1})$ are white Gaussian noise. From the above discussion, the white Gaussian noise assumption is compatible with the assumptions made for LTS. For simplicity, we assume $t' = 0$. Now there are at least two approaches to jointly estimate A and ρ_x:

- We can choose the time difference $t_1 - t_0$ so large that the magnitude of $w(t_0)$ and $w(t_1)$ is very likely negligible as compared to $\rho_x(t_1 - t_0)$. Hence,

$$O(t_1) - O(t_0) = A + \rho_x(t_1 - t') + w(t_1) - A - \rho_x(t_0 - t') - w(t_0) \approx \rho_x(t_1 - t_0)$$

which gives us ρ_x from the observations $O(t_0)$ and $O(t_1)$. In a second step, we can subtract $\rho_x(t_1 - t')$ from $O(t_1)$ to obtain an estimate of A.
- If more estimates are available, we have conceptually the problem of line fitting with noisy data. The probability density function of making the observations $O(t_0), \ldots, O(t_{n-1})$ for assumed known parameters A and ρ_x can be written as:

$$p(O(t_1), \ldots, O(t_{n-1}); A, \rho_x) = \prod_{i=0}^{n-1} \frac{1}{\sqrt{2\pi\sigma_x^2}} \exp\left(-\frac{(O(t_i) - A - \rho_x t_i)^2}{2\sigma_x^2}\right)$$

$$= \left(2\pi\sigma_x^2\right)^{-\frac{n}{2}} \cdot \exp\left(-\frac{1}{2\sigma_x^2} \sum_{i=0}^{n-1} n(O(t_i) - A - \rho_x t_i)^2\right).$$

KAY [414, Chap. 3] shows that the minimum-variance unbiased estimator for this type of estimation problem is given by:

$$\widehat{A} = \frac{2(2n-1)}{n(n+1)} \sum_{k=0}^{n-1} O(t_k) - \frac{6}{n(n+1)} \sum_{k=0}^{n-1} k \cdot O(t_k)$$

$$\widehat{\rho_x} = \frac{-6}{n(n+1)} \sum_{k=0}^{n-1} O(t_k) + \frac{12}{n(n^2-1)} \sum_{k=0}^{n-1} k \cdot O(t_k)$$

(the least-squares estimator is the same). More advanced estimation techniques are needed when $w(t_i)$ is known to be not correlated or non-Gaussian.

Another interesting case is when node x's clock is readjusted upon every resynchronization by the estimated offset $O(t_k)$. In the following, we derive a simple maximum-likelihood estimator for the drift rate ρ_x. According to our data model from above, for the first observation $O(t_0)$, we have:

$$O(t_0) = A + \rho_x(t_0 - t') + w(t_0)$$

and immediately after this observation node x corrects its clock by subtracting $O(t_0)$. The "new" phase offset $A(t_0)$ immediately after clock correction is just $-w(t_0)$. From now on, we do not have to consider the initial offset $A = A(t')$ anymore. For $O(t_1)$ we have:

$$O(t_1) = A(t_0) + \rho_x(t_1 - t_0) + w(t_1) = \rho_x(t_1 - t_0) + w(t_1) - w(t_0)$$

and the new phase offset immediately after correction is $A(t_1) = -w(t_1)$. By repeating this argument, we obtain

$$\begin{aligned} O(t_k) &= A(t_{k-1}) + \rho_x(t_k - t_{k-1}) + w(t_k) \\ &= \rho_x(t_k - t_{k-1}) + w(t_k) - w(t_{k-1}) \\ &= \rho_x(t_k - t_{k-1}) + w_*(t_k) \end{aligned}$$

where the $w_*(t_k)$ are independent Gaussian random variables with zero mean and variance $\sigma_*^2 = 2 \cdot \sigma_x^2$. The log-likelihood of making the observations $O(t_1), \ldots, O(t_{n-1})$ for given ρ_x is given by:

$$\begin{aligned} L(\rho_x) &= \log\left[\prod_{k=1}^{n-1} \frac{1}{\sqrt{2\pi\sigma_*^2}} \exp\left(\frac{1}{2\sigma_*^2}(O(t_k) - \rho_x(t_k - t_{k-1}))^2\right)\right] \\ &= -\frac{n}{2}\log(2\pi\sigma_*^2) + \frac{1}{2\sigma_*^2}\sum_{k=1}^{n-1}(O(t_k) - \rho_x(t_k - t_{k-1}))^2. \end{aligned}$$

Setting the derivative $\frac{\mathrm{d}}{\mathrm{d}\rho_x} L(\rho_x)$ to zero and solving for ρ_x yields

$$\rho_x = \frac{\sum_{k=1}^{n-1} O(t_k) \cdot (t_k - t_{k-1})}{\sum_{k=1}^{n-1} (t_k - t_{k-1})^2}.$$

8.2.3 Timing-sync protocol for sensor networks (TPSN)

The Timing-Sync Protocol for Sensor Networks (TPSN) [271, 272] is another interesting sender/receiver based protocol. Again, we first explain the approach to pair-wise synchronization before turning to the multihop case.

Pair-wise synchronization

The pair-wise synchronization protocol of TPSN has some similarities with LTS. It operates in an asymmetric way: Node i synchronizes to the clock of another node j but not vice versa (Figure 8.4). The operation is as follows:

- Node i initiates resynchronization at time t_0. It formats a **synchronization pulse packet** and hands it over to the operating system and networking stack at time t_1.
- The networking stack executes the MAC protocol and determines a suitable time for transmission of the packet, say t_2. Immediately before transmission, the packet is timestamped with $L_i(t_2)$. By timestamping the packet immediately before transmission and not already when the packet has been formatted in the application layer, two sources of uncertainty are removed: the operating system/networking stack and the medium access delay. The remaining uncertainty is the small time between timestamping the packet and the true start of its transmission. This delay is created, for example, by the need to recompute the packet checksum immediately before sending it.
- After propagation delay and packet transmission time, the last bit arrives at the receiver at time t_3, and some time after this the packet receive interrupt is triggered, say at time t_4. The receiver timestamps the packet already in the interrupt routine with $L_j(t_4)$.
- Node j formats an **acknowledgement packet** and hands it over at time t_5 to the operating system and networking stack. Again, the networking stack executes the MAC protocol and sends the packet at time t_6. Immediately before transmission, the packet is timestamped with $L_j(t_6)$, and the packet carries also the other timestamps $L_i(t_2)$ and $L_j(t_4)$.

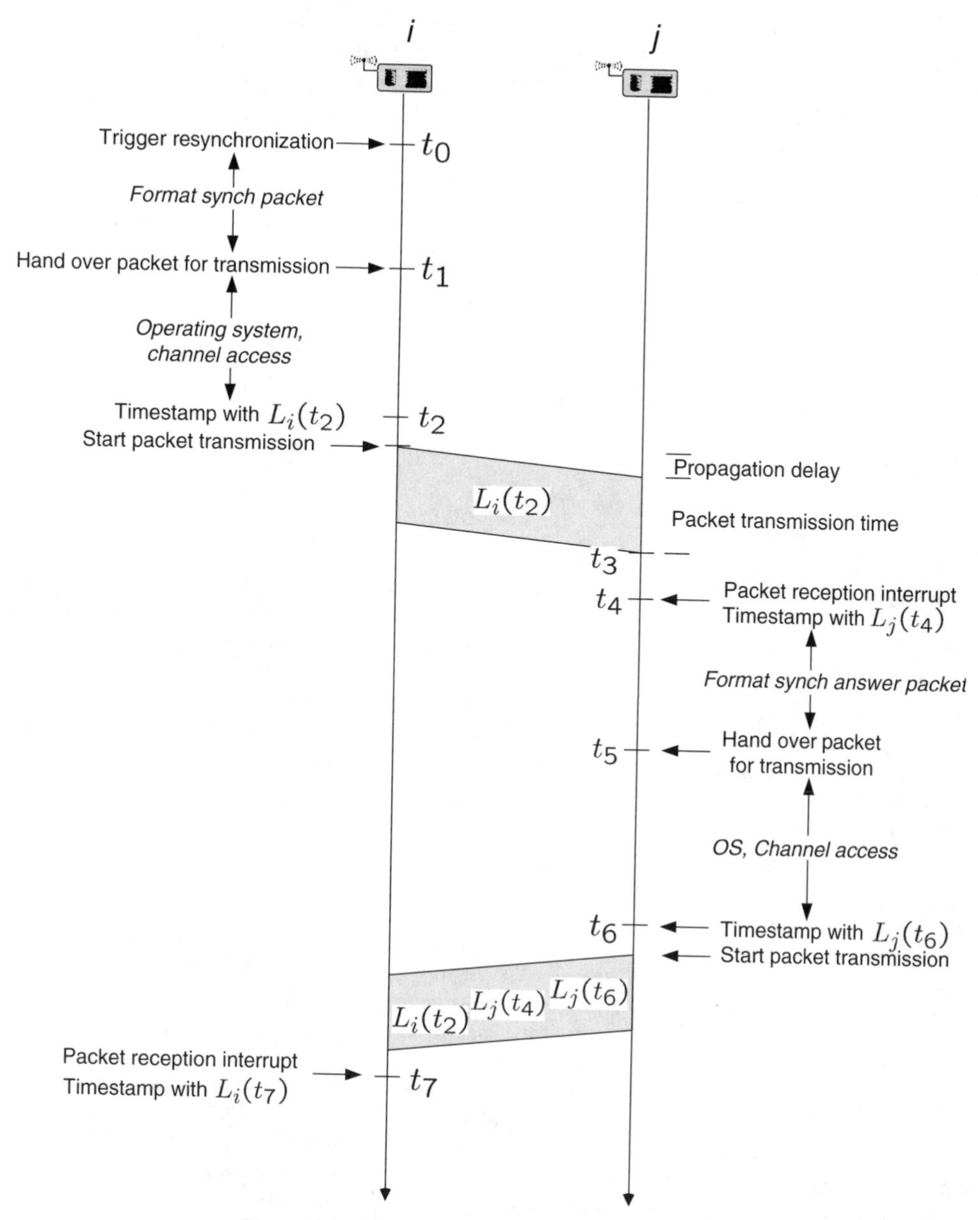

Figure 8.4 Sender/receiver synchronization in TPSN

- Finally, node i timestamps the incoming acknowledgement packet as early as possible with $L_i(t_7)$.

If O is the true offset of node i's clock to node j's clock, we have for the synchronization pulse packet

$$L_j(t_4) = L_i(t_2) + O + \tau + t_p + \delta_{s,i} + \delta_{r,j},$$

where τ is the propagation delay, t_p is the packet transmission time (packet length divided by channel bitrate), $\delta_{s,i}$ is the (small) uncertainty at the transmitter side between timestamping and actual start of transmission, and $\delta_{r,j}$ is the **receiver uncertainty** at j. In the other direction, for the acknowledgement packet, we have

$$L_i(t_7) = L_j(t_6) - O + \tau + t_p + \delta_{s,j} + \delta_{r,i}$$

(assuming that acknowledgement packet and synchronization pulse packet have the same length and the propagation delay is the same in both directions). Now

$$\begin{aligned}
&(L_j(t_4) - L_i(t_2)) - (L_i(t_7) - L_j(t_6)) \\
&= (O + \tau + t_p + \delta_{s,i} + \delta_{r,j}) - (-O + \tau + t_p + \delta_{s,j} + \delta_{r,i}) \\
&= 2 \cdot O + (\delta_{s,i} - \delta_{s,j}) + (\delta_{r,j} - \delta_{r,i}).
\end{aligned}$$

Therefore,

$$O = \frac{(L_j(t_4) - L_i(t_2)) - (L_i(t_7) - L_j(t_6))}{2} - \frac{\delta_{s,i} - \delta_{s,j}}{2} - \frac{\delta_{r,j} - \delta_{r,i}}{2}.$$

The key feature of this approach is that node i timestamps the outgoing packet as lately as possible and node j timestamps the incoming packet as early as possible. This requires support from the MAC layer, which is easier to achieve in sensor nodes than with commodity hardware like IEEE 802.11 network interface cards. For an implementation of this protocol on MICA motes with a 115 kbps transceiver, the average magnitude of the transmitter uncertainty $\delta_{s,i} - \delta_{s,j}$ is 1.15 μs (the expectation is clearly zero) [272]. The average magnitude of the total synchronization error was found to be ≈17 μs.

This protocol allows arbitrary jumps in node i's local clock. The relative performance of TPSN compared with the RBS protocol is discussed in Section 8.3.1.

Networkwide synchronization

The networkwide synchronization algorithm of TPSN essentially builds a **spanning tree** where each node knows its level in the tree and the identity of its parent. The **root node** is assigned level 0 and it is its responsibility to trigger the construction of the tree. All reachable nodes in the network synchronize with the root node. If the root node has access to a precise external timescale like UTC, all nodes therefore synchronize to UTC.

The protocol works as follows. To start the tree construction, the root node sends a *level_discovery* packet containing its level 0. All one-hop neighbors of the root node assign themselves a level of one plus the level indicated in the received *level_discovery* packet and accept the root as their parent. Subsequently, the level 1 nodes send their own *level_discovery* packets of level 1 and so forth. The level 1 nodes choose a random delay to avoid excessive MAC collisions. Once a node has received a *level_discovery* packet, the packet originator is accepted as parent and all subsequent packets are dropped. After a node has found a parent, it periodically resynchronizes to the parent's clock.

A node might fail to receive *level_discovery* packets because of MAC collisions or because it is deployed after initial tree construction. If a node i does not receive any *level_discovery* packet within a certain amount of time, it asks its one-hop neighborhood about an already existing tree by issuing a *level_request* packet. The neighboring nodes answer by sending their own level. Node i collects the answers from some time window and chooses the neighbor with the smallest level as its parent.

The tree maintenance is integrated with resynchronization. To account for drift, a node i must run the pair-wise algorithm with its parent j periodically. If this fails subsequently for a number of times, node i concludes that its parent has moved or passed away. If the level of i is two or larger, it sends a *level_request* packet, collects the answers for some time and assigns itself a new level from the lowest-level answer packet. If i is at level one, it concludes that the root node has died. There are several possibilities to resolve this situation. One of them is to run a leader election protocol among level 1 nodes.

This approach has the following properties:

- The resulting spanning tree is not necessarily a minimal one, since MAC collisions and random delays may lead to a situation where a node receives a *level_discovery* from a higher-level node first. However, there is a trade-off between the synchronization accuracy (longer paths imply larger average error) and the overhead for tree construction. Algorithms for finding minimal spanning trees are more elaborate.
- If two nodes i and j are geographically close together and receive the same level ν *level_discovery* in the tree setup phase, both assign themselves level $\nu + 1$ and try to resend the *level_discovery* packet. One of them wins contention. Since both are close together, their one-hop neighborhoods are almost identical. As a result, all so-far-unsynchronized neighbors accept node i as their parent and create significant resynchronization load for i, whereas node j spends almost no energy because it has no children. To avoid unfairness, the tree construction should be repeated periodically, which in turn creates network load.
- The average magnitude of the synchronization error between a level ν node and the root node increases with ν, but gracefully. For one hop, the average synchronization error is ≈ 17 μs and for five hops ≈ 23 μs.
- It is possible to achieve post-facto synchronization. In this case, no spanning tree is constructed. Consider a scenario in which a node i_0 wants to communicate an event (which happened at time t) to another node i_n over a number of intermediate hops $i_1, i_2, \ldots, i_{n-1}$. Node i_0 sends the packet with its local timestamp $L_{i_0}(t)$ to i_1. Subsequently, node i_1 synchronizes its clock to that of node i_0 and forwards the packet to node i_2, and so forth. Finally, all nodes including node i_n have synchronized to node i_0 and i_n has the packet with timestamp $L_{i_0}(t)$ and can thus decide about the age of the event.

8.3 Protocols based on receiver/receiver synchronization

In sender/receiver based synchronization, the receiver of a timestamped packet synchronizes with the sender of the packet. In receiver/receiver synchronization approaches, multiple receivers of the same timestamped packet synchronize with each other, but not with the sender. We discuss two protocols belonging to this class.

8.3.1 Reference broadcast synchronization (RBS)

The **reference broadcast synchronization (RBS)** protocol [236] consists of two components. In the first one, a set of nodes within a single broadcast domain (i.e. a set of nodes that can hear each other) estimate their peers' clocks. The second component allows to relate timestamps between distant nodes with several broadcast domains between them. We explain each component in turn.

Synchronization in a broadcast domain

The basic idea is as follows. A sender sends periodically a (not necessarily timestamped) packet into a broadcast channel and all receivers timestamp this packet. The receivers exchange their

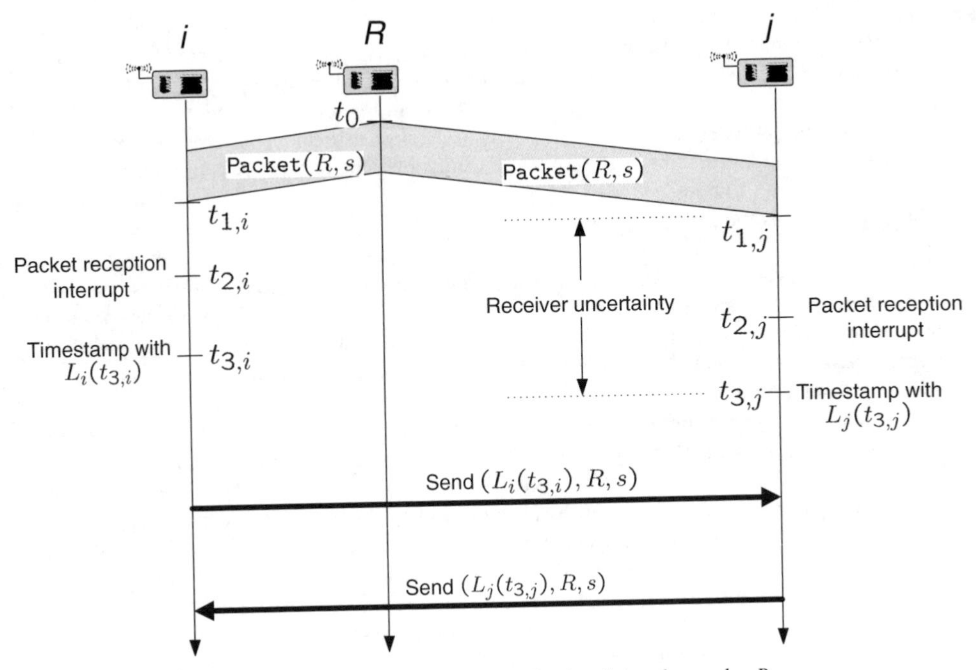

Figure 8.5 RBS example: Two nodes i and j and a sender R

timestamps and can use this data to learn about their neighbors' clocks. By repeating this process, the nodes can not only learn about their mutual phase offsets but also about their drift rates. The nodes do not adjust their local clocks but construct a table that, for each neighbor, stores the necessary parameters to convert clock values.

An example

An example is shown in Figure 8.5. Two nodes i and j want to synchronize. At time t_0, another node R broadcasts a **pulse packet**, which includes its identification R and a sequence number s.[9] Since nodes i and j have a different distance to node R, the propagation delays differ: The last bit reaches node i at real time $t_{1,i}$ and node j at time $t_{1,j}$, and the propagation delays are τ_i and τ_j, respectively. In both nodes, a packet reception interrupt is generated, say, at times $t_{2,i}$ and $t_{2,j}$ for nodes i and j. Some short time later, at time $t_{3,i}$, node i timestamps the packet with its local timestamp $L_i(t_{3,i})$; node j behaves similarly, producing a timestamp $L_j(t_{3,j})$. After this, nodes i and j exchange their timestamps and the identity (sender address, sequence number) of the corresponding pulse packet. Both nodes can easily compute the relative phase shifts of their clocks by assuming $t_{3,i} = t_{3,j}$. Specifically, node i stores the value $O(t_{3,i}) = L_i(t_{3,i}) - L_j(t_{3,j})$ as the phase offset in a local table without readjusting its clock. Clearly, this scheme can benefit from having receivers timestamp incoming packets as quickly as possible in the interrupt routine.

Adopting the terminology introduced for TPSN, the time between receiving the last bit and timestamping the packet is called receiver uncertainty. This is denoted by $\delta_{r,i}$ and $\delta_{r,j}$ for nodes i and j, respectively.

[9] The pulse packet does not need to be a dedicated time synchronization packet, normal data packets carrying sequence numbers suffice.

Achievable precision for a single pulse

Ultimately, the whole computation is perfectly precise when the assumption $t_{3,i} = t_{3,j}$ is really true. Let us briefly analyze the sources of possible synchronization errors.

- The *propagation delay*: In a sensor network, the broadcast domain is typically small and the propagation delay of the packet is negligible. Furthermore, it is only the *difference* in propagation delay between i and j that matters and this difference tends to be even smaller.
- The delay between receiving the last bit and generating the packet reception interrupt: This might be due to hardware processing delays (like checksum computations) and also short-term blocking of interrupts in case of critical sections or servicing interrupts of higher priority. Again, it is the *difference* between i and j that counts for synchronization errors.
- The delay between receiving the packet interrupt and timestamping the packet: If the timestamp is already generated in the interrupt routine, this delay is small.
- The drift between timestamping and exchanging the observed timestamps also contributes to synchronization errors. The more the time that elapses, the larger this error will be.

Compared to sender/receiver based approaches, the time required for R to format its packet, move it through its operating system and networking software, as well as the medium access delay are completely irrelevant since the common point of reference for i and j is the time instant t_0 where the packet appears on the medium.

ELSON et al. [236] have characterized the differences between the times where receivers timestamp a pulse packet among a set of five Berkeley motes. All motes are equipped with a 19.200 b/s transceiver and run TinyOS. For the measurements, the motes raise an I/O pin at the same time they timestamp the packet, and the I/O signals were picked up by an external logic analyzer. Another node sent 160 pulse packets at random times, and for each pulse packet and for each of the 10 possible receiver pairs, the difference of the signal transition times was captured. The results, shown in Figure 8.6, indicate that the error between receivers seem to have a normal distribution with sample mean zero and a standard deviation of $\sigma = 11.1$ μs (according to a Chi-square test with confidence level 99.8 %), which is significantly smaller than the time needed to transmit a single bit at a bitrate of 19.200 bps.

Comparison of RBS and TPSN

GANERIWAL et al. [272] compare RBS and TPSN, both analytically and by comparing implementations on MICA motes. Both protocols timestamp received packets already in the receiver interrupt.

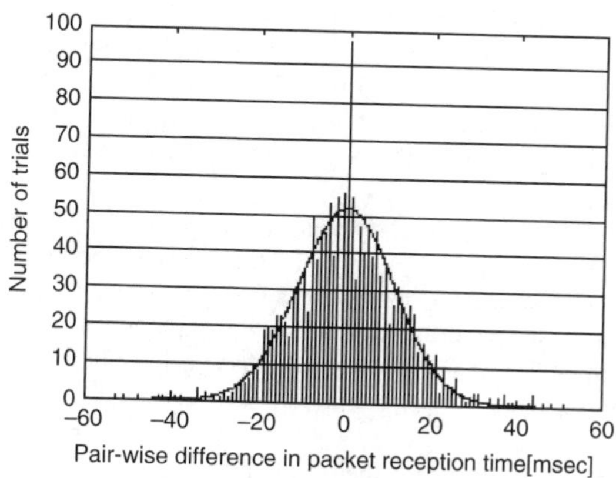

Figure 8.6 Distribution of differences in packet reception time. Reproduced from [236, Fig. 2] by permission of Jeremy Elson

As compared to other sender/receiver based protocols, TPSN removes much of the uncertainty at the sender (operating system and networking stack, medium access delay) by timestamping an outgoing packet immediately before transmission. For the case of TPSN, we have derived the equation

$$O = \frac{(L_j(t_4) - L_i(t_2)) - (L_i(t_7) - L_j(t_6))}{2} - \frac{\delta_{s,i} - \delta_{s,j}}{2} - \frac{\delta_{r,j} - \delta_{r,i}}{2}$$

where $L_x(t_y)$ are the timestamps and $\delta_{s,i}$ and $\delta_{r,i}$ are transmitter and receiver uncertainty at node i, respectively.

Now, let us consider RBS. Say, we have two receivers i and j, and node i wants to estimate the offset to node j. Let the true offset be O. From Figure 8.5, we can write (with $t_1 = t_{1,i}$ being the time where i timestamps the pulse packet):

$$L_i(t_1) = L_j(t_1) + O - (\delta_{r,i} - \delta_{r,j}) - (\tau_i - \tau_j)$$

resulting in

$$O = L_i(t_1) - L_j(t_1) + (\delta_{r,i} - \delta_{r,j}) + (\tau_i - \tau_j).$$

Assuming that (i) the difference in propagation delay is negligible, and (ii) he transmitter uncertainty is small (Section 8.2.3), the error in estimating the offset is dominated by the term $\frac{\delta_{r,j}-\delta_{r,i}}{2}$ in the case of TPSN and by $\delta_{r,i} - \delta_{r,j}$ for RBS.

For neither TPSN nor RBS, the possible clock drift during exchange of the synchronization packets has been considered.

Using multiple pulses

The sender can send pulse packets regularly, and for each pulse packet, the nodes i and j exchange their local observations. The least-squares linear regression proposed for RBS is almost equivalent to the minimum-variance unbiased estimator discussed in Section 8.2.2. An outlier-rejection technique is used additionally. By disregarding those observations older than a certain threshold (a few minutes in RBS), time-varying drift rates can be accommodated.

Multiple nodes and RBS costs

It is straightforward to extend this technique to $m > 2$ nodes in a broadcast domain: The pulse packets are picked up by all m nodes and all nodes exchange their observations. This way, a single one of these nodes, say node i, can estimate drift and phase offsets for all its peers in the broadcast domain. By storing these values for each peer in a table, node i can convert its local timestamps to the time base of any of its peers. As for the pulse senders, there are at least the following choices:

- In networks where stationary infrastructure nodes are present (like, for example, IEEE 802.15.4 in the beaconed mode, see Section 5.5), these nodes can send the pulse packets. They can do even more – they can also collect the observations from the sensor nodes, compute for each node pair the offsets and drift rates with the least-squares method, and send the results back.
- Each node acts as both a receiver and a sender of pulse packets. In this scenario, both the sender identification and the sequence numbers in pulse packets are really needed.

The precision goal in a broadcast domain is then to reduce the maximum of the phase errors between all node pairs, called **group dispersion**. It is shown that the group dispersion depends both on the group size (larger groups have larger dispersion) and on the number of observations used to estimate phase shift and drift rate (increasing the number of observations decreases the group dispersion). For two nodes, it is shown that increasing the number of observations beyond 30 provides no additional gains in precision. The RBS protocol is also compared with NTP, which uses a sender/receiver based scheme. In the investigated scenario, RBS has a much smaller group dispersion than what could be achieved with NTP. Furthermore, the performance of RBS is almost insensitive to the background load in the broadcast domain, whereas NTPs group dispersion increases with increasing load. This can be explained by the influence of medium access delays in NTP. In the given scenario, RBS is approximately eight times better than NTP without background load.

This technique of synchronizing introduces some overhead, which is not yet fully characterized:

- To exchange the observations of a group of m nodes in a broadcast domain, a number of $2m$ packets is needed when the observations are collected and evaluated at a central node and afterward sent back to the receivers. In scenarios without central nodes, exchanging observations takes $m \cdot (m-1) = O(m^2)$ packets. In dense sensor networks, a node can be a member in several (partially overlapping) broadcast domains, and the workload increases accordingly.
- The rate of pulse packets needed to keep the group dispersion below a certain level depends on the group size. A more precise characterization of this relationship needs to be developed.
- The least-squares regression requires significant computation.

RBS and post-facto synchronization

When looking at the overhead, one gets the feeling that RBS can become quite expensive in terms of the number of exchanged packets and computation. However, one of the most important features of RBS is that it can participate in **post-facto synchronization**. In this mode, the nodes do not synchronize with each other until an external event of interest happens. Each node i timestamps this event with a local timestamp $L_i(t_0)$. This event triggers synchronization among nodes and the nodes learn about their relative phase shifts and drift rates. If synchronization is acquired quickly enough after the event occurred, it is safe to assume that the relative drift has not changed. Therefore, node j can use its estimation of node i's phase shift and drift rate to accurately estimate $L_j(t_0)$, that is, to make a *backward extrapolation.*

Network synchronization over multiple hops

What we have seen so far is local synchronization in a broadcast domain. In most cases, a broadcast domain covers only a tiny fraction of a whole sensor network. So the question is: How can an a node learn at which time, measured with respect to its local timescale, a remote event occurred?

Timestamp conversion approach

The basic idea is to not produce a global timescale, but to convert a packet's timestamp at each hop into the next hop node's timescale until it reaches the final destination. Figure 8.7 serves as an example. Let us assume that node 1 has observed an event that it wants to report to the sink node. The human operator at the sink node wants to know when the event happened and he needs this information in an accepted timescale like UTC. Node 1 timestamps the event at time $L_1(t)$, and the packet has to pass through nodes $1 \rightarrow 3 \rightarrow 5 \rightarrow 9 \rightarrow 8 \rightarrow 14 \rightarrow 15 \rightarrow$ Sink. Furthermore, assume that nodes form broadcast domains as indicated by the circles. Observe that nodes 5, 8, and 9 are members of multiple broadcast domains. The source node 1 is in the same broadcast domain as the next node 3 and thus node 1 can use its estimates of the phase shift and drift rate of node 3 to convert the timestamp $L_1(t)$ into $L_3(t)$. Node 1 timestamps the event packet not with its own timestamp but with $L_3(t)$. Node 3, upon receiving this packet, makes a routing decision and infers that node 5 is the next hop. Fortunately, node 5 is in the same broadcast domain as node 3 and node 3 can thus convert the received timestamp $L_3(t)$ into $L_5(t)$. This process is continued until the packet reaches the sink. The last node before the sink, node 15, is in the same broadcast

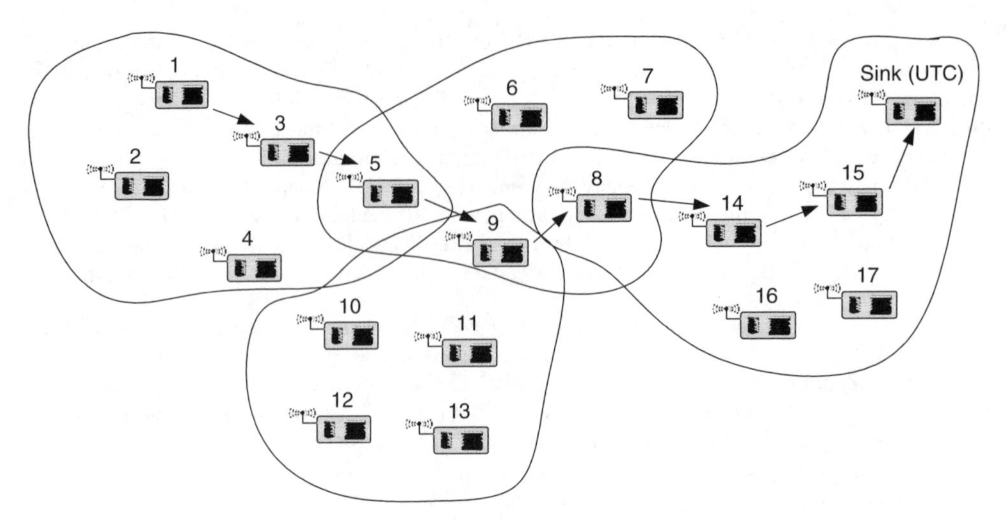

Figure 8.7 RBS example: Multiple broadcast domains

domain as the sink, which keeps UTC time. Again, node 15 knows its phase shift and drift rate with respect to the sink and can thus convert the timestamp $L_{15}(t)$ to UTC time.

The following points are noteworthy:

- The sink node gets an event timestamp in the UTC timescale.
- Each node keeps its local timescale and timestamp conversion can be integrated with packet forwarding. In fact, when node i has to decide about a possible forwarder, only candidates j that have a broadcast domain in common with i are suitable candidates. This requires that for node i and a possible forwarder j there exists a third node whose pulse packets can be heard by both i and j. These might not always exist in sparse sensor networks.
- There is no information related to time synchronization that must be globally available.
- If the synchronization error over a single hop is characterized by a normal distribution with zero mean and standard deviation σ and if the synchronization errors of different hops are independent, then after n hops the synchronization error is Gaussian with zero mean and standard deviation $\sigma \cdot \sqrt{n}$. This growth rate is also confirmed by measurements.

How to create the broadcast domains?

One of the open points in RBS is how the broadcast domains are constituted and how it can be ensured that a time conversion path actually exists between two desired nodes, say, a sensor and a sink node.

Let us look at two different scenarios (Figure 8.8). In the first scenario, a number of static nodes act as dedicated pulse senders and also as packet forwarders; these are shown as circles in the figure. Ordinary sensor nodes (displayed as rectangles) in the range of a pulse sender do not necessarily have to be immediate neighbors to be synchronized. The sensor nodes timestamp the pulse packets, transmit their observations to the pulse sender, which computes for each pair of sensor nodes the

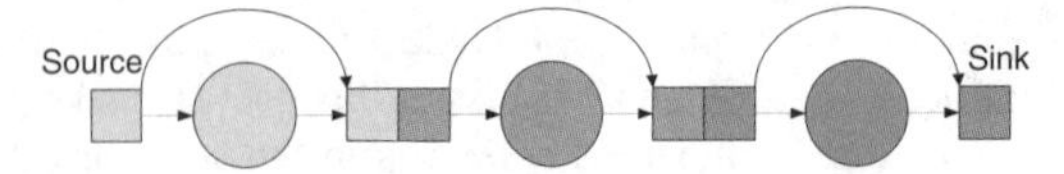

Figure 8.8 RBS: Integrating packet forwarding and timestamp conversion

relative phase shifts and drift rates, and distributes the results back. A broadcast domain is thus formed by the transmission range of the pulse sender. A sensor node i about to forward a packet has two options:

- It can by itself determine another sensor node j in the same broadcast domain to forward the packet to. If both are one-hop neighbors, node i can convert the timestamp and send the packet directly to j. This is indicated by solid arcs in Figure 8.8. In the other case, node i can ask the pulse sender to forward the packet to node j without manipulating it. In both cases, the pulse sender is not required to keep the table with conversion information.
- It can forward the packet to the pulse sender. It is the pulse sender's responsibility to look up a successor node j and to convert the timestamp (indicated by the dashed lines in Figure 8.8). Accordingly, the pulse sender is required to keep the table with conversion parameters.

In either case, the pulse sender *does not make use of its local clock* and thus need not be synchronized with its sensor nodes.[10] In both cases, the broadcast domains should be chosen as follows:

- Each sensor node is a member of at least one broadcast domain.
- When a packet is to be delivered from a source node to a sink node in a different broadcast domain, there must be a series of neighboring broadcast domains on the path between source and sink and two neighboring broadcast domains must overlap in at least one node. These **gateway nodes** (the rectangular nodes with two different grey levels in Figure 8.8) know the conversion parameters for all nodes in both broadcast domains and can convert timestamps accordingly. There needs to be enough overlap between domains to ensure the presence of gateway nodes. Ideally, between two domains there is more than one gateway node to allow for some balancing of forwarding load.
- The number of conversions necessary between a source and a sink node should be small to avoid loss of precision; this calls for large broadcast domains. On the other hand, in larger domains, more packets have to be exchanged and a larger transmit power has to be used to reach all neighbors, draining batteries more quickly.

This can be considered as a clustering problem (see Chapter 10), with the goal to find **overlapping clusters**. A dedicated clustering protocol for the purposes of RBS is presented by [567]. This protocol allows to adjust the cluster size to find a trade-off between minimizing the number of hops and minimizing the number of packets/transmit power.

If there are no dedicated nodes available or when really all nodes including pulse senders need to be synchronized, the broadcast domains must be smaller and all members of a broadcast domain must be one-hop neighbors to be able to exchange the pulse observations as well as to forward data packets. A broadcast domain is thus a **clique** of nodes. In such a scenario each node acts both as a receiver as well as a sender of pulse packets, and forwarding decisions are restricted to those one-hop neighbors that share at least one broadcast domain with the forwarding node.

8.3.2 Hierarchy referencing time synchronization (HRTS)

The TSync protocol presented by Dai and Han [189] combines a receiver/receiver and a sender/receiver based technique. It consists of two subprotocols, which may act independent of each other. The Hierarchy Referencing Time Synchronization (HRTS) protocol discussed below is a receiver/receiver technique synchronizing all nodes in a broadcast domain and additionally constructing a synchronization tree to synchronize a whole network. The synchronization is triggered periodically

[10] To achieve such a synchronization by means of RBS, *another* pulse sender would be needed in range of the original pulse sender and its associated nodes ...

by the root of the tree. The other protocol, called ITR (Individual-based Time Request) protocol, is discussed briefly in Section 8.4.

Synchronization in a broadcast domain

We explain the approach by an example, shown in Figure 8.9. A dedicated node, called a **root node** or **base station** (node R in Figure 8.9) triggers time synchronization at time t_1 (corresponding to local time $L_R(t_1)$) by broadcasting a **sync_begin** announcement packet. This packet includes two parameters, a **level** value (explained below) and the identification of one of node R's one-hop neighbors, say node i. The selected node i timestamps the received packet at time t_2 with its local time $L_i(t_2)$. Another node j timestamps the same packet at time t_2' with $L_j(t_2')$. Similar to RBS, t_2 and t_2' can differ slightly because of random receiver uncertainties. Since node i has found its identification in the packet, it formats an answer packet and timestamps it for transmission at time t_3 with its local time $L_i(t_3)$. Both timestamps $L_i(t_2)$ and $L_i(t_3)$ are included in the answer packet. The root node R receives the packet and timestamps it at time t_4 with $L_R(t_4)$. [11] The root node R

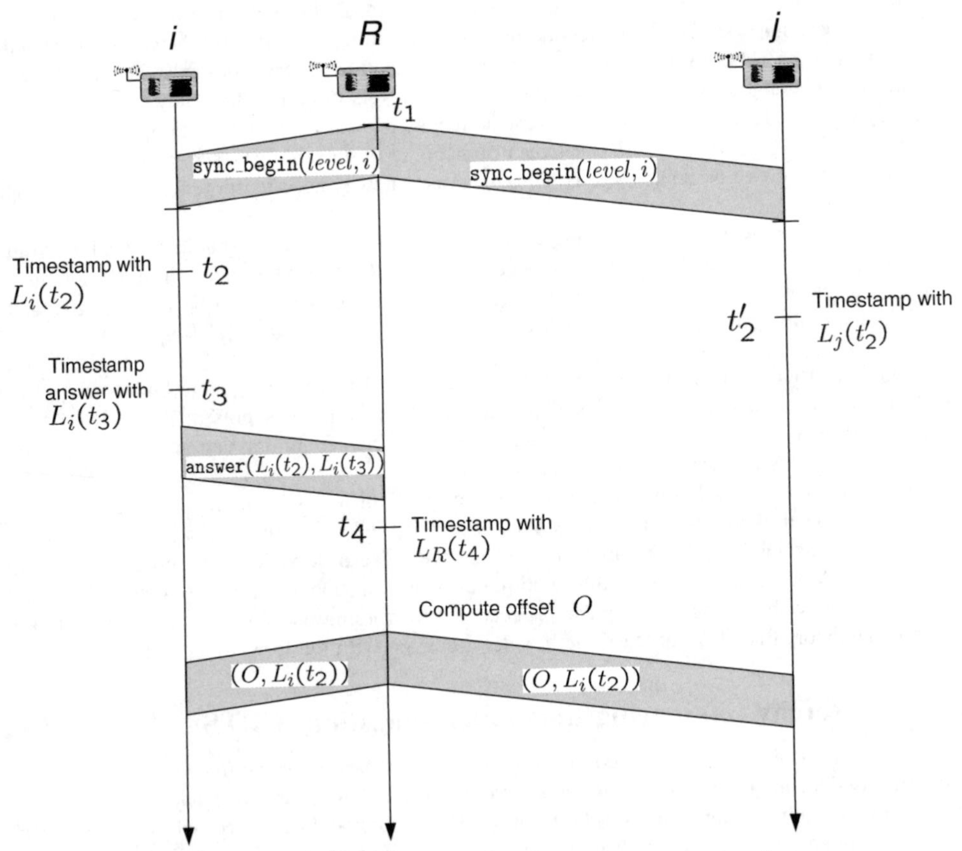

Figure 8.9 TSync: Synchronization in a single broadcast domain

[11] The explicit inclusion of a node identification requires node R to know its neighborhood. Alternatively, node R can omit the identification and all neighbors start a timer with a random value. Node R then just picks the first answer.

can now estimate the offset $O_{R,i}$ between its own local clock and the local clock of node i in a similar fashion as in the pair-wise synchronization protocol of LTS and TPSN as

$$O_{R,i} = \frac{(L_i(t_2) - L_R(t_1)) - (L_R(t_4) - L_i(t_3))}{2}.$$

In the next step, the root node R broadcasts the values O and $L_i(t_2)$ to all nodes. Now, let us look how the individual nodes adjust their clocks:

- Node i simply subtracts the offset $O_{R,i}$ from its local clock.
- Assume that another node j in the same broadcast domain has phase offset $O_{j,i}$ to node i. Under the assumption that i and j receive the **sync_begin** at the same time, we have $t_2 = t_2'$ and $L_i(t_2) = L_j(t_2') + O_{j,i}$. Upon receiving the clock value $L_i(t_2)$ from the root node's final broadcast, node j can compute $O_{j,i}$ directly as $O_{j,i} = L_i(t_2) - L_j(t_2')$. On the other hand, since

$$L_i(t_2) = L_R(t_2) + O_{R,i}$$
$$L_j(t_2) = L_r(t_2) + O_{R,j}$$

we obtain

$$L_R(t_2) + O_{R,i} = L_j(t_2') + O_{j,i}$$

which gives

$$\begin{aligned} O_{R,j} &= L_j(t_2) - L_R(t_2) \\ &\approx L_j(t_2') - L_R(t_2) \\ &= L_R(t_2) + O_{R,i} - O_{j,i} - L_R(t_2) \\ &= O_{R,i} - O_{j,i} \\ &= O_{R,i} - (L_i(t_2) - L_j(t_2')). \end{aligned}$$

Therefore, node j can compute the phase offset to node R directly from its own observation $L_j(t_2')$ and the values from the final broadcast, *without exchanging any packets with other nodes.*

The important property of this scheme is that three packets suffice to synchronize *all* of R's neighbors to R's clock, no matter what their number is and whether they are in mutual range or not. In contrast to RBS, the receivers of the **sync_begin** packet do not have to exchange their observations in a pair-wise fashion and the protocol is therefore insensitive to the network density.

Dai and Han [189] have shown experimentally that with respect to a single broadcast domain as well as over multiple hops the HRTS protocol and RBS have approximately the same distribution of synchronization errors when taken over a large number of repetitions. Similar to the other protocols, the receive and transmit uncertainties can be reduced by timestamping outgoing packets as lately as possible and timestamping incoming packets as early as possible.

Although not necessary for this approach to work, Dai and Han [189] propose to run this protocol over a separate MAC channel. The goal is to separate synchronization-related traffic from user data traffic and thus to reduce the probability of collision and the medium access delays when the selected node i answers to the root node's **sync_begin** packet. Specifically, the root node sends the **sync_begin** on a dedicated **control channel** and includes the specification of another dedicated channel, the **clock channel**. Both the root node and the selected node i switch to the clock channel and i transmits its answer packet. Finally, the root node broadcasts the packet containing $L_i(t_2)$ and O, again on the control channel. All other nodes can continue forwarding data packets as R and i exchange their packets.

Network synchronization over multiple hops

Let us assume that the whole network is static and connected and that there is at least one **reference node** having access to an external time source and, thus, to UTC time; there can be multiple reference nodes. These reference nodes are assigned level 0 and become root nodes for their one-hop neighborhoods. They trigger resynchronization periodically, according to the protocol described in the previous section. The level is included in the **sync_begin** packet. All nodes behave according to a simple rule. They maintain a local level variable, initialized with a sufficiently large value. If a node receives a **sync_begin** packet with a level value being truly smaller than its own level variable, the node accepts the packet, sets its own level to the received level value plus one, and becomes a root node of its own, starting synchronization with its neighbors after the triggering synchronization round has finished. In the other case, if the received level is larger than or equal to the own level variable, the **sync_begin** packet is simply dropped. The process continues in a recursive fashion until the fringe of the network is reached.

This way, nodes always synchronize over multiple hops with the closest reference node, and their level variable indicates the hop distance. If an already-synchronized node learns later about a closer reference node, it assigns itself a new level and starts resynchronization of its children.

The overhead (expressed as number of exchanged packets) of HRTS and RBS are compared by DAI and HAN [189] for a scenario with 200 nodes placed in an area of $400 \times 400\ \text{m}^2$ square meters when varying the average node degree. RBS creates much higher overhead than HRTS, however, the overhead of HRTS increases when the node density increases. One explanation for this might be the observation that the broadcast domains of higher-level nodes are smaller than those of reference nodes because of overlapping transmission ranges of neighboring children. This way they have relatively fewer children.

The protocol has a further mechanism to restrict the depth of a synchronization tree rooted in a reference node. In addition to the level parameter and the identity of the answering node, each **sync_begin** packet carries also a depth parameter, which is decremented in every level of the tree. The tree construction stops upon reaching a depth of zero.

8.4 Further reading

- The TinySync and MiniSync protocols are presented by SICHITIU and VEERARITTIPHAN [764]. They belong to the class of sender/receiver protocols and consist of schemes for pair-wise synchronization and synchronizing a network. Similar to RBS, the nodes do not adjust their local clocks but compute conversion parameters for their respective phase shifts and drift rates. Let us assume that node i wants to synchronize with node j. At time $t_{0,1}$ (and local time $L_i(t_{0,1})$), node i sends a packet to node j, which timestamps it as $L_j(t_{1,1})$. Node j answers immediately and i receives the answer at $L_i(t_{2,1})$. Under the assumption of constant drift rate, we need to find coefficients $a_{i,j}$ and $b_{i,j}$ such that $L_j(t) = a_{i,j} \cdot L_i(t) + b_{i,j}$ holds for all t. From only a single three-way handshake, one can constrain the parameters $a_{i,j}$ and $b_{i,j}$ by considering that node j must have received the first packet *after* it has been sent by i. By the same token, node j must have sent the answer *before* node i has received it. To summarize, it must hold that

$$L_j(t_{1,1}) > a_{i,j} \cdot L_i(t_{0,1}) + b_{i,j}$$
$$L_j(t_{1,1}) < a_{i,j} \cdot L_i(t_{2,1}) + b_{i,j}.$$

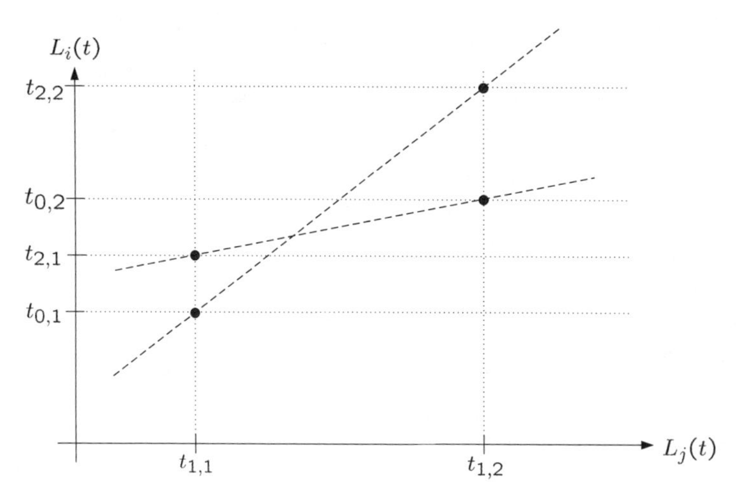

Figure 8.10 Deriving constraints for drift rate and phase offset in TinySync/MiniSync (adapted from [764, Fig. 2])

An estimate of $a_{i,j}$ and $b_{i,j}$ can be obtained by taking more three-way handshakes at later times $t_{0,k}$, $t_{1,k}$ and $t_{2,k}$ into account, compare Figure 8.10 in which an example involving two handshakes is shown. The slope of the dashed line between the points $(t_{1,1}, t_{0,1})$, and $(t_{1,2}, t_{2,2})$ gives an upper bound $\overline{a_{i,j}}$ on the drift rate $a_{i,j}$. The other dashed line indicates a lower bound $\underline{a_{i,j}}$; finally, an estimate $a_{i,j}$ is taken from between these bounds. SICHITIU and VEERARITTIPHAN [764] demonstrate how these bounds can be tightened; TinySync and MiniSync differ in the respective set of handshake operations taken into account (observation set) and in the update policies for this observation set.

- The concept of sender/receiver synchronization and the approach of using timestamps taken immediately after reception of a packet (thus eliminating the uncertainties coming from medium access and the sender's operating system) are combined in reference [568, 569]. In the proposed protocol, a set of IEEE 802.11 stations synchronize their times independently with an access point. The necessary time synchronization pulses are actually provided by beacon packets sent by the access points in regular intervals. The trick is that not only the receiver station i timestamps the beacon packet immediately after arrival (say, with $L_i(t_1)$) but the access point a is also forced to receive its own beacon packet and to timestamp it with $L_a(t_1')$ (Figure 8.11). The access point piggybacks its timestamp onto the *next* beacon packet (actually, a history of timestamps is piggybacked to provide some fault tolerance). Having two such measurements – say, $L_i(t_1)$, $L_i(t_2)$ and $L_a(t_1')$, $L_a(t_2')$ – the receiver can estimate the drift with respect to the access point by equating t_1 and t_1' as well as t_2 and t_2' and computing the slope $\frac{L_i(t_2)-L_i(t_1)}{L_a(t_2)-L_a(t_1)}$. The actual clock adjustment policy used in reference [569] avoids sudden jumps: Instead, of modifying a receiver's phase shift ϕ_i (for example, by adding $L_a(t_2) - L_i(t_2)$ to $L_i(t_2)$) the drift rate θ_i is controlled instead. Node i modifies its drift rate at t_2 such that its local clock will coincide with the access points clock at a carefully selected time in the future. Because of its focus on the infrastructure mode of IEEE 802.11, the multihop case is not discussed.
- KARP et al. [410] discuss some problems of RBS. The first problem is that the pair-wise synchronization approach of RBS based on least-squares linear regression is, in general, not consistent.[12] The second problem is that for synchronizing two nodes i and j, which are n hops apart, the variance of the synchronization error increases linearly with n. KARP et al. [410] argue that this

[12] To explain consistency, assume that node i has learned its phase shift $\phi_{i,j}$ and drift rate $\theta_{i,j}$ with respect to another node j; thus, $L_j(t) = \theta_{i,j} \cdot L_i(t) + \phi_{i,j}$. Additionally, node j has learned about node k, resulting in $L_k(t) = \theta_{j,k} \cdot L_j(t) + \phi_{j,k}$.

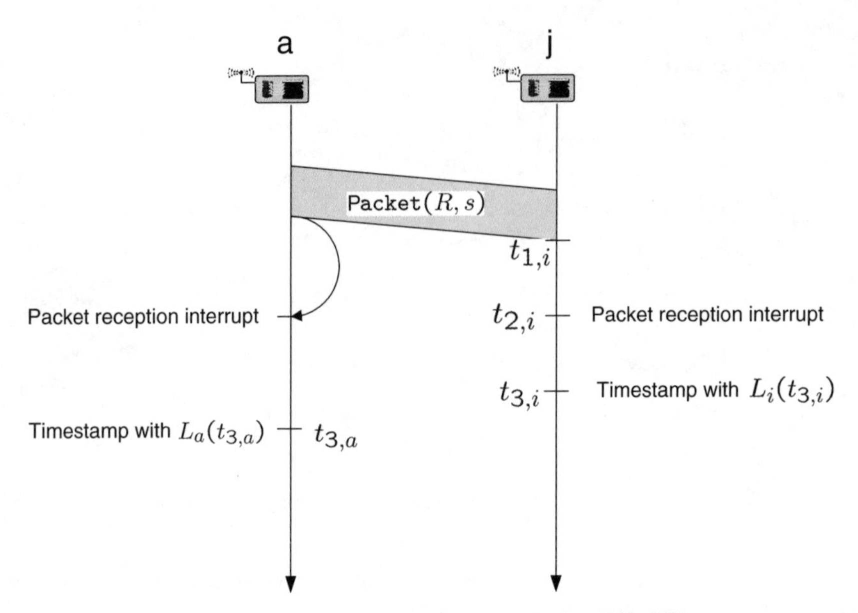

Figure 8.11 Continuous clock synchronization [568, 569]

is an artifact of considering only a single path between i and j. Theoretically, by increasing the amount of information available to a pair of nodes (which can exchange their observations regarding *multiple* common pulse senders instead of only a single one) and by considering multiple synchronization paths between remote nodes i and j, a minimum-variance global estimator can be produced that is consistent and for which the synchronization error grows only logarithmically in the number of hops (at least in selected scenarios). However, without optimizations, this global estimator requires observations from *all* nodes and all pulse senders.

- A scheme with an approach similar to RBS is CesiumSpray system from VERISSIMO et al. [841]. CesiumSpray is designed for a local network with a single broadcast domain and contains no provisions for multiple hops.
- The MultiHop Time Synchronization Protocol (MTSP) [777] uses a pair-wise synchronization protocol similar to LTS and TPSN. It requires that transmitted packets are timestamped as lately as possible and received packets are timestamped as early as possible. Both the pair-wise protocol and the networkwide protocol are designed such that nodes with smaller local clock values synchronize to other nodes with higher clock values but not vice versa. Eventually, all nodes in the network synchronize to the fastest node.
- The Individual-based Time Request (ITR) protocol proposed by DAI and HAN [189] is supposed to run in parallel to the HRTS protocol. ITR allows a node to acquire synchronization on demand, for example, when it has missed some of HRTSs synchronization packets or after waking up from a long sleep period. ITR is also useful for implementing post-facto synchronization. The protocol is built on a similar pair-wise synchronization approach as TPSN (Section 8.2.3) or Simple NTP [555]. An interesting feature of ITR is its way to synchronize two nodes that are $n \geq 2$ hops

Consistency would require that

$$L_k(t) = \theta_{j,k} \cdot \left[\theta_{i,j} \cdot L_i(t) + \phi_{i,j}\right] + \phi_{j,k}$$

and hence, $\theta_{i,k} = \theta_{i,j} \cdot \theta_{j,k}$ as well as $\phi_{i,k} = \theta_{j,k} \cdot \phi_{i,j} + \phi_{j,k}$.

away: The intermediate nodes just forward synchronization request and answer packets without acquiring synchronization on the fly!

- SWOL HU and SERVETTO [801], HU and SERVETTO [802] propose an interesting technique for global synchronization in dense sensor networks, which we explain briefly by an analogy. Let us assume that you are in a large football stadium and somewhere someone else (the **seed**) starts to clap hands rhythmically. The neighbors of the seed (**the first generation**) get interested, listen to the seed, independently estimate the seed's clapping period and start to clap at the estimated period. The second generation neighbors hear the seed's and the first generation neighbors' clapping. The first generation neighbors can have small random deviations from the seed; however, these will be independent. The second generation neighbors try to estimate the clapping frequency from the composite signal and start clapping too. All further generations behave in the same way. Under suitable assumptions (one of them being a high node density and thus a large number of nodes in the respective generations), the peak intensity of the composite signal perceived by second and further generation neighbors coincides with the seed's clapping period.

9

Localization and positioning

Objectives of this Chapter

This chapter gives an overview of the methods to determine the symbolic location – "in the living room" – and the numeric position – "at coordinates (23.54, 11.87)" – of a wireless sensor node. The properties of such methods and the principal possibilities for a node to determine information about its whereabouts are discussed. The mathematical basics for positioning are introduced and the single-hop and multihop positioning case are described using several example systems.

At the end of the chapter, the reader will understand the principal design trade-offs for positioning and gain an appreciation for the overhead involved in obtaining this information.

Chapter Outline

In many circumstances, it is useful or even necessary for a node in a wireless sensor network to be aware of its location in the physical world. For example, tracking or event-detection functions are not particularly useful if the WSN cannot provide any information *where* an event has happened. To do so, usually, the reporting nodes' location has to be known. Manually configuring location information into each node during deployment is not an option. Similarly, equipping every node with a Global Positioning System (GPS) receiver [354] fails because of cost and deployment limitations (GPS, e.g. does not work indoors).

This chapter introduces various techniques of how sensor nodes can learn their location automatically, either fully autonomically by relying on means of the WSN itself or by using some assistance from external infrastructure.

9.1 Properties of localization and positioning procedures

The simple intuition of "providing location information to a node" has a number of facets that should be classified to make the options for a location procedure clear. The most important properties are (following the survey papers [349, 350]):

Physical position versus symbolic location Does the system provide data about the *physical position* of a node (in some numeric coordinate system) or does a node learn about a *symbolic location* – for example, "living room", "office 123 in building 4"? Is it, in addition, possible to match physical position with a symbolic location name (out of possibly several applicable ones)?

While these two concepts are different, there is no consistent nomenclature in the literature – position and location are often used interchangeably. The tendency is to use "location" as the more general term. We have to rely on context to distinguish between these two contexts.

Absolute versus relative coordinates An absolute coordinate system is valid for all objects and embedded in some general frame of reference. For example, positions in the Universal Transverse Mercator (UTM) coordinates form an absolute coordinate system for any place on earth. Relative coordinates, on the other hand, can differ for any located object or set of objects – a WSN where nodes have coordinates that are correct with respect to each other but have no relationship to absolute coordinates is an example.

To provide absolute coordinates, a few **anchors** are necessary (at least three for a two-dimensional system). These anchors are nodes that know their own position in the absolute coordinate system. Anchors can rotate, translate, and possibly scale a relative coordinate system so that it coincides with the absolute coordinate system. These anchors are also commonly called *"beacons" or "landmarks"* in the literature.

Localized versus centralized computation Are any required computations performed locally, by the participants, on the basis of some locally available measurements or are measurements reported to a central station that computes positions or locations and distributes them back to the participants? Apart from scaling and efficiency considerations (both with respect to computational and communication overhead), privacy issues are important here as it might not be desirable for a participant to reveal its position to a central entity.

Accuracy and precision The two most important figures of merit for a localization system are the accuracy and the precision of its results. Positioning accuracy is the largest distance between the estimated and the true position of an entity (high accuracy indicates a small maximal mismatch). Precision is the ratio with which a given accuracy is reached, averaged over many repeated attempts to determine a position. For example, a system could claim to provide a 20-cm accuracy with at least 95 % precision. Evidently, accuracy and precision values only make sense when considered together, forming the accuracy/precision characteristic of a system.

Scale A system can be intended for different scales, for example – in indoor deployment – the size of a room or a building or – in outdoor deployment – a parking lot or even worldwide operation. Two important metrics here are the area the system can cover per unit of infrastructure and the number of locatable objects per unit of infrastructure per time interval.

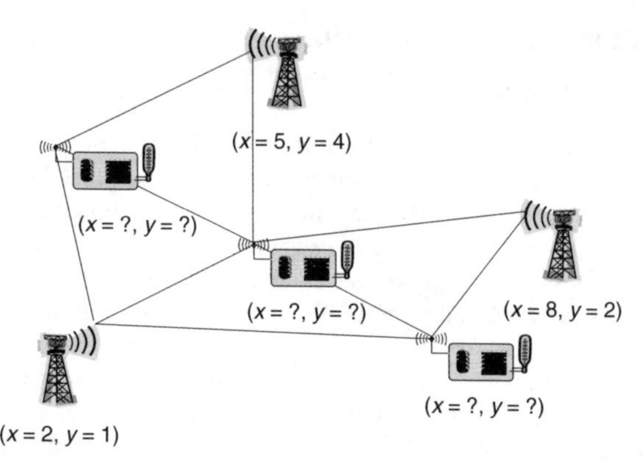

Figure 9.1 Determining the position of sensor nodes with the assistance from some anchor points; not all nodes are necessarily in contact with all anchors

Limitations For some positioning techniques, there are inherent deployment limitations. GPS, for example, does not work indoors; other systems have only limited ranges over which they operate.

Costs Positioning systems cause costs in time (infrastructure installation, administration), space (device size, space for infrastructure), energy (during operation), and capital (price of a node, infrastructure installation).

Figure 9.1 illustrates the positioning problem. The figures in this chapter use the "access point" icon to indicate anchors for easy distinction. It should be pointed out, however, that normal sensor nodes can just as well be used as anchors, as long as they have are aware of their position.

In addition, a positioning or localization system can be used to provide the recognition or classification of objects; this property is less important in the WSN context or, if used, usually not considered a part of the localization system.

9.2 Possible approaches

Three main approaches exist to determine a node's position: Using information about a node's neighborhood (proximity-based approaches), exploiting geometric properties of a given scenario (triangulation and trilateration), and trying to analyze characteristic properties of the position of a node in comparison with premeasured properties (scene analysis). The overview given here again considerably follows reference [350].

9.2.1 Proximity

The simplest technique is to exploit the finite range of wireless communication. It can be used to decide whether a node that wants to determine its position or location is in the proximity of an anchor. While this only provides coarse-grain information, it can be perfectly sufficient. One example is the natural restriction of infrared communication by walls, which can be used to provide a node with simple location information about the room it is in.

Proximity-based systems can be quite sophisticated and can even be used for approximate positioning when a node can analyze proximity information of several overlapping anchors (e.g. [106]). They can also be relatively robust to the uncertainties of the wireless channel – deciding whether a node is in the proximity of another node is tantamount to deciding connectivity, which can happen on relatively long time scales, averaging out short-term fluctuations.

9.2.2 Trilateration and triangulation

Lateration versus angulation

In addition to mere connectivity/proximity information, the communication between two nodes often allows to extract information about their geometric relationship. For example, the distance between two nodes or the angle in a triangle can be estimated – how this is done is discussed in the following two subsections. Using elementary geometry, this information can be used to derive information about node positions. When distances between entities are used, the approach is called **lateration**; when angles between nodes are used, one talks about **angulation**.

For lateration in a plane, the simplest case is for a node to have precise distance measurements to three noncolinear anchors. The extension to a three-dimensional space is trivial (four anchors are needed); all the following discussion will concentrate on the planar case for simplicity. Using distances and anchor positions, the node's position has to be at the intersection of three circles around the anchors (Figure 9.2).

The problem here is that, in reality, distance measurements are never perfect and the intersection of these three circles will, in general, not result in a single point. To overcome these imperfections, distance measurements form more that three anchors can be used, resulting in a **multilateration** problem. Multilateration is a core solution technique, used and reused in many concrete systems described below. Its mathematical details are treated in Section 9.3.

Angulation exploits the fact that in a triangle once the length of two sides and two angles are known the position of the third point is known as the intersection of the two remaining sides of the triangle. The problem of imprecise measurements arises here as well and can also be solved using multiple measurements.

Determining distances

To use (multi-)lateration, estimates of distances to anchor nodes are required. This **ranging** process[1] ideally leverages the facilities already present on a wireless node, in particular, the radio communication device. The characteristics of wireless communication are partially determined by the distance between sender and receiver, and if these characteristics can be measured at the receiver,

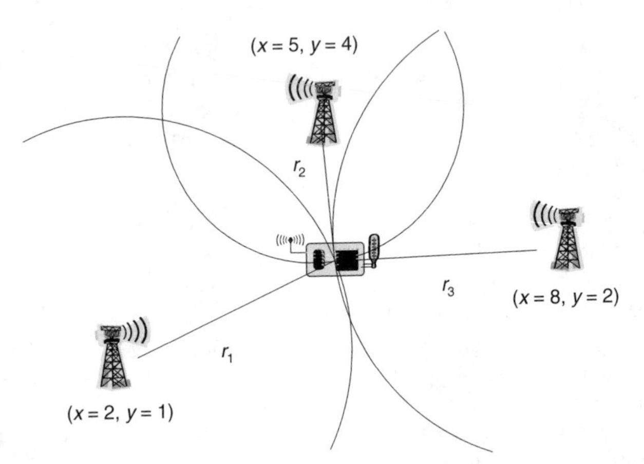

Figure 9.2 Triangulation by intersecting three circles

[1] Because of this name, proximity-based approaches are sometimes also called "range-free" approaches.

they can serve as an estimator of distance. The most important characteristics are Received Signal Strength Indicator (RSSI), Time of Arrival (ToA), and Time Difference of Arrival (TDoA).

Received signal strength indicator

Assuming that the transmission power P_{tx}, the path loss model, and the path loss coefficient α are known, the receiver can use the received signal strength P_{rcvd} to solve for the distance d in a path loss equation like

$$P_{rcvd} = c\frac{P_{tx}}{d^{\alpha}} \Leftrightarrow d = \sqrt[\alpha]{\frac{cP_{tx}}{P_{rcvd}}}.$$

This is appealing since no additional hardware is necessary and distance estimates can even be derived without additional overhead from communication that is taking place anyway. The disadvantage, however, is that RSSI values are not constant but can heavily oscillate, even when sender and receiver do not move. This is caused by effects like fast fading and mobility of the environment – ranging errors of $\pm 50\,\%$ are reported, for example, by SAVARESE et al. [724]. To some degree, this effect can be counteracted by repeated measurements and filtering out incorrect values by statistical techniques [864]. In addition, simple, cheap radio transceivers are often not calibrated and the same actual signal strength can result in different RSSI values on different devices (reference [873] considers the calibration problem in detail); similarly, the actual transmission power of such a transceiver shows discrepancies from the intended power [725]. A third problem is the presence of obstacles in combination with multipath fading [104]. Here, the signal attenuation along an indirect path, which is higher than along a direct path, can lead to incorrectly assuming a longer distance than what is actually the case. As this is a structural problem, it cannot be combated by repeated measurements.

A more detailed consideration shows that mapping RSSI values to distances is actually a random process. RAMADURAI and SICHITIU [674], for example, collected, for several distances, repeated samples of RSSI values in an open field setup. Then, they counted how many times each distance resulted in a given RSSI value and computed the density of this random variable – Figure 9.3(a) shows this probability density function for a single given value of RSSI, Figure 9.3(b) for several. The information provided in particular by small RSSI values, indicating longer distances, is quite limited as the density is widely spread.

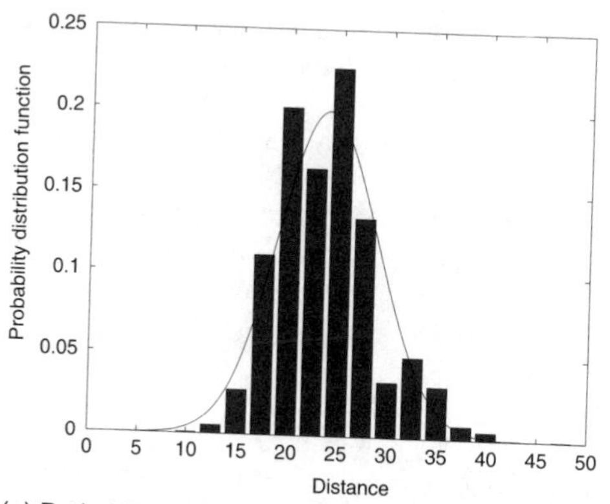

(a) Probability density function of distances resulting in a given RSSI value

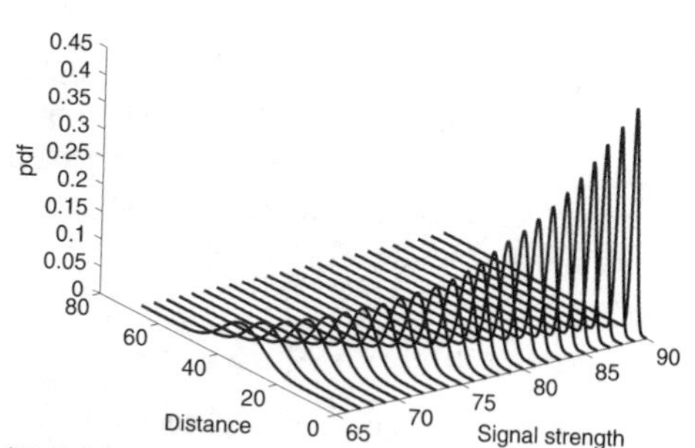

(b) Several probability density functions of distances for various given RSSI values

Figure 9.3 Random nature of mapping RSSI values to distances [674]. Reproduced by permission of V. Ramadurai and M. L. Sichitiu

Hence, when using RSSI as a ranging technique, it is necessary to accept and deal with considerable ranging errors or to treat the outcome of the ranging process as a stochastic result to begin with.

Time of arrival

Time of Arrival (ToA) (also sometimes called *"time of flight"*) exploits the relationship between distance and transmission time when the propagation speed is known. Assuming both sender and receiver know the time when a transmission – for example, a short ultrasound pulse – starts, the time of arrival of this transmission at the receiver can be used to compute propagation time and, thus, distance. To relieve the receiver of this duty, it can return any received "measurement pulse" in a deterministic time; the original sender then only has to measure the round trip time assuming symmetric paths.

Depending on the transmission medium that is used, time of arrival requires very high resolution clocks to produce results of acceptable accuracy. For sound waves, these resolution requirements are modest; they are very hard for radio wave propagation.

One disadvantage of sound is that its propagation speed depends on external factors such as temperature or humidity – careful calibration is necessary but not obvious.

Time difference of arrival

To overcome the need for explicit synchronization, the Time Difference of Arrival (TDoA) method utilizes implicit synchronization by directly providing the start of transmission information to the receiver. This can be done if two transmission mediums of very different propagation speeds are used – for example, radio waves propagating at the speed of light and ultrasound, with a different in speed of about six orders of magnitude.[2] Hence, when a sender starts an ultrasound and a radio transmission simultaneously, the receiver can use the arrival of the radio transmission to start measuring the time until arrival of the ultrasound transmission, safely ignoring the propagation time of the radio communication.[3]

The obvious disadvantage of this approach is the need for two types of senders and receivers on each node. The advantage, on the other hand, is a considerably better accuracy compared to RSSI-based approaches. This concept and variations of it have been used in various research efforts [659, 725] and accuracies of down to 2 cm have been reported [725].

Discussion

There is a clear trade-off between ranging error and complexity/cost. RSSI-based approaches are simpler and work with hardware that is required anyway; TDoA-based approaches provide superior ranging results but need more complex and additional hardware, which also adds to energy consumption. The open question is thus whether it is possible to solve the actual positioning or localization problem based on the error-prone measurements provided by RSSI-based approaches or whether the overhead for TDoA is unavoidable.

There is a number of additional work on making ranging measurements more reliable, for example, by using acoustic ranging as opposed to radio frequency attenuation [291] or by considering the effects introduced by quantizing RSSI values (in effect, proximity systems can be regarded as yes/no quantizations of RSSI) [624].

Determining angles

As an alternative to measuring distances between nodes, angles can be measured. Such an angle can either be the angle of a connecting line between an anchor and a position-unaware node to a

[2] Speed of light in vacuum: 299,792,458 m/s, ultrasound in air about 344 m/s at 21 °C.

[3] In fact, this is precisely what happens when one estimates the distance to a thunderstorm by counting the seconds between lightning and thunder.

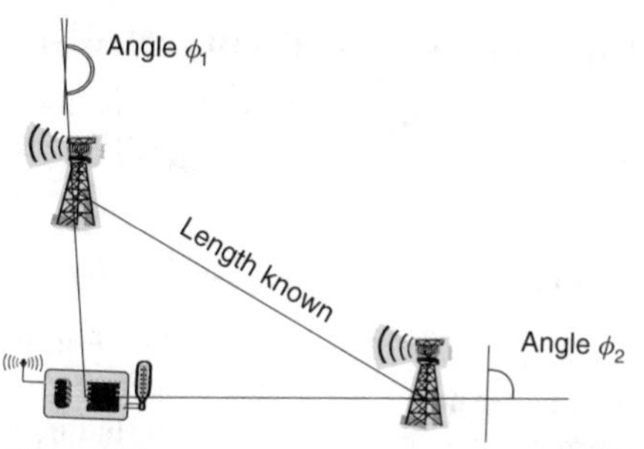

Figure 9.4 Angulation based on two anchors

given reference direction ("0° north"). It can also be the angle between two such connecting lines if no reference direction is commonly known to all nodes (Figure 9.4).

A traditional approach to measuring angles is to use directional antennas (antennas that only send to/receive from a given direction), rotating on their axis, similar to a radar station or a conventional lighthouse. This makes angle measurements conceptually simple, but such devices are quite inappropriate for sensors nodes; they can be useful for supporting infrastructure anchors.

Another technique is to exploit the finite propagation speed of all waveforms. With multiple antennas mounted on a device at known separation and measuring the time difference between a signal's arrival at the different antennas, the direction from which a wavefront arrived at the device can be computed. The smaller the antenna separation, the higher the precision of the time differences has to be, which results in strenuous timing requirements given the desirable small size of sensor nodes.

Overall, angulation is a less frequently discussed technique compared to lateration; Section 9.4 discusses some examples.

9.2.3 Scene analysis

A quite different technique is **scene analysis**. The most evident form of it is to analyze pictures taken by a camera and to try to derive the position from this picture. This requires substantial computational effort and is hardly appropriate for sensor nodes.

But apart from visual pictures, other measurable characteristic "fingerprints" of a given location can be used for scene analysis, for example, radio wave propagation patterns. One option is to use signal strength measurements of (one or more anchors) transmitting a known signal strength and compare the actually measured values with those stored in a database of previously off-line measured values for each location – the RADAR system [35] is one example that uses this approach to determine positions in a building. Using other physical characteristics such as multipath behavior is also conceivable.

While scene analysis is interesting for systems that have a dedicated deployment phase and where off-line measurements are acceptable, this is not always the case for WSNs. Hence, this approach is not the main focus of attention.

9.3 Mathematical basics for the lateration problem

Since (multi)lateration is one of the most popular techniques for positioning applied in WSNs and serves as a primitive building block for some of the approaches discussed later, it is worthwhile to have a closer look at the mathematics behind it.

9.3.1 Solution with three anchors and correct distance values

Assume that there are three anchors with known positions (x, y_i), $i = 1, \ldots, 3$, a node at unknown position (x_u, y_u), and perfect distance values r_i, $i = 1, \ldots, 3$. From the Pythagoras theorem, a set of three equations follows:

$$(x_i - x_u)^2 + (y_i - y_u)^2 = r_i^2 \text{ for } i = 1, \ldots, 3. \tag{9.1}$$

To solve this set of equations, it is more convenient to write it as a set of linear equations in x_u and y_u. To do so, the quadratic terms x_u^2 and y_u^2 have to be removed. This can be achieved by subtracting the third equation from the two previous ones, resulting in two remaining equations:

$$(x_1 - x_u)^2 - (x_3 - x_u)^2 + (y_1 - y_u)^2 - (y_3 - y_u)^2 = r_1^2 - r_3^2 \tag{9.2}$$

$$(x_2 - x_u)^2 - (x_2 - x_u)^2 + (y_2 - y_u)^2 - (y_2 - y_u)^2 = r_2^2 - r_3^2. \tag{9.3}$$

Rearranging of terms results in

$$2(x_3 - x_1)x_u + 2(y_3 - y_1)y_u = (r_1^2 - r_3^2) - (x_1^2 - x_3^2) - (y_1^2 - y_3^2) \tag{9.4}$$

$$2(x_3 - x_2)x_u + 2(y_3 - y_2)y_u = (r_2^2 - r_2^2) - (x_2^2 - x_3^2) - (y_2^2 - y_3^2), \tag{9.5}$$

which can be easily rewritten as a linear matrix equation

$$2\begin{bmatrix} x_3 - x_1 & y_3 - y_1 \\ x_3 - x_2 & y_3 - y_2 \end{bmatrix}\begin{bmatrix} x_u \\ y_u \end{bmatrix} = \begin{bmatrix} (r_1^2 - r_3^2) - (x_1^2 - x_3^2) - (y_1^2 - y_3^2) \\ (r_2^2 - r_2^2) - (x_2^2 - x_3^2) - (y_2^2 - y_3^2) \end{bmatrix}, \tag{9.6}$$

where the matrix on the left side and the right hand side only consists of known constants.

Example 9.1 (Position determination) Using the example positions of Figure 9.2 – $(x_1, y_1) = (2, 1)$, $(x_2, y_2) = (5, 4)$, and $(x_3, y_3) = (8, 2)$ – with the distances between anchors and node of unknown position $r_1 = \sqrt{10}$, $r_2 = 2$, $r_3 = 3$, Equation 9.6 becomes

$$2\begin{bmatrix} 6 & 1 \\ 3 & -2 \end{bmatrix}\begin{bmatrix} x_u \\ y_u \end{bmatrix} = \begin{bmatrix} 64 \\ 22 \end{bmatrix}, \tag{9.7}$$

resulting in $x_u = 5$ and $y_u = 2$ as the position of the unknown nodes.

9.3.2 Solving with distance errors

The real challenge for triangulation arises when the distance measurements are not perfect but only estimates $\tilde{r}$ with an unknown error ε are known. Solving the above equations with $\tilde{r}_i = r_i + \varepsilon_i$ will in general not yield the correct values for the unknown positions (x_u, y_u).

The intuitive solution to this problem is to use more than three anchors and redundant distant measurements to account for the error in each individual measurement. Mathematically, this turns the above equations into an overdetermined system of equations, written in matrix form as

$$2\begin{bmatrix} x_n - x_1 & y_n - y_1 \\ \vdots & \vdots \\ x_n - x_{n-1} & y_n - y_{n-1} \end{bmatrix}\begin{bmatrix} x_u \\ y_u \end{bmatrix} = \begin{bmatrix} (r_1^2 - r_n^2) - (x_1^2 - x_n^2) - (y_1^2 - y_n^2) \\ \vdots \\ (r_{n-1}^2 - r_n^2) - (x_{n-1}^2 - x_n^2) - (y_{n-1}^2 - y_n^2) \end{bmatrix}. \tag{9.8}$$

For such an overdetermined system of linear equation, a solution can be computed that minimizes the mean square error, that is, the solution is the pair (x_u, y_u) that minimizes $\|\mathbf{Ax} - \mathbf{b}\|_2$, where 0.5**A** is the left-hand matrix (an $n - 1 \times 2$ matrix), $\mathbf{x} = (x_u, y_u)$ a shorthand for the vector describing the unknown position, and **b** the right hand side (an $n - 1$ row vector) from Equation (9.8). Since $\|\cdot\|_2$, the 2-norm of a vector (the square root of the sum of the squares of the vector elements), is minimized, this reflects solving for the position that best satisfies, with minimum average error, all the position constraints from all n anchors.

To find a solution for this minimization problem, look at an expression for the square of the 2-norm from above. Observe that for any vector $\mathbf{v}$, $\|\mathbf{v}\|_2^2 = \mathbf{v}^{\mathrm{T}}\mathbf{v}$. Hence,

$$\|\mathbf{Ax} - \mathbf{b}\|_2^2 = (\mathbf{Ax} - \mathbf{b})^{\mathrm{T}}(\mathbf{Ax} - \mathbf{b}) = \mathbf{x}^{\mathrm{T}}\mathbf{A}^{\mathrm{T}}\mathbf{Ax} - 2\mathbf{x}^{\mathrm{T}}\mathbf{A}^{\mathrm{T}}\mathbf{b} + \mathbf{b}^{\mathrm{T}}\mathbf{b}. \tag{9.9}$$

Minimizing this expression is equivalent to minimizing the mean square error. Regarding this as a function in $\mathbf{x}$, its gradient has to be set equal to zero:

$$2\mathbf{A}^{\mathrm{T}}\mathbf{Ax} - 2\mathbf{A}^{\mathrm{T}}\mathbf{b} = 0 \Leftrightarrow \mathbf{A}^{\mathrm{T}}\mathbf{Ax} = \mathbf{A}^{\mathrm{T}}\mathbf{b}. \tag{9.10}$$

Equation (9.10) is called the **normal equation** for the linear least squares problem. This equation has a unique solution under certain conditions (**A** has to have full rank). There are various methods to solve such an equation, for example, Cholesky or QR factorization (by substituting $\mathbf{A} = \mathbf{QR}$, $\mathbf{Q}$ an orthonormal and **R** an upper triangular matrix, directly into the normal equation and simplifying the resulting term), which differ in overhead and numeric stability (refer to any decent book on numerical mathematics for details).

Example 9.2 (Positions with imprecise information) To illustrate this concept, look at the previous example, assuming that for the three anchors only incorrect position estimates $\tilde{r}_1 = 5$, $\tilde{r}_2 = 1$, and $\tilde{r}_3 = 4$ are available. Solving the resulting equation corresponding to Equation 9.7 gives the incorrect position (5.2, 4.8) with a distance of $\sqrt{(5.2 - 5)^2 + (4.2 - 2)^2} \approx 2.2$ between estimated and correct position.[4]

Adding additional anchors at $(x_4, y_4) = (3, 1)$, $(x_5, y_5) = (7, 5)$, $(x_6, y_6) = (2, 8)$, and $(x_7, y_7) = (4, 6)$ with distance estimates $\tilde{r}_4 = 2$, $\tilde{r}_5 = 3$, $\tilde{r}_6 = 7$, and $\tilde{r}_7 = 4$, respectively should improve this estimate. The resulting matrix **A** and right hand side **b** are

$$\mathbf{A} = \begin{bmatrix} 2 & 5 \\ -1 & 2 \\ -4 & 4 \\ 1 & 5 \\ -3 & 1 \\ 2 & -2 \end{bmatrix} \quad \mathbf{b} = \begin{bmatrix} 56 \\ -4 \\ -16 \\ 30 \\ -29 \\ 17 \end{bmatrix}. \tag{9.11}$$

Solving $\mathbf{A}^{\mathrm{T}}\mathbf{Ax} = \mathbf{A}^{\mathrm{T}}\mathbf{b}$ for $\mathbf{x}$ results in $\mathbf{x} = (5.5, 2.7)$, with a distance error of $\sqrt{(5.5 - 5)^2 + (2.7 - 2)^2} \approx 0.86$.

---

[4] Note that this is not necessarily the mean square solution!

Hence, this formalism allows computing of the position with the smallest mean square error out of $n \geq 3$ anchors in the presence of errors in the distance measurements. The generalization to three dimensions is obvious.

It is possible to extend this formalism even further when other parameters have to be estimated as well. SAVVIDES et al. [725], for example, discuss how to include the estimation of the unknown ultrasound propagation speed (in the context of a TDoA ranging system) into this optimization problem.

9.4 Single-hop localization

Using these basic building blocks of distance/range or angle measurements and the mathematical basics, quite a number of positioning or locationing systems have been developed. This section concentrates on systems where a node with unknown position can directly communicate with anchors – if anchors are used at all. The following section contains systems where, for some nodes, multihop communication to anchors is necessary. These single-hop systems usually predate wireless sensor networks but provide much of the basic technology upon which multihop systems are built.

9.4.1 Active Badge

The "Active Badge Location System" [863] is the first system designed and built for locating simple, portable devices – badges – within a building. It uses diffused infrared as transmission medium and exploits the natural limitation of infrared waves by walls as a delimeter for its location granularity. A badge periodically sends a globally unique identifier via infrared to receivers, at least one of which is installed in every room. This mapping of identifiers to receivers (and hence rooms) is stored on a central server, which can be queried for the location of a given badge. HARTER and HOPPER [333] describe an appropriate software environment for the Active Badge system. It is possible to run additional queries, such as which badge is in the same room as a particular given badge. As soon as badges are directly connected to persons, privacy issues play a crucial role as well.

9.4.2 Active office

After the Active Badge system introduced locating techniques, WARD et al. [864] targeted the positioning of indoor devices. Here, ultrasound is used, with receivers placed at well-known position, mounted in array at the ceiling of a room; devices for which the position is to be determined act as ultrasound senders.

When the position of a specific device shall be determined, a central controller sends a radio message, containing the device's address. The device, upon receiving this radio message, sends out a short ultrasound pulse. This pulse is received by the receiver array that measures the time of arrival and computes the difference between time of arrival of the ultrasound pulse and the time of the radio pulse (neglecting propagation time for the radio wave). Using this time, a distance estimate is computed for every receiver and a multilateration problem is solved (on the central controller), computing a position estimate for the mobile device. Sending the radio pulse is repeated every 200 ms, allowing the mobile devices to sleep for most of the time.

The system also compensates for imprecision in the distance estimates by discarding outliers based on statistical tests. The obtained accuracy is very good, with at least 95 % of averaged position estimates lying within 8 cm of the true position. With several senders on a mobile device, the accuracy is even high enough to provide orientation information.

9.4.3 RADAR

The RADAR system [35] is also geared toward indoor computation of position estimates. Its most interesting aspect is its usage of scene analysis techniques, comparing the received signal characteristics from multiple anchors with premeasured and stored characteristic values. Both the anchors and the mobile device can be used to send the signal, which is then measured by the counterpart device(s). While this is an intriguing technique, the necessary off-line deployment phase for measuring the "signal landscape" cannot always be accommodated in practical systems.

9.4.4 Cricket

In the Active Badge and active office systems described above, the infrastructure determines the device positions. Sometimes, it is more convenient if the devices themselves can compute their own positions or locations – for example, when privacy issues become relevant. The "Cricket" system [659] is an example for such a system. It is also based on anchors spread in a building, which provide combined radio wave and ultrasound pulses to allow measuring of the TDoA (signal strength information had been found to be not reproducible enough to work satisfactorily). From this information, symbolic location information within the building is extracted. Reference [659] also discusses interference and collision considerations that are necessary when devices/anchors are not synchronized with each other. A simple randomized protocol is used to overcome this obstacle.

9.4.5 Overlapping connectivity

BULUSU et al. [106] describe an example for an outdoor positioning system that operates without any numeric range measurements. Instead, it tries to use only the observation of connectivity to a set of anchors to determine a node's position (Figure 9.5). The underlying assumption is that transmissions (of known and fixed transmission power) from an anchor can be received within a circular area of known radius. Anchor nodes periodically send out transmissions identifying themselves (or, equivalently, containing their positions). Once a node has received these announcements from all anchors of which it is in reach (typically waiting for a few periods to smooth out the effect of random packet losses), it can determine that it is in the intersection of the circles around these anchors. The estimated position is then the arithmetic average of the received anchors' positions. Moreover, assuming that the node knows about all the anchors that are deployed, the fact that some anchor announcements are not received implies that the node is outside the respective circles. This information further allows to restrict the node's possible position.

The achievable absolute accuracy depends on the number of anchors – more anchors allow a finer-grained resolution of the area. At 90 % precision, the relative accuracy is one-third the separation distance between two adjacent anchors – assuming that the anchors are arranged in a regular mesh and that the coverage area of each anchor is a perfect circle. In a 10 m × 10 m area, the average error is 1.83 m; in 90 % of the cases, positioning error is less than 3 m. Accuracy degrades if the real coverage range deviates from a perfect sphere (as it usually does in reality). In

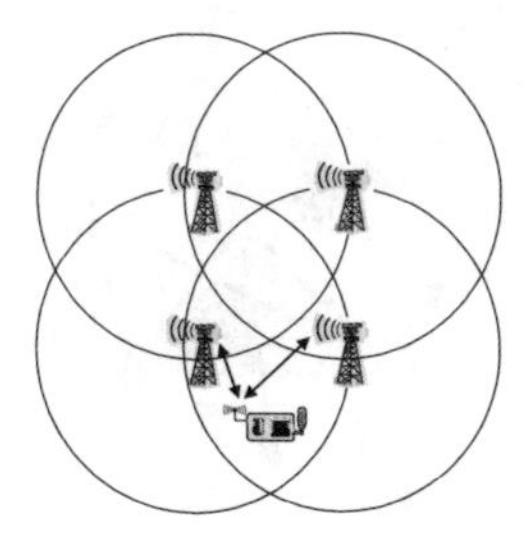

Figure 9.5 Positioning using connectivity information to multiple anchors [106]

addition, the transmission range has to be chosen carefully to result in a minimal positioning error, given a set of anchors.

9.4.6 Approximate point in triangle

The previous approach has used a range-free connectivity detection to decide whether a node is inside or outside a circle around a given anchor. In fact, more information can be extracted from pure connectivity information. The idea is to decide whether a node is within or outside of a triangle formed by any three anchors [339]. Using this information, a node can intersect the triangles and estimate its own position, similar to the intersection of circles from Section 9.4.5.

Figure 9.6 illustrates the idea. The node has detected that it is inside the triangles BDF, BDE, and CDF and also that it is outside the triangle ADF (and ABF, AFC, and others). Hence, it can estimate its own position to be somewhere within the dark gray area – for example, this area's center of gravity.

The interesting question is how to decide whether a node is inside or outside the triangle formed by any three arbitrarily selected anchors. The intuition is to look at what happens when a node inside a triangle is moved: Irrespective of the direction of the movement, the node must be closer to at least one of the corners of the triangle than it was before the movement. Conversely, for a node outside a triangle, there is at least one direction for which the node's distance to all corners increases.

Moving a sensor node to determine its position is hardly practical. But one possibility to approximate movements is for a node to inquire all its neighbors about their distance to the given three corner anchors, compared to the enquiring node's distance. If, for all neighbors, there is at least one corner such that the neighbor is closer to the corner than the enquiring node, it is assumed to be inside the triangle, else outside – this is illustrated in Figure 9.7. Deciding which of two nodes is closer to an anchor can be approximated by comparing their corresponding RSSI values.

Both the RSSI comparison and the finite amount of neighbors introduce errors in this decision. For example, for a node close to the edge of the triangle, there is a chance that the next neighbor in the direction toward the edge is already outside the triangle, incorrectly leading the enquiring node to assume this also – reference [339] gives more cases and details. Therefore, the approach is likely to work better in dense networks where the probability of such kinds of errors is reduced. Note that it can still be regarded as a range-free algorithm since only relative signal strength received by two nodes is compared, but no direct relationship is presupposed between RSSI values and distance. Nonetheless, nonmonotonic RSSI behavior over distance is a source of error for this approach. Because of these potential errors, it is only an Approximate Point in Triangle (APIT) test.

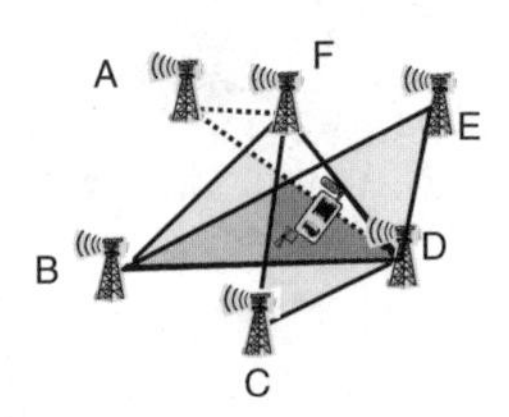

Figure 9.6 Position estimates using overlapping triangles

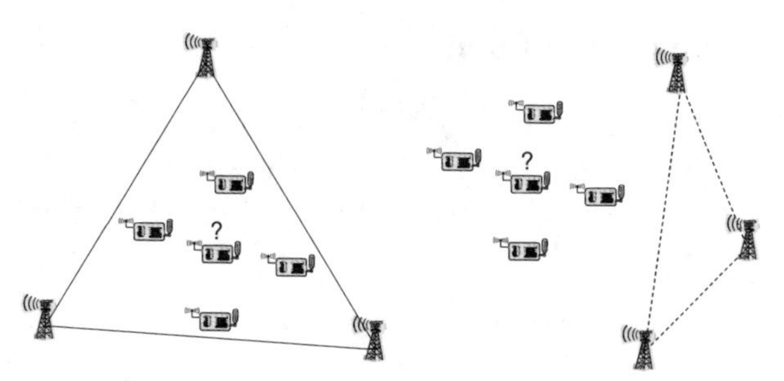

Figure 9.7 Testing whether a node is in a triangle or not using APIT (enquiring node is marked with a "?")

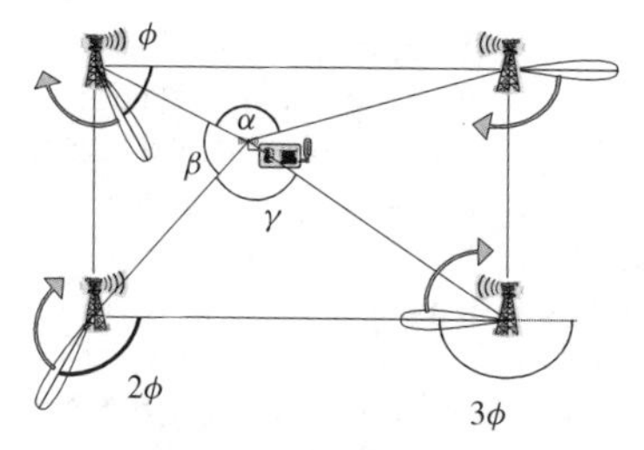

Figure 9.8 Rotating beacons provide angle of arrival information via timing offsets [586]

9.4.7 Using angle of arrival information

One example method to obtain angular information in a sensor network is described by Nasipuri and Li [586]. They use anchors nodes that use narrow, rotating beams where the rotation speed is constant and known to all nodes. Nodes can then measure the time of arrival of each such beam, compute the differences between two consecutive signals, and determine the angles α, β, and γ from Figure 9.8 using straightforward geometric relationships. The challenge here is mainly to ensure that the beams are narrow enough (less than 15° are recommended) so that nodes have a clear triggering point for the time measurements and to handle effects of multipath propagation. An advantage of this approach is that it is unaffected by the network density and causes no traffic in the network; the sensor nodes themselves can remain quite simple. In simulations, excellent accuracy is reported, limiting the positioning error to about 2 m in a 75 m × 75 m area [586].

9.5 Positioning in multihop environments

All the approaches and concepts described in Section 9.4 assume that a node trying to determine its position can directly communicate with – in general – several anchor nodes. This assumption is not always true in a wireless sensor network – not every node is in direct contact with at least three anchors. Mechanisms are necessary that can somehow cope with the limited geographic availability of (relatively) precise ranging or position information. Such mechanisms and approaches are described here. In some form or another, they rest upon the fact that for a sufficiently connected graph with known length of the edges, it is possible to reconstruct its embedding in the plane (or in three-dimensional space).

9.5.1 Connectivity in a multihop network

A semidefinite program feasibility formulation

A first approach [206] to the multihop positioning problem is (predominantly) based upon connectivity information and considers the position determination as a feasibility problem. Assume that the positions of n anchors are known and the positions of m nodes is to be determined, that connectivity between any two nodes is only possible if nodes are at most R distance units apart, and that the connectivity between any two nodes is also known. The fact that two nodes are connected introduces a constraint to the feasibility problem – for two connected nodes, it is impossible to choose positions that would place them further than R away. For a single node, multiple such constraints can exist that have to be satisfied concurrently – akin to the overlapping circles from above, which restrict the possible positions of a node.

On the basis of this formulation, DOHERTY et al. [206] give a formulation of the feasibility problem as a semidefinite program (a generalization of linear programs). This problem can be solved, but only centrally, requiring all connectivity information at one point.

The main observation here is that in this formulation, the fact that two nodes are not connected does not provide any additional information – it is impossible to write down a constraint that two nodes are *at least* a given distance away from each other in a semidefinite program; nodes cannot be "pushed apart". As an example consequence, a linear chain of nodes with only one anchor in it cannot be distinguished from a situation where all nodes are clustered around the anchor. This implies that anchor nodes should preferably be placed at the borders of the network, to impose as many "pull apart" constraints as possible. Such controlled placement considerably reduces the average positioning error compared to random anchor placements.

DOHERTY et al. [206] also discuss variations and extensions of this basic idea, in particular, using estimates of the actual distance instead of only the upper bound derived from connectivity; angular information instead of ranges (inspired by directed, optical communication); and computing bounds on the position error by solving multiple feasibility problems. Because of its essentially centralized character, however, this concept is only of limited applicability to WSNs.

MultiDimensional scaling

The same basic problem of range-free, connectivity-based locationing is solved by SHANG et al. [755] using the mathematical formalism of MultiDimensional Scaling (MDS). On the basis of connectivity between nodes, an all-pair shortest path algorithm roughly estimates positions of nodes. This initial estimate is improved by MDS, and if nodes with absolute position information are available, the resulting coordinates are properly normalized.

The details of this mathematical technique are somewhat involved; the reader is referred to reference [755]. The main advantage to this approach is that it is fairly stable with respect to anchor placement, achieving good results even if only few anchors are available or placed, for example, inside the network. JI and ZHA [383] show, in addition, that MDS is also suitable for anisotropic networks (networks where the distance between neighbors is not uniform over the extent of the network).

9.5.2 Multihop range estimation

The basic multilateration approach requires a node to have range estimates to at least three anchors to allow it to estimate its own position. NICULESCU and NATH [597] consider the problem when anchors are not able to provide such range estimates to all nodes in the network, but only to their direct neighbors (because of, for example, limits on the transmission power). The idea is

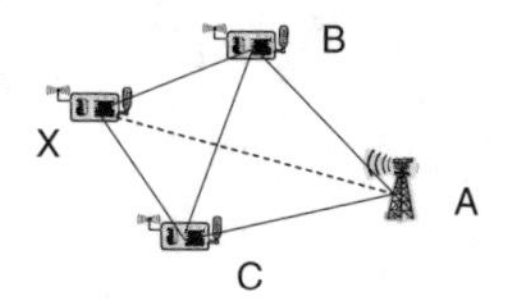

Figure 9.9 Euclidean distance estimation in the absence of direct connectivity [597]

to use indirect range estimation by multihop communication to be able to reuse the well-known multilateration algorithm.

To compute range estimates between a node and a far-off anchor via multiple intermediate hops, NICULESCU and NATH [597] describe three different possibilities. All of them are based on flooding the network with information, independently starting from each anchor, similar to the operation of a distance vector (DV) routing protocol.

The simplest possibility is the "DV-Hop" method. The idea is to count the number of hops (along the shortest path) between any two anchors and to use it to estimate the average length of a single hop by dividing the sum of the distances to other anchors by the sum of the hop counts. Every anchor computes this estimated hop length and propagates it into the network. A node with unknown position can then use this estimated hop length (and the known number of hops to other anchors) to compute a multihop range estimate and perform multilateration. Note that this is, in fact, a range-free approach as there is no need to estimate internode distances.

When range estimates between neighboring nodes are available, they can be directly used in the same framework, resulting in the "DV-Distance" method.

In presence of range estimates and a sufficient number of neighbors, a node can actually try to compute its true Euclidean distance to a faraway anchor. Figure 9.9 illustrates the idea: Assuming that the distances AB, AC, BC, XB, XC are all known, it is possible to compute the unknown distance XA (actually, there are two solutions, one where X is on the other side of the line BC – node X can potentially distinguish these two solutions based on local information). This way, actual positions can be forwarded between nodes.

The obtainable accuracy here depends on the ratio of anchors relative to the total number of nodes. The "Euclidean" method increases accuracy as the number of anchors goes up; the "distance vector"-like methods are better suited for a low ratio of anchors. As one would expect, the distance vector methods perform less well in anisotropic networks than in uniformly distributed networks; the Euclidean method, on the other hand, is not very sensitive to this effect.

9.5.3 Iterative and collaborative multilateration

The previous approach tried to estimate distances between nodes with unknown position and the anchors in order to apply multilateration with the anchors themselves. An alternative approach is to use normal nodes, once they have estimated their positions, just like anchor nodes in a multilateration algorithm. Figure 9.10 shows an example: Nodes A, B, and C are unaware of their position. Node A can triangulate its own position using three anchors. Once node A has a position estimate, node B can use it and two anchors for its own estimate, in turn providing node C with the missing information for its own triangulation. The basic idea for such **iterative multilateration** has been proposed by SAVVIDES et al. [725] and similarly by SAVARESE et al. [723, 724].

A centralized implementation is fairly trivial, typically starting with the as-yet-undetermined node that has the most connections to anchors/nodes with already-determined position estimates and iteratively computing more position estimates. In a distributed implementation, nodes can compute a position estimate once at least three neighbors can provide position information, resulting in an

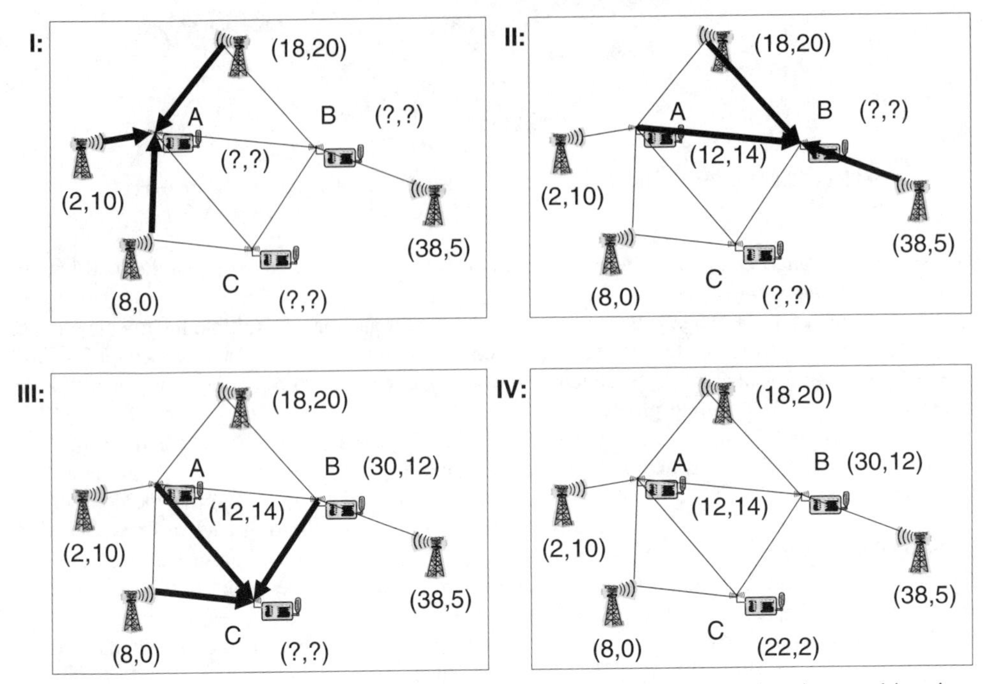

Figure 9.10 Iterative multilateration: Nodes A, B, and C can determine their positions in several iterations (bold arrows indicate the links whose range estimates are used in a given iteration)

initial estimate of a node's position. When more information becomes available – for example, because more neighbors have estimated their own position – it is possible to use it to improve the position estimate and propagate an updated estimate to a node's neighbors. The hope is that this algorithm will converge to the correct set of positions for all nodes. It should be pointed out that the initial position estimates for such an iterative refinement can also be computed by other means, for example, the DV-hop or DV-distance approaches from Section 9.5.2.

The average position error after such an iterative refinement depends on the accuracy of the range estimation, the initial position estimate, the average number of neighbors, and on the number of anchors [724]. Also, it is not guaranteed that the refinement algorithm converges at all; there can be situations where the position error increases the longer the algorithm runs. In fact, the straightforward refinement algorithm does not result in acceptable performance. This is mainly due to error propagation through the entire network. As one improvement, Savarese et al. [724] suggest to add confidence weights to all position estimates and to solve a modified weighted optimization problem, resulting in the convergence of almost all scenarios.

One particular challenge to this class of algorithm occurs when not all nodes in the network will have three nodes with position estimates. This is easy to detect for a single node, but difficult to do for an entire group of nodes. Depending on the topology, however, it might still be possible to estimate at least some positions – Savvides et al. [725] call this **collaborative multilateration**.

Figure 9.11 illustrates two problematic cases. The scenario on the left side is still fully determined as sufficient information is available to solve the equation system for the two nodes with unknown position. For the right scenario, there are two solutions (X and X′) for position of node X, which cannot be distinguished, but the position of the other unknown node can still be determined.

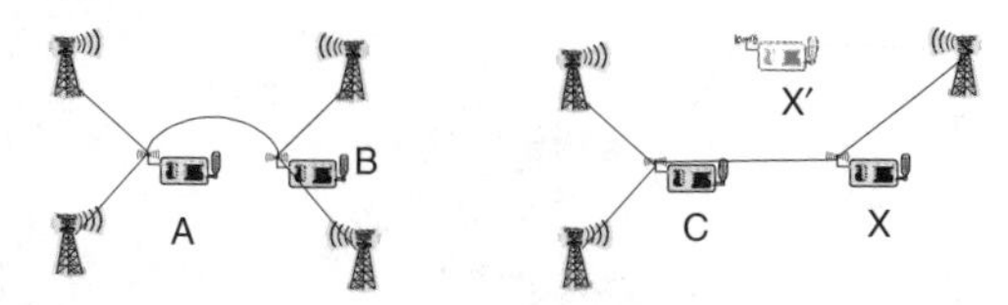

Figure 9.11 Problematic scenarios for iterative multilateration (adapted from [725])

Savvides et al. [725] approach these problematic cases by defining "participating nodes" – nodes that have at least three anchors or other participating nodes as neighbors, making nodes A and B in Figure 9.11 participating nodes. For such participating nodes, positioning can be solved.

Savarese et al. [724] take a slightly different approach here. The crucial observation is that a node, in order to determine its position, needs at least three *independent* references to anchor nodes – the paths to the anchors have to be edge-disjoint. Such nodes are called *sound*. In Figure 9.11, nodes A, B, and C are all sound. Soundness can be detected during the initial position estimation, for example, by recording over which neighbor the shortest path to a given anchor extends. If three or more such paths are detected, the node declares itself sound and enters the refinement phase. Node X from Figure 9.11, for example, will not be able to declare itself sound. As a consequence, the "soundness" procedure will be able to locate more nodes than the participating node concept from [725]. The algorithm of reference [724] is also better suited to low anchor ratios and can determine the positions of more nodes.

As quite a number of parameters influence precision and accuracy of these mechanisms, the reader is referred to the original publications for details; reference [465] also presents an extensive quantitative comparison of different mechanisms. General guidelines that can be derived are to ensure a high connectivity (more than 10 neighbors on average), to employ at least 5 % anchors, and to place anchors toward the edge of a network [724].

9.5.4 Probabilistic positioning description and propagation

The previous approaches all have described the position of a node, once it has been determined, by an explicit set of coordinates. This "deterministic" description collapses the inherent randomness, caused, for example, by uncertainties in range estimates. Ramadurai and Sichitiu [674] try to explicitly take into account this randomness by describing the position of a node by a probability function of the node's possible location, describing the amount of information that is available for this node's location.

Initially, a node can be at all locations with equal probability. A concrete distance measurement, for example, an RSSI value, gives rise to a probability density function, relating each distance to a certain probability with which it corresponds to the RSSI value (Figure 9.3). Hence, a node that measures a certain RSSI value from an anchor knows that it has a high probability of being somewhere in a circle around the anchor (Figure 9.12(a)). Once information from a second anchor becomes available (Figure 9.12(b)), the two density functions can be convoluted and an improved description of the node's position probabilities results (Figure 9.12(c)).

Ramadurai and Sichitiu [674] describe the mathematical details of these operations and the proper formalism for describing these probability functions. The resulting positioning error varies considerably between different nodes, ranging from excellent to quite large.

9.6 Impact of anchor placement

Several references discussed so far have already pointed out the importance of properly placed anchor nodes [206, 724], expressing a preference for anchors to be placed on the perimeter of a

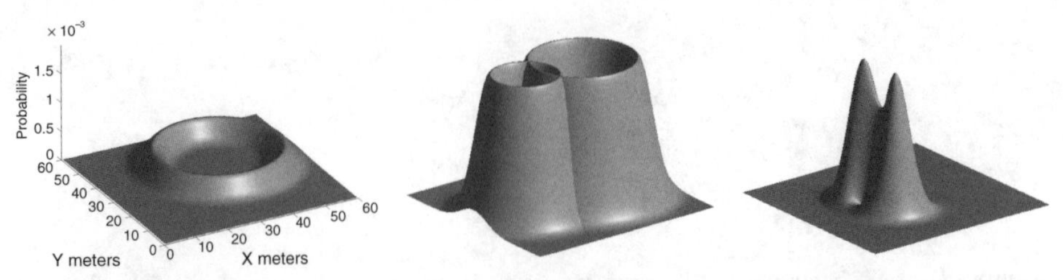

(a) Probability density function of a node positions after receiving a distance estimate from one anchor (b) Probability density functions of two distance measurements from two independent anchors (c) Probability density function of a node after intersecting two anchor's distance measurements

Figure 9.12 Probabilistic treatment of node positions [674]. Reproduced by permission of V. Ramadurai and M. L. Sichitiu

given area. Also, accuracy and precision improve if more anchors are available. The question is, hence, where to deploy such anchors and how many are necessary?

An up-front planning of anchor deployment is usually not possible in combination with most WSN deployment scenarios. Rather, an adaptive deployment scheme where anchors are added when and where necessary appears more promising. One simple such approach is described in reference [107] where a mobile entity is wandering around the given area, measuring positioning error compared against an external positioning source (e.g. differential GPS). Two different algorithms are suggested to decide where to deploy an additional anchor – either at the point of maximum location error or at the center of grids with maximum cumulative location error.

The obvious drawback of this approach is the need to have an absolute measure of positioning error. Bulusu et al. [105, 108] present an adaptive algorithm that tries to replace the mobile agent with a localized estimate of positioning errors. Already-existing anchors collect information about each other. Then, an anchor tries to estimate the positioning error that a hypothetical node would encounter at some given point; the anchor uses information about existing anchors and their reachability for this estimate. On the basis of these estimates, there are several schemes (analogous to those in reference [107]) that decide where an additional anchor should be placed. Clearly, this concept will be less precise than the actual measurements but incurs considerably less overhead. In addition to these adaptive placement algorithms, suited mostly for low-density networks, reference [105] also considers the possibility to selectively turn on/off anchors to reduce energy consumption while providing a given level of positioning accuracy.

9.7 Further reading

There is a lot of additional basic and advanced material on the topic of localization and positioning. Some pointers for going more into depth are:

GPS The practically most important system is certainly the Global Positioning System (GPS) [229, 400], which is based on some of the same ranging and multilateration principles as described here, despite being based on satellites as anchors.

Angulation Used in aviation, the VHF Omnidirectional Ranging (VOR) system is a practically important example for an angulation-based system. Angulation is also used by Niculescu and Nath [598] to amend the APS system with angulation information in a multihop context.

Error impact Some basic design considerations for positioning systems are discussed in reference [104]. It considers the tuning of density for quality/lifetime trade-offs, multiple sensor modalities to make ranging more robust, and exploiting environmental characteristics. It also considers some basic problems and the impact of various error sources – the non-line-of-sight problem for ranging, correlation between RSSI and distance, and the impact of anchor number. The impact of errors and their modeling is also discussed in detail in reference [775]. Both papers are somewhat more general, using positioning as a working example for more general considerations.

Anchor-free systems In the absence of anchors, only relative coordinate systems can be computed. References [581, 658] are two example papers dealing with this problem. An interesting result from reference [581] is the fact that an average neighborhood size of 15 is necessary for good accuracy.

Performance LANGENDOEN and REIJERS [465] compare important positioning algorithms in a uniform simulation environment under a set of different assumptions. They find that no singe algorithm performs best in all situations. Reference [726] compares centralized and distributed algorithms. WANG et al. [857] complement simulation-based work with an analysis of lower bounds on positioning accuracy.

Nonstandard approaches The lighthouse system [698] is an interesting alternative approach to localization – it uses a rotating, broad beam from anchor nodes to let nodes measure the start and end time of the beam, from which the distance to the anchor can be computed at high precision. NICULESCU and NATH [599] describe a positioning system where only nodes that are engaged in a "conversation" are arranged in a common coordinate system; this restriction allows to save overhead compared to a general solution.

9.8 Conclusion

Determining positions – and, to a lesser degree, also locations – in a wireless sensor network is burdened with considerable overhead and the danger of inaccuracies and imprecision. A non-negligible amount of anchors is necessary for global coordinate systems, and the time and message overhead necessary to compute positions if no direct communication between anchors and nodes is available should not be underestimated. Nonetheless, it is possible to derive out of erroneous measurements an often satisfactory degree of position estimates.

10

Topology control

Objectives of this Chapter

In a densely deployed wireless network, a single node has many neighboring nodes with which direct communication would be possible when using sufficiently large transmission power. This is, however, not necessarily beneficial: high transmission power requires lots of energy, many neighbors are a burden for a MAC protocol, and routing protocols suffer from volatility in the network when nodes move around and frequently form or sever many links.

To overcome these problems, *topology control* can be applied. The idea is to deliberately restrict the set of nodes that are considered neighbors of a given node. This can be done by controlling transmission power, by introducing hierarchies in the network and signaling out some nodes to take over certain coordination tasks, or by simply turning off some nodes for a certain time.

Chapter Outline

10.1 Motivation and basic ideas

One perhaps typical characteristic of wireless sensor networks is the possibility of deploying many nodes in a small area, for example, to ensure sufficient coverage of an area or to have redundancy present in the network to protect against node failures. While these are clear advantages of a *dense* network deployment – density as measured, for example, by the average number of neighbors that

Protocols and Architectures for Wireless Sensor Networks H. Karl and A. Willig

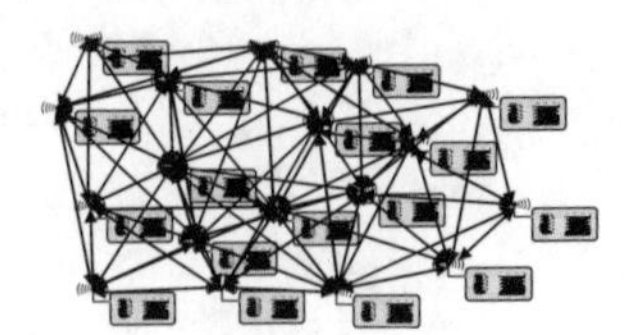

Figure 10.1 Topology in a densely deployed wireless sensor network

a single node has – there are also disadvantages. In a relatively crowded network (Figure 10.1), many typical wireless networking problems are aggravated by the large number of neighbors: many nodes interfere with each other, there are a lot of possible routes, nodes might needlessly use large transmission power to talk to distant nodes directly (also limiting the reuse of wireless bandwidth), and routing protocols might have to recompute routes even if only small node movements have happened.

Some of these problems can be overcome by **topology-control** techniques. Instead of using the possible connectivity of a network to its maximum possible extent, a deliberate choice is made to restrict the topology of the network. The topology of a network is determined by the subset of active nodes and the set of active links along which direct communication can occur. Formally speaking, a topology-control algorithm takes a graph $G = (V, E)$ representing the network – where V is the set of all nodes in the network and there is an edge $(v_1, v_2) \in E \subseteq V^2$ if and only if nodes v_1 and v_2 can directly communicate with each other – and transforms it to a graph $T = (V_T, E_T)$ such that $V_T \subseteq V$ and $E_T \subseteq E$.

10.1.1 Options for topology control

To compute a modified graph T out of a graph G representing the original network G, a topology-control algorithm has a few options:

- The set of active nodes can be reduced ($V_T \subset V$), for example, by periodically switching off nodes with low energy reserves and activating other nodes instead, exploiting redundant deployment in doing so.
- The set of active links/the set of neighbors for a node can be controlled. Instead of using all links in the network, some links can be disregarded and communication is restricted to crucial links.

  When a *flat* network topology (all nodes are considered equal) is desired, the set of neighbors of a node can be reduced by simply not communicating with some neighbors. There are several possible approaches to chose neighbors, but one that is obviously promising for a WSN is to limit the *reach* of a node's transmissions – typically by *power control*, but also by using adaptive modulations (using faster modulations is only possible over shorter distances) – and using the improved energy efficiency when communicating only with nearby neighbors.

  Figure 10.2 illustrates how the dense topology from Figure 10.1 can be reduced by applying power control. In essence, power control attempts to optimize the trade-off between the higher likelihood of finding a (useful) receiver at higher power values on the one hand and the increased chance of collisions/interference/reduced spatial reuse on the other hand [357].
- Active links/neighbors can also be rearranged in a *hierarchical* network topology where some nodes assume special roles. One example, illustrated in Figure 10.3, is to select some nodes as a "backbone" (or a "spine") for the network and to only use the links within this backbone and direct links from other nodes to the backbone. To do so, the backbone has to form a **dominating**

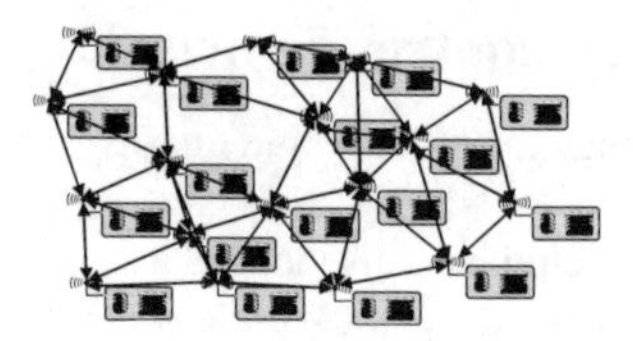

Figure 10.2 Sparser topology after reducing transmission power

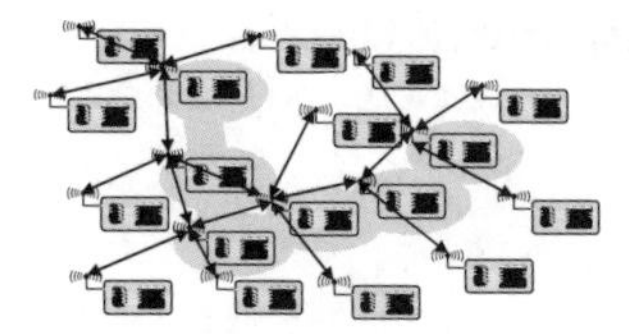

Figure 10.3 Restricting the topology by using a backbone

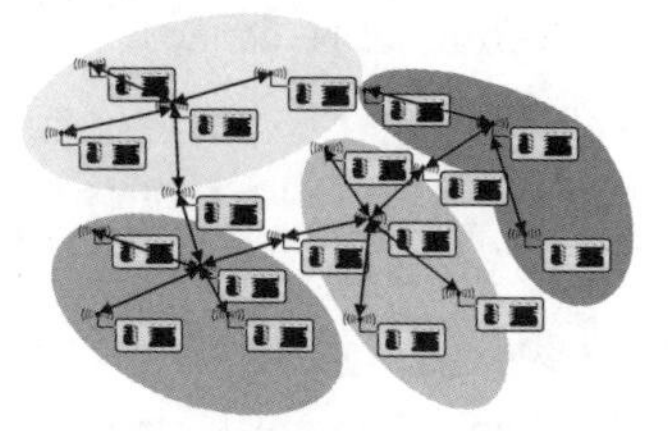

Figure 10.4 Using clusters to partition a graph

set: a subset $D \subset V$ such that all nodes in V are either in D itself or are one-hop neighbors of some node $d \in D$ ($\forall\, v \in V : v \in D \ \vee\ \exists\, d \in D : (v, d) \in E$). Then, only the links between nodes of the dominating set or between other nodes and a member of the active set are maintained. For a backbone to be useful, it should be **connected**.

A related, but slightly different, idea is to partition the network into **clusters** (Figure 10.4). Clusters are subsets of nodes that together include all nodes of the original graph such that, for each cluster, certain conditions hold (details vary). The most typical problem formulation is to find clusters with **clusterheads** – a representative of a cluster such that each node is only one hop away from its clusterhead. When the (average) number of nodes in a cluster should be minimized, this is equivalent to finding a maximum (dominating) **independent set** (a subset $C \subset V$ such that $\forall\, v \in V - C : \exists\, c \in C : (v, c) \in E$ and no two nodes in C are joined by an edge in E – $\forall\, c_1, c_2 \in C : (c_1, c_2) \notin E$). In such a clustered network, only links within a cluster are maintained (typically only those involving the clusterhead) as also selected links between clusters to ensure connectivity of the whole network".

Both problems are intrinsically hard and various approximations and relaxations have been studied.

These three main options for topology control – flat networks with a special attention to power control on the one hand, hierarchical networks with backbones or clusters on the other hand – will be treated in more detail in Sections 10.2 10.3, and 10.4, respectively. First, a few desirable aspects of topology-control algorithms should be discussed.

10.1.2 Aspects of topology-control algorithms

There are a few basic metrics to judge the efficacy and quality of a topology-control algorithm [671]:

Connectivity Topology control should not disconnect a connected graph G. In other words, if there is a (multihop) path in G between two nodes u and v, there should also be some such path in T (clearly, it does not have to be the same path).

Stretch factors Removing links from a graph will likely increase the length of a path between any two nodes u and v. The **hop stretch factor** is defined as the worst increase in path length for any pair of nodes u and v between the original graph G and the topology-controlled path T. Formally,

$$\text{hop stretch factor} = \max_{u,v \in V} \frac{|(u, v)_T|}{|(u, v)_G|} \tag{10.1}$$

where $(u, v)_G$ is the shortest path in graph G and $|(u, v)|$ is its length.

Similarly, the **energy stretch factor** can be defined:

$$\text{energy stretch factor} = \max_{u,v \in V} \frac{E_T(u, v)}{E_G(u, v)} \tag{10.2}$$

where $E_G(u, v)$ is the energy consumed along the most energy-efficient path in graph G.

Clearly, topology-control algorithms with small stretch factors are desirable. It particular, stretch factors in $O(1)$ can be advantageous.

Graph metrics The intuitive examples above already indicated the importance of a small number of edges in T and a low maximum degree (number of neighbors) for each node.

Throughput The reduced network topology should be able to sustain a comparable amount of traffic as the original network (this can be important even in wireless sensor networks with low average traffic, in particular, in case of event showers). One metric to capture this aspect is **throughput competitiveness** (the largest $\phi \leq 1$ such that, given a set of flows from node s_i to node d_i with rate r_i that are routable in G, the set with rates ϕr_i can be routed in T), see reference [671] for details.

Robustness to mobility When neighborhood relationships change in the original graph G (for example, because nodes move around or the radio channel characteristics change), some other nodes might have to change their topology information (for example, to reactivate links). Clearly, a robust topology should only require a small amount of such adaptations and avoid having the effects of a reorganization of a local node movement ripple through the entire network.

Algorithm overhead It almost goes without saying that the overhead imposed by the algorithm itself should be small (low number of additional messages, low computational overhead). Also, distributed implementation is practically a condition sine qua none.

In the present context of WSNs, connectivity and stretch factors are perhaps the most important characteristics of a topology-control algorithm, apart from the indispensable distributed nature and low overhead. Connectivity as optimization goal, however, deserves a short caveat.

A caveat to connectivity

Consider a simple example of power control. Five thousand nodes are uniformly, randomly deployed over a an area of 1000 by 1000 m. The transmission range of each node can be set to a precise radius of r m (i.e. all nodes at most r m apart can communicate directly, and no other nodes can). This model is known as the **disk graph model**; the special case of $r = 1$ is called the **Unit disk graph**. For one such example network and a given transmission range, the network is either connected or not. Determining connectivity for 100 different, randomly generated networks gives a rough estimate of the probability of connectivity as a function of the transmission range, shown as the dotted line in Figure 10.5.

As expected, the probability of connectivity is zero for small transmission ranges and raises relatively sharply, until it levels off and slowly approaches probability 1 at about 30 m transmission range.

The same experiment allows to consider an additional metric. For each repetition, the size of the largest connected component can be computed, and this size, averaged over the 100 repetitions, is shown in Figure 10.5, also as a function of the transmission range. Clearly, even for relatively small transmission ranges, *almost all* nodes are connected into a single component, even though the probability of connectivity is still practically zero because there are three nodes in an unfortunate position. For example, for transmission range 25 m, the average size of the largest component is 4997 – that is, only 3 out of 5000 nodes are not connected – whereas the probability of connectivity is still practically zero. Evidently, for WSN, connectivity might not be the relevant metric, but rather, a large value of the maximum component size would be more important.

That being said, the overwhelming part of research has gone into studying connectivity properties, with the importance of component sizes being only slowly realized.

Another important aspect treated later (Chapter 13) is **coverage** – making sure that all points in the plane are covered by an observing node. If the few nodes missing for connectivity are important for coverage as well, it might actually be inevitable to invest the energy required to connect them.

Figure 10.5 also shows another effect. The average size of the largest component and, to a slightly smaller degree, also the probability of having a connected network do not slowly increase with the maximum transmission range (or, equivalently, the density of the network). Rather, both metrics increase sharply from zero to their maximum values once a certain critical threshold for the transmission range is exceeded. This effect is known for a large number of aspects of (random) graphs in general and called a **phase transition**. The existence of such thresholds is provable for purely random graphs and plausible for (unit) disk graphs as used to model wireless networks. KRISHNAMACHARI et al. [441] discuss various examples for such thresholds and recommend to set operational parameters of a network just slightly larger than the critical threshold to obtain the

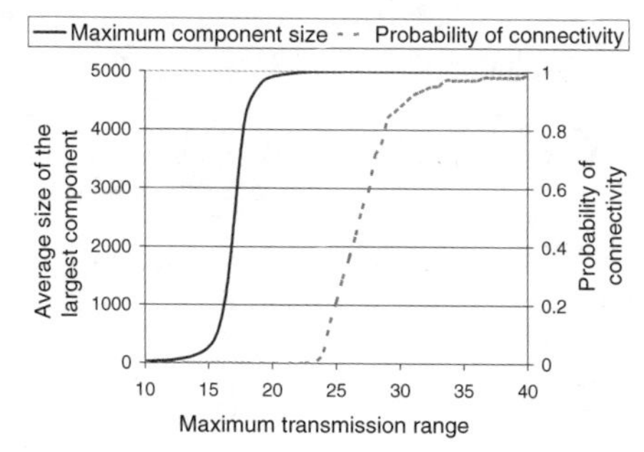

Figure 10.5 Optimization goal of topology control: Probability of connected network versus average size of the largest component

desired behavior without wasting resources. Phase transition phenomena are often characterized using percolation theory techniques [545].

10.2 Controlling topology in flat networks – Power control

Controlling the set of neighbors to which a node talks to is the basic approach of topology control. In this section, a *flat* topology is considered where all nodes are operational and have the same tasks. This problem is closely linked to controlling the transmission power of nodes and can often be found in the literature under this headline. For brevity, power control will here also be used as the concise, albeit somewhat imprecise, term for this problem class.

10.2.1 Some complexity results

The possible variations of the power control problem are numerous. To give a rough orientation, some basic complexity results are summarized here before some power control algorithms are discussed in more detail below.

To describe a power control problem, a four-tuple $(\mathcal{M}, \mathcal{P}, \mathcal{O}, \mathcal{I})$ is used (adapted and extended from reference [517]):

- $\mathcal{M} \in$ {dir, undir} describes a directed or undirected graph model (the definition given above is a directed graph (V, E), the corresponding undirected graph is $(V, E \cap E^{-1})$, that is, asymmetric links are removed).
- $\mathcal{P}$ is the property of the graph that is to be guaranteed in the topology-controlled graph T. Examples for properties include but are not limited to:
  - Strongly connected.
  - k-node connected (k-NC), that is, there are k nodes in the graph the deletion of which disconnects the undirected graph; the graph stays connected when deleting any $k-1$ nodes.
  - k-edge connected (k-EC) (analogous for edges). Note that k-node connectivity implies k-edge connectivity, but not necessarily vice versa.

  Particularly important are so-called *monotone* properties that continue to hold when transmission powers of some nodes are increased [517]. Being k-NC connected, for example, is monotone (increasing range only adds new connections), whereas being acyclic is not a monotone graph property (increasing range can lead to cycles).
- $\mathcal{O}$ is the objective function that is to be minimized. Examples include minimizing the maximum power assigned to any node ($\mathcal{O} =$ max P) or minimizing the sum of the power values assigned to all nodes ($\mathcal{O} =$ total P).
- $\mathcal{I}$ is any additional information that the topology control can use. The most prominent example is information about the geographic positions of nodes ($\mathcal{I} =$ pos).

Lloyd et al. [517] give a summary of some basic complexity results for topology control:

- (undir, 1-NC, total P, –) is NP-hard, there is an approximation algorithm with a performance guarantee[1] of 2 [146]. Geometric versions of the problem stay NP-hard in 2 or 3 dimensions.
- (undir, 1-NC, max P, –) and (undir, 2-NC, max P, –) are solvable in polynomial time [678].

[1] An approximation algorithm has a *performance guarantee* ρ if, for every problem instance, the approximated solution is within a factor of ρ of the optimal solution.

- For any *monotone* property $\mathcal{P}$ testable in polynomial time on an (un)directed graph, (undir, $\mathcal{P}$, max P, –) and (dir, $\mathcal{P}$, max P, –) can be solved in polynomial time. This is not necessarily the case for all $\mathcal{P}$ if also the number of nodes using maximum power should be minimized [517].
- There are nonmonotone but still efficiently testable properties $\mathcal{P}$ (e.g. graph is a tree) such that minimizing max P is NP-complete [517].
- LLOYD et al. [517] also describe a generic approach to design approximation algorithms for total P problems, which in general are NP-hard.
- There are several distributed, polynomial time constructions for subgraphs T of G with a constant stretch factor – these graphs are called **spanner graphs** (the distance between any two nodes in T is at most a given constant times the distance of the two nodes in G) [671].

10.2.2 Are there magic numbers? – bounds on critical parameters

Given this range of different problems, how do we design concrete algorithms that achieve a given optimization goal by controlling the graph's topology? A first idea is for every node only to communicate with its k nearest neighbors, where the (expected) number k is to be determined such that it optimizes a given goal, for example, throughput or connectivity. The simplest situation would be if it were possible to determine a value of k that would not depend on the actual graph to be controlled. Such a constant value of k could justifiably be called a *"magic number"* and there have been several attempts in the literature to proof the existence of such a number, starting already in the 1970s.

The probably best-known of these papers is by KLEINROCK and SILVESTER [421]. They consider the problem of a set of nodes uniformly distributed on a square and try to maximize the expected progress that a packet can make toward its destination, assuming a slotted ALOHA MAC protocol and several concurrent packet transmissions. They suggested that $k = 6$ would indeed maximize the progress per hop. This result was soon corrected to $k \approx 8$ by TAKAGI and KLEINROCK [804], using basically the same line of reasoning. Similar results, in slightly different settings, were achieved by HOU and LI [357].

While these are interesting results, it is important to realize that these papers do *not* address the problem of connectivity. Rather, their optimization goal is throughput. In other words, when using these "magic numbers", it is not guaranteed that the resulting graph will be connected. It fact, it is most likely not the case.

When looking at the connectivity problem, there are two options to approach the problem. One considers the *transmission range* of a node, and the other considers the *number of neighbors*. Under certain assumptions, these two options are equivalent, but lead to slightly different styles of proofs and results.

Controlling transmission range

The first option, controlling the transmission range, is purely geometric and assumes a unit disk graph model and a uniform distribution of nodes in a given area of size A. This model assumption corresponds to the theory of *geometric random graphs*, which can be used here to great benefit (see e.g. reference [205] for an overview).

For such a network, it is known that its probability of being connected goes to zero if $r(|V|) \leq \sqrt{\frac{(1-\varepsilon)A \log |V|}{\pi |V|}}$, for any $\varepsilon > 0$ [642].[2] If, on the other hand, $r(|V|) \geq \sqrt{\frac{A(\log |V| + \gamma_{|V|})}{\pi |V|}}$, then the graph

[2] This paper also conjectures that connectivity and coverage have similar bounds, but this conjecture was refuted by PIRET [645], who showed a counterexample for a linear network (nodes arranged on a line).

is connected with probability converging to one if and only if $\gamma_{|V|} \to \infty$ as a function of the number of nodes in the network [315].

A similar approach, based on geometric random graphs, is taken by BETTSTETTER [69] to determine an expression for the probability of a (dense) graph being k connected (making these results more general than the previous ones), depending on the transmission range r of the nodes and on the node density ρ (assuming uniform distribution of nodes over a given area). The key result used here is by PENROSE [630], who shows that a graph with a large number of nodes is k connected as soon as the transmission power becomes large enough to ensure that the smallest degree in the graph is at least k (with high probability). BETTSTETTER uses this result to develop a formula for the probability of the minimum node degree in a graph, resulting in the following main result:

$$P(G \text{ is } k \text{ connected}) \approx \left(1 - \sum_{l=0}^{k-1} \frac{(\rho\pi r^2)^l}{l!} \mathrm{e}^{-\rho\pi r^2}\right). \tag{10.3}$$

The paper also shows that this formula holds for infinite areas but not for finite areas as used in simulations and explains how to account for this **edge effect**. This treatment of edge effects and the practical approximation formulas make this a valuable and practical paper for use in simulation-based investigations as well. It is unclear, however, how these results would generalize to nonuniform node deployment and, in particular, to sparse networks as the theory of geometric random graphs does not apply there. In another paper, BETTSTETTER [70] also provides an extension of this work to mobile networks; reference [71] considers the multihop case.

The problem of k connectivity is also addressed by LI et al. [495] who show that the probability that a network of n nodes is at least $(k+1)$ connected is at least $\mathrm{e}^{\mathrm{e}^{-\alpha}}$ when the transmission radius r satisfies $n\pi r^2 \geq \ln n + (2k-1)\ln\ln n - 2\ln k + 2\alpha$ for $k > 0$ and n sufficiently large. As the k-connectivity problem is NP-hard, HAJIAGHAYI et al. [324] provide some additional heuristics, some of which achieve an $O(k)$ approximation of the optimal solution.

To also address **sparse networks**, SANTI and BLOUGH [722] formulate the problem slightly differently. They consider the random, uniform deployment of a fixed number of (possibly mobile) nodes n in a d-dimensional deployment region $R = [0, l]^d$, also including the hitherto not considered case of $d = 3$. There is no restriction on the node density $\rho = n/l^d$. The goal is to assign a minimal range r to all nodes such that the resulting graph is connected.

For the one-dimensional case $d = 1$, it is shown that the graph is connected with high probability if $rn \geq 2l \ln l$ and that it is disconnected with high probability if $rn < l \ln l$. For the $d = 2$ and $d = 3$ case, the following results are given:

- Assume $r^d l = k l^d \ln l$ for some constant $k > 0$, $r = r(l) \ll l$ and $n = n(l) \gg 1$. If $k > d \cdot 2^d d^{d/2}$ (or $k = d \cdot 2^d d^{d/2}$ and additionally $r = r(l) \gg 1$), then the communication graph is connected with high probability.
- Assume $r = r(l) \ll l$ and $n = n(l) \gg 1$. If $f^d \in O(l^d)$, then the communication graph is disconnected with high probability.

It is important to point out that these results, being formulated as constraints on the transmission range, can just as well be seen as a constraint on the minimum number of nodes that are necessary to cover a given region when there is an externally (e.g. hardware) imposed maximum transmission range.

SANTI and BLOUGH [722] also validated their analysis by simulation and provided some results on the dependence of the size of the largest component on the transmission range. These results are consistent with the caveat of Section 10.1.2. Moreover, they considered the mobile version of the connectivity problem as well. The transmission range can be substantially reduced if short-term disconnections can be tolerated (a temporal analog to only requiring a large component to be connected).

Controlling the number of neighbors

The second option is not to look at the area that a node's transmission range must cover, but rather directly at the number of nodes. The previous results already suggest that the expected number of neighbors of a node should grow logarithmically, and this conjecture has indeed been proven by XUE and KUMAR [904]. More precisely, they show the following (assuming that links are symmetric, i.e. an originally asymmetric link is complemented by the missing link in the reverse direction): (i) For the network to be connected, the number of neighbors of a node needs to grow like $\theta(\log |V|)$, with the constants being bounded as follows. (ii) For less than $0.074 \log |V|$ neighbors, the network is asymptotically disconnected. (iii) For more than $5.1774 \log |V|$ neighbors, the network is asymptotically connected. BLOUGH et al. [83] point out that these results continue to hold when a strictly asymmetric link is dropped.

These results show that there are no magic numbers defining the number of neighbors required to obtain a connected network. The large degree (logarithmic in the number of nodes) required for connectivity is actually somewhat inconvenient for practical network operation.

10.2.3 Some example constructions and protocols

Sparsing a topology can be efficiently done locally if information about distances between nodes or their relative positions is available. Several constructions for such *proximity graphs* exist with different properties – some are described here (partially following the description given in reference [671]). Much of this work relies on results from computational geometry; EPPSTEIN [242], for example, provides an overview.

The concrete results are sometimes in the form of a geometric construction on a given graph, which would – usually – correspond to a centralized protocol. Sometimes, distributed protocols are also given.

The relative neighborhood graph

The Relative Neighborhood Graph (RNG) [830] T of a graph $G = (V, E)$ is defined as $T = (V, E')$ where there is an edge between nodes u and v if and only if there is no other node w (a "witness") that is closer to either u or v than u and v are apart from each other – formally, $\forall\, u, v \in V : (u, v) \in E'$ iff $\nexists\ w \in V : \max\{d(u, w), d(v, w)\} < d(u, v)$, where $d(u, v)$ is the Euclidean distance between two nodes. Put another way, the RNG construction removes the longest edge from any triangle (Figure 10.6).

The RNG is easy to determine with a local algorithm. It is also necessarily connected if the original graph G is connected. However, its worst-case spanning ratio is $\Omega(|V|)$ (hence, nodes that are only a few hops apart in the original graph can become very distant from each other in the RNG), and its energy stretch is polynomial. Its average degree is 2.6.

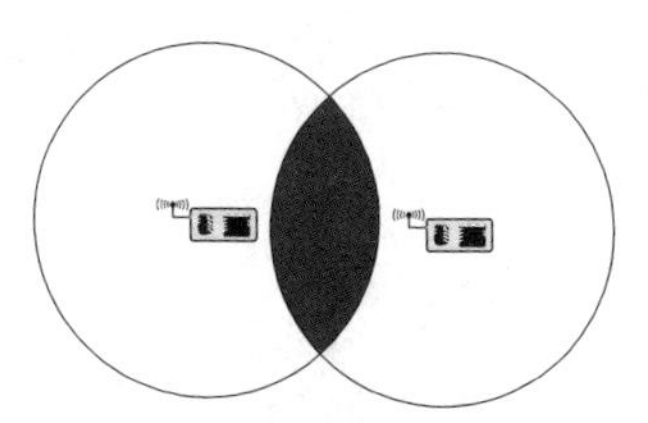

Figure 10.6 Constructions of the RNG: Shaded region must not contain another node for two nodes to be connected

The Gabriel graph

The Gabriel graph (GG) is defined similarly to the RNG; the formal definition for its edges is $\forall\, u, v \in V : (u, v) \in E'$ if f $\nexists\, w \in V : d^2(u, w) + d^2(v, w) < d^2(u, v)$. Equivalently, u and v are connected if and only if the circle with diameter $d(u, v)$ and nodes u and v on its circumference contains no other nodes but u and v. It also maintains connectivity, its worst-case spanning ratio is $\Omega(\sqrt{|V|})$, its energy stretch is $O(1)$ (depending on the precise energy-consumption model), its worst-case degree is $\Omega(|V|)$.

A distributed construction for the Gabriel graph is described, for example, in references [93, 409]. A node simply has to test, for all its neighbors, whether the circle definition of the edge holds – this is easy to do if all nodes exchange their position with their neighbors.

Delaunay triangulation

Another sparsing construction leverages a classical structure from computational geometry, the Delaunay triangulation. To construct it, imagine that to each node all the points in the plane for which it is the closest node are assigned. The resulting structure (Figure 10.7) is called the **Voronoi diagram** of a node set with the Voronoi region around each node (it can extend to infinity) [605]; it can be constructed in $O(|V| \log |V|)$ time. Then, connect any two nodes for which the Voronoi regions touch to obtain the **Delaunay triangulation**. Another way of constructing this triangulation is the "empty circle" rule: there is an edge between nodes u and v if and only if there exists a circle that does contain no other nodes except u and v.

Using of the Delaunay triangulation of a graph for topology control has been suggested by various authors, for example, [279, 362, 494]. First, it is known to be a spanner (see references in [279]). Its construction via the empty circle rule, however, requires global knowledge. What is worse, the Delaunay construction might produce very long links, longer than the maximum transmission range. GAO et al. [279] show that the restricted Delaunay graph, which only contains links up to the maximum transmission range, is still a spanner with constant stretch factor. Please refer to reference [279] for details on the construction as well as a performance comparison with the RNG and the Gabriel graph. LI et al. [494] present another distributed construction of a localized Delaunay graph, resulting in a spanner of factor 2.5.

Spanning tree–based construction

Another construction is based on local minimum spanning trees [487]. The idea is that each node will collect information about its neighboring nodes (at maximum transmission power) and then construct (using Prim's algorithm, for example) a minimum spanning tree for these nodes, with energy costs used as link weights (links with same costs are distinguished by adding node identifiers

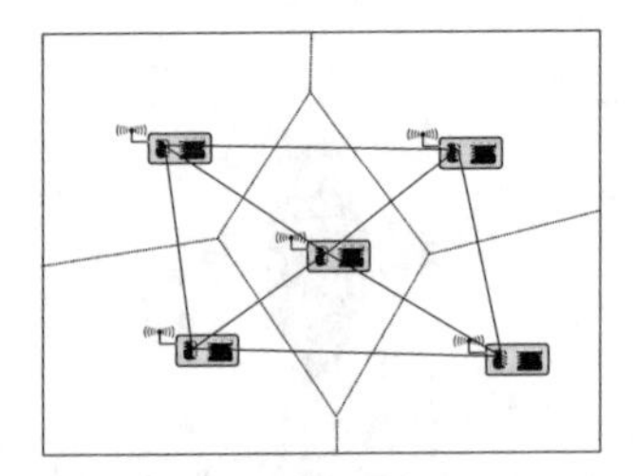

Figure 10.7 Voronoi diagram (dotted lines) and Delaunay triangulation (straight lines) for five nodes

as tiebreaker). The key is then to only maintain those edges in the reduced topology that correspond to direct neighbors in the minimal spanning tree.

This construction results in interesting properties. It preserves the connectivity of the original graph, and the maximum degree of each node will be six. It is possible to restrict to bidirectional links, and power control can be easily added. Moreover, the average node degree is small (close to theoretical bounds). It compares favorably with several other constructions discussed here.

Relay regions and enclosures

A crucial part of constructing a topology is deciding which neighbors to use. For wireless networks, this decision should be based (at least to a large degree) on the energy consumption that results from using a given node as a direct neighbor. Rodoplu and Meng [697] start from this point of view and first consider the notion of a **relay region**: Given a node i and another node r, for which points in the plane would i use r as a relay node in order to reduce the total power consumption, compared with the direct communication case? Formally, the relay region $R_{i \to r} = \{(x, y) | P_{i \to (x,y)} > (P_{i \to r} + P_{r \to (x,y)})\}$, where $P_{a \to b}$ is the minimum power required to communicate directly from a to b (identifying nodes with their position in the notation). Depending on the used power consumption model, the shape of the relay region is different; taking into account that the relay also requires power itself, the relay region is "bended" around the relay node as sketched in Figure 10.8 and asymptotically approaches a line orthogonal to the shortest line between nodes i and r.

On the basis of this power-aware definition of the set of points to which a relay node should be used, the next step is to define the nodes with which a given node i should communicate directly – in reference [697], the area where these nodes are located is called the **exposure**. A first intuition would be to simply intersect the complements of all relay regions and to only consider nodes that lie in this intersection as direct neighbors. For a node outside this intersection, there is at least one other node that can provide a less power costly route than direct communication. This intuition holds for some cases, as exemplified by the left half of Figure 10.9.

To ensure that a sufficient number of edges are preserved in the graph, Rodoplu and Meng [697] deviate from this simple definition. The example in the right part of Figure 10.9 highlights a case where both nodes x and z are maintained as neighbors, since node y, which would "dominate" node z, is in turn dominated by node x and thus not considered. Formally, this characterization of the neighboring set is expressed as a fixed point equation. On top of the thus defined graph, a routing algorithm is used to find minimum energy paths in the network.

Several papers have followed up on this definition of relay regions. Li and Halpern [484], for example, consider an enclosure definition closer to the intuition and characterize conditions under which this definition is applicable; they also describe a simple algorithm for determining it based on a stepwise increment of the transmission power that a node uses.

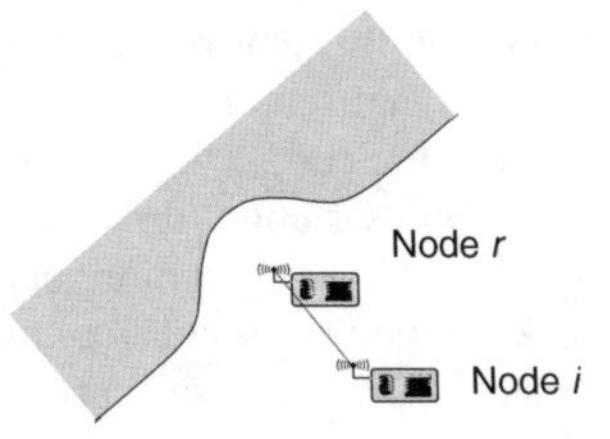

Figure 10.8 Illustration of the relay region of node i with node r as possible relay (shaded area indicates points where it is more power efficient to relay instead of direct communication) [697]

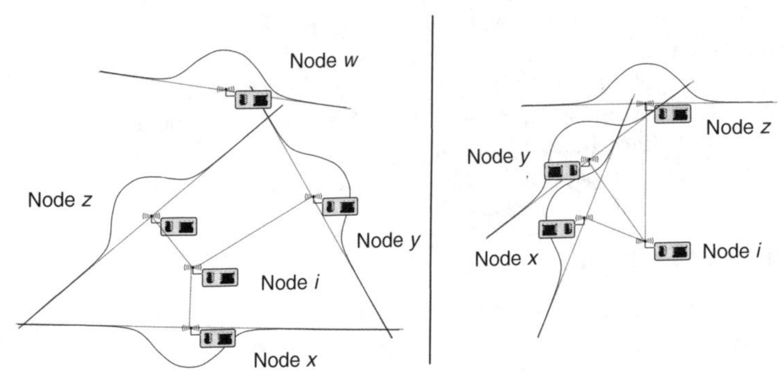

Figure 10.9 Two scenarios of how to construct the neighbors for node i out of the relay regions with other nodes [697]

Cone-based topology control

When not only distances to neighboring nodes but their *directions* are also available (with respect to some arbitrary absolute orientation, using some angle-of-arrival technique discussed in Section 9.1), even more powerful topology-control algorithms become possible. One out of several ideas based on such direction information – cone-based topology control – is introduced by WATTENHOFER et al. [865] and further analyzed in reference [485] (the "Yao graph", another, related idea is described in reference [905]). They set out to obtain a connected subgraph that minimizes power consumption by finding minimum power paths; moreover, the topology should have a small degree.

Cone-based topology control is a two-phase algorithm. The first phase constructs a connected topology with each node starting from a very small transmission power and increasing it until it has "sufficiently many" neighbors. In the second phase, redundant edges are removed. The creative insight here is a clever solution to the "sufficiently many" decision problem.

An arbitrary node u starts looking for neighbors by sending broadcast messages with increasing power; these packets are answered by neighbors. Any discovered neighbor is recorded by u in a neighbor list. Node u continues this process until there is a neighbor v in every cone of angle α or until u uses the maximum transmission power. Node u can easily detect this condition as each discovered neighbor v "covers" a cone of angle α around itself. When the superposition of these cones covers the full angle 2π, the process terminates.

The important observation is now the choice of α. WATTENHOFER et al. [865] show that when $\alpha = 2\pi/3$ (and nodes of course only use the smallest power to reach all their neighbors), then the resulting graph is connected if the graph obtained by using maximum power is also connected, that is, their topology algorithm maintains connectivity. LI et al. [485] improve this result by showing that taking $\alpha = 5\pi/6$ is a necessary and sufficient condition to preserve connectivity (in other words, using an $\alpha > 5\pi/6$ does no longer guarantee connectivity).

The second phase of the algorithm simply removes all neighbors w of u for which there is another neighbor v of both u and w such that sending via v is more energy efficient than sending directly (or up to a given constant more efficient).

In addition to this appealingly simple construction, the resulting graph also shows good performance compared with an optimal graph obtained by a more complicated algorithm.

An extension to the Yao graph makes this type of construction even k connected (if the original graph was k connected in the first place) [495]: adding links to the closest $k + 1$ nodes per cone, not only the closest node per cone.

Centralized algorithms for (bi)connectivity, minimizing maximum power

RAMANATHAN and ROSALES-HAIN [678] were likely the first authors to seek topology-control algorithms for node (bi)connectivity that minimize the *maximum* power: the (undir, 1-NC, max P, –) and the (undir, 2-NC, max P, –) problems in the notation of Section 10.2.1. In reference [678], they present two centralized algorithms to solve this problem.

For both algorithms, the network is represented as a graph $G = (V, L, P, \gamma, E)$, where V is the set of nodes, $L : V \rightarrow C$, $C = \mathbb{R} \times \mathbb{R}$, represents the locations of all nodes in the plane, $P : V \rightarrow \mathbb{R}$ the transmission power used by each node (this is easily generalized to other tunable parameters like antenna direction, but is not pursued in the paper), $\gamma : C \times C \rightarrow \mathbb{R}$ the path loss between any two sets of coordinates, and $E = \{(u, v) : P(u) - \gamma(L(u), L(v)) \geq S\}$ (power and path loss given in dB) is the set of edges that exist if and only if the transmission power of node u suffices to result, after path loss, in a received power at v larger than a given, constant receiver sensitivity S. The paper only assumes that path loss increases monotonically with distance and uses distance orderings as equivalent to path-loss orderings. In that only these orderings are used in the algorithms, the approach is nongeometric.

The paper then looks for a mapping P such that the resulting (V, E) is connected and $\max_{u \in V} P(u)$ is minimized over all possible mappings P. This problem is solved by using a centralized, greedy algorithm (similar to minimum spanning tree algorithms) that starts out with each node using transmission power 0, forming its own connected component. The algorithm iteratively connects those two components that have the "cheapest" link between them. It stops when only one connected component remains and the graph is therefore connected. The simplest implementation traverses a list of all node pairs (u, v) in order of increasing costs and checks whether these two nodes belong to different components; if so, they are connected by assigning them the smallest transmission power that allows them to communicate.

This problem formulation (and the algorithm) does not have a unique solution as it does not constrain the power levels of the nodes that use less than the maximum power. Here, preference is given to *per-node optimal* solutions, in which it is not possible to reduce the transmission power of any node without sacrificing connectivity. In particular, it may be possible to reduce the transmission power of some nodes because the links that they had to form early on in the greedy algorithm are not needed in the final graph as other nodes provide paths that ensure connectivity. An example for such nodes is shown in Figure 10.10.

The biconnectivity problem is formulated as finding a set of minimal power increments necessary to make a connected graph biconnected. A similar greedy algorithm, with a corresponding postprocessing phase, achieves this goal.

In addition to these two centralized algorithms, RAMANATHAN and ROSALES-HAIN [678] also provide two distributed heuristics. One tries to keep a node's degree at a specified value (results from Section 10.2.2 apply here), and the other also exploits information from a routing protocol.

TSENG et al. [831] extend the basic ideas of this concept to account for nodes with unequal initial energy supplies. They are interested in maximizing the lifetime of the network (time until first node failure) and consider a "fixed power regime", where nodes have to change their transmission power levels once at initialization time, and a "variable power" one, where nodes can change their transmission power under certain conditions. The principal construction in this paper is, similar to reference [678], also based on an adaption of a minimal spanning tree.

A distributed, common power protocol – COMPOW

The motivation behind the COMPOW ("common power") protocol [582] is twofold: The first observation is that when assigning the identical transmission power to all nodes the resulting

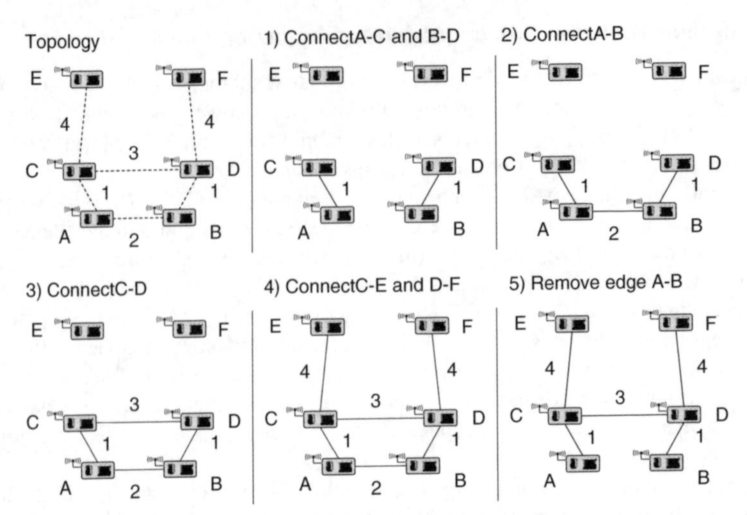

Figure 10.10 Greedy algorithm for connectivity at minimal transmission power with postprocessing phase. Nodes use the largest transmission power indicated on any adjacent edge. Postprocessing phase 5 allows nodes A and B to reduce their transmission power back to 1. [678]

per-node throughput is only negligibly worse than when every node has its individual power level – the difference is only a factor of $\frac{1}{\sqrt{|V|}}$. The second observation is the need to keep the transmission power level as low as possible to just ensure connectivity, lest valuable area would be needlessly consumed by long-range transmissions. Both observations are based on arguments adapted from reference [316].[3]

The heuristic used by the COMPOW protocol to determine the smallest power that results in a connected network is fairly simple. It is based on the assumption that a finite number of different power level is available and that the COMPOW protocol can be tightly integrated with a routing protocol. Then, each node determines routing tables for each transmission power level. A node will use the smallest transmission power for which the associated routing table has the same number of entries (i.e. reachable nodes) as the table for the maximum transmission power.

While this idea is relatively simple to implement, the need to maintain routing tables with all potential neighbors in the network makes it hardly appropriate for wireless sensor networks.

The K-NEIGH protocol

The results of Section 10.2.2 are used also by Blough et al. [83] to develop the K-NEIGH protocol. The idea is to keep the number of neighbors per node at or around a number k determined by the formulas of Xue and Kumar [904].

The proposed protocol is distributed, based on distance estimates between nodes, and requires a total of $2|V|$ message exchanges. It is based on nodes announcing their identifier at high transmission power, collecting their observed neighbors, sorting neighbors by distance, and computing the k nearest neighbors that can mutually reach each other. Each node then uses the smallest transmission power that suffices to reach all its neighbors thus computed. In the design of this protocol, care has to be taken to wait long enough for nodes waking up at random times and also to properly account for potentially asymmetric links.

[3] The arguments in the paper about power-optimal routes rest on a over-simplified power consumption model.

10.2.4 Further reading on flat topology control

Further sparsing constructions WANG et al. [861] describe another sparsing construction, the Yao-Yao graph. It is unknown whether this is actually a spanner. Other extensions of the Yao graph are discussed in references [308, 710].

Hardness results CLEMENTI et al. [171, 172] and BLOUGH et al. [83] provide further hardness results.

Percolation theory Many of the problems discussed here can also profitably be approached with techniques from percolation theory. XUE and KUMAR [904] provide some references that can serve as a starting point.

Distributed power control KUBISCH et al. [448] describe two distributed algorithms for controlling the number of neighbors to certain values and evaluate their performance by simulation.

Asymmetric maximum power LIU and LI [512] are concerned with the problem of nodes having different maximum transmission ranges, resulting in the formation of asymmetric links. They describe a distributed topology-control algorithm that minimizes maximum power and maintains the reachability of every node.

Power control and mobility ROYER et al. [705] have studied how power control interacts with ad hoc routing protocols – namely, Ad hoc On-demand Distance Vector (AODV) – performance. They show that there is no single optimum density but that density should increase with movement.

Power control and IEEE 802.11 When not considering topology control as such, but only using power control as a means to improve energy efficiency and interference in an IEEE 802.11 context, the mechanisms described by AGARWAL et al. [11] are useful. They allow each node to choose a separate power level per neighbor, put explicit RSSI information into the RTS/CTS exchange packets, and adapt power levels according to this information, which is also cached. The authors show some modest improvements in energy efficiency and throughput. Roughly similar goals are also targeted by references [571, 572].

Power control and code assignment HUANG et al. [363] consider the combined problem of power control for topology control and the resulting, variable code assignment problem (where code can be a time slot, a frequency band, or an orthogonal CDMA code).

Impact on network performance metrics Much of the work mentioned so far has treated the network purely as a graph, abstracting away from the need to carry traffic. LI and SINHA [486], for example, evaluate the impact of power control on actual network performance metrics like throughput and energy per delivered packet and show considerable improvements in these crucial metrics. Similarly, ZUNIGA and KRISHNAMACHARI [947] have looked at the time it takes to distribute a query in a WSN and derived an equation for the optimum transmission radius to minimize the time when the last node has received the query.

Cross-layer aspects As some of the previous items here have shown, considering power control in isolation of other network layers is not necessarily a good approach. As another, fairly general example, consider the work by CRUZ and SANTHANAM [185], who have described a combined link schedule and power control algorithm, minimizing total power under constraints of minimal data rate that each link has to carry; it also solves the routing problem.

10.3 Hierarchical networks by dominating sets

10.3.1 Motivation and definition

The previous Section 10.2 has looked at controlling the transmission range and hence the *number of neighbors* of a node. This and the following section look at approaches that choose *which specific* nodes should be neighbors of a given nodes, and which other nodes (and the links to these nodes) should be ignored. While this implicitly also influences the number of active neighbors, some of these neighbors can be far away, and some nearby.

Usually, but not necessarily, this selection of neighbors/links implies some form of hierarchy among nodes. In this section, some nodes are selected to from a "virtual backbone" or, formally, a dominating set – a set of nodes $D \subset V$ is a dominating set if all nodes in V are either in D itself or are one-hop neighbors of some node $d \in D$ ($\forall\, v \in V : v \in D \vee\; \exists\, d \in D : (v, d) \in E$). Figure 10.3 has already visualized this notion. HAYNES et al. [337] provide an extension treatment of domination issues in general graphs.

Having such a dominating set at hand simplifies routing, for example, by allowing to restrict the actual routing protocol to the backbone nodes only; all "dominated" nodes can simply forward a nonlocal packet to (one of) their adjacent backbone node(s) that will then take care of forwarding the packet towards its destination.

To be useful, such a dominating set should be in some sense "small" or even minimal – the above definition admits $V = U$ as a solution, which evidently does not provide any advantages. A typical metric in which to measure the size and/or minimality of a dominating set U is its number of nodes; other options are conceivable as well.This is called the *Minimum Dominating Set* (MDS) problem. Moreover, a backbone should intuitively be connected using nodes only within itself; it should not be necessary to recur to other, nonbackbone nodes to route a packet from one place to the other.

Hence, the problem at hand is the following: Devise a preferably distributed algorithm that determines a Minimum Connected Dominating Set (MCDS). In addition, each node should know whether it is in this set or not and, if not, which of its neighbors are in the dominating set.

10.3.2 A hardness result

Not surprisingly, the MDS problem is NP-hard (this and the following results are taken from reference [182]). It is even a hard problem to approximate in general graphs as it not even approximable within $c \log |V|$ for some $c > 0$ (i.e. there is no polynomial time algorithm that always finds a dominating set that has at most $c \log |V|$ more nodes than a minimal dominating set would have). It is approximable within $1 + \log |V|$. For the case of unit disk graphs, it is possible to find a Polynomial Time Approximation Scheme (PTAS).

When also requiring connectivity in solving the MCDS problem (which is also NP-hard), it is possible to approximate it within $\ln \Delta + 3$, where Δ is the maximum degree of the original graph [182]. This observation is encouraging for the design of practical algorithms (and also an incentive to solve MCDS on top of a power-controlled topology which limits the degree of the nodes).

A further important result is proven by WAN et al. [851]. They show that any algorithm that computes a nontrivial connected dominating set solution must send at least $\Omega(m \log n)$ messages, where the message size is on the order of the number of bits required to express unique identifiers for all nodes $O(\log |V|)$.

10.3.3 Some ideas from centralized algorithms

While centralized algorithms are usually not directly applicable to WSNs, they can provide some ideas on how to design distributed algorithms for a given problem. Two such centralized examples are described here; both are taken from reference [309].

Growing a tree

A naïve approach

The idea of the first algorithm is to construct the dominating set as a spanning tree, iteratively adding nodes and edges to this tree until all the nodes are covered. This algorithm is formulated using "colors" of nodes: white nodes have not yet been processed, black nodes belong to the dominating set, and gray nodes are the dominated nodes.

The algorithm is shown in pseudocode in Listing 10.1. It takes a graph $G = (V, E)$ and produces a set of edges T that belong to the tree-shaped dominating set and a coloring of the nodes. A run of this algorithm is illustrated in Figure 10.11.

The resulting graph is indeed a tree as there is never an edge between two gray nodes or between a gray node and two different black nodes; the black nodes also form a tree. The set of black nodes

Listing 10.1: A naïve dominating set algorithm based on growing a tree

```
initialize all nodes' color to white
pick an arbitrary node and color it gray

while (there are white nodes) {
  pick a gray node v that has white neighbors
  color the gray node v black
  foreach white neighbor u of v {
    color u gray
    add (v,u) to tree T
  }
}
```

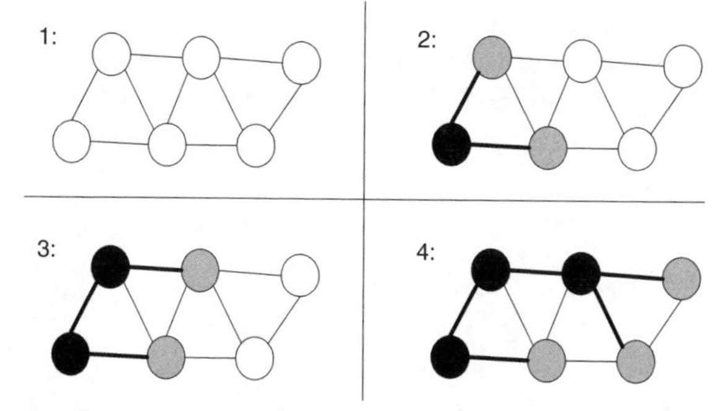

Figure 10.11 Illustration of a naïve dominating set algorithm based on growing a tree (thick lines indicate edges in the tree, black nodes form the dominating set)

(the nonleaf nodes of the spanning tree) is also a dominating set, as each gray node has a black node as a neighbor and there are no white nodes left over. The size of this dominating set, however, need not be particularly small or even minimal. In the example of Figure 10.11, the two nodes at the bottom right would also have formed a (smaller) dominating set. Clearly, picking the gray node to be turned into a black node is the crucial step in this algorithm.

Judiciously choosing gray nodes

One intuitively appealing heuristic choice for the next gray node is that node which would turn the whitest nodes gray [309]. This number can be regarded as the "yield" of choosing a particular gray node.

This greedy heuristic does indeed find the optimal solution for the example in Figure 10.11. But there are simple graphs for which this heuristic fails; one such graph is shown in Figure 10.12. Here, the algorithm would start at node u, add any one of the nodes directly connected to u, and then has to break ties between node u's peer nodes and chooses one node connected to node v. If node u has d neighbors, the worst case for such a greedy yield-based algorithm is a dominating set size of $O(d)$, where the optimal size is 4 (u, v, and two nodes from any of the vertical lines – nodes u and v would suffice if connectivity of the dominating set were not required).

The reason for this nonoptimal performance is the fact that the algorithm cannot distinguish between choosing a gray node from the first or the second row – each operation reduces the number of white nodes by one. This myopic operation hinders the algorithm from realizing that once a node from the second row has been chosen, the bottom node v is gray and could be turned black, producing the required dominating set as node v would cover all the remaining white nodes in the second row.

This shortsightedness can be overcome by allowing the algorithm to look ahead one step [309]. In addition to considering the yield of turning individual gray nodes black, the algorithm can also look at pairs of a gray node g and an adjacent white node w, speculating about what would happen if this node w would be also turned black in the following step. For each combination of nodes g and w, the yield of this operation can be computed. The heuristic is then to choose the individual gray node or the pair of a gray and a white node that produces the highest yield.

This idea is illustrated in Figure 10.13. The upper part shows the plain yield-based heuristic, failing by turning gray nodes in the upper row black one after the other without making essential progress – picking any gray node individually gives a yield of 1 without favoring a node in the second row. In the lower part, the lookahead for nodes marked g and w (after the first step of the algorithm) gives a yield of $1 + (d - 1)$ white nodes turned gray, clearly favoring this operation over choosing individual gray nodes.

This particular example shows that this lookahead heuristic is promising. It is rather powerful: GUHA and KHULLER [309] have shown that it produces connected dominating sets at most a factor

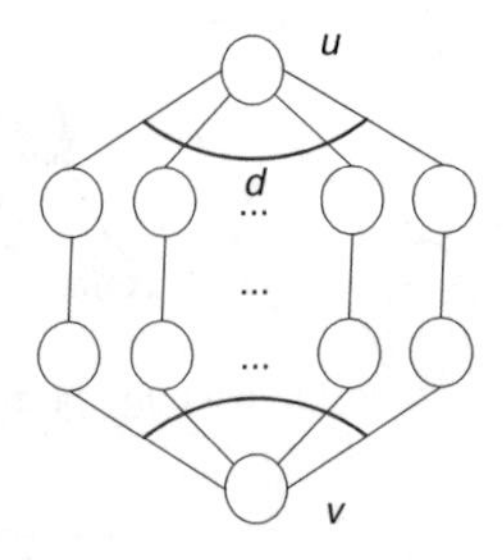

Figure 10.12 Example graph for which a greedy, one-step yield-based heuristic fails [309]

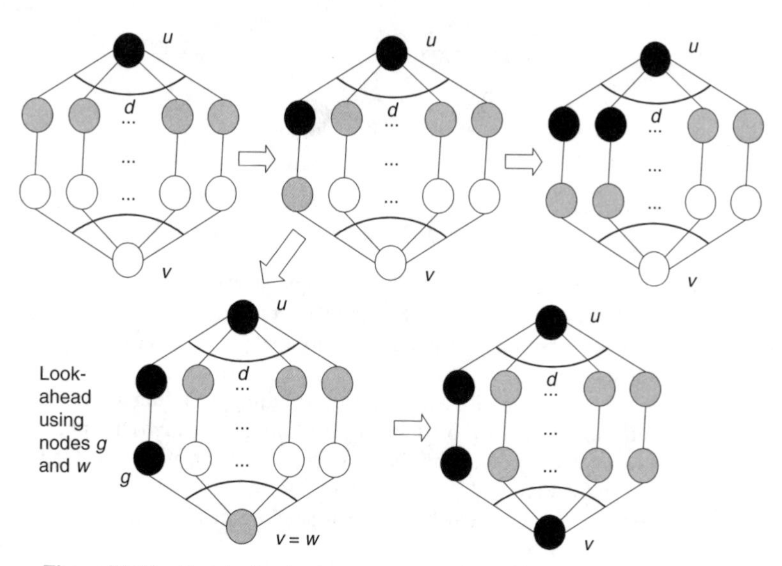

Figure 10.13 Greedy dominating set construction without and with lookahead

of $2(1 + H(\Delta))$ larger than an optimal one ($H(\cdot)$ is the harmonic function $H(k) = \sum_{i=1}^{k} 1/i \leq \ln k + 1$ and Δ the maximum degree of the graph). Moreover, there is empirical evidence that this algorithm works well in practice.

Connecting separate components

In the previous approach, the set of nodes that will eventually form the dominating set is always connected. An alternative idea is to first construct a not necessarily connected dominating set and then, in a second phase, explicitly connect the nodes in this set.

Finding a nonconnected dominating set

In a centralized algorithm, finding *some* (nonconnected) dominating set is fairly easy. Using the white/gray/black coloring codes from above, pick a gray or white node, color it black and color all of its white neighbors gray. Proceed until there is no white node left. Like in the previous algorithm, care has to be taken in choosing which node to turn black in order to obtain a small dominating set.

One simple heuristic would be to pick that node that turns most white nodes gray. Having in mind the need to later on connect the black nodes into a connected set, it also intuitively makes sense to choose gray nodes that lie in between two black nodes since there is a good chance that such gray nodes might have to be added to the dominating set anyway. Figure 10.14 shows an example where this consideration breaks the tie between nodes A and B, resulting in B being added to the dominating set.

Formally, the heuristic proposed by Guha and Khuller [309] is thus to select that white or gray node as the next black node that most reduces the sum of the number of the white nodes and the number of connected components of black nodes. When there is no node left that reduces this number, there is also no white node left and the first phase can terminate.

Ensuring connectivity – building a Steiner tree

Once this nonconnected dominating set has been found, it has to be connected, that is some more gray nodes have to be turned black so as to ensure that two black nodes can reach each other with

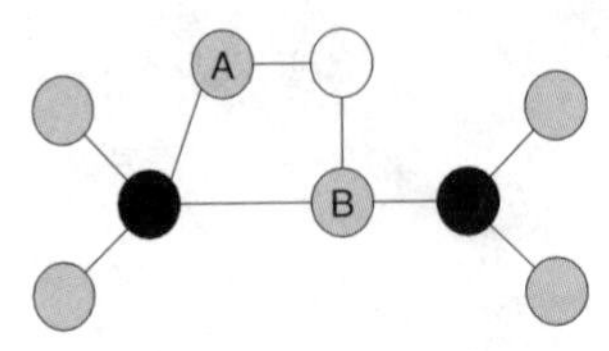

Figure 10.14 Taking into account the need to connect two black nodes breaks the tie between nodes A and B

only other black nodes as intermediaries. This is fairly simple as well, since at most two gray nodes can separate two "adjacent" black components. Recursively connecting two black components with one or two gray nodes in between will eventually lead to a single connected component of black nodes, solving the problem.

As a side remark, this problem can also be regarded as finding a **Steiner tree**: find a minimum spanning tree that contains all nodes of a predefined set of nodes, adding other nodes as required. This is yet another NP-complete problem, but has a constant approximation factor [182]. Steiner tree heuristics are treated in Section 11.4.2 in the context of multicasting.

Overall, the dominating sets determined by this heuristic are at most $\ln \Delta + 3$ larger than the optimal possible ones.

10.3.4 Some distributed approximations

Distributedly growing a tree

The heuristic of growing a tree [309] can be straightforwardly implemented in a distributed fashion as well, as described by Das and Bharghavan [192]. Essentially, all gray nodes explore their two-hop neighborhood, determining the biggest yield that each node could achieve. Then, the largest yield is chosen distributedly (in effect, a leader election takes place along the already existing backbone) and the next black node is determined.

The dominating set C determined by this algorithm is at most $2H(\Delta)$ large than the optimal one; it takes $O(|C|(\Delta + |C|))$ time and $O(n|C|)$ messages to compute [192].

Connecting a dominating set

Das and Bharghavan [192] have also considered how to adapt the other centralized algorithm described in Section 10.3.3, determining a small dominating set and then connecting it in a separate step.

Suppose the degree of the graph is bounded by some constant c, $\Delta \leq c$. Let every node broadcast its degree to all of its neighbors. Using this degree information from all neighbors, every node marks the neighboring node with the highest degree as its dominating node (or itself). The resulting set is dominating (obviously, since each node chooses one dominator), but not necessarily connected; a Steiner tree–like connection algorithm takes care of ensuring connectivity in a second phase.

The resulting dominating set has a performance ratio of $\Delta + 1$ in $O(n\Delta)$ time. Since $\Delta \leq c$, this is still a constant performance ratio. Guaranteeing a maximum degree for a random deployment of sensor nodes is nontrivial, though.[4]

Overall, the algorithms presented by Das and Bharghavan [192] have a logarithmic approximation factor [851] (see this reference also for an example).

[4] Das and Bharghavan [192] also discuss a similar algorithm for the case of unbounded degree; however, the details of this algorithm are not fully spelled out in their paper. Reference [851] has some additional analysis on this algorithm, showing a logarithmic performance ratio.

Marking nodes with unconnected neighbors

While both the algorithms discussed so far tried from the outset to limit the size of the dominating set, another idea is to first look for some, possibly large, connected dominating set that can be easily constructed and then reduced in size. If a such constructed set had some other favorable properties, even a nonminimal dominating set might be acceptable.

A nonoptimal connected dominating set construction

A simple observation [894] does allow to quickly, with low message overhead, construct a connected dominating set. Assume again that all nodes are initially unmarked (colored white). Mark any node (make it a member of the dominating set) if it has two neighbors that are not directly connected. This decision can be made locally after nodes have exchanged their neighbor sets – any node simply has to check whether each of its neighbors is included in all the received neighbor lists.

This simple construction entails some interesting properties:

- If the original graph is a connected but not fully connected graph, the resulting set of marked nodes is a dominating set. This can be shown by considering several cases. The crucial case is a nonmarked node v with only nonmarked neighbors and at least one nondirectly connected node u. Recognizing that the first node on a shortest path between v and u would have to be marked by construction shows the contradiction.
- The resulting set of marked nodes is connected. The proof is also by contradiction, using a shortest path between two marked nodes from two different connected components of the dominating set. It is shown that either all nodes along this path have to be marked or it is not the shortest path, contradicting the assumption that the two components were disconnected.
- The shortest path between *any* two nodes does not include any nonmarked nodes. The proof again goes by contradiction about the property of nodes on a shortest path. This property is not necessarily enjoyed by the previous constructions and makes this particular dominating set much more amenable for using as a routing backbone.
- The dominating set is not minimal; in fact, it can be trivial (encompassing all nodes of the original graph). Wu and Li [894] claim that it performs well for practical graphs.

Figure 10.15 shows an example graph where the resulting set of marked nodes is not minimal (both u, v and u, w would suffice).

Two pruning heuristics

Wu and Li [894] propose two heuristics to reduce the size of the dominating set in such situations. The first one is to unmark any node v such that both v itself and its neighborhood are included in the neighborhood of some other marked node u; also, the unique identifier of v must be smaller than u's. The second heuristic unmarks a node v if its neighborhood is included in the neighborhoods of two marked neighbors u and w and v has the smallest identifier. Figure 10.15 also shows that the distribution of the identifiers has considerable impact on which nodes can be removed. The case of mobile nodes is also treated in reference [894].

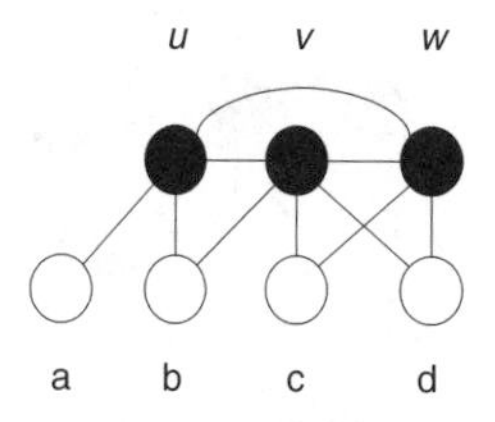

Figure 10.15 Example graph for marking nodes with unconnected neighbors [894]

Compared to the algorithms of DAS and BHARGHAVAN [192], the algorithm here only requires $O(\Delta^2)$ time to exchange neighborhood sets and constant time for the reduction phase. Moreover, simulation results indicate that the dominating set (after reduction heuristics) tends also to be smaller than that produced by the previous algorithms. The approximation factor of this algorithm even with both heuristics, however, is only linear [851].

A degree and position-based pruning heuristic

In the previously described pruning heuristics, the unique identifiers of nodes are used to decide which node can be excluded from the dominating set. STOJMENOVIC et al. [796] propose to modify these heuristics by replacing the identifier by a key that comprises the degree of a node and its x and y coordinates (alternatively, identifiers can be used instead of the coordinates if they are not known). Node u's key is larger (and will more likely remain in the dominating set) than node v's if either u's degree is larger or, if both degrees are equal, nodes' x and y coordinates are (lexicographically) larger.

This heuristic gives preference to nodes with large degrees, which makes intuitive sense for a dominating set. However, the authors evaluated their CDS construction only in the context of broadcasting, so a direct comparison is not available. WAN et al. [851] show that this improved heuristic is not enough to improve the linear approximation factor of the original algorithm.

Span

Another distributed approximation to the backbone problem is Span [141]. An important motivation for the Span design is the need to carry traffic in a backbone-based network at about the same rate as the underlying, flat network topology could accomplish – essentially, capacity should not be sacrificed for topology control. The practical consequence is that paths that "could operate without interference in the original network should be represented in the backbone" [141].

This desire to have a backbone that preserves capacity and has interference-free paths means that paths from the original network should not be overly dilated in the backbone network. More concretely, if any two nodes can communicate in the original network via at most one intermediate node, an extended path with at most two intermediaries is acceptable, but if a two-hop path extends to a five-hop path, additional nodes are required in the backbone – they should become "coordinators". Evidently, this rule is a relaxed version of the heuristic to mark nodes with unconnected neighbors. Figure 10.16 shows a simple example case. Note that the question whether the resulting backbone structure is actually connected is not explicitly discussed in the Span paper [141].

This rule can be easily implemented: A node v observes its neighbors and their broadcast traffic. If node v has two neighbors u and w that cannot communicate via at most two backbone nodes, this node v should become a member of the backbone itself to give a shorter route to nodes u and w. Typically, such a condition will be observed by multiple nodes and avoiding that all of them announce their willingness to join the backbone at the same time is one of the main concerns of the Span design. Such an "announcement contention" is combated using simple random backoff delays. This delay depends on (i) the relative number of not satisfactorily connected nodes that would be connected by this new node becoming a coordinator and (ii) the relative remaining battery

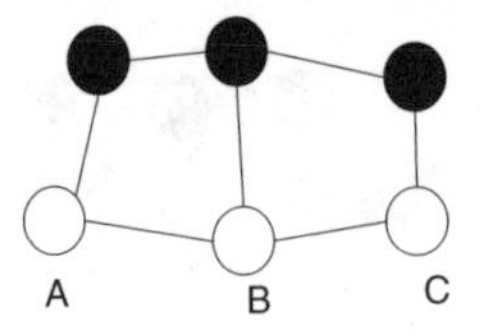

Figure 10.16 A case where Span would extend the backbone by node B as nodes A and C could have a two-hop path via B

capacity of the prospective coordinators. Both these numbers are used to bias the backoff delay, giving precedence to potential coordinators that connect many additional node pairs and that have a relatively full battery left. Reference [141] gives detailed formulas. Implicitly, this consideration of remaining battery capacity will result in rotating the role of coordinators among all nodes as the backbone membership is periodically reelected.

Role-based hierarchical self organization

All the dominating set approaches discussed so far were purely graph theoretical ones; they did not take into account the particular tasks and roles that the nodes in this graph have to fulfill. KOCHHAL et al. [427] propose a scheme to construct a dominating set that explicitly considers such information.

In particular, the degree of fault tolerance in sensing a given area that is implied by redundant deployment of sensors, resulting in multiple sensors surveying a given point, is used to decide the roles played by a given sensor. This degree of fault tolerance is expressed as an attribute of any given sensor, the Cumulative Sensing Degree (CSD) value. Reference [427] describes the technicalities necessary to define this value (which are somewhat involved because the surveillance space is discretized into small rectangles); intuitively, a high CSD value implies that the chances of an event being detected among its neighboring sensors is high.

This property of the CSD values is then used to assist in constructing a Connected Dominating Set (CDS). The initial construction marks nodes with unconnected neighbors [894], as described in the previous section. The pruning heuristics, however, use the CSD values to select nodes to be removed from the dominating set (among other values like energy level or node degree). On the basis of the resulting connected dominating set, sensor coordinators are finally elected.

This solution is attractive in that it combines a simple graph theoretic approach with knowledge about the actual sensing tasks that individual nodes have to perform. Reference [427] goes into further detail about how to use the CSD metric also as a basis for a more general QoS discussion.

10.3.5 Further reading

Weakly connected dominating sets CHEN and LIESTMAN [148] relax the requirement of finding a connected dominating set and are only looking for *weakly* connected dominating sets instead. A subset $S \subset V$ of nodes in a graph $G = (V, E)$ is weakly connected if the weakly induced subset $S_{<w>}$ is connected; $S_{<w>}$ consists of S and all the neighbors of S, and the edges of $S_{<w>}$ are all those edges in E that have at least one endpoint in S. Weakly connected dominating sets can be smaller than CDSs but retain most of their attractive properties. While it is still NP-complete to find a minimal weakly connected set, CHEN and LIESTMAN [148] present distributed approximation algorithms with a performance ratio of $O(\ln(\text{maximum degree of } G))$.

Nontrivial approximation in constant time None of the previous algorithms managed to produce a nontrivial approximation ratio in a constant number of rounds. The first algorithm that did achieve this has been presented by KUHN and WATTENHOFER [450]. More precisely, given a parameter k and maximum degree Δ of the input graph, their algorithm produces a dominating set of expected size $O(k\Delta^{2/k} \log \Delta |DS_{\text{opt}}|)$ in $O(k^2)$ rounds; each node sends $O(k^2\Delta)$ messages of size $O(\log \Delta)$. The approximation is based on a linear programming relaxation.

Generalized pruning heuristics DAI and WU [188] describe another heuristic, to remove any "gateway" node that is already covered by k other gateways. This rule formulation generalizes the two separate heuristics proposed in reference [894].

Backbones and databases LIANG and HAAS [497] describe a scheme how a virtual backbone can assist in the maintenance of location data bases for mobile nodes.

Exploiting node heterogeneity In many practical deployments, nodes will be heterogeneous in their capabilities and energy resources, for example, because some nodes can be powered from electrical outlets. Such more powerful nodes obviously lend themselves to act as backbone nodes (or as clusterheads, see below). As an example of such an approach, CONNER et al. [177] describe the ReOrg protocol along with a proper routing protocol and give an experimental evaluation.

"Virtually minimal" dominating sets When using a minimal dominating set in a very dense network (with a lot of nodes in one radio range) to retrieve information, the resulting resolution of the retrieved information can be too low to be useful. For such cases, DEB et al. [197] introduce the notion of minimal dominating sets, which are constructed on a "virtual" graph that, for the purpose of deciding domination relationships, use a radius r smaller than the actual communication range R. In addition, distance information between nodes already part of that minimal virtual dominating set and candidate nodes is used for probabilistic delays in forwarding join requests and the actual joining of nodes into the dominating set. The result of this construction is a backbone that contains more nodes than necessary; the process is parameterized to result in a desired density of active nodes.

Backbones with many leaves In combination with data aggregation as described in Chapter 12, it may be desirable to have a dominating set with many leaves. BOUKERCHE et al. [94] describe an appropriate, distributed heuristic.

Energy efficiency and broadcasting The problem of finding a backbone is clearly related to the broadcasting problem. This is discussed in more detail in Section 11.4.

10.4 Hierarchical networks by clustering

10.4.1 Definition of clusters

The previous Section 10.3 has introduced a hierarchy into a network by designating some nodes as belonging to a backbone, a dominating set. Another idea for a hierarchy is to locally mark some nodes as having a special role, for example, controlling neighboring nodes. In this sense, local groups or **clusters** of nodes can be formed; the "controllers" of such groups are often referred to as **clusterheads**. The hoped-for advantages of such clustering are similar to that of a backbone, but with additional emphasis on local resource arbitration (e.g. in MAC protocols), shielding higher layers of dynamics in the network (making routing tables more stable since all traffic is routed over the clusterheads), and making higher-layer protocols more scalable (since the size and complexity of the network as seen by higher layers is in a sense reduced by clustering). In addition, clusterheads are natural places to aggregate and compress traffic converging from many sensors to a single station.

Formally, given a graph $G = (V, E)$, clustering is simply the identification of a set of subsets of nodes V_i, $i = 1, \ldots, n$ such that $\cup_{i=1,\ldots,n} V_i = V$. A number of questions about the detailed properties required from these sets distinguish various clustering approaches:

Are there clusterheads? The partitioning of V into several clusters does not mandate anything about the internal structure of a cluster; in principle, all nodes can be equal (one example is described in reference [501]). Typically, however, for each set V_i there is a unique node c_i, the clusterhead, that represents the set and can take on various tasks. We shall almost exclusively deal with examples using clusterheads.

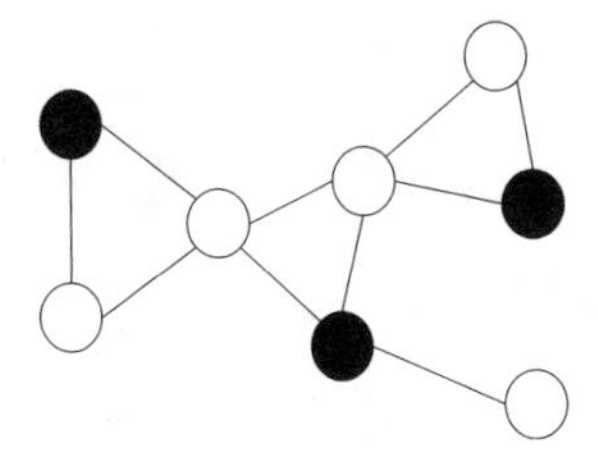

Figure 10.17 An example graph with a maximum independent set [59]

May clusterheads be neighbors? In principle, again, it is perfectly acceptable for two clusterheads (of two different clusters) to be direct neighbors. It is, however, often desirable to have clusterheads separated. Formally, clusterheads should form an **independent set**: a subset $C \subset V$ such that no two nodes in C are joined by an edge in E – $\forall c_1, c_2 \in C : (c_1, c_2) \notin E$.

Finding an arbitrary such set is trivial; the interesting case is **maximum independent sets**, which contain as many nodes as possible without violating the independence property, resulting in as many clusters around these clusterheads as possible. Figure 10.17 shows an example graph with one maximum independent set; others are possible (and easy to find) as the maximum independent set is, in general, not unique.

An important property of such a maximal independent set is that it is also dominating (easily proven via the contraposition). This property essentially justifies the importance of this problem formulation: maximum independent sets naturally partition the network and also form a subset of nodes that can control the network.

Determining maximum independent sets is, as expected, NP-complete. It does admit a PTAS for scalar graphs and unit disk graphs [182]. For bounded degree graphs, it is approximable within $(\Delta + 3)/5$ for small Δ, and within $O(\Delta \log\log \Delta / \log \Delta)$ for larger values [182].

While the maximum independent set formulation is elegant and simple, it does not necessarily reflect the actually desired configuration of the clusters. Consider, for example, a graph $G = (\{v_0, \ldots, v_n\}, \{(v_0, v_i)|i = 1, \ldots, n\})$ (i.e. one node connected to n other nodes that are not connected with each other). The maximum independent set for this graph is $v_1, \ldots, v_n$, resulting in n clusters, one of size 2, the others of size 1. Much more practical, in most circumstances, would be to use node 0 as the head of only a single cluster. Such an intuition about networks is reflected in most of the later-on described heuristics even though the actual optimization objectives are usually not fully formalized. Objectives like uniform spread of clusters over a given area are often considered important [59].

May clusters overlap? When forming clusters out of the maximal independent set shown in Figure 10.17, the question arises to which cluster to assign nonclusterhead nodes, particularly those nodes that are adjacent to two clusterheads. One option would be to assign these nodes to both clusters, resulting in overlapping clusters. If that is not desirable, some decision rule is required to unambiguously assign nodes to clusterheads. Figure 10.18 highlights these possibilities.

How do clusters communicate? Whether clusters overlap or not, a node that is adjacent to two clusterheads can naturally assist in the communication between two clusters – it forms a **gateway** (other names are bridge, boundary node, or similar terms). The idea is that intracluster communication can be routed via the clusterheads, who then use the gateways for any intercluster communication.

There may be cases, however, where two clusterheads are separated by two nodes, and no single node can fulfill the duties of a gateway. In such a situation, two nodes from each

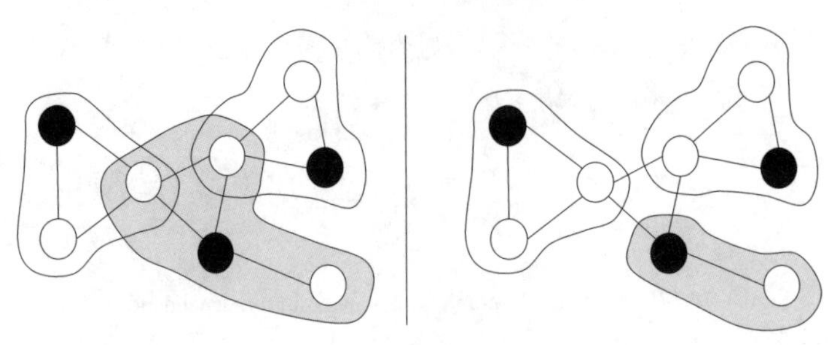

Figure 10.18 Maximum independent set induces overlapping or nonoverlapping clusters (example adapted from reference [59])

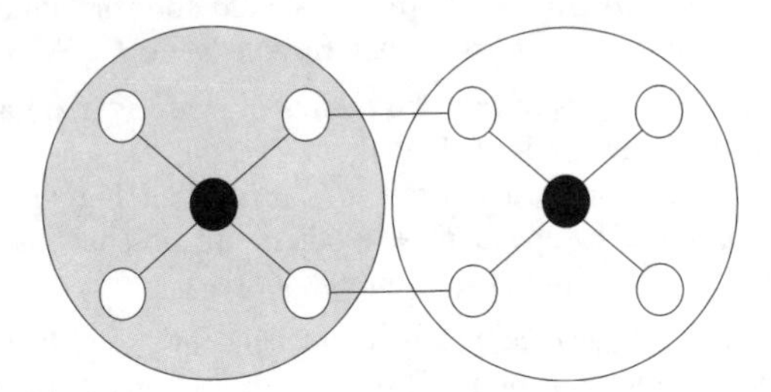

Figure 10.19 Two clusters connected by two distributed gateways

cluster together can act as a so-called **distributed gateway** to enable the communication between clusters. This idea is shown in Figure 10.19.

The clusterheads together with the (distributed) gateways again form a connected dominating set and thus a backbone of the entire network. This equivalence can also steer the gateway selection, as for example it might not be necessary to connect all neighboring clusters via gateways (although this is often done regardless of global optimization opportunities) or the choice between different gateways can be optimized by preferring nodes to serve as gateways that can connect more than two clusters, being in the intersection of several clusters. Choosing the optimal set of gateways to connect the given clusterheads into a connected sets is again a Steiner tree problem.

How many gateways exist between clusters? There can be several options to connect two clusterheads via several (distributed) gateways (examples are described in reference [40]). Depending on the optimization goal for the eventual connected dominating set, some degree of redundancy in the intercluster communication may be desirable.

What is the maximal diameter of a cluster? The presence of clusterheads and the goal of constructing a maximum independent set point to a maximum cluster diameter of two – each node in a cluster is at most two hops away from any other node. This is not necessarily the case: sometimes, one-hop clusters are considered (which often do not have clusterheads); sometimes, multihop clusters with larger diameters are used.

Is there a hierarchy of clusters? Clusterheads impose a hierarchy of nodes onto the network. Usually, such a two-level hierarchy is considered sufficient. Nonetheless, it is possible to consider the clusters as such as nodes in a new, induced graph, along with the links between clusters as edges in this graph. To this graph, again, clustering (or other dominating set approaches) can be applied.

10.4.2 A basic idea to construct independent sets

A first, simple idea to construct – hopefully large – independent sets exploits the inherently local nature of being independent – if selected nodes can restrain all their neighbors from being selected as well, independence ensues. The idea is thus for every node to communicate with its neighbors and to locally select nodes to join the set of independent nodes (to become clusterheads in the end).

To do so, all nodes need a property that can be locally determined, easily exchanged with all neighbors, and unambiguously ranked by each node (ties can be broken locally). A simple example for such a property is a unique identifier of each node, sorted for example in ascending order, where ties cannot happen at all. Using the identifier has actually been the first proposal for a distributed clustering algorithm [38, 39, 40].

Irrespective of the precise choice of the property used for ranking nodes, a basic distributed algorithm to compute independent sets starts out by marking all nodes as being ready to become clusterheads, but as yet undecided. During the course of the algorithm, this status is switched to either "clusterhead" or "cluster member" (comparable to the colors white, black, and gray in Section 10.3). In the first step, each node determines its local ranking property and exchanges it with all of its neighbors. Once this information is available, a node can decide to become a clusterhead if it has the largest rank (or the smallest, depending on definition) among all its as-yet-undecided neighbors. It changes its state accordingly and announces its new state to its neighbors. Nodes that learn about a clusterhead in their neighborhood switch to cluster member state and in turn announce that to their neighbors. Note that this is the crucial step: Once a node with a large rank becomes a cluster member to some other node, it can "unblock" nodes with lower rank in its vicinity to become clusterheads on their own. The algorithm terminates once all nodes have decided to become either a clusterhead or a cluster member.

This algorithm is illustrated with a simple linear network in Figure 10.20. Note how, in step 1, nodes 2 and 5 cannot become clusterheads because their neighboring nodes 3 and 6 have not yet decided and would, potentially, take precedence over them. Once nodes 3 and 6 have learned about node 7 being a clusterhead in their vicinity, they decide to become cluster members and propagate this information to nodes 2 and 5. Then, these nodes can become clusterheads in step 3.

This essential algorithm has been considered with several small variations. One variation is whether to actually hold back nodes from forming clusters as long as the clusterhead decision might still be revised, or to allow intermediate clusters to be formed, which will later be reclustered

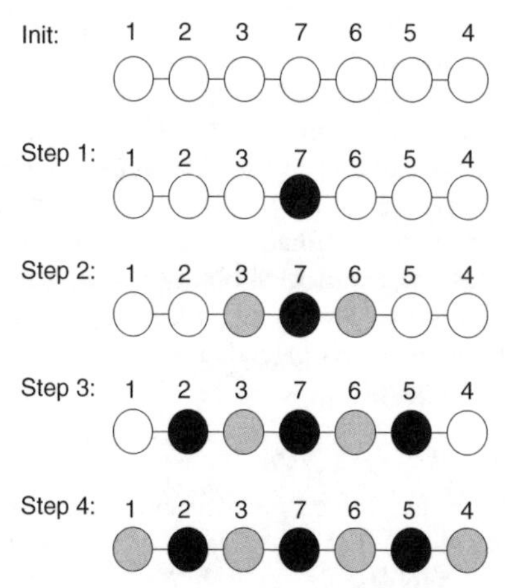

Figure 10.20 Basic algorithm for determining independent sets, using node identifiers as rank (white nodes are undecided, black nodes are clusterheads, gray ones are cluster members)

and nodes might join another clusterhead. This variation might be particularly useful in mobile networks (compare e.g. reference [502]).

Another important variation is how to rank nodes. The using of smallest (or largest) identifiers has been the first proposal [38, 39, 40] (describing, for example, the "linked cluster architecture", with some provisions made for how to exchange connectivity information between nodes).

Ranking nodes according to their degree, using the identifiers only to break ties, was proposed and investigated in references [288, 501, 616]. Essentially the same idea has been used in reference [501], where clusters are grown around nodes with the highest degree, but no clusterhead is elected.

10.4.3 A generalization and some performance insights

Other rankings besides identifiers or node degrees are conceivable. Basagni et al. [59] generalize these approaches by introducing weights for each node and formulate the clustering problem as the Maximum Weight Independent Set (MWIS) problem. Here, the goal is to find an independent set of nodes such that the sum of the weights of the nodes in this set is maximized. As it generalizes the maximum independent set problem, MWIS is NP-hard as well.

The algorithm described in reference [59] is straightforward and quite similar to the algorithm described above. It is actually a centralized algorithm, in each round choosing the node with the largest weight as a clusterhead and assigning all neighbors to this clusterhead; all these nodes are removed from the set of nodes that have to be considered. The algorithm terminates when all nodes have been assigned to some cluster. The result is a set of independent, dominating clusterheads, with nonoverlapping clusters. The algorithms of the previous section are obtained through proper choice of node weight.

The concrete performance of this algorithm depends on the actual choice of weights. It is possible to provide a lower bound on its performance, as long as all the weights are nonnegative. To do so, the notion of "performance" of a clustering algorithm has to be made more precise; the obvious choice here is how well it approximates the maximum weighted set that it is supposed to find. In other words, what is the ratio between the maximum weight of the best independent set and the weight of the set found by the algorithm, given a graph G and a node weighting w. Using this performance definition, Basagni et al. [59] show that this generalized algorithm always finds independent sets at least as heavy as maximum weight/Δ, where Δ is the maximum degree of the graph. This is nontrivial, as it holds irrespective of the actually used node weighting. [5]

What is more, they also show that this is the best bound on a performance ratio that can be proven for any polynomial time algorithm for nontrivial classes of graphs, as long as $P \neq NP$. In this sense, these simple algorithms are actually optimal.

Nonetheless, the actual performance of an algorithm (and *not* the worst-case bound) does considerably depend on the concrete weighting in use. The authors compare a "lowest ID" weighting with a weighting that gives preference to slowly moving nodes in a mobile ad hoc network; the metrics of interest are the number of reaffiliations of nodes to new clusters and the number of elections of new clusterheads as the result of mobility. In both these metrics, a mobility-aware weighting outperforms an identifier-based weighting (degree-based approach are known to perform not well in such situations). Reference [58] extends upon this work.

10.4.4 Connecting clusters

Once the clusterheads have been determined, by whatever algorithm, it is usually also necessary to determine the (possibly distributed) gateways between the clusters. Put simply, this problem is reduced again to the Steiner tree problem.

[5] They actually also prove another, sharper bound in their paper.

But the situation here is simpler than in the general Steiner tree setting as some properties of the clusterheads are known. In particular, they form a dominating, independent set where all nodes are separated by at most three hops (in case of two ordinary cluster members meeting at the edge of two clusters). For such a setting, CHLAMTAC and FARAGO [159] have shown that a connected backbone results if each clusterhead connects to all other clusterheads that are at most three hops away. While for some networks, this might mean more connections than necessary, but there are networks where all this links are needed to ensure connectivity.

In addition to this basic connectivity consideration, other aspects like load balancing between multiple gateways can be considered. Various approaches have been proposed here; reference [53], for example, treats this topic in more detail.

10.4.5 Rotating clusterheads

Being a clusterhead means taking over additional tasks: organizing medium access within the cluster or participating in routing decisions. Hence, the battery of clusterheads will tend to be exhausted sooner. Often, it is considered desirable that all nodes have roughly equal battery capacities at any point in time.[6] Hence, the duty of being a clusterhead should be shared among all nodes. Such sharing is in fact a viable option as there is usually not only a single solution to a maximum independent set problem but rather a number of different, (nearly) equally good ones.

To be able to rotate the clusterheads, the clustering algorithm cannot run only once but must be repeatedly executed. These repetitions can happen periodically or can be triggered by node mobility, for example. Of course, choosing periods and triggers judiciously is an important optimization problem, depending for example, on average node speed, battery draining rate, and so on.

Using virtual identifiers for rotation

As an example for clusterhead rotations, consider the extensions to the node-identifier-based or node-degree-based algorithms introduced by AMIS and PRAKASH [21]. To enable the ID-based algorithm to rotate clusterheads, the identifier is replaced, on each node independently, by a queue of virtual identifiers that are used in a round-robin fashion in the actual clustering algorithm. The node degree heuristic is adapted by forcing a clusterhead to step down if its degree has changed more than a given threshold in between two runs of the clustering algorithm.

Low-Energy Adaptive Clustering Hierarchy (LEACH)

Another early and popular example for rotating clusterheads is the Low-Energy Adaptive Clustering Hierarchy (LEACH) [344, 346] protocol. Its target scenario is a sensor network with a known number of nodes and known area, with a dedicated data sink to which all data is to be reported. As the data can be aggregated (e.g. by averaging), the introduction of clusterheads stands to reason. These nodes shall collect data readings from their cluster members and transmit it directly, at high transmission power, to the data sink in a single hop. As this is an energy-intensive operation, it makes sense to protect the clusterheads from being drained by rotating their role among all nodes.

A simple, lightweight protocol for clusterhead election and rotation is desirable, and the idea here is to use a simple random choice of clusterheads, foregoing the entire overhead for determining optimal clusterheads as their role is a temporary one anyway. Nodes independently decide to act as clusterheads and announce this to their neighboring nodes. These nodes then join that clusterhead in their vicinity with minimal communication costs (if there is more than a single one); nodes that do not hear a clusterhead announcement but do not want to become clusterheads themselves have to communicate with the data sink directly.

[6] An alternative is to deliberately burn out some nodes early on and to preserve other nodes for a long time. This is a principal design decision.

For such a random choice of clusterheads, the optimal ratio of clusterheads out of the total number of nodes is required. Taking into account the high costs for communication with a remote data sink, operation without clusterheads will result in low energy efficiency. Adding even a few will quickly improve overall energy efficiency, despite the additional effort for aggregation that these clusterheads incur. When increasing the ratio further, the advantages of clustering slowly diminish (in the extreme case, each node is a clusterhead for itself, voiding any aggregation or multihopping benefits). Hence, there is an optimal number at a relatively low ratio; for a typical example scenario, HEINZELMAN et al. [346] determine an optimal number of 5 %, but this does depend on the particular setup and has to be determined beforehand.

Once such an optimal percentage P of clusterheads is known, the actual LEACH algorithm proceeds in $1/P$ rounds (assuming, for simplicity, that $1/P$ is an integer value). In each round, a set of clusterheads of expected size nP (n the total number of nodes) of nodes is elected from the set G of nodes that have not yet served as a clusterhead (initially, and after every $1/P$ rounds, G encompasses all nodes). At the beginning of round r, each node in G becomes a clusterhead with probability $P/(1 - P \cdot (r \mod 1/P))$. This probability increases with every round, such that in round $1/P - 1$, all as-yet-unelected nodes will become a clusterhead with probability 1, ensuring that every node is serving as a clusterhead exactly once in some round. In round $1/P$, the process starts afresh.

HEINZELMAN et al. [346] further discuss the suitability of the resulting clustering structure for transmission scheduling and how this scheme can be used to determine multiple levels of clustering. Overall, this is a simple and elegant solution to the rotation problem, but requiring that all clusterheads can directly talk to a data sink should (and can) be replaced by some more elaborate mechanisms.

10.4.6 Some more algorithm examples

On the basis of these principal considerations, a few more algorithms shall be described in slightly more detail.

A Weighted Clustering Algorithm

The node weights (or ranks) discussed so far have been fairly simple: identifiers, node degree, or (inverse) node speed. None of these parameters can fully express all aspects of a node's suitability to serve as a clusterhead. Moreover, there might be constraints imposed by other system layers on the topology selection; for example, Bluetooth only allows a clusterhead (a master) to control clusters of at most seven members (slaves). In general, it might be desirable to prescribe a desirable *size* of a cluster in number of nodes that a clusterhead can efficiently control.

CHATTERJEE et al. [140] describe a clustering algorithm that takes the following aspects into account to compute node weights:

- A cluster should not exceed a maximum size δ
- Battery power (being a clusterhead means increased effort, which should be balanced over all nodes)
- Mobility (slow nodes are preferred)
- Closeness of neighbors (clusters with short distances between members are preferred).

Formally, the weight of a node v is expressed as

$$W_v = w_1 |d_v - \delta| + w_2 \left(\sum_{u \in N(v)} \text{dist}(v, u) \right) + w_3 S(v) + w_4 T(v)$$

where the w_i are nonnegative weighting factors, $N(v)$ are the neighbors of v (at maximum power), $S(v)$ is the average speed of node v, and $T(v)$ is the time node v has already served as a clusterhead (since system start). Tuning the weights will give importance to different system aspects.

The actual algorithm is then essentially identical to the ones discussed above where small weights take precedence (ties are broken arbitrarily). An interesting aspect of this algorithm is that it will, all else being equal, rotate the role of clusterheads among several nodes to ensure sharing of the load between several nodes.

Several other papers consider similar problems; reference [918] is a more recent example.

An emergent algorithm for cluster establishment

Most of the algorithms described so far were distributed in that there was no central entity that knew about the complete state of the network and computed the final solution. They were, in this sense, localized – nodes only drew upon information known to themselves or to their neighbors – but they still had a clear goal explicitly incorporated into the algorithm (e.g. nodes with highest degree become clusterheads).

An alternative approach to construct localized algorithms does away with such explicit goals. These are so-called **emergent algorithms** or protocols. To quote from reference [130]: "...an emergent protocol for a sensor network is a localized protocol in which the desired global property is neither explicitly encoded in the protocol nor organized by a central authority, but emerges as a result of repeated local interaction and feedback between the nodes."

Chan and Perrig [130] apply this design principle to the clustering problem. They define an emergent algorithm for clustering that essentially moves around clusterheads until an even spread of clusters has been achieved, without explicitly writing down this as a goal of the algorithm.

In this algorithm, every node can be in three states: unclustered (unaware of any cluster), clusterhead, or follower (to potentially more than one clusterhead; only at the end a node finally decides for a single clusterhead). Unclustered nodes turn themselves into clusterheads spontaneously (after random delays) if there is no cluster in their vicinity; these clusterheads recruit their neighbors as followers. The interesting idea now is that clusterheads can abdicate if there is a follower node that would make a better clusterhead, for example, one that would have more followers and less overlap with other clusters. Such a superior node will be promoted to clusterhead status by the old, abdicating clusterhead. In effect, the clusterhead role moves around in the network. Nodes terminate the algorithm after a predefined time.

The interesting property of this algorithm is that it achieves a packing efficiency that approaches closest hexagonal packing of clusters in a given area. Its runtime is constant, independent of the size of the network (as enough clusterheads are spawned in a distributed fashion). It does outperform algorithms like "lowest ID".

10.4.7 Multihop clusters

The clusters discussed so far have all been derived from the maximum independent set formulation, with clusterheads forming a dominating set as well. Consequently, the maximum diameter of a cluster is two, resulting in relatively small clusters. Depending on the purpose of clustering, larger clusters can be useful even though not every node is a neighbor of a clusterhead then – routing or aggregation protocols, for example, can profit even from larger clusters whereas cluster support for MAC protocols is mostly based on the dominance property of the clusterheads.

A crucial problem here is to limit the cluster size from both above and below. An early treatment of this topic can be found in reference [676], where an expanding ring search is used. In this search, the depth limit is successively increased until the cluster exceeds a given size threshold. Bannerjee and Khuller [52] also discuss the problem; their contribution is discussed in

Section 10.4.8. Another example for such multihop clusters, which also discusses the relationship to MAC protocols, is described in reference [186].

NP-completeness and a heuristic

When using multihop clusters, the definition of the dominating set is extended to allow a node to dominate nodes that are up to $d \geq 1$ hops away. The resulting problem is then to find a minimal set of nodes D such that all nodes of the graph are either in D or at most d hops away from some node in D. Amis et al. [22] show that this problem is NP-complete.

Amis et al. [22] also provide a heuristic solution for this problem. Their algorithm only requires $O(d)$ message exchanges between nodes and forms a backbone between the clusterheads by determining gateways between clusters. It is fair in that it tries to distribute the load of being a clusterhead; it is also a stable algorithm since clusterheads are reelected when possible.

The details of the heuristic are somewhat complex. The basic idea is to have two phases of flooding of node identifiers, with the flooding limited to d hops. In the first case, the largest identifiers are propagated and nodes that have not heard from nodes with larger identifiers within these d rounds can safely elect themselves as clusterheads. However, there may be cases of nodes that have to serve as clusterheads since they have the largest identifier in the d-hop neighborhood of some node but are themselves dominated by some other node in their vicinity. To allow these nodes to learn about the need to become clusterheads, a second round of d-hop-limited flooding is undertaken, but this time, the smallest identifiers are propagated. The information exchanged in these rounds lets nodes find out whether they should become clusterheads. The clusters themselves are then formed by nonclusterheads starting convergecasts toward their clusterheads and adopting nodes in the wake of this convergecast into a cluster. For details, please refer to reference [22].

Fixing the size of clusters by growth budgets

A slightly different tilt is given to the clustering problem when trying to prescribe the *size* of a cluster, that is, the number of nodes within it, rather than its maximum depth or diameter. One example of how to achieve this goal is "growth budgets," introduced by Krishnan and Starobinski [444]. The basic idea (incorporated in their "rapid algorithm") is quite simple. Given a cluster target size B, a clusterhead asks its neighbors to adopt $B_i \geq 0$ nodes into the cluster, where $B - 1 = \sum B_i$ (the clusterhead itself counts as a member as well, thus $B - 1$). Each node, on being asked to adopt x nodes, becomes a member of the cluster and again asks its neighbors to find another set of nodes of size $x - 1$ (implicitly, a spanning tree of the cluster is formed as well). Such a search terminates when the budget has been used up or when there are no more nodes to adopt into a cluster.

In the latter case, the growth budget has not been used fully and the resulting cluster will be too small. The "persistent" algorithm [444] will try to repair such a situation by reporting unused budget to the parent node in the spanning tree, which can then try to allocate the budget to other neighbors. This readjustment can percolate up to the clusterhead and shift budget to other parts of the cluster.

Evidently, the first algorithm uses fewer messages, namely $O(B)$, than the persistent algorithm, which has a polynomial message complexity. Other algorithms, like expanding ring search, can achieve similar goals but have worse complexities.

When to use multihop clusters

The question when to actually use multihopping within a cluster is considered, for example, by Mhatre and Rosenberg [550]. They assume a heterogenous system model where clusterheads communicate directly (over longer distances) with a remote data collection entity and sensors send their data to the clusterheads, where they can possibly be aggregated. These sensors can communicate with their clusterhead either directly or via multihop communication. The authors provide an expression for the critical distance beyond which multihopping should be used; this distance only depends on radio parameters (in particular, path-loss coefficient) and is independent of the network characteristics. Moreover, a scheme to compute the optimum number of clusterheads in such a scenario is also provided.

10.4.8 Multiple layers of clustering

Once clusters and their gateways have been determined, they induce a new graph where clusters are the nodes of the graph and any two nodes are connected if there exists a gateway between the clusters. To this induced graph, again a clustering algorithm can be applied, electing new clusterheads and connecting neighboring nodes by gateways. Evidently, this process can be repeated recursively. One hoped-for advantage of such multiple layers of clustering is to contain topology changes, for example, relevant for routing protocols, better and only to modify information in a local vicinity.

One of the first papers to describe such multilayer clustering is reference [679]. There, a heterogeneous setup is assumed where only some nodes can relay traffic. Naturally, these nodes form the first-level clusterheads, attempting to control the size of each cluster by forming, merging, and splitting clusters. These clusters in turn can elect clusterheads, forming higher-layer clusters. The height of the hierarchy should be kept small, that is, clusters should be of uniform size. An interesting aspect is how the gateways between clusters are formed. Relay-capable nodes that are at the edge of a cluster and detect such nodes of another cluster in their vicinity can invite them to form a "virtual gateway". To increase redundancy and stability of the topology, these gateways can also incorporate additional relay-capable nodes moving in range, merge with another gateway, or split up into two if nodes move out of range.

Bannerjee and Khuller [52] extend upon this setting and provide a distributed solution to the problem of hierarchical clustering. They start out by considering a standard clustering problem in a graph, with the following three additional requirements: (i) Cluster size $|V_i|$ for any cluster V_i is bounded by a given constant k, $k \leq |V_i| < 2k$ (one cluster is allowed to be smaller than k to avoid some special cases), (ii) two clusters should only have a small, constant number of nodes in common, (iii) each node should only belong to a small set of different clusters. In fact, there are graphs where these requirements cannot be met but they are feasible for unit disk graphs (and similar graphs). These requirements are satisfied by an algorithm that traverses a breadth-first spanning tree of a given graph and connects nodes in subtrees of a node u into clusters, possibly using u to connect these subtree clusters together if they are too small. This process continues up the tree.

Bandyopadhyay and Coyle [49] suggest to the use of multiple levels of clustering to save energy in a scenario where sensor nodes are to report sensor readings to a remote processing center. They start out by a simple, randomized clusterhead election protocol where a node volunteers as a clusterhead with probability p. Clusters are of size k. Any node that is not covered by such a cluster also becomes a "forced" clusterhead. On the basis of quite standard assumptions about energy consumption and node deployment, closed-form solutions for both p and k are analytically derived. This randomized algorithm can be fairly easily extended to multiple levels: From the (level 1) clusterheads, again some of them elect themselves as level 2 clusterheads and announce this fact to their level 1 clusterhead neighbors at most k_2 hops away; these then join such a level 2 cluster. The extension to more layers is obvious. Again, optimal values for p_i and k_i are determined, minimizing the energy spent to communicate data readings to a processing center. It is assumed that each node sends its data to a next level clusterhead, which aggregates the data from all its children before forwarding them. The authors claim that this scheme outperforms other clustering schemes in the resulting energy efficiency and that the algorithm has lower complexity than most other ones.

10.4.9 Passive clustering

In terms of energy consumption, one of the most expensive operations in a network is flooding: disseminating a particular piece of operation to all nodes. Flooding happens, for example, in routing protocols when routes have to be computed, but it also occurs when a new data sink announces its interest in certain kinds of observable data.

Usually, flooding is implemented by every node repeating every packet that it has received, with the exception of already received ones to avoid cycles. But not *every* node would have to retransmit a packet; because of the broadcast nature of the wireless channel, retransmission by a minimum dominating set – such as clusterheads and gateways connecting them – would suffice. The methods discussed above would be amenable to compute such a set; however, they can incur considerable overhead.

This clustering overhead can be reduced if the information flow that is happening anyway during a flooding operation is leveraged to compute a clustering structure on the fly. Actively sending out any message for clustering as such is avoided; the approach discussed here is hence called **passive clustering** [462]. The necessary information exchange is achieved by adding state information about each sender into any packet that is sent anyway, namely "initial", "clusterhead", "gateway", and "ordinary node". This distributes information about the state of neighboring nodes; it suffices to build a clustering structure that well approximates maximum independent sets with optimal gateway choice and is competitive with ID-based or degree-based algorithms.

The procedure works as follows. Suppose a node starts a flood; it will be stamped as coming from an "initial" node. The first node receiving and forwarding this packet will become a clusterhead and announces this fact by appropriately stamping the forwarded packet. Any initial nodes receiving such a packet will turn into "ordinary nodes" or into gateways.

The decision to become a gateway depends on the number of clusterheads and other gateways that a node has already heard from. Intuitively, a node that has heard from two or more clusterheads should become a gateway to connect these two clusterheads but only if there is no other gateway nearby already fulfilling this role. This intuition is formalized by two system parameters a and β: A nonclusterhead hearing from clusterheads or gateways becomes a gateway if and only if $a(\text{\# adjacent clusterheads}) + \beta > \text{\# adjacent gateways}$; a and β control the degree of redundancy of gateways between two clusters.

Hence, after the initial declaration of a clusterhead, up to two nodes can declare themselves as gateways (depending on the choice of a and β) and forward the flood packet; all other nodes (hearing from these two gateways) will declare themselves as ordinary nodes and stop forwarding the packet.

In effect, a set of clusterheads and gateways is constructed while performing the flooding operation, limiting the required overhead. While there is no means to guarantee a nearly optimal performance, simulations show that for practical networks the resulting clustering structure is quite similar to that produced by active clustering schemes.

Combining this passive clustering with WSN-typical protocols is also appealing. HANDZISKI et al. [331] discuss an example how to combine passive clustering with directed diffusion (described in Chapter 12).

10.4.10 Further reading

1-hop clusters One of the few papers dealing with 1-hop clusters (cliques, every node in a cluster can communicate with every other one) is reference [438]. A first fit heuristic is proposed to find largest cliques but cannot guarantee that optimal clusters are always found. It requires three passes for each change of the network topology.

Clustering and mobility In general, clustering in a mobile network is difficult because the clustering structure can undergo "ripple through" changes as nodes move around, resulting in network-wide, high-overhead consequences of local activities. Most of the schemes described above have been evaluated to some smaller or larger degree for mobile networks. Few of them, however, have been designed on the basis of a concrete mobility model. McDonald and Znati [544] present such a mobility-based framework for clustering where clusters are constructed such that there is a specified lower probability bound α on the mutual availability of paths between all nodes of the cluster over a given time interval t. They also provide an analysis of link errors and intra and intercluster routing issues.

Gao et al. [278] also consider this problem and present an algorithm that determines a constant-factor approximation of the smallest possible number of clusterheads covering all mobile nodes (which is an NP-complete problem). They describe a kinetic data structure to update the set of clusterheads in a mobile network.

Energy efficiency Safwat et al. [714] describe a scheme where the residual battery capacities of nodes is taken into account in the election process. Another paper that considers energy efficiency in cluster formation is reference [325].

10.5 Combining hierarchical topologies and power control

Both hierarchal approaches (backbones, clusters) and power control are effective means to influence the topology of a wireless network. Several approaches that combine these mechanisms exist and some of them are briefly described here.

10.5.1 Pilot-based power control

One early proposal [461] used the clusterheads to perform power control in a way akin to the power control mechanisms in cellular networks. After an initial clustering structure has been set up (by some arbitrary mechanism), clusterheads use power control on both pilot signals and on normal data packets. The pilot signal power control is used to control the cluster membership as nodes only join a cluster based on these pilots. The data packet power control is used to ensure adequately low errors for faraway nodes and efficient transmission for nearby nodes; it also combats unusually bad transmission conditions. The main advantage is that the power control logic can be "centralized" in the clusterheads, simplifying the problem of a fully distributed power control.

10.5.2 Ad hoc Network Design Algorithm (ANDA)

Allowing the clusterheads to control the size of their cluster by power control is also used in the Ad hoc Network Design Algorithm (ANDA) system [151] and concrete rules are derived to maximize the network lifetime. The assumption is that network lifetime is mostly determined by the clusterheads as they have to fulfill the most demanding tasks. Hence, energy drainage should be balanced across clusterheads.

The underlying assumptions for this approach are that (i) the positions of ordinary nodes and of (preselected) clusterheads are known, (ii) the traffic load is evenly distributed over ordinary nodes, (iii) the lifetime of a clusterhead is proportional to its initial energy supply and inversely proportional to $cr^{\alpha} + dn$, where r is the coverage radius of a clusterhead, n is the number of cluster members, α is the path-loss coefficient, and c, d are constants. The optimization goal is to maximize the lifetime of all clusterheads or, equivalently, maximize the minimum lifetime over all clusterheads.

This maximization task is an optimization problem where the decision variables describe the membership of ordinary node i in cluster j; the required radio range is implied. This problem can be solved optimally for static networks by a simple greedy algorithm: Assign node i to that clusterhead that gives the longest lifetime, and repeat for all nodes. For dynamic network scenarios, an additional reconfiguration procedure is necessary and optimality can no longer be guaranteed; practical performance is still good.

10.5.3 CLUSTERPOW

The CLUSTERPOW [413] protocol grew out of the observation that the COMPOW protocol (described in Section 10.2.3) is not well suited for a scenario where node distribution is nonhomogeneous as the lowest common power level to ensure connectivity would be far too high for most communications between nodes.

The basic idea is simply to assume a discrete set of transmission power levels as given, for example, 1, 10, and 100 mW. Clusters are formed independently at each power level and there are separate routing tables for each power level. Packet transmissions take place at the lowest power level that guarantees that the destination is reachable (such that an appropriate entry exists in the routing table) and the power level is reduced once the packet enters clusters of lower power levels containing the destination.

Replacing the initial transmissions at high power levels can often be useful. To do so, the "tunneled CLUSTERPOW" protocol is also described in reference [413]. To avoid infinite routing loops in this context, also the addresses of intermediate nodes used to route at lower power levels must be encapsulated into the data packet.

10.6 Adaptive node activity

There are some additional approaches to topology control that do not fit strictly under the headline of backbone/dominating set computation or clustering. All of them influence the topology of a graph by selecting certain nodes to be turned on or off – an operation that of course also fits well into the context of clustering or backbone mechanisms. Evidently, nodes that are sources or sinks of data are always kept active.

10.6.1 Geographic Adaptive Fidelity (GAF)

One option to exploit redundancy in the network is to declare certain subsets of node as equivalent from a the perspective of a higher-layer (e.g. for routing) protocol. In this sense, two nodes can be classified as equivalent if they (i) do not play any special role from an application perspective, that is, they are neither source nor sink of data, and (ii) can communicate with the exact same set of neighbors and can hence replace each other in a routing protocol. As the second point clearly is a topology-control issue, this equivalence opens a lever to alternatively turn off equivalent nodes.

This idea of equivalent nodes and a position-based realization has been described by Xu et al. [903] in the Geographic Adaptive Fidelity (GAF) protocol. The idea is to divide the area into rectangles that are small enough such that any node in one rectangle can communicate with any other node in an adjacent rectangle (rectangle touching only at the corner are not considered). The critical positions are those that are at diametrically opposed corners of two rectangles; two such

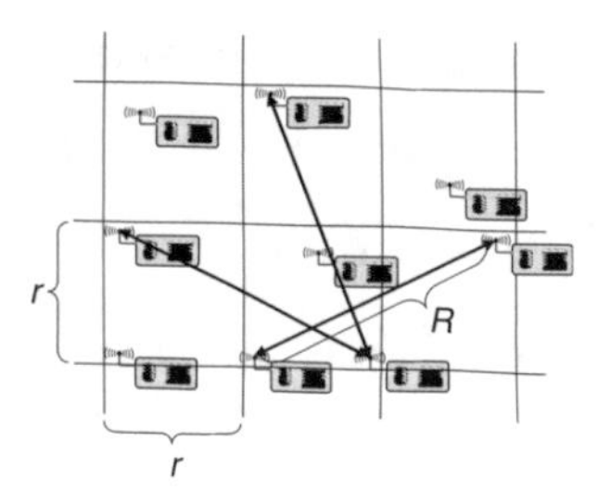

Figure 10.21 Relationship between maximum radio range R and rectangle length r in the GAF protocol

nodes should still be able to communicate at maximum radio range. Figure 10.21 illustrates this relationship for a rectangle length r and a maximum radio R.

The distance between two such critical nodes is $\sqrt{r^2 + (2r)^2}$. As this distance has to be smaller than R, it follows that $r < R/\sqrt{5}$. Since nodes are assumed to know their location, they can easily construct such equivalency rectangles, determine nodes in their own rectangle, and collaboratively determine a sleeping pattern, taking turns in participating in a routing protocol. Xu et al. [903] show that using this scheme, existing ad hoc routing protocol energy efficiency can be improved between 40 and 60 %.

On top of GAF (or other approaches that have a notion of equivalent nodes), the Sparse Topology and Energy Management (STEM) protocol [733, 737] – described from a MAC layer perspective in Section 5.2 – can be added. GAF turns off nodes but maintains the network's forwarding capacity; STEMs mechanism of a virtual wakeup channel increases path setup delay in return for an increase in energy efficiency. Schurgers et al. [737] suggest to combine these two approaches by applying STEM to a set of nodes equivalent from the GAF perspective.

GAF has also been extended and amended by Xu et al. [902] by removing its dependence on actual location information. The protocol proposed in this reference works directly on radio connectivity information but follows otherwise similar design goals in determining redundancy among nodes.

10.6.2 Adaptive Self-Configuring sEnsor Networks' Topologies (ASCENT)

The Adaptive Self-Configuring sEnsor Networks Topologies (ASCENT) approach [127] advocates a slightly different take on the topology-control problem. Rather than trying to construct a backbone or a clustering structure, the network tries to adapt itself to the needs of ongoing communications. This adaption takes place by incorporating nodes that are normally passive into a link (turning a one-hop into a multihop communication); it is triggered by communications that are undertaken over long distances, at high energy costs, with high error rates.

In such a situation, the receiver can send out "help" messages to the network. A node receiving such a help message will test out whether its joining the network will actually be beneficial. It does so by temporarily participating in packet forwarding and observing whether its own assistance improves network performance. If so, it becomes active; if not, it becomes passive again. Nodes can also switch themselves off but have to periodically enter the passive state again to monitor for pleas for help.

This probing of the network's performance, depending on the topology-control decisions, distinguishes ASCENT from most other protocols. However, it has a large number of parameters that have to be carefully optimized.

10.6.3 Turning off nodes on the basis of sensing coverage

Unlike "traditional" ad hoc networks consisting of laptops or PDAs, merely ensuring connectivity of the network is not sufficient for a WSN. Rather, a primary task of a WSN is to sense and measure something about its environment. To do so, the observed area has to be covered by a sufficient number of nodes, irrespective of (possibly lower) connectivity needs. In general, the problem of sensing coverage provided by a WSN is a difficult one. One example approach should suffice here.

TIAN and GEORGANAS [816] point out that to turn off nodes, it has to be assured that the area for which a certain node provides sensing coverage is taken care of by some other nodes. Only then is a node "eligible" for going to sleep for some time.

The protocol described in reference [816] assumes, in its basic form (some extensions are described in the paper), that nodes know their position and their sensing ranges. After exchanging this information with their neighbors (e.g. at the beginning of a round), it is a simple geometric problem for a node to decide whether its own sensing area is covered by its neighbors. If that is the case, a node declares itself eligible for sleeping, announces this fact to its neighbors, and goes to sleep.

If all the eligible nodes did sleep simultaneously and turned themselves off at once, there is a danger of "blind spots" forming. This can happen if two neighboring nodes are both eligible, based on the assumption that the respective other node were to stay active (the remaining coverage at the opposite side would have to be provided by some further nodes). To avoid such a case, the eligibility announcements are randomly delayed, using common random backoff approaches. Once a node receives such a message, it removes this soon-to-be-sleeping node from its neighbor list and reevaluates its own eligibility status. Repeating this process periodically at the beginning of a round will tend to distribute sleeping possibility over multiple nodes by virtue of the random backoff.

The protocol rests upon the fact that nodes are redundantly deployed; otherwise there is no possibility to sleep in the first place. It is a simple yet effective approach that takes the actual WSN characteristics into account.

10.7 Conclusions

Topology control – namely, power control, backbones, and clustering – is a powerful means to change the appearance and properties of a network for other protocol layers: MAC layers see reduced contention, routing protocols work on a different graph, changes in neighborhood relationships can be hidden. Judicious use of topology control can significantly improve operational aspects of a network, such as lifetime. However, determining an optimal topology is usually prohibitively expensive and appropriate approximations and heuristics have to be used instead.

11

Routing protocols

Objectives of this Chapter

In a multihop network, intermediate nodes have to relay packets from the source to the destination node. Such an intermediate node has to decide to which neighbor to *forward* an incoming packet not destined for itself. Typically, *routing tables* that list the most appropriate neighbor for any given packet destination are used. The construction and maintenance of these routing tables is the crucial task of a distributed *routing protocol*.

This chapter discusses mechanisms for routing and forwarding when the destination of a packet is identified by a unique node identifier, by a set of such identifiers, or when all the nodes in the network shall receive a packet. One particular type of identifier will be position information, which can identify both individual nodes and groups of nodes.

This chapter does *not* consider the dissemination or collection of *data* in the general sense. These tasks are crucial for WSNs and are also clearly routing problems. To give them due credit, they are treated separately in Chapter 12.

Chapter Outline

11.1 The many faces of forwarding and routing

Whenever a source node cannot send its packets directly to its destination node but has to rely on the assistance of intermediate nodes to forward these packets on its behalf, a multihop network

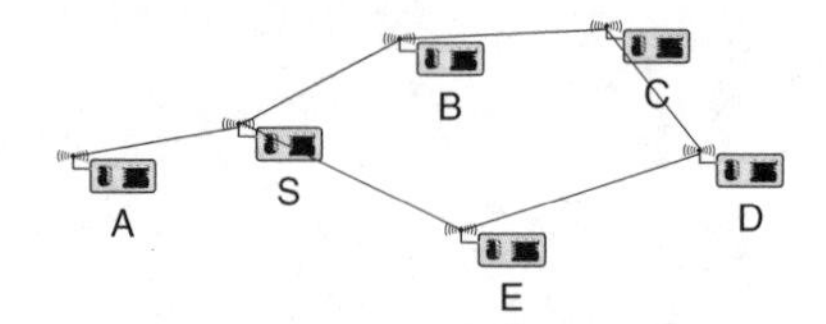

Figure 11.1 A simple example of routing in a multihop network – node S sends packets to node D

results – an example is shown in Figure 11.1. In such a network, an intermediate node (as well as the source node) has to decide to which neighboring node an incoming packet should be passed on so that it eventually reaches the destination – for example, node S sending to node A would not do. This act of passing on is called **forwarding**, and several different options exist how to organize this forwarding process.

The simplest forwarding rule is to **flood** the network: Send an incoming packet to all neighbors. As long as source and destination node are in the same connected component of the network, the packet is sure to arrive at the destination. To avoid packets circulating endlessly, a node should only forward packets it has not yet seen (necessitating, for example, unique source identifier and sequence numbers in the packet). Also, packets usually carry some form of expiration date (time to live, maximum number of hops) to avoid needless propagation of the packet (e.g. if the destination node is not reachable at all).

An alternative to forwarding the packet to all neighbors is to forward it to an arbitrary one. Such **gossiping** results in the packet randomly traversing the network in the hope of eventually finding the destination node. Clearly, the packet delay can be substantially larger. Flooding and gossiping are two extreme ends of a design spectrum; alternatively, the source could send out more than a single packet on a random walk or each node could forward an incoming packet to a subset of its neighbors – for example, as determined by a topology-control algorithm, equivalent to flooding on a reduced topology. This last option is sometimes called **controlled flooding**.

While these forwarding rules are simple, their performance in terms of number of sent packets or delay, . for example, is likely poor. These shortcomings are due to ignoring the network's topology. In the example of Figure 11.1, without knowing that node A is even further away from the destination node D, the source node S has no means of avoiding it when forwarding its own packet. Hence, some information about the suitability of a neighbor in the forwarding process would be required. A neighbor's suitability is captured by the **cost** it incurs to send a packet to its destination via this particular neighbor. These costs can be measured in various metrics, for example, the minimal number of hops or the minimal total energy it requires to reach the destination via the given neighbor. Each node collects these costs in **routing tables**; Table 11.1 shows two examples.

Table 11.1 Routing tables for some nodes from Figure 11.1, using hop count as cost metric

Destination	Next-hop neighbor	Cost
A	A	1
D	A	3
D	B	3
D	E	2
E	E	2

(a) Node S's routing table

Destination	Next-hop neighbor	Cost
A	S	2
D	C	2
D	S	3
E	A	2
E	C	3

(b) Node B's routing table

Determining these routing tables is the task of the **routing algorithm** with the help of the **routing protocol**. In wired networks, these protocols are usually based on link state or distance vector algorithms (Dijkstra's or Bellman–Ford). In a wireless, possibly mobile, multihop network, different approaches are required. Routing protocols here should be distributed, have low overhead, be self-configuring, and be able to cope with frequently changing network topologies. This question of **ad hoc routing** has received a considerable amount of attention in the research literature and a large number of ad hoc routing protocols have been developed. A commonly used taxonomy [707] classifies these protocols as either (i) **table-driven** or **proactive** protocols, which are "conservative" protocols in that they do try to keep accurate information in their routing tables, or (ii) **on-demand** protocols, which do not attempt to maintain routing tables at all times but only construct them when a packet is to be sent to a destination for which no routing information is available. As usual, the borders are not sharp between these classes and there are some ideas for hybrid solutions. Examples for table-driven protocols are Destination-Sequenced Distance Vector (DSDV) [633], Clusterhead Gateway Switch Routing (CGSR) [149], and Wireless Routing Protocol (WRP) [673]. Popular on-demand protocols are, among others, Dynamic Source Routing (DSR) [536], AODV [634], Temporally Ordered Routing Algorithm (TORA) [619], Associativity-Based Routing (ABR) [825], and Signal Stability Routing (SSR) [212]. Overviews of this topic can be found in various books on ad hoc networks [373, 635, 827] and survey papers [356, 680, 707]. A common problem for many of these ad hoc routing protocols is that they require flooding of control messages to explore the network topology and to find destination nodes.

The full range of ad hoc networking is too broad to be covered here in full detail and not all the research in this context is relevant to the case of wireless sensor networks (e.g. routing of multimedia traffic in ad hoc networks). Rather, the exposition in this chapter will concentrate on the most crucial aspect: energy efficiency. This pertains both to the selection of energy-efficient routes as well as to the overhead imposed by the construction of the routing tables themselves. Secondary aspects that are briefly touched upon are stability and dependability of the routes as well as routing table size (nodes with limited memory cannot store large routing tables). In particular, the issues related to mobile ad hoc networks where all nodes move around will be considered at best superficially; the case of a mobile sink is briefly discussed at the end of this chapter. Energy-efficient unicast without and with the help of routing tables is described in Sections 11.2 and 11.3; overviews can be found, for example, in references [17, 273, 298, 385, 639].

In addition to energy efficiency, resiliency also can be an important consideration for WSNs [276]. For example, when nodes rely on energy scavenging for their operation, they might have to power off at unforeseeable points in time until enough energy has been harvested again. Consequently, it may be desirable to use not only a single path between a sender and receiver but to at least explore multiple paths. Such multiple paths provide not only redundancy in the path selection but can also be used for load balancing, for example, to evenly spread the energy consumption required for forwarding. Multipath routing schemes have been considered in the ad hoc literature as well [585, 618, 641, 837]; they are treated in Sections 11.2 and 11.3 where appropriate.

Apart from the unicast case, where one node sends packets to another, uniquely identified node, both **broadcasting** (sending to all nodes in a network)[1] and **multicasting** (sending to a specified group of nodes) are important tasks in WSNs. Both these tasks are treated in Section 11.4. One special way to define such a group is by specifying a geographic region such that all nodes in the region should receive the packet. This requires nodes to know about their positions, and once such knowledge is available, it can be used both to assist conventional routing and as a definition for target groups in a multicast sense. Section 11.5 describes such **geographic routing** approaches.

[1] The difference to flooding, as the terms are used in this book, is that broadcasting *intentionally* distributes a packet to all nodes, whereas flooding – every nodes forwards every new, incoming message – is often only used for lack of better options to distribute a message. Evidently, flooding is one implementation option for broadcasting, but not the only one.

All these options discussed so far are in a sense *node-centric* in that certain nodes are addressed by source nodes and packets should be delivered to these nodes. An alternative view on routing is enabled by *data-centric* network where the set of target nodes is only implicitly described by providing certain attributes that these nodes have to fulfill (geographic routing can indeed be conceived of as data-centric routing in this sense). These routing approaches are very important in WSNs as they reflect natural usage cases – in particular, collection of data and dissemination of events to interested nodes. These approaches are treated separately in Chapter 12 along with in-network processing schemes for aggregating information in the network.

11.2 Gossiping and agent-based unicast forwarding

11.2.1 Basic idea

This section deals with forwarding schemes that attempt to work without routing tables, either because the overhead to create these tables is deemed prohibitive (when a node only issues a command, for example, and does not expect any answers) or because these tables are to be constructed in the first place. The simplest option is flooding – forwarding each new, incoming message – but more efficient schemes are desirable. The topology-control discussion in Chapter 10 has already shown that reducing the forwarding set can considerably improve efficiency. The approaches taken here try to find a forwarding set without recurring to topology-control mechanisms but try to solve it strictly locally.

The perhaps earliest paper [201] along these lines draws a parallel between the distribution of data in a replicated database system and epidemics occurring in human populations. Various options are described; one is "rumor mongering": Once a site receives an update, it periodically, randomly chooses another site to propagate this update to; it stops doing so after the update has already been received by a sufficient number of sites (supposedly similar to the way rumors or epidemics are propagating in a population). The goal is to spread updates to all nodes as fast as possible while minimizing the message overhead. The question is to select neighbors for gossiping the rumor at hand (how often, which neighbors, etc.).

This same idea of *randomly* choosing forwarding nodes can also be applied to wireless sensor networks. There is, in fact, one advantage of wireless communication over wired communication that comes to bear in this context: a single transmission can be received by all neighboring nodes in radio ranges, thus incurring transmission costs only once for many neighbors. This property has been called the **wireless multicast advantage** [874]. Evidently, whether this advantage is actually relevant heavily depends on the deployed MAC protocol and on the relative costs of sending and receiving.

11.2.2 Randomized forwarding

On the basis of this consideration, HAAS et al. [320] look at the question how information spreads in a wireless network by such a gossiping mechanism. The key parameter of their mechanism is the probability with which a node retransmits a newly incoming message. In the simplest case, this probability is constant. They show that there is a critical probability value below which the gossip – typically – dies out quickly and reaches only a small number of nodes. If, on the other hand, nodes use a probability larger than the critical threshold to retransmit messages, then most of the gossips reach (almost) all of the nodes in the network. Typical value for the critical threshold are about 65 to 75 %. The existence of this threshold shows that gossiping exhibits a typical phase transition behavior, in accordance with what can be expected from a percolation-theoretical treatment of the problem.

HAAS et al. [320] also point out that nodes near the boundary of the sensor network's deployment region are critical as they have, on average, a smaller number of neighbors than nodes in the center of the region. They discuss various possible remedies, for example, (i) to have the neighbors of a node with few neighbors retransmit with higher probability, (ii) to prevent a gossip from dying out too fast by retransmitting messages over the first few hops with probability 1, or (iii) to retransmit a message (despite having decided not to do so) if the node does not overhear the message repeated from at least one of its neighbors (the actual minimum number is an optimization problem). Using such enhancements, the ratio of nodes that receive a gossip is considerably increased.

NI et al. [595, 596], TSENG et al. [833] also look at a similar problem, under the perspective of implementing broadcasting. They propose a couple of heuristics that let a node decide when to repeat a received or overheard packet. They look at rules that are based on counters (do not retransmit when a message has been overheard a certain number of times), distance based (do not retransmit if the distance to the sender is small), or location based (determine the additional coverage that could be obtained by retransmitting, based on the location of the nodes that have already sent the message).

11.2.3 Random walks

Limiting flooding by only probabilistically forwarding a packet is only one option. Another approach is to think of a data packet as an "agent" that wanders through the network in search of its destination. In the simplest form, this is a purely random walk, where a packet is randomly forwarded to an arbitrary neighbor. Hence, the agents are sent via unicast, not via local broadcast, to their next hop. Instead of a single "agent", several of them can be injected into the network by the source to shorten the time to arrival by parallelism.[2] The probabilistic properties of random walks have been extensively studied, but without any additional measures, a purely random walk is too inefficient to be useful for WSNs. Two examples of such extensions to random walks shall be briefly discussed.

Rumor routing

BRAGINSKY and ESTRIN [99] consider this approach in the context of event notification: Assume some sensors are interested in certain events (e.g. temperature exceeding a given value) and a sensor can observe it. Classical options are to flood either the query for the event or the notifications that an event has occurred through the entire network. The "rumor routing" approach proposed here does not flood the network with information about an event occurrence but only installs a few paths in the network by sending out one or several agents. Each of these agents propagates from node to node and installs routing information about the event in each node that it visited. This is illustrated in Figure 11.2(a) where the node in the middle detects an event and installs two event paths in the network (shaded areas). Once a node tries to query an event (or to detect whether an event actually exists), it also sends out one or more agents. Such a search agent is forwarded through the network until it intersects with a preinstalled event path and then knows how to find an event. In Figure 11.2(b), the node in the lower left corner sends out such a search, which happens to propagate upward until it intersects with one event path. All these agent propagations are limited to avoid endless circling of data.

The rationale behind this technique is the relatively high probability that two random lines in a square intersect each other; BRAGINSKY and ESTRIN [99] state a probability of about 69 %. While neither the event paths nor the search paths will in reality be straight lines, the approximation is claimed to be good enough. Using five instead of one event paths increases this probability to about

[2] In a sense, gossiping is a form of random walk where each node has the additional option to copy the agent arbitrarily often; flooding corresponds to copying the agent to all neighbors.

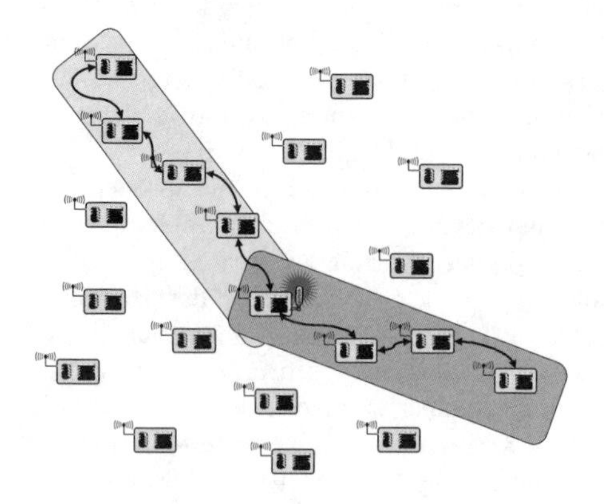

(a) Setting up two event paths from the event source in the middle

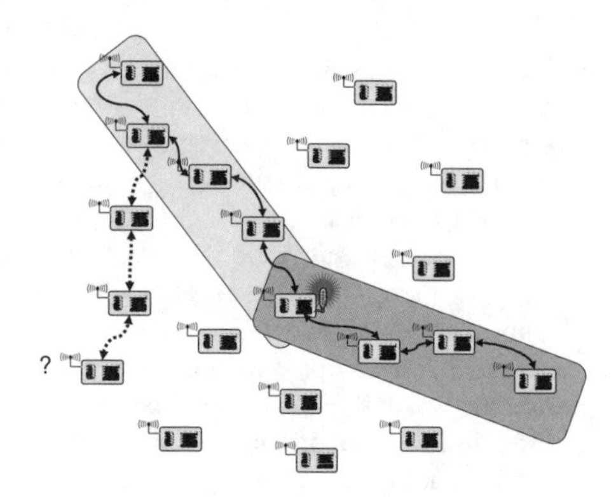

(b) Node starts a query (in the lower left corner, marked with "?"), propagating until it meets a pre-installed event path

Figure 11.2 Rumor routing example

99.7 %. In effect, rumor routing allows to trade off effort in path creation and/or search against probability of detecting an event.

There are a few more functionalities included in rumor routing. For example, agents spread information about more than one event if they have crossed an event path for another event. Also, an agent uses opportunities to shorten existing event paths if they know about shorter paths. Please refer to reference [99] for more details.

Random walks with known destination

A different perspective on random walks is taken by Servetto and Barrenechea [750]. They consider the problem of a WSN where lots of nodes are redundantly deployed but some of these nodes are randomly turned off and later on again (e.g. due to energy scavenging), giving rise to a dynamic graph. The idea is to use random walks to ensure that all possible paths in the network are used with equal probability, spreading the forwarding burden over all nodes. To do so, only local computations should be required for each node.

The concrete scenario under investigation in reference [750] is a rectangular grid of nodes where the source is in the upper left corner and the destination in the lower right corner; nodes in between are randomly active. For such a situation, formulas are developed to compute the probability of passing an incoming packet either down or to the right, based on a distributed computation of the number of paths from the source to an intermediate node and from the intermediate node to the destination. Compared to assigning both the lower node and the node to the right a probability of 50 % each, the random walks based on these formulas indeed result in a much more uniform traffic density in the network.

11.2.4 Further reading

- The basic ideas of random walks are related to biologically inspired algorithms, for example, based on the behavior of ants or other swarm insects. Eberhart and Kennedy [220] or Bonabeau et al. [86] give overviews of these topics.

- The idea of mobile agents originally contains the idea of sending code, through the network, that is executed at each node (active networks). Borcea et al. [90] present an example of this idea in the context of pervasive computing.
- Barrett et al. [56] describe another example for a multipath approach where an intermediate node makes a probabilistic decision about whether to forward a packet. This probability depends on, for example, distance between sensor and destination or number of hops that a packet has already traveled. They provide a fairly extensive comparison with other gossiping-type approaches.

11.3 Energy-efficient unicast

11.3.1 Overview

At a first glance, energy-efficient unicast routing appears to be a simple problem: take the network graph, assign to each link a cost value that reflects the energy consumption across this link, and pick any algorithm that computes least-cost paths in a graph. An early paper along these lines is reference [747], which modified Dijkstra's shortest path algorithm to obtain routes with minimal total transmission power. What qualifies as a good cost metric in general is, however, anything but clear and depends on the precise intention of energy-efficient unicast routing.

In fact, there are various aspects how energy or power efficiency can be conceived of in a routing context. The presentation here mostly follows reference [768]; acronyms are taken from reference [826]. Figure 11.3 shows an example scenario for a communication between nodes A and H including link energy costs and available battery capacity per node.

Minimize energy per packet (or per bit) The most straightforward formulation is to look at the total energy required to transport a packet over a multihop path from source to destination (including all overheads). The goal is then to minimize, for each packet, this total amount of energy by selecting a good route.

Minimizing the hop count will typically not achieve this goal as routes with few hops might include hops with large transmission power to cover large distances – but be aware of distance-independent, constant offsets in the energy-consumption model. Nonetheless, this cost metric can be easily included in standard routing algorithms. It can lead to widely differing energy consumption on different nodes.

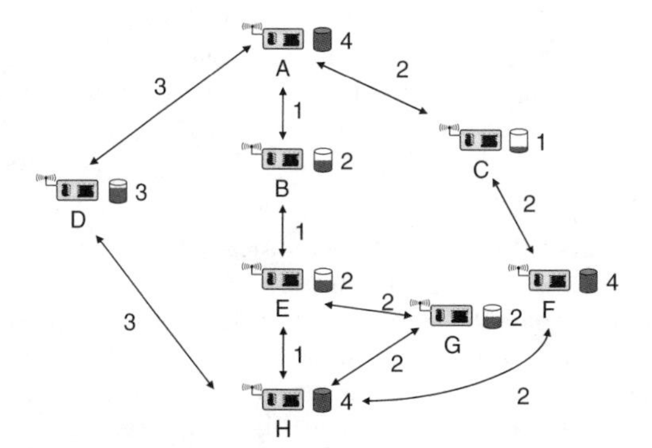

Figure 11.3 Various example routes for communication between nodes A and H, showing energy costs per packet for each link and available battery capacity for each node (adapted from reference [17])

In the example of Figure 11.3, the minimum energy route is A-B-E-H, requiring 3 units of energy. The minimum hop count route would be A-D-H, requiring 6 units of energy.

Maximize network lifetime A WSN's task is not to transport data, but to observe (and possibly control). Hence, energy-efficient transmission is at best a means to an end and the actual end should be the optimization goal: the network should be able to fulfill its duty for as long as possible.

Which event to use to demarcate the end of a network's lifetime is, however, not clear either. Several options exist (already discussed at length in Section 3.2.2):

- Time until the first node fails.
- Time until there is a spot that is not covered by the network (loss of coverage, a useful metric only for redundantly deployed networks).
- Time until network partition (when there are two nodes that can no longer communicate with each other) [136, 768].

While these aspects are related, they require different solutions. For the network partition, for example, nodes in the graph's minimal cut set should have equal energy consumption (or rather, supplies) to ensure maximum time to partition. Also, their solutions can be infeasible – for example, maximizing the time to network partition is reported as NP-complete [768]. Moreover, Li et al. [489] state that maximizing the time until the first node runs out of energy does not have a constant competitive ratio with the optimal off-line algorithm that knows the arrivals of future packets (when optimizing the number of messages the network can successfully carry, a competitive ratio logarithmic in the number of nodes can be shown [402]). Because of these theoretical limitations, only approximative solutions are practically relevant.

Routing considering available battery energy While maximizing the network lifetime is clearly a useful goal, it is not immediately obvious how to reach this goal using observable parameters of an actual network. As the finite energy supply in nodes' batteries is the limiting factor to network lifetime, it stands to reason to use information about battery status in routing decisions. Some of the possibilities are:

Maximum Total Available Battery Capacity Choose that route where the sum of the available battery capacity is maximized, without taking needless detours (called, slightly incorrectly, "maximum available power" in reference [17]).

Looking only at the intermediate nodes in Figure 11.3, route A-B-E-G-H has a total available capacity of 6 units, but that is only because of the extra node G that is not really needed – such detours can of course arbitrarily increase this metric. Hence, A-B-E-G-H should be discarded as it contains A-B-E-H as a proper subset. Eventually, route A-C-F-H is selected.

Minimum Battery Cost Routing (MBCR) Instead of looking directly at the sum of available battery capacities along a given path, MBCR instead looks at the "reluctance" of a node to route traffic [768, 826]. This reluctance increases as its battery is drained; for example, reluctance or routing cost can be measured as the reciprocal of the battery capacity. Then, the cost of a path is the sum of this reciprocals and the rule is to pick that path with the *smallest* cost. Since the reciprocal function assigns high costs to nodes with low battery capacity, this will automatically shift traffic away from routes with nodes about to run out of energy.

In the example of Figure 11.3, route A-C-F-H is assigned a cost of $1/1 + 1/4 = 1.25$, but route A-D-H only has cost $1/3$. Consequently, this route is chosen, protecting node C from needless effort.

Min–Max Battery Cost Routing (MMBCR) This scheme [768, 826] follows a similar intention, to protect nodes with low energy battery resources. Instead of using the sum of reciprocal battery levels, simply the largest reciprocal level of all nodes along a path is used as the cost for this path. Then, again the path with the smallest cost is used. In this sense, the optimal path is chosen by minimizing over a maximum.

The same effect is achieved by using the smallest battery level along a path and then maximizing over these path values [17]. This is then a maximum/minimum formulation of the problem.

In the example of Figure 11.3, route A-D-H will be selected.

Conditional Max–Min Battery Capacity Routing (CMMBCR) Another option is to conditionalize upon the actual battery power levels available [826]. If there are routes along which all nodes have a battery level exceeding a given threshold, then select the route that requires the lowest energy per bit. If there is no such route, then pick that route which maximizes the minimum battery level.

Minimize variance in power levels To ensure a long network lifetime, one strategy is to use up all the batteries uniformly to avoid some nodes prematurely running out of energy and disrupting the network.[3] Hence, routes should be chosen such that the variance in battery levels between different routes is reduced.

Minimum Total Transmission Power Routing (MTPR) Without actually considering routing as such, Bambos [47] looked at the situation of several nodes transmitting directly to their destination, mutually causing interference with each other. A given transmission is successful if its SINR exceeds a given threshold. The goal is to find an assignment of transmission power values for each transmitter (given the channel attenuation metric) such that all transmissions are successful and that the sum of all power values is minimized.

MTPR is of course also applicable to multihop networks.

A direct performance comparison between these concepts is difficult as they are trying to fulfill different objectives. Moreover, while these objectives are fairly easy to formulate, it is not trivial to implement them in a distributed protocol that judiciously balances the overhead necessary to collect routing information with the performance gained by clever routing choices. The following section describes some concrete protocols that tackle this challenge; a good overview is also included in reference [916]. Care has to be taken about the details here – Safwat et al. [715] show that a non-power-aware protocol can actually have (in many circumstances) a better energy-consumption behavior than some straightforward power-aware solutions (although it is not clear to what degree this conclusion is owing to the use of an IEEE 802.11-type MAC protocol in the paper).

11.3.2 Some example unicast protocols

Attracting routes by redirecting

An early proposal by Gomez et al. [300] uses the idea that nodes can overhear packet exchanges between other nodes. If, in these packets, information about the energy required to communicate

[3] It is by no means obvious that this in fact maximizes network lifetime; other factors like deployment pattern, event patterns, and battery discharge/recharge mechanisms also have to be considered.

between two adjacent nodes X and Z is included, a third node Y can decide whether it can offer a more energy-efficient route by breaking the direct communication X-Z into a two-hop communication X–Y–Z. If so, Y can "attract" this route by sending route redirect messages to X, Z, or both.

The advantage of this scheme is that its administrative overhead regarding explicit message exchanges is small. The need to overhear traffic is, however, not quite as appealing and makes this scheme not particularly suitable for WSNs. Further details can be found in reference [301].

Distance vector routing on top of topology control

The relay regions concept described in Section 10.2.3 also lends itself to a formulation of an energy-efficient routing problem. In reference [697], a Bellmann–Ford–type algorithm is used to find paths with minimal power consumption in the enclosure graph.

Maximizing time to first node outage as a flow problem

Chang and Tassiulas [137] attempt to maximize the time until the first node runs out of energy. To do so, they use a centralized, flow-based modeling approach. Given is a directed graph to represent the network annotated with the initial battery capacity of each node and, for each link, the energy costs to transmit a fixed-size packet. Moreover, the rates of data flows coming from certain nodes in the network, and their destination nodes are known. The goal is to find assignments of flows to forwarding nodes such the time until the first node runs out of energy is maximized.

This problem can be formulated as a linear programming problem with certain conditions on flow conservation. As the forwarding energy differs at each node, the normal maximum flow algorithms are actually not applicable to solve this problem. Therefore, two approximation algorithms are proposed.

The core idea of the first algorithm is to find a generalized description of the "costs" of a link. The observation is that both the actual energy cost e_{ij} of a link from node i to j as well as the initial and residual battery capacity E_i and E_i' of node i should be taken into account. Hence, a generalized link cost $c_{ij} = e_{ij}^{\alpha} E_i^{\beta} / E_i'^{\gamma}$ is used, where α, β, and γ are nonnegative weighting factors. Setting some of these factors to 0 and computing the "lowest cost" paths results in the routing strategies described above; it also allows to generalize these diverse approaches into a single metric.

The second algorithm is a flow redirection algorithm. Both of them are distributed and base on locally available information. The core result is that system lifetime can be extended up to 60 % in the scenarios investigated here in comparison to simple minimum energy routing when battery capacity also is taken into account. Further information on this approach can be found in references [136, 138]; Zussman and Segall [948] discuss an extension of the same basic techniques to an anycast routing problem.

Maximizing time to first node outage by a max–min optimization

Li et al. [489] approach the network lifetime maximization problem as a max–min optimization problem. They propose two algorithms. One of them has to know the battery power level of each node in the network and the other can work without this information at only slightly reduced performance.

The max min $zP_{\min}$ approximation

This heuristic starts out from the intuition to use paths that have a large residual energy, that is, that path where the minimal remaining power in all nodes is the largest. This heuristic, however, can result in arbitrarily bad performance (compare Fig. 3 in reference [489]). Moreover, this approach does not take into account the total power consumption of a given path, possibly giving preference to very expensive paths (in absolute terms). Hence, a proper compromise must be found.

The idea is to use a max–min path but limit its maximum power consumption. This limit cannot be chosen in absolute terms but is best defined as a ratio to the path with the smallest possible power consumption. Thus, the path to be chosen should have at most a power consumption of a factor z times the power consumption along the most efficient path, $P_{\min}$; among the paths that fulfill these constraints, the max–min path with respect to battery power will be selected as the actual path to be used.

Evidently, proper choosing of $1 < z < \infty$ will determine the efficiency of this approximation. In fact, z should adapt itself to the residual power levels in the network. This adaptation can be based on estimates of the shortest remaining lifetime of all nodes in the network, estimated over some period T. Parameter z will be additively increased or decreased after each period and the sign of the change depends on whether the estimated shortest lifetime has increased or decreased compared to the previous period. The magnitude of the adaption is reduced over time.

This heuristic is shown to have good performance by both simulations and analysis.

The zone routing approximation

The disadvantage of $\max\min zP_{\min}$ is that knowledge of battery power levels is required. The "zone routing" heuristic removes this need. This is done by partitioning the network in geographical zones where nodes within the zones are responsible for routing in the zone. Routing among zones is organized hierarchically. The reader is referred to reference [489] for a full description. The main point is that the resulting performance loss is relatively small, showing that the approximative maximum lifetime routing can be implemented on the basis of locally available information (assuming location information is available).

Maximizing number of messages

A slightly different optimization goal is to maximize the number of messages that can be sent over a network before it runs out of energy. In practice, this can be more important than maximizing the time until the first node runs out of energy, depending on what can be assumed about the actual data sensing process and energy consumption.

Kar et al. [402] consider this problem. Interestingly enough, they can prove a competitive ratio that is logarithmic in the network size if admission control (the routing algorithm is allowed to reject messages although there would be a path to carry the message) is assumed. What is more, the constructed "CMAX" routing algorithm does not depend in practice on admission control but performs well nonetheless (even when using network lifetime as figure of merit, for which the algorithm is not actually designed).

The crucial insight for this property is the choice of the link weights. Given an edge between nodes i and j with energy costs e_{ij} to transmit a message of unit size, the weight w_{ij} is chosen as $w_{ij} = e_{ij}(\lambda^{\alpha_i} - 1)$, where λ is a constant and α_i is the fraction of battery capacity that node i has already used up when the present routing decision is to be made. Admission control is then formulated as disregarding paths the total weight of which exceeds a given threshold. In practice, this path weight threshold can be ignored. Using any standard algorithm, the lowest cost path (based on these link weights) is then selected to transmit the packet.

The information needed for this heuristic and the $\max\min zP_{\min}$ heuristic are identical. In fact, the same zone-based variant should apply to this heuristic as well. An obvious improvement over the $\max\min zP_{\min}$ heuristic is that only a single shortest path computation is necessary.

Reference [402] discusses performance properties of the CMAX routing algorithm in detail. In short, it performs well with respect to maximum number of messages and, with respect to network lifetime, outperforms even specialized algorithms like the $\max\min zP_{\min}$ heuristic.

Note that both this and the previous scheme – and similarly all other schemes that depend on battery capacity – have to recompute routing tables relatively frequently to mirror the change in available capacity. This can be a considerable burden.

Bounding the difference between routing protocols

The previous sections have discussed several possible approaches to energy-efficient unicast routing. It seems like the choice of the routing has a considerable impact on the chosen energy efficiency metric. But the remaining question is: What is the maximal improvement that can be gained from an improved routing protocol? What is the biggest possible difference between a stupid and a perfect routing protocol?

ALONSO et al. [19] answer this question (Figure 11.4). They consider a class of networks where a all nodes transmit with identical power and all nodes continuously have data to deliver to a base station (possibly over multiple hops). Apart from these assumptions, their results apply to a wide range of real networks, irrespective, for example, of the concrete topology; data aggregation is not considered. The key question is the energy consumption of nodes during the data exchange, *not* the overhead of the routing protocol itself.

To approach this problem, the graph is partitioned into "spheres" S_i that include all the nodes that are reachable from the base station in at most i hops. The interesting case is then networks where "most" nodes are more than a single hop away from the base station (otherwise, the network is not particularly interesting anyway). Then, all traffic has to go through the nodes of sphere S_1, and because there are relatively few of these nodes, they limit the lifetime of the network.

For such networks, the authors show that no routing will have an energy efficiency worse than a factor of $2|S_1| - 1$ than the best possible one – and $|S_1|$ is a small constant compared to the network size. Put the other way around, unless the number of direct neighbors of the base station is large, the possible impact of routing is limited. For example, for a base station with four neighbors, even the worst possible routing protocol will only reduce the energy efficiency of data delivery

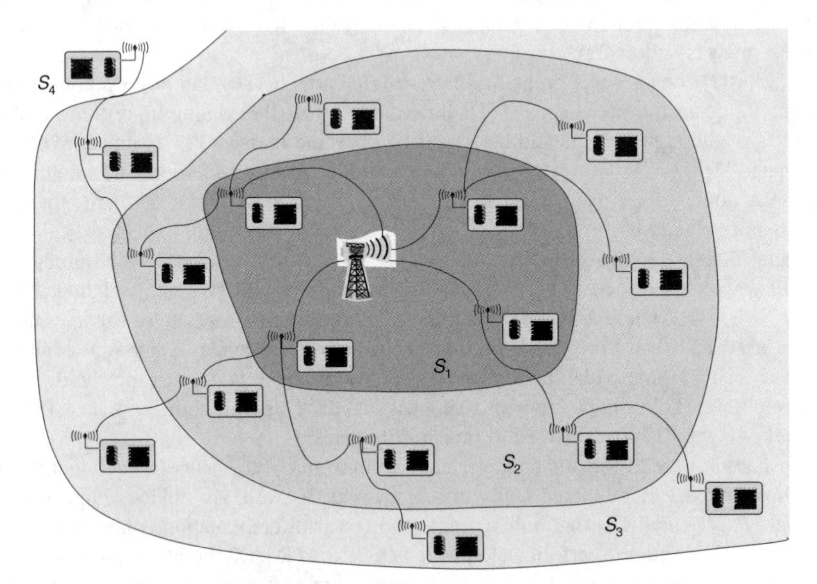

Figure 11.4 Spheres and balls as used by ALONSO et al. [19]

by a factor of seven compared to the optimal routing. Nonetheless, while such factors may not be impressive theoretically, they obviously have a large practical relevance.[4]

11.3.3 Further reading

The range of possible alternatives is large. The reader is particularly encouraged to check the general literature on mobile ad hoc networking that can provide, despite some fundamental differences, valuable inspiration. An example for a proposal that is related to an ad hoc networking protocol is the cost-field-based approach described in reference [910]: The cost field is a "height" assigned to a node, measuring the smallest possible energy to get to the source. This idea is clearly related to the older TORA ad hoc routing protocol [619], which assigns heights in hop counts. An interesting aspect of the cost field is that it is updated by randomly delaying the rebroadcasting of advertisement messages.

11.3.4 Multipath unicast routing

Overview

The unicast routing protocols discussed so far tried to construct a single energy-efficient path (with whatever interpretation of this term) between a sink and a receiver, typically by giving a clever meaning to the "cost" of a link. These costs try to balance, for example, energy required for communication across this link against the battery capacity of the nodes involved. Focusing on choosing the *best* possible path, however, limits the opportunities for making such trade-offs. Extending the focus to *multiple* paths and trying to balance, for example, energy consumption across multiple path is therefore an option worthwhile exploring. Moreover, multiple paths provide redundancy in that they can serve as "hot standbys" to quickly switch to when a node or a link on a primary path fails.

Such multipath routing protocols construct several paths between a given sender and receiver. The basic goal is to find k paths that do not have either links or nodes in common (apart from source and destination node, of course). Some basic references on finding multiple paths in general networks are [475, 604, 765]. Once the paths have been established by the routing protocol, the forwarding phase can then dynamically decide which path (or even paths) to choose to transmit a packet. This can increase the robustness of the forwarding process toward link or node failures.

Applying multipath routing to wireless networks, both general ad hoc and sensor networks, is a well-studied problem; some example references include [32, 50, 473, 584, 585, 612, 627, 929]. Some of the more WSN-relevant papers are briefly described here. But even some routing schemes discussed in other chapters, for example, directed diffusion, have multipath characteristics, even though it might not be their most prominent feature.

Sequential Assignment Routing (SAR)

As a basic rule of thumb, computing such k-disjoint paths requires about k times more overhead than a single-path routing protocol [778]. Sohrabi et al. [778] try to reduce the multipath-induced overhead by focusing the disjointness requirements to that part of a network where they truly matter – near the data sink, as the nodes close to the sink are (often) those that likely are going to fail first because of depleted battery resources. Hence, they only require paths to use different neighbors of the sink. The Sequential Assignment Routing (SAR) algorithm achieves this objective

[4] In addition, the MAC protocol is likely to impose a background load onto each node that can be considerably more important than the additional small fluctuations caused by different routing protocols.

by constructing trees outward from each sink neighbor; in the end, most nodes will then be part of several such trees. A packet's actual path is then selected by the source on the basis of information about the available battery resources along the path and the performance metrics (e.g. delay) of a given path.

Constructing energy-efficient secondary paths

When using multiple paths as standby paths to quickly switch to when the primary path fails, an obvious concern is that of the energy efficiency of these secondary paths compared to the (hopefully) optimal primary path. GANESAN et al. [276] consider the question how to construct the secondary paths from this perspective, without worrying about battery capacity or similar metrics along the various paths.

Their first observation is that strictly requiring node disjointness between the various paths tends to produce rather inefficient secondary paths as large detours can be necessary. To overcome this problem and yet retain the robustness advantages of multiple paths, they suggest the construction of so-called "braided" paths (sometimes also called "meshed" multipaths [195]). These braided paths are only required to leave out some (even only one) node(s) of the primary path but are free to use other nodes on the primary path. This relaxed disjointness requirement results in paths that can "stay close" to the primary path and are therefore likely to have a similar, close to optimal energy efficiency as the primary path. Figure 11.5 illustrates these two redundant paths' concepts.

Constructing these two different types of redundant paths is simple in a centralized fashion; a distributed construction is described in reference [276] as a modification to the reinforcement mechanism (popularized by directed diffusion, Section 12.2.2). For disjoint paths, the data sink not only reinforces the primary path via its best neighbor toward the data source but also sends out an "alternate path" reinforcement to its second-best neighbor (or several such neighbors, for multiple standby paths). This alternate path reinforcement is then forwarded toward the best neighbor that is not already on the primary path. For braided paths, each node on the primary path (including the

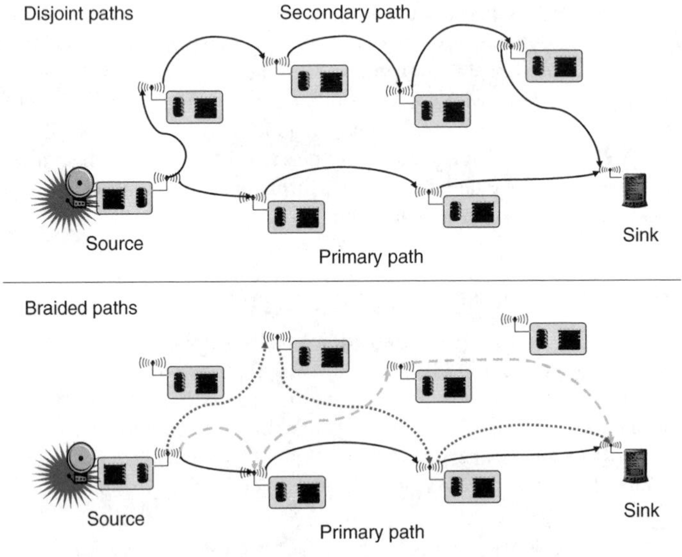

Figure 11.5 Disjoint and braided paths around a primary path

sink) sends out such an alternate path reinforcement, which only has to avoid the next upstream node on the primary path but is then free to use nodes on the primary path.

Which of these two schemes is advantageous clearly depends on the node failure patterns. The authors look at both independent node failures and so-called "patterned" failures (all nodes within a circle of known radius around randomly selected points fail; the appearance of points follows a Poisson distribution). The main figure of merit is the "resilience", the percentage of cases where the failure of the primary path is compensated for by an alternative path. Reference [276] discusses the relative performance in detail; in summary, braided paths tend to have a better overall resilience.

Simultaneous transmissions over multiple paths

When using multiple paths as a standby for the primary path, failover times might be improved compared to strictly single-path solutions. Nevertheless, there is some delay in detecting the need to use a secondary path. Depending on which node makes this decision – only the source node or any node on the primary path – there can be more or less overhead involved.

To further shorten the time to delivery and to increase the delivery ratio of a given packet, it is also conceivable to use all or several of the multiple paths simultaneously. The simplest idea is to assume node-disjoint paths and to send several copies of a given packet over these different paths to the destination. Clearly, this trades off resource consumption against packet error rates. De et al. [195] provide a performance comparison of such a packet replication scheme with other multipath schemes, for example, one that uses additional FEC to protect against packet errors. Dulmann et al. [214] combine the basic idea of sending packet replicas with FEC by proposing to split a packet and its error correction redundancy over several paths, to be recombined at the receiver. The degree of redundancy and the number of paths can be tuned to the expected error behavior, trading off overhead against residual packet error rate.

Randomly choosing one of several paths

When maintaining multiple paths, it actually makes sense also to use paths that are less energy efficient than the optimal one. One reason to do so is to share the load among all nodes in order to use the available battery capacity in the network better.

A relatively straightforward way of doing so is described by Shah and Rabaey [752]. Each node maintains an energy cost estimate for each of its neighbors (toward the destination, packets are not routed "away" from their destination). When forwarding a packet, the next hop is randomly chosen proportional to the energy consumption of the path over this neighbor. To the upstream node, the appropriately weighted average of these costs (i.e., the harmonic mean of the costs) is reported.

More formally: suppose node v has neighbors v_1 to v_n that advertise cost $c_1, \ldots, c_n$, respectively. Node v will advertise $c = n/\sum_{i=1}^{n} c_i$ as its own cost and will forward an incoming packet to neighbor i with probability $(1/c_i)/(1/\sum_{i=1}^{n} c_i)$.

This routing approach is extended by Willig et al. [877] by introducing the notion of **altruists**. An altruistic node is one that is willing to do more work on behalf of its neighbors, for example, because it has a tethered power supply. Willig et al. show that such asymmetric nodes can be efficiently exploited by the routing protocol, simply by occasionally broadcasting "altruistic announcements" into the network.

Trade-off analysis

Clearly, supporting such multiple paths in a network implies a trade-off between robustness (the probability that paths are available even after node failure) and energy efficiency (as both the management of these paths and the nonoptimal choices made for packet forwarding decisions imply increased energy expenditure) – irrespective of the concrete routing protocol in use. This tradeoff is analyzed by KRISHNAMACHARI et al. [440], who compare the robustness gained by multiple paths with those owing to simply increasing transmission power.

Their basic observation, made in a simplified scenario of five nodes, is that it is not possible to simultaneously optimize both robustness and energy efficiency of a given set of paths, but rather that only the notion of Pareto optimality can be applied.[5] They do observe, however, that single-path solutions that require a larger transmission power tend to dominate multipath solutions with low transmission power.

To test these basic observations, the authors conducted a set of simulation experiments, comparing various degrees of redundancy via braided multipaths. As one might expect, for low failure rates, the robustness of even two paths is perfectly sufficient. The two controlled parameters are the degree of redundancy via additional paths and the maximum transmission power, enabling the system to bridge across failed nodes if necessary. Using these two factors to influence Pareto optimality with respect to the robustness and energy efficiency objectives shows, interestingly, that the single-path schemes actually perform "best". Overall, the results of this paper highlight the need to carefully choose between various sources of redundancy.

11.3.5 Further reading

Section 11.3 could only provide a tiny glimpse – unicast routing is perhaps the broadest and most widely investigated research topic in the context of ad hoc and wireless sensor networks. Some further interesting aspects are briefly mentioned here.

Routing and topology control Chapter 10 has discussed topology control in a wireless network. In a clustered network or in a network where the topology is based on a dominating set, the routing problem has to be solved as well. Superficially, the problem is simple owing to a reduced network topology. However, it is not clear how, for example, information about battery capacity can be taken into account. A few references dealing with this topic are [150, 150, 356, 488, 628, 771]; LEACH [344, 346], for example, also belongs to this class of approaches.

Maximizing data flow for multiple source/destination pairs In reference [781], the authors look at a situation where several sources of data are distributed in the network, each one trying to send as much energy to a dedicated sink, possibly using multiple routes, each one equipped with a utility function. The optimization is to select routes such that the total utility of the network is maximized before the first node runs out of energy. The authors derive a flow control algorithm; the techniques used here are interesting and should be applicable to similar problems as well.

[5] Given an optimization problem with two objectives A and B, it is usually not possible to find "the" optimum. Rather, among the set of possible solutions, one can talk about "dominance" of one solution over another solution. S_1 dominates S_2 if either

- S_1's value in objective A is at least as good as that achieved by solution S_2 and S_1 performs strictly better with respect to objective B than does solution S_2,
- or vice versa, S_1 is strictly better with respect to objective A and at least as good as S_2 with respect to objective B.

In other words, A improves in one objective without loosing in the other one. The set of Pareto-optimal solutions is that set of solutions that is not dominated by any other solution. Usually, this set contains more than one element (as dominance is only a partial order on the set of solutions).

Consider all costs Some of the previously discussed papers have tried to find minimum (energy) cost paths but did not necessarily take into account all possible sources of energy consumption. BANERJEE and MISRA [51] argue, for example, that costs for retransmissions have to be taken into account as well (normally, residual error rates over wireless links cannot be neglected) and compare paths resulting from local retransmission schemes with end-to-end retransmission schemes.

Integrate scheduling and power control To obtain optimal solutions, CRUZ and SANTHANAM [185] describe a scheme that takes into account link scheduling and power control jointly with the computation of routes. The result is a relatively complicated optimization problem that minimizes total average power consumption, based on traffic requirements for all links. The resulting scheme has some strong assumptions, though.

A similar approach is taken by BERGAMO et al. [65], who use the results of a power control algorithm as an input to classical routing algorithms.

Routing and link quality A related issue is the quality of the underlying links. While most routing protocols are formulated in a graph-theoretical manner, it is often by no means clear which nodes are connected by a link. Links fluctuate in reliability and can have relatively high packet error rates. Using flooding-based protocols over such links can result in rather convoluted routing tables where nodes are considered to be neighbors only because a flooding packet happened to go through despite actually bad link quality. To overcome these problems, WANG et al. [859] advocate a careful selection of actual neighbors (in their case, parents for a routing tree toward a data sink), using information that the link layer can provide.

Routing and lifetime guarantees The determined routes evidently influence the lifetime of the network and several of the discussed protocols address this issue. References [716, 717] go a step further in that they attempt to provide guarantees on the lifetime of the network.

Routing for one-shot queries In WSN, a typical routing problem is that of a query that is to be routed to the place where it can be evaluated and then the routing of the answer back to the place where the query originated. Depending on the dynamics of the network, such a query can be regarded as a "one-shot" event if the structure of the network has changed sufficiently by the next query so that any topology information that the first query might have acquired is already outdated.

In such a situation it is not clear how to route a query and the answer. A typical assumption is that the location of the target is known and that geographical routing can be used or some form of data-centric routing comes into play. HELMY [348] proposes a scheme that handles such one-shot queries without recurring to location knowledge. The intuition behind it is that nodes know about their R-hop neighborhood; for queries outside this immediate neighborhood, so-called contact nodes are involved. These nodes are selected by nodes at the border of a given vicinity when the query has not been satisfied locally.

11.4 Broadcast and multicast

11.4.1 Overview

The protocols described in Sections 11.2 and 11.3 were trying to find efficient means to transmit data from one node to another one, possibly over multiple hops. In doing so, some of them had to collect or distribute information to all nodes in the network; they had to perform a **broadcast** operation. In fact, broadcasting can be a common operation in many wireless network applications.

Similarly, it is often necessary to distribute some data to a given, typically known, subset of all nodes in the network; this is called **multicast**.

The question of broadcasting has been treated rather extensively in Chapter 10 already. Essentially, the question is how to restrict the set of forwarding nodes as much as possible while still ensuring that all nodes receive the data. Multicast bases on the same principal ideas but both these tasks deserve some specific treatment from the routing perspective.

The multicast problem in a graph $G = (V, E)$ can be described by a set of sources $S = (s_1, \ldots, s_n)$ and, for each source, a set of destinations $D_i = (d_{i1}, \ldots, d_{im_i})$, for each $i = 1, \ldots, n$. In general, $D_i \subset V$. A frequent simplification is to assume that all destination sets are identical. Also, the edges of the graph are annotated with communication costs. There are several possibilities how to construct routing structures for multicast (overviews can be found, for example, in references [147, 474, 706, 827, 876, 917]).

Source-based tree The first idea is to construct, for each source, a tree, rooted at the given source, that contains all the destinations for this source and, if necessary, additional nodes of V to ensure that the tree can be constructed.

Which tree to select (out of, in general, many possible ones) is determined by the optimization goal, which reflects the link costs:

For each source, minimize the *total* cost Try to find a tree for which the *sum* of all link costs is minimal (over all possible trees rooted at the source). This is the **Steiner tree** problem[6] and matches best the intuitive expectation how a multicast routing structure should look like. It is, however, NP-complete (yet approximable within $1 + (\ln 3)/2 \approx 1.55$ [182]). Note that the minimal cost *broadcast* problem is equivalent to the minimal cost spanning tree problem (unless "wireless advantage" is assumed, see below), which is in fact solvable in polynomial time.

For each source, minimize the *maximum* cost to each destination Owing to the complexity of the previous optimization problem, a different optimization goal can be considered: Instead of trying to minimize the *total* cost of the tree, one can minimize the costs to each individual destination separately. In effect, this maps the multicast problem to repeated unicast shortest path problems, which can be solved by any routing algorithm, for example, Dijkstra's [147, 874].

Figure 11.6 illustrates the difference between these two optimization problems and the resulting trees.

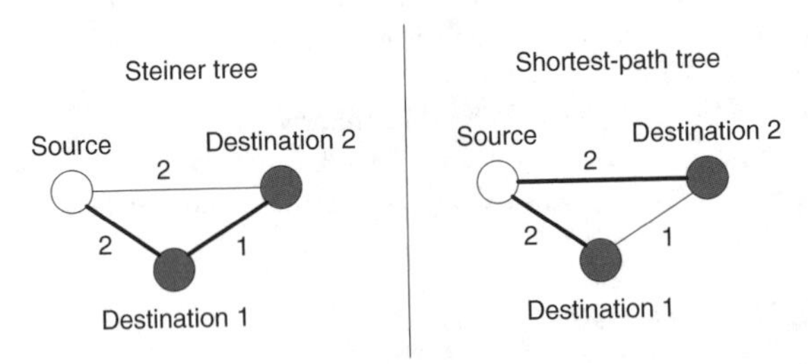

Figure 11.6 Difference between Steiner tree and shortest-path tree (thick lines indicate links that are part of the tree)

[6] In fact, Fermat was the first to ask a related problem: Given three points in the plane, how to find a point that minimizes the sum of the distance to the three given points. This is today known as the Euclidean Steiner problem; Steiner only later worked on the general problem of n nodes in a graph.

Shared, core-based tree Constructing and maintaining a dedicated tree for each source incurs considerable overhead. This overhead can be reduced if only a single tree is maintained – typically, this is promising when the destination sets for all sources are identical. The obvious downside is that for a given source the paths to its destinations can, in general, no longer be as short as with a dedicated, source-based tree.

To share a tree among several sources, a representative node in the network (not necessarily a source or a destination node) is selected and, from this node, a tree is constructed to contain all destination nodes; this tree is shared among all the sources. Selecting this **core** node is again NP-complete and optimization goals similar to the source-based trees can be considered.

In such a shared tree concept, the core node evidently becomes a single point of failure. To overcome this shortcoming, **multicore** shared trees are also considered in the literature.

Mesh While trees represent the overhead-optimal routing structures, they are not redundant – failure of even a single link will disconnect the tree. Adding additional links to the tree to obtain redundancy, however, will alter its essential properties, in particular the absence of cycles. The resulting routing structure will be a **mesh** [283] and requires more complicated forwarding structures than does a simple tree.

Orthogonal to the resulting structure of the routing tables is the assumption about the forwarding behavior of a node. Does a node use local unicast or can it actually, by virtue of the broadcast nature of the wireless channel, reach several or all of its neighbors with a single transmission? The first option directly maps to the standard graph model (separate transmissions are required to reach each neighbor). The second option is attractive since only a single transmission can spread information to many neighbors, but whether this is realistic depends on the assumptions made about hardware, MAC protocol, and, in particular, sleeping cycles of nodes. Ultimately, the ratio between transmission power and idle power consumption comes into play here. When this **wireless multicast advantage** [874] is assumed, the design of protocols (and of the resulting trees or meshes) changes considerably as now not the sum of the costs of outgoing links has to be considered but only the cost of the most expensive link that has to be invested in to transmit data to all neighbors. In practice, when a link is activated in the course of a routing protocol, all the cheaper links of that node are implicitly contained in the routing graph as well. CLEMENTI et al. [171] and also CAGALJ et al. [113] show that even under this assumption the problem of computing the optimal multicast graph is still NP-complete; some approximation results are proved in references [852, 853] – the best known, provable approximation factor is 12, making this a harder problem than the Steiner tree construction, for example.

It is interesting to note that other broadcast optimization problems are NP-hard as well. GANDHI et al. [270], for example, show that minimizing the broadcast latency (finding a broadcast tree such that the time until the latest node has received the message is minimized) is NP-hard, as is the problem of minimizing the number of retransmissions (they also provide heuristics with a constant approximation factor). FERREIRA and JARRY [256] consider the problem in the context of **evolving graphs** (graphs where the set of available links changes over time) and show that computing minimum spanning trees in a planar, mobile network is NP-complete.

Figure 11.7 summarizes these principal options how to organize multicasting in a wireless network. The following sections will discuss typical representatives of these categories in some more detail.

For completeness, it should be pointed out that *flooding* the network is also a means to implement broadcast and/or multicast. Efficient flooding structures (minimum connected dominating sets) have been discussed in Chapter 10. References [287, 610] are some examples where this concept is investigated further. Also, some of the gossiping techniques from Section 11.2 can also be amenable to implement a (probabilistic) broadcast.

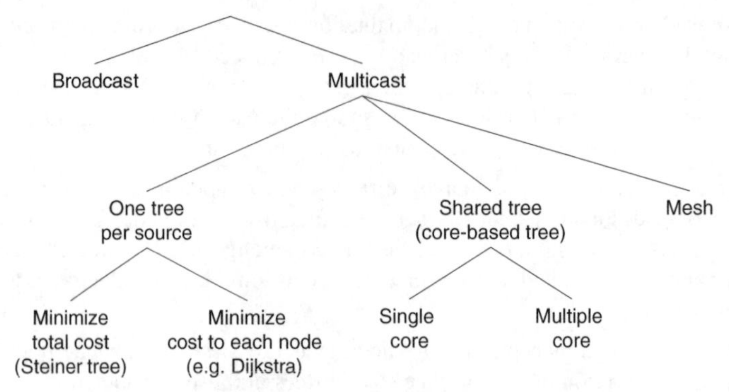

Figure 11.7 Overview of possible multicast approaches

11.4.2 Source-based tree protocols

Source-based trees can be constructed in a number of different ways. First, a simple heuristic is explained that ("simplifyingly") maps the tree construction onto the problem of finding shortest paths. Then, an essential algorithm for broadcast trees is introduced and the multicast problem is solved by various Steiner tree approximations.

These approaches work on a basic graph interpretation of the network. Solutions that exploit the wireless multicast advantage – in particular, BIP – are described later.

A greedy heuristic – Shortest Path Tree

A trivial heuristic for broadcasting and/or multicasting is to compute, to each destination, the shortest (or rather, cheapest) path and overlay all these paths onto a tree (described, for example, as Shortest Path Tree (SPT) in reference [874]). WAN et al. [852] show that this greedy heuristic does not have a good approximation ratio. Consider a network with the source node at the center, m nodes $p_1, \ldots, p_m$ distributed on the unit circle, and another m nodes $q_1, \ldots, q_m$ each placed on the line connecting the origin with node p_i at distance ε from the origin. SPT will send one broadcast packet (using the wireless multicast advantage) to the inner ring at cost ε^2 (assuming quadratic path loss), and each node q_i forwards at cost $(1-\varepsilon)^2$. The total cost is thus $\varepsilon^2 + m(1-\varepsilon)^2$. Having the source node send at power 1 will, on the other hand, distributed the information to all nodes. Thus, the approximation ratio is $(\varepsilon^2 + m(1-\varepsilon)^2)/1$, which converges to m as $\varepsilon \to 0$.

WAN et al. [852] also describe another greedy heuristic, broadcast average incremental power, which has an unsatisfying approximation ratio. The heuristics described below actually have constant ratios, outperforming greedy heuristics.

Broadcasting using minimum cost spanning tree – Prim's algorithm

A simple broadcasting algorithm can be based on a minimum cost spanning tree. One possible algorithm to compute it is due to Prim [657].[7] It starts with a tree consisting of the source node, and in $|V|-1$ steps adds one node per step to the tree. The node that is added is the one that has the lowest-cost link to any node already in the tree (evidently, this is not an immediately distributed algorithm).

[7] Kruskal's algorithm is of course also applicable. Prim's algorithm is described here as a later discussed one is structurally similar.

More formally, given a graph $G = (V, E)$ with edge weights $W : E \to \mathbb{R}$ (for example, energy cost for the edge), the algorithm maintains the tree (V_T, E_T) itself and a set of candidate nodes V_C along with one candidate edge per candidate node. Listing 11.1 shows this algorithm in more detail.

Listing 11.1: Prim's minimum-cost spanning tree algorithm

```
V_T = E_T = ∅
V_C = source node
while (V_T ≠ V) {
  Select v ∈ V_C with smallest candidate edge weight
  Add v to V_T
  foreach neighbor u of v in V \ {V_T ∪ V_C} {
    // new candidate u found
    add u to V_C
    add (v,u) as candidate edge
  }
  foreach neighbor u of v in V_C {
    if (W(v,u) < weight of u's existing candidate edge) {
      replace u's candidate edge with edge (v,u)
    }
  }
}
```

An important newer algorithm is a randomized one with linear-time complexity [403].

Some Steiner tree approximations for multicasting

A simple approximation

As computing the optimal Steiner tree is an NP-complete problem, quick approximations are required (references [371, 607] provide overviews). One simple approximation arbitrarily orders all the destination nodes as well as the source and then successively adds these nodes to the tree: For the first two nodes, a shortest path between these two nodes is constructed. Then, for every node, construct the shortest path to any node already on the Steiner tree (where it is not a trivial task to determine what the best connection point on the tree is). The result will be a tree including all the required nodes; the quality of the approximation depends on the order in which nodes are added to the tree. In practice, this approximation tends to perform reasonably well.

Takahashi Matsuyama

Instead of fixing the order in which nodes are to be added to the tree a priori, it stands to reason to let the algorithm find the next best node to be added. Thus, start with the source node. In each step, determine the next so-far-unconnected destination that has the shortest distance to the already existing tree; add this node via a shortest path to the tree. Repeat for all destination nodes. This is the Takahashi–Matsuyama heuristic [806].

KMB heuristic

The KMB heuristic (after Kou et al. [434]), on the other hand, maps the Steiner tree construction in the original graph G to the finding of a minimum spanning tree in another graph. Let D be the set of all nodes to be connected by the Steiner tree (including the source node). Construct the complete graph K (all nodes are directly connected) with D as the node set; assign edge weights in K as the cost of the shortest path between the respective nodes in the original graph. Construct a minimum spanning tree in K. Then, transform back this spanning tree to the original graph, that is, replace the artificial links in K with the (in general) multihop paths in G, obtaining another graph T'. This graph T' need not necessarily be a tree, so compute another minimum spanning tree on T', obtaining T''. This is almost the sought-after Steiner tree; it remains now only to remove any possibly unnecessary leaves. KMS is attractive since its competitive ratio is at most 2 and in practice KMS often comes within 5 % of the best possible solution [607].

A heuristic for multiple rates

The Steiner tree is well suited for a situation where a data source periodically sends data updates to multiple sinks. The problem becomes slightly different if these sinks require data more or less frequently. In such a situation, branching points in the tree only forward data with the necessary maximum rate per child to cover all the sinks in the respective subtree. Adding a source to such a tree can then require an increase in the sending rate of a branching point and will thus have an impact on the energy consumption.

KIM et al. [416] suggest a heuristic to handle this case, exploiting knowledge about location of nodes. When adding a new sink, they perform a recursive computation of the required additional power when using a given node as a branching point, taking into account possible needs to increase rates in a branch of the tree. As a result, paths might be used that are not strictly the most energy efficient ones but are those paths that already carry a high rate anyway. In addition, the tree can be locally modified to merge existing branches that run in parallel, pushing the branching point deeper toward the sinks.

Another heuristic that uses node locations is a modification of the Takahashi–Matsuyama heuristic and is described in reference [144].

Broadcasting/multicasting with a finite set of power levels

The Steiner tree is also amenable to solve the broadcast or multicast problem when each node only has a finite set of k different power levels at its disposal. A simple helper graph construction [498] can solve this problem: Replace each node v in the original graph $G = (V, E)$ with a small subgraph, consisting of v itself and k nodes $u_{v1}, \ldots, u_{vk}$. Add an edge (v, u_{vi}) and annotate it with a weight representing the transmission costs at power level i, $i = 1, \ldots, k$. The helper graph is then the union of all these replacement graphs for each v, with the additional edges (u_{vi}, w) (with $w \in V$ an original node) if and only if node v can communicate with node w using transmission power level i (for each v and each $i = 1, \ldots, k$). On this helper graph, solve a Steiner tree problem, with all or some of the original nodes as destinations. Further optimizations are possible if all nodes use the same transmission power levels.

Exploiting wireless multicast advantage for broadcast: Broadcast incremental power

In all the algorithms described so far, a node that wants to transmit to multiple neighbors (because it has multiple children in the tree, for example) experiences a proportional cost. The Broadcast Incremental Power (BIP) algorithm [874] differs here in that it exploits the wireless multicast advantage to compute a heuristic for a broadcast tree. The core idea is that a node that is already transmitting to some other node would only have to raise its transmission power in order provide data also to further nodes, without incurring cost for another transmission. Hence, the *additional* cost for a node to supply a further node with data is only the *difference* between the current and the needed (higher) transmission power.

On the basis of this idea, a modification of Prim's algorithm is possible. Like in Prim's algorithm, one node is added to the tree per round and each as-yet-not-added node maintains a "candidate edge" that represents its current best option to be added to the tree (updated each round). In each round, the node with the lowest-cost candidate edge is chosen. The difference lies mainly in the computation of the cost assigned to the candidate edge of a node and, since each node is reachable from each other node with sufficiently high power, there is no notion of a set of "candidate" nodes representing the fringe of the growing tree (alternatively, all nodes are in the fringe since the underlying graph is complete). An obvious modification of the algorithm assumes maximum transmission power per node, thus reintroducing possible candidates for transmission and speeding up the computing; this and other improvements are left out here for simplicity. The interesting part is the candidate edge weight computation: Unlike in Prim's algorithm, the currently used transmission power of a node is *subtracted* from the actual edge weight to reflect the fact that only *additional* costs are incurred if the wireless multicast advantage can be exploited.

Listing 11.2 shows the raw form of this algorithm (without possible efficiency improvements) and Figure 11.8 illustrates one example execution – black and gray nodes are nodes part of the tree, black nodes are transmitting themselves, dotted lines show the candidate edge with the candidate edge weight (note how this weight is reduced when the source nodes, for example, increases its transmission power), thick lines are the edges in the final tree. In the example, node A's candidate edge switches from S to B in round 2 and then, in round 3, back to S as source node S increases its transmission power to 3 to cover node C, making it cheaper to add A directly from S at an additional cost of 2 instead of having node B transmit at an additional cost of 3.

Listing 11.2: The broadcast incremental power algorithm for exploiting the wireless multicast advantage

```
// Initialize
V_T = {source node}
P(source node) = 0 // transmission power assigned to a node
foreach (v in V \ V_T) {
  Set candidate edge to (source node, v)
  Set candidate edge weight to transmission power to
    reach v from source node
}
// Compute tree
while (V_T ≠ V) {
  Select v ∈ V \ V_T with smallest candidate edge weight
  Add v to V_T using its candidate edge (u, v)
  Increase P(u) to smallest power that reaches v
  // Recompute candidate edges and their weights
  foreach (v in V \ V_T) {
    Select u which minimizes P'(u) - P(u)
    // where P'(u) ≥ P(u) is smallest power to reach v from u
    Set candidate edge to (u, v)
    Set candidate edge weight to P'(u) - P(u)
  }
}
```

This algorithm tends to produce good broadcast graphs. In some situations, however, it can unnecessarily assign high transmission power levels to nodes to cover neighbors that are already covered by some third node. Figure 11.9 shows such an example where nodes are placed at coordinates $A = (0, 0)$, $B = (3, 0)$, $S = (4, 0)$, and $C = (6, 0)$, the node at $(4, 0)$ is the source, and the transmission costs between two nodes are assumed to be proportional to the square of the distance (path loss coefficient of 2). Since B has to use transmission power 9 anyway to cover node A, it

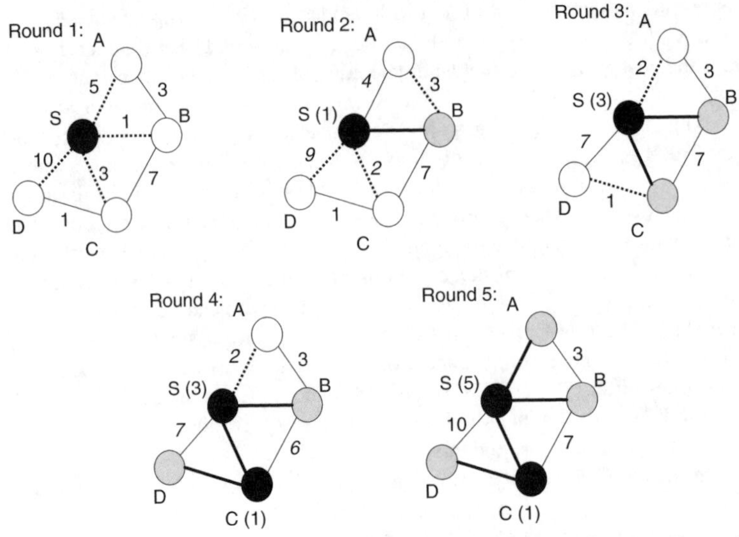

Figure 11.8 Illustration of the BIP algorithms's operation (in parentheses, the currently used transmission power is shown)

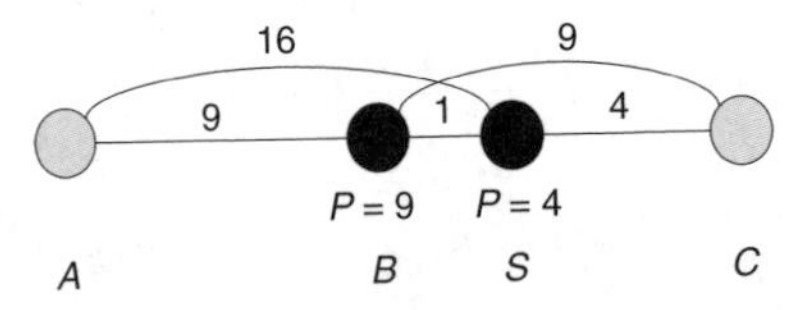

Figure 11.9 Example of opportunity for "sweeping" in the BIP algorithm

also implicitly supplies node C, avoiding the need for S to send at more than transmission power 1. Detecting such opportunities to reduce transmission power is called a sweep operation [874]. Such sweeping is of course applicable to other broadcast/multicast algorithms as well.

Not only does this algorithm work well in practice, but also its approximation ratio is provably better than that of the Minimum Spanning Tree (MST): WAN et al. [852, 853] show that the MST has an approximation ratio between 6 and 12, and BIP has one between 13/3 and 12.

Exploiting wireless multicast advantage for multicast: Pruning broadcast trees by Multicast Incremental Power (MIP)

On the basis of the computation of a broadcast tree, one simple heuristic for the multicast case is evident. Given a broadcast tree, prune it by removing all nodes from the tree that have no members of the destination set as "downstream" nodes; in addition, the sweep operation can be applied to reduce transmission power further if high power is only needed to supply subtrees without any destination nodes. WIESELTHIER et al. [874] study this pruning heuristic on the basis of several broadcast tree heuristics, in particular their BIP algorithm.

Embedded wireless multicast advantage – Transforming existing graphs

A different approach to leverage the wireless multicast advantage is followed by CAGALJ et al. [113]. They start from a "traditional", link-oriented broadcast tree, for example, the minimum-cost spanning tree. From such a starting point, they look for opportunities to increase transmission power levels of certain trees such that the additionally covered nodes can stop transmitting and the resulting, modified tree consumes less energy than the original one that did not take into account the wireless multicast advantage.

The algorithm proceeds in rounds, building up the final broadcast tree along the lines of the preliminary, MST-based tree. The first round starts from the source node. In each round, nodes are added as covered nodes to the final tree and one node is selected as a transmitting tree; some nodes that were transmitting in the preliminary tree can be demoted to normal, nontransmitting nodes (they are "excluded"). The decision of which node to select as a transmitting node and at which power to operate it depends on the possible *gain* that such a decision incurs. The gain is computed with respect to the possibility to exclude certain nodes, thus saving energy: For a potential transmission candidate u and a to-be-excluded node v, the transmission power for u is computed as that power required to cover both v itself and all of its children. This requires more energy on behalf of u but saves energy in v and also in potentially other transmitting nodes (which will then also be excluded). Eventually, the node that results in the biggest gain is chosen as transmission node. This algorithm is repeated until all nodes are covered.

Cagalj et al. [113] also describe a distributed version of this heuristic. It is also based on a preliminary, distributed determination of a minimum spanning tree. The final tree's computation is again divided into rounds, which are moreover structured into three phases. In these phases, information about neighboring nodes and the structure of the transmission chain from the source to a given node is used to modify the original tree – the details are somewhat involved and the reader is referred to reference [113].

Marks et al. [541] follow similar lines as Cagalj et al. [113] in that they also are interested in transforming existing trees to obtain ones with better energy consumption. Marks et al., however, start from stochastically generated trees (rather than MIPs) and their search operations are different as well. They base their heuristics on the viability lemma.

A distributed, position-based approach to the wireless multicast advantage

Another construction [120] for energy-efficient broadcasts does not start from the minimum spanning tree but starts from the Relative Neighborhood Graph (RNG) (already introduced in Section 10.2.3). The RNG is defined as that subgraph of the original graph where two nodes u and v are only connected if there is no third node w that would be closer to each of u and v. While the RNG typically has a higher degree than the MST, it is easily constructed using locally available information about positions of neighboring nodes or distances between neighbors' neighbors.

The RNG as such is a unicast graph as it does not take into account the wireless multicast advantage. To correct this, the proposed heuristic first determines, for each node v, the transmission range as the smallest range that connects v to all of its neighbors in the RNG. Then, the RNG* is defined as the subgraph induced by these (local broadcast) transmission ranges. Neighbors in this graph can know about joint neighbors in the underlying RNG.

A broadcast is then performed on the RNG*, using a "neighbor elimination" procedure. The source node s broadcasts a message using its assigned transmission range. A receiving neighbor v of the source can then decide which of its neighbors in the RNG have already been covered by the first broadcast, eliminates them from its list, and, if this list is nonempty, rebroadcasts the received message. Nodes that redundantly receive a message can still extract information about which neighboring nodes have already been supplied with the message and can eliminate them from their neighbor list. To take advantage of this information from other nodes' forwarding, the rebroadcasting of a received message is randomly delayed for some time. Reference [120] describes the procedure in more detail. It also compares this schemes with others, in particular, with BIP. The authors show that this RNG-based scheme is advantageous, especially in dense networks.

11.4.3 Shared, core-based tree protocols

The challenge in core-based tree multicast protocols lies in finding a good core node. Once this node is determined, essentially the problem can be reformulated as a source-based tree protocol with the core node as the source (although better optimization is possible). An overview of such protocols and a performance comparison can be found in reference [474]; further examples are contained in references [147, 917].

To give an idea how such a core search could work, the "merge point formation" from reference [564] is briefly described here. Assume there are a few sinks in a network to which data shall be distributed via a core-based multicast tree. A "merge point" for this tree is to be found. To do so, each sink broadcasts advertisement messages indicating its presence; each node in the network collects these advertisements along with sink identifier and number of hops that the advertisement took. After a certain time, each node that has received more than one sink advertisement broadcasts merge advertisement messages. These messages are only forwarded by nodes that have heard from fewer sinks or whose cumulative distance to all sinks is larger. Eventually, only one node (depending on network topology and timer values) will not have heard other merge advertisements overruling its own and will declare itself the merge point. This leader election result is then spread through the network.

When dropping the assumption that the broadcast or multicast tree is actually rooted in and constructed from the actual source of the broadcast, optimality is evidently sacrificed. On the other hand, having to maintain only a single tree for all sources instead of one tree per source is attractive as well. The question is thus how big the performance penalty is that has to be paid by using only a single (e.g. core-based) tree for all multicast operations.

This question is answered by Papadimitriou and Georgiadis [613]. They show that, given a tree and exploiting the wireless multicast advantage, the difference in total power consumption between any two sources on this tree is at most a factor of two. Moreover, they give a construction for a single broadcasting tree (SBT) such that the total power using this tree from any arbitrary source s is at most $2H(|V|-1)$ times that of the optimal tree for the given source (where $H(\cdot)$ is the harmonic function).

11.4.4 Mesh-based protocols

To overcome scalability (in number of sources) and robustness issues of tree-based protocols, a structure with higher connectivity is necessary that can connect multiple sources to their destinations. The first proposal in this sense was the Core-Assisted Mesh Protocol (CAMP) [283]. The mesh, a subgraph of the original graph, has to contain all sources and destinations and provide at least one path from each source to each destination. The redundancy of the mesh can actually enable shorter paths in the mesh than would be possible in a core-based tree; it is, however, up to the forwarding procedure to actually be able to exploit these shortcuts without resorting to flooding the entire mesh with data. Reference [917] summarizes several mesh-based protocols, mostly from an ad hoc background, and compares them against tree-based protocols. The energy consumption of mesh-based protocols is, on the average, $(f+1)/2$ times larger than that of tree-based multicast protocols (where $f \leq 2$ is the node connectivity, analyzed for nodes laid out in a grid pattern). Hence, their use in sensor networks requires careful deliberation.

As an example of a (in a sense) mesh-based protocol, let us consider the Two-Tier Data Dissemination (TTDD) described by Ye et al. [911] (Figure 11.10). Here, source nodes detect certain events that have to be forwarded to several mobile sinks. To do so, each source node, after detecting an event, starts to build a regularly spaced mesh of "dissemination points", resulting in rectangular cells of known size l (geographical unicast routing is assumed to be available). Sinks,

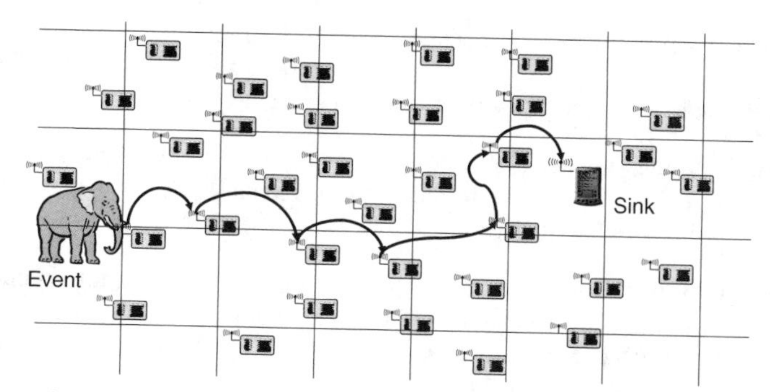

Figure 11.10 Two-tier data dissemination as an example of a mesh-based protocol

on the other hand, flood their queries in their local vicinity with a bounded radius to live (significantly more efficient than unrestricted flooding). This radius is selected large enough so that at least one dissemination point must be found within it. In this sense, two tiers of forwarding functionality are deployed – global multicasting on a mesh and local query broadcasting. Once a dissemination node receives a query, it forwards it upstream to the source of the data via the multicast mesh; it does *not* repeat that for multiple queries, reducing the query overhead. Dissemination nodes keep on forwarding (aggregated) queries toward the source and store information from where the queries arrived. This information is used to then send the actual data back to the sinks, implicitly setting up a multicast tree on top of the multicast mesh.

11.4.5 Further reading on broadcast and multicast

Gossiping for multicast Gossiping can be used for multicast and broadcast as well. CHANDRA et al. [131], for example, describe a scheme where gossiping is used to improve the reliability of multicasting. Reference [523] is a similar example.

Directed antennas for multicasting When directed antennas are available, robustness and traffic-carrying capacity of a network should improve considerably (given a fixed amount of available power). WIESELTHIER et al. [875], for example, extend their BIP/MIP protocols to take advantage of such antennas, assuming that directed antennas can be used to concentrate power to the neighbors in the multicast tree and thus reduce power consumption. This results in considerable improvements in network lifetime and total delivered traffic.

Relationship to topology control There is clearly a tight relationship to topology-control protocols. A difference is often the source-based nature of multicast/broadcast protocols, whereas topology-control protocols try to optimize the network as a whole. There are also some papers that explicitly combine these aspects, for example, references [629, 796].

Optimal solutions by linear programming An optimization technique that has not been discussed so far is linear programming where the problem is cast as a linear equation that is to be maximized (or minimized) subject to certain constraints. While this modeling approach obviously does not change the complexity of the problem (integer linear programming is NP-hard), it can often lead to good approximations of the optimal solutions by standard relaxation techniques. DAS et al. [191] describe three different integer linear programming problems to capture the minimum power broadcast problem. The solution to these problems can act as benchmarks for practical algorithms. Reference [20] is another example of this concept.

Optimal solution for tree networks How to optimally collect and distribute data in a tree network is considered in reference [261].

Time to complete a multicast So far, mostly the energy required for a multicast or broadcast has been considered. But the time necessary to do so can also be important. Reference [261] gives an example of how this problem can be approached.

Data placement A further variant of Steiner tree approximations is investigated by BHATTARCHARYA et al. [77]. They are interested in placing data caches in a network, which boils down to choosing good Steiner points.

Cooperative multihop broadcast In a broadcast flood, a node can overhear the same packet transmitted by several different senders. Using advanced signal processing, a node might be able to reconstruct the correct packet even if each individual reception is erroneous. Such cooperative broadcasting is described in reference [540].

11.5 Geographic routing

The idea behind the relatively large class of geographic routing protocols is twofold:

- For many applications, it is necessary to address physical locations, for example, as "any node in a given region" or "the node at/closest to a given point". When such requirements exist, they have to be supported by a proper routing scheme.
- When the position of source and destination is known as are the positions of intermediate nodes, this information can be used to assist in the routing process. To do so, the destination node has to be specified either geographically (as above) or as some form of mapping – a **location service** [483] – between an otherwise specified destination (e.g. by its identifier) and its (conjectured) current position is necessary.
  The possible advantage is a much simplified routing protocol with significantly smaller or even nonexisting routing tables as physical location carries implicit information to which neighbor to forward a packet to.

The first aspect – sending data to arbitrary nodes in a given region – is usually referred to as **geocasting**. It was originally introduced in an Internet context [589]; a survey can be found in reference [533]. The second aspect is called **position-based routing** (in particular in combination with a location service); it was probably first introduced by FINN [258] as "Cartesian routing". MAUVE et al. [543] provide an overview. The necessary techniques to make nodes aware of their position have been described in Chapter 9.

In wireless sensor networks, usually the geocasting aspect of geographic routing is considerably more important. Since nodes are considered as interchangeable and are only distinguished by external aspects, in particular their position, a location service is usually not necessary. Hence, this chapter concentrates on the geocasting aspect, with position-based routing aspects treated where necessary, and are briefly surveyed in Section 11.5.1. The presentation given here partially follows, in its broad structure, the survey papers [533, 543]

11.5.1 Basics of position-based routing

Some simple forwarding strategies

Most forward within *r*

Assume a node wants to send a data packet to a node at known position and assume also that every node in the network knows its own position and that of its neighbors. In a simple greedy forwarding approach, the packet is forwarded to that neighbor that is located closest to the destination (the

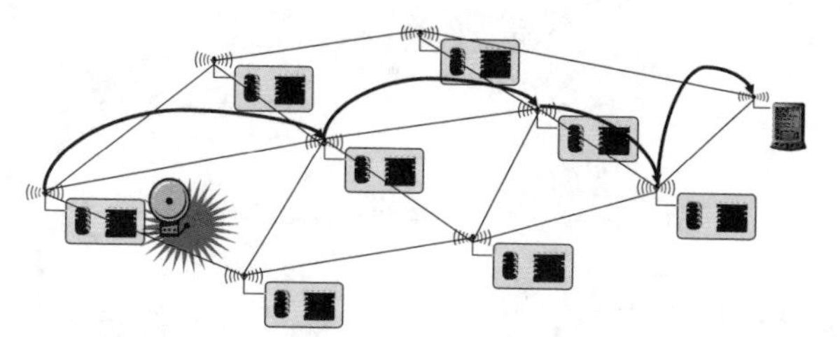

Figure 11.11 Simple greedy geographic forwarding

destination's position is included in the packet), minimizing the *remaining* distance that the packet has to travel.[8] Formally, the next hop of node v toward destination d is chosen as

$$\text{next hop}(v) = \text{argmin}_{u \in N(v)}\{|ud|\},$$

where $|ud|$ indicates the distance between nodes u and d and $N(v)$ is the set of neighbors of node v. This scheme is called **most forward within** r [804], where r indicates the maximum transmission range and thus the neighborhood. As Stojmenovic and Lin [794] show, this method is necessarily loop free.

Figure 11.11 illustrates this scheme and immediately shows one principal shortcoming: by ignoring topology information, geographic routing is, in general, not able to find the shortest possible path (in hop count). This trade-off between simplified routing scheme and reduced efficiency is, in general, unavoidable.

Nearest with forward progress

An alternative to the greedy forwarding is to choose the *nearest* neighbor that still results in some progress toward the destination [357]. The rationale is to reduce the collision rate and thus to maximize the expected progress per hop; it is not clear how this scheme would interact with an actual MAC layer.

Directional routing

Yet another possibility is to forward to nodes that are closer in direction rather than closer in distance. Compass routing [437] is an example, where that neighbor is chosen that is closest to the direct line between transmitter or destination. (A variation would be to choose the *angularly* closest node; this is not identical.)

Distance Routing Effect Algorithm for Mobility (DREAM) [61] is another example of this idea. Unlike the most progress within r scheme, however, a direction-based scheme like DREAM is not necessarily loop free; Stojmenovic and Lin [794] give an example. To ensure loop freeness in direction-based algorithm, memory about which nodes have already been forwarded by a node has to be used in the nodes.

The problem of dead ends

What is more, these simple strategies also cannot deal with dead ends. Figure 11.12 illustrates how an obstacle that blocks the direct path between source S and destination D interrupts communication even though S and D are actually connected by the network.

An apparently simple fix for a situation where no forward progress can be made is to use the "least unappealing" node, that is, the neighboring node that loses the least progress [804]. However,

[8] A common misconception is to use the node farthest away from the sender.

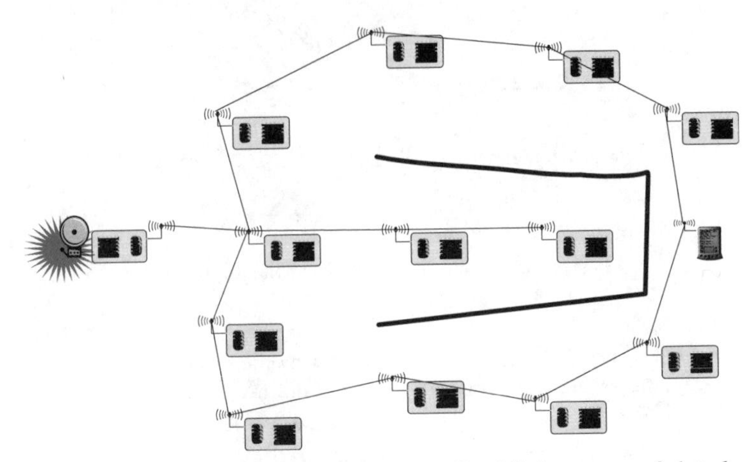

Figure 11.12 Simple greedy geographic forwarding fails in presence of obstacles

as Figure 11.12 shows, this heuristic can lead to packets looping back and forth between the nodes near the obstacle.

The obstacle problem is also not solved by randomly choosing a node that is closer to the destination than the transmitter is, as proposed in reference [590] (randomly forwarding to *any* node results in random walks). Hence, improvements over these simple schemes are required.

Restricted flooding

Figure 11.12 also shows that even an extended greedy forwarding where a source forwards to some or all of the nodes that are closer to the destination than itself (so-called **geographically restricted flooding**) will not remedy the shortcoming – the scheme will not be able to find detours.

Restricted flooding is, on the other hand, quite suited to compensate for mobility of the destination. Assume that the destination moves at a given speed v and that the distance between transmitting node and destination is known, it is a question of simple trigonometry to find an angle α such that D will receive the packet (with given probability) when all neighbors in this angle, centered around the line between transmitter and destination, will receive the packet. BASAGNI et al. [61], for example, provide the required formulas.

Right-hand rule to recover greedy routing – GPSR

Figure 11.12 not only illustrates the problem of greedy forwarding in dead ends but also gives an intuition about a possible solution. When being stuck in a dead end, or even in a labyrinth, one certain way of escaping from the labyrinth is to keep the right hand to the wall and keep walking. This way, all the walls of the labyrinth will be eventually visited [87]. The practical consequence is to backtrack the packet out of the dead end, counterclockwise around the obstacle; it will eventually find a node closer to the destination.

This intuition has been turned into various protocols, for example, Compass Routing II [437], "face-2" [93], or, later, the Greedy Perimeter Stateless Routing (GPSR) protocol [409]. GPSR forwards a packet as long as possible using greedy forwarding with the "most forward" rule. If a packet cannot make any more progress, the packet is switched to another routing mode: perimeter routing. A perimeter is a set of nodes defining a **face** (the largest possible region of the plane that is not cut by any edge of the graph; faces can be exterior or interior). Perimeter routing essentially consists of sending the packet around the face using the right-hand rule. To do so, the packet carries

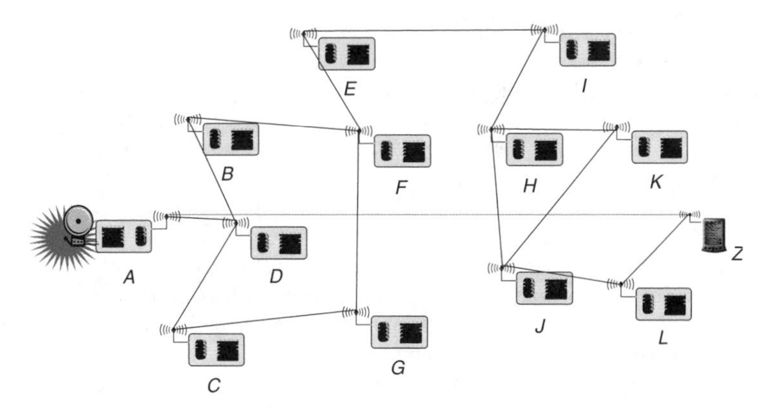

Figure 11.13 Example for GPSR

information where it entered a given face. This node v and the connecting line between v and the destination are used to decide whether the packet should leave the face and proceed to the next one (when the edge from the current node to the next node on the face intersects the connecting line between v and the destination node). Also, the packet can return to greedy forwarding if the distance of the current node to the destination and node v has been effectively reduced (but see the next section regarding performance guarantees of such fallback heuristics).

Figure 11.13 illustrates how a packet would be routed from node A to node Z. While at node A, the packet can be greedily forwarded to node D. At node D, greedy forwarding fails (both B and C are further away from Z than D itself), so the packet has to be routed around the perimeter of the interior face defined by *BFGCD*. That is, it is forwarded to B and from there to F. Here, edge FG intersects line DZ and routing can proceed to the next face (note that greedy forwarding to G would not help here). The packet proceeds around the perimeter of the exterior face via E and I to H, from there via K to J and thence to L and Z (the last steps via greedy forwarding).

Since this face-based procedure is based on properties of the plane, it only applies to planar graphs. In general, wireless network graphs are not planar, requiring the construction of a planar subgraph first. Reference [93] suggests to use a Gabriel graph; reference [409] discusses both Relative Neighborhood Graph (RNG) and Gabriel graph. Both these subgraphs can be constructed in a distributed fashion assuming that node positions are known and have been discussed in Section 10.2.3 already.

Performance guarantees of combined greedy/face routing

When combining face routing and greedy routing, face routing is tasked with routing around obstacles or out of dead ends while greedy routing tries to make quick progress toward the destination. One would thus like to switch to greedy routing as soon as possible once the obstacle has been cleared. It is, however, nontrivial to select this face-to-greedy switching point correctly or even to provide performance guarantees about the behavior of such an algorithm. A simple heuristic for such a fallback like switching to greedy mode whenever a node has been found that is closer to the destination than the node where face routing started is in fact not worst-case optimal [452].

In fact, the first combined greedy/face routing algorithm that is provably worst-case optimal was described in reference [452], but in order to show the worst-case optimality, quickly switching back to greedy routing could not be used. The proved performance bound was that face routing reaches the destination in $O(c^2)$ steps, where c is the cost of the optimal path from source to destination. The idea here is to adaptively grow an area in which next hops are searched. This performance is worst-case optimal since a graph can be constructed on which no geometric algorithm (without routing tables) can do any better.

The result from reference [452] has been improved by the same authors in reference [451] by presenting the Greedy and (Other Adaptive) Face Routing (GOAFR)+ algorithm that is worst-case optimal and at the same time efficient in the average case. The crucial point is when to fall back to greedy mode – too soon loses worst-case optimality, too late wastes average-case performance. Two techniques realize this behavior:

- The algorithm maintains a bounding circle, centered at the destination node, that prevents the face search from needlessly exploring in the wrong direction. This circle is reduced at every step in the greedy forwarding phase and can be enlarged in face routing if, with the current circle restrictions, no progress toward the destination can be made.
- A packet maintains two counters, p and q. When switching to face-based forwarding, both counters are set to 0. Counter p contains the number of nodes on the face perimeter that are closer to the destination than is the node where face search started; q counts nodes farther away. The algorithm falls back to greedy search if $p > \sigma q$ (for some properly chosen constant σ), that is, when substantially more nodes are closer to the destination on this face than are further away.

As shown in reference [451], this algorithm is worst-case optimal. It is also efficient in the average case as shown by simulation-based comparison against other algorithms, notably GPSR. An interesting observation is that the difference between these algorithms is largest in the phase transition from a barely connected to a very dense network (where it is either trivial or impossible to find good paths).

Reference [451] also explores the consequences of different cost metrics; the reader is referred to the paper for details.

Combination with ID-based routing, hierarchies

Purely position-based routing can be problematic in the immediate vicinity of the destination node, for example, when the destination has moved around or the location information is not very accurate. Identity-based routing protocols solve this issues relatively smoothly but have difficulties maintaining state information over long distances. Hence, a natural combination would use (even coarse-grained) position information to forward a packet into the vicinity of the destination where then an identity-based protocol (like any mobile ad hoc networking protocol) would take over. An example for such a hybrid, hierarchical approach is the "Terminodes" project's routing protocol [81].

Randomized forwarding and adaptive node activity – GeRaF

Zorzi and Rao [940, 941] investigate the combination of position-informed, random forwarding and nodes that switch on and off to save energy. Their scenario for investigation is the following: Assume nodes are uniformly distributed over the plane, each node knows its position, and each node turns on or off at arbitrary times; nodes also know their position and that of their neighbors. The goal is to transmit a message, usually over multiple hops, to a destination node the position of which is also known; the challenge is the constantly changing topology.

Their basic idea is then to use **receiver-initiated** forwarding: A node S forwarding a message simply broadcasts it *without* specifying – in the packet or otherwise – which node shall forward it. Ideally, the node closest to the destination node T and in range of S will pick up the message and forward it onward. However, since the neighbors of S do not know which of the other nodes is currently asleep, a deterministic rule to pick the forwarder would be difficult. Hence, the problem is solved by a position-informed randomization. Define N annuli with inner radius

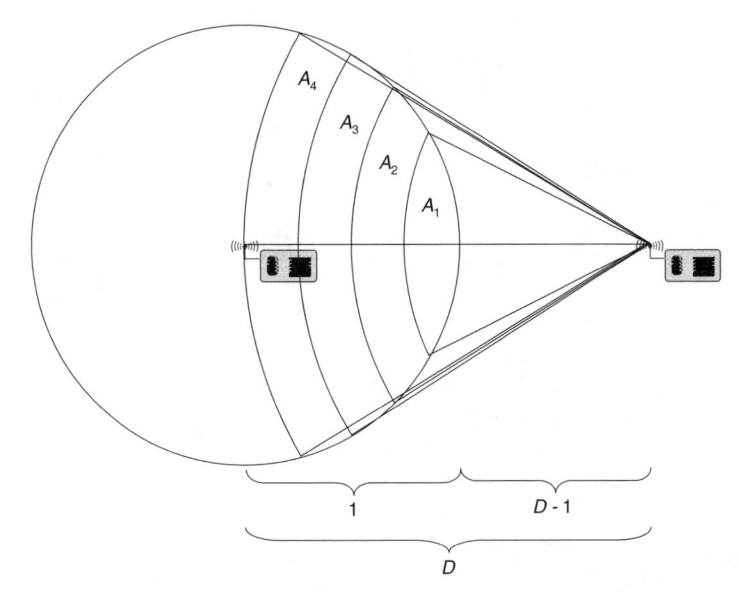

Figure 11.14 Contention regions for selecting the next hop node in GeRaF [941]

$D - 1 + (i - 1)/N$, $i = 1, \ldots, N$ and width $1/N$ (where $D = |ST|$ is the distance between nodes S and T and the radio range is normalized to 1); let A_i be the intersection of these annuli with the radio range of S (Figure 11.14).

After S broadcasts its packet, nodes in A_1 contend for forwarding (treating all nodes in A_1 as equivalent from the point of view of distance to T). If one node forwards the packet, the forwarding problem is solved. If several nodes attempt forwarding, the collision has to be resolved using a standard resolution algorithm (backoff or similar). If no node answers, then, in a next time slot, nodes in A_2 should attempt to forward the packet, and so on. If there is not even a node in A_N that is awake, node S simply can wait for some time and reattempt transmission, hoping that some nodes have woken up in the meantime.

In this scheme, a larger N will better approximate the geographically closest node but increases latency of the forwarding. Interestingly enough, Zorzi and Rao [941] show that even $N = 3$ achieves an average number of hops to the destination that is close to the optimal one. Hence, the scheme is feasible. Moreover, references [940, 941] contain a detailed analysis of multihop, energy, and latency performance, deriving bounds for the most important metrics.

It is also interesting to compare this scheme to GAF, discussed in Section 10.6.1. GeRaF is similar in spirit but since it does not require all nodes to be identical from a routing perspective, it need not artificially restrict the range of the nodes and can thus achieve a substantially reduced average number of hops to the destination.

Geographic routing without positions – GEM

What seems like a contradiction – using principles of geographic routing without position information – is actually a rather ingenious idea. Newsome and Song [593] describe a method that uses *virtual* instead of actual physical coordinates of nodes to facilitate routing (Figure 11.15). The method has two essential parts: Routing using a *polar* coordinate system (given in radius and angle from a center, as opposed to the usual Cartesian coordinates) and an efficient, distributed construction of such virtual polar coordinates that does not depend upon actual physical coordinates.

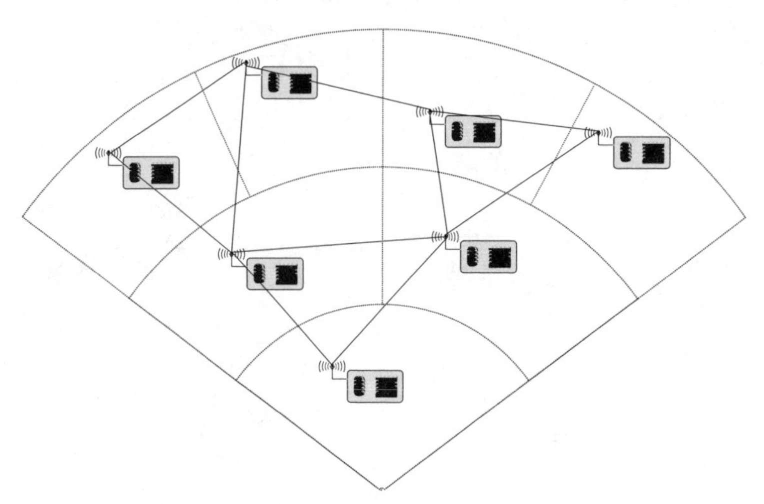

Figure 11.15 Geographic routing with positions [593]

To start with, let us construct the Virtual Polar Coordinate Space (VPCS). Pick a node in the center of the network and construct a standard spanning tree with that node as the route. The spanning tree immediately defines the radius of a node – it is simply the number of hops between a node and the root node in that spanning tree. The angle is more complicated. First, each node is assigned an *angle range*, which it can use to assign angles to nodes in its subtree; the root node has the angle range $[0, 2\pi]$. Larger subtrees need a bigger angular range so that the given range is split up onto child nodes, proportional to the size of the subtree of each child. Formally: If node v has n children $v_1, \ldots, v_n$, each with a subtree size of $s_1, \ldots, s_n$, and node v is assigned range $[\alpha, \beta]$, then child v_i is assigned a range of size $(\beta - \alpha)s_i/(s_1 + \cdots + s_n)$. The remaining challenge is which child is assigned which particular range.

Intuitively, if child nodes knew their position, they could be easily sorted in ascending angles and proper ranges could be assigned to them (child v_i would be assigned angle range $[\alpha + (\beta - \alpha)(s_1 + \cdots + s_{i-1})/(s_1 + \cdots + s_n), \alpha + (\beta - \alpha)(s_1 + \cdots + s_i)/(s_1 + \cdots + s_n)]$). Such position information could be determined using any of the methods described in Chapter 9. The authors show, however, that the resulting accuracy tends to be insufficient for a virtual coordinate system. Instead, they propose a method based purely on hop count, working without any physical location information.

Determining angular information using hop counts works as follows. Choose two nodes in addition to the original root; the three nodes should be far apart and not collinear. Determine, for each node in the network, the hop count of the shortest path between each of these three nodes (using two additional spanning trees rooted at the two additional nodes, for example). In addition, each node in the network has to know the distances (in hops) between the three reference nodes. On the basis of this information, each node can triangulate its own position in the hop count metric. So far, this method is similar to the DV-Hop method discussed in Section 9.5.2. Using the angular information resulting from these positions would still be fairly inaccurate. Hence, for each subtree, its *center of mass*, the average position of all nodes in a subtree, is computed and propagated to the root of the respective subtree. As it turns out, determining the ordering of child nodes using this center of mass of their subtrees results in a fairly accurate virtual polar coordinate system.

As a result, a tree has been computed and nodes have been labeled with virtual coordinates that correspond well to the topological situation of the network. This tree is embedded in the original graph, whence the name Graph EMbedding (GEM) for this approach.

Routing in a tree from one node to another, both identified by their virtual polar coordinates, is actually trivial – route up in the tree until a node has been found that is a common parent of both source and destination (which is easily decidable via the angular range for which a node is responsible) and then route down to the destination node.

This simple scheme is rather inefficient as it does not exploit possible "shortcuts" between physical neighbors (in radio range) belonging to different subtrees, in particular, when far away from the root. To remove this inefficiency, NEWSOME and SONG [593] propose a suitable routing algorithm, Virtual Polar Coordinate Routing (VPCR), that can take advantage of "circular" links (connecting nodes of same or similar radius but of different angle) in the network. Instead of always routing up the tree, VPCR checks to see if there is a neighboring node that is angularly closer to the destination than the current node. If so, that node is given preference; if not, the packet is routed up the tree. Once an ancestor of the destination is reached, the packet is routed downward just like in the simple routing scheme.

Overall, the procedure is remarkably simple once the virtual coordinates have been constructed and even their construction in a stable network can be done with acceptable overhead. NEWSOME and SONG [593] also discuss options to repair the graph embedding when nodes fail or move around. This scheme works especially well when geographic information is not available or imprecise or when geographic and topological proximity do not coincide. Moreover, it lends itself to implementing data-centric routing and storage, as discussed in Section 12.4.

11.5.2 Geocasting

Geocasting – sending data to a subset of nodes that are located in an indicated region – is evidently an example of multicasting and thus would not require any further attention. Similar to the case of position-based routing, position information of the designated region and the intermediate nodes can be exploited to increase efficiency. Thus, a few dedicated geocasting protocols shall be briefly described in the following.

A broad classification can be made into protocols that are essentially based on some form of geographically restricted flooding even outside the destination region and protocols that are based on some unicast routing protocol to transport a packet into the destination region. Within that region, clearly some form of flooding is required as all nodes in that region are supposed to receive the data. Most of the examples discussed here are based on restricted flooding; GeoTORA is one example based on unicast routing.

Location Based Multicast

A simple way to implement geocasting is to base it on flooding but somehow restrict the area where packets are forwarded. The Location-Based Multicast (LBM) protocol [425] does just that. There is a **forwarding zone** such that only nodes within the forwarding zone forward a received data packet. This zone can be defined in various ways:

Static zone The smallest rectangle that contains both the source and the entire destination region, with its sides parallel to the axes of the coordinate system. (Alternative geometric definitions are of course possible as well, for example, the destination region and two tangents to it defined by the source node's location [423].)

Adaptive zone Each forwarding node recalculates the zone definition, using its own position as the source. This way, nodes that would be included in the static zone but would represent a detour once the intermediate node has been reached are excluded from forwarding. Since this can, however, again lead to dead end situations, this rule is only applied if an intermediate node actually has neighbors within its newly calculated forwarding zone; otherwise it forwards the packet to all neighbors.

Adaptive distances While the previous two schemes contained the forwarding zone explicitly in each packet, this scheme recomputes it in each step, on the basis of information about the destination region and coordinates of the previous hop (or the source). The idea here is that a node u forwards a packet to its neighbors if its distance to the center of the destination region is smaller than the distance of the previous hop v to the center (the packet has made progress). If not, the packet is only forwarded if the node is actually within the destination region (to ensure that all destinations receive the packet).

Ko and Vaidya [425] point out the importance of not only looking at the overhead caused by a geocasting protocol but also at its accuracy, defined as the ratio of the nodes in the geocast region that actually received the packet. The adaptive algorithms, in fact, achieve a good trade-off between reduced overhead and maintained accuracy.

Finding the right direction: Voronoi diagrams and convex hulls

To correctly decide which neighbors of a forwarding node are the "right" direction is not an immediately obvious task for directional routing approaches like Compass routing or LBM. Stojmenovic et al. [795] suggest to use Voronoi diagrams:[9] Given a node S that has to forward a message, the destination region D (or the region of uncertainty where the destination node is located), and the set $N(S)$ of neighbors of S. Construct the Voronoi diagram for $N(S)$ (not including S itself). Then, a given neighbor $A \in N(S)$ is closest to some node in D if and only if its Voronoi polygon intersects D. Hence, these neighbors should be selected as next hops. Figure 11.16 shows an example.

Since Voronoi diagrams can be constructed in $|N(S)| \log |N(S)|$ time and $|N(S)|$ is likely to be relatively small, the overhead is acceptable. Reference [795] also proposes some approximation constructions. The advantage of this construction is that it will also find nodes that deviate from the immediate direction if they are necessary to forward the packet, improving upon the heuristics given in the LBM protocol.

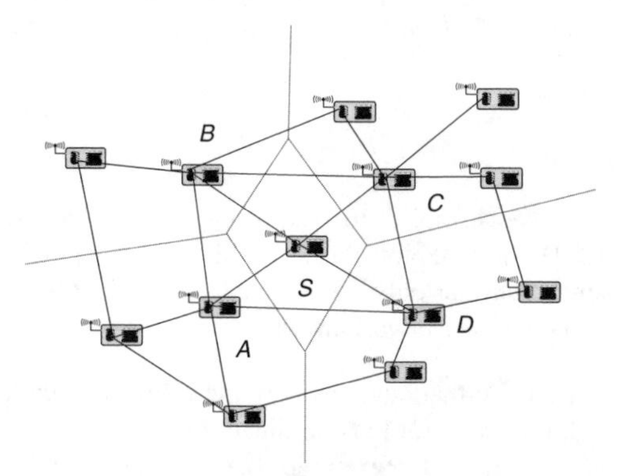

Figure 11.16 Illustration of the Voronoi diagram-based neighbor selection scheme [795] – node S uses the Voronoi cells to decide which neighbor to use for a given destination area

[9] A Voronoi diagram of a set N of nodes is a tessellation of the plane into $|N|$ convex polygons, one for each node, such that the polygon associated with node v contains all the points that are closest to node $v \in N$ than to any other node in N. It has been discussed in the context of the Delaunay triangulation in Section 10.2.3.

A similar rule can be developed to improve the "most forward within r" rule [795]. The problem is to find those neighbors that make most progress toward some point in the destination region D. To this end, construct the two tangents from S to the region D. Call the intersection points of the tangents with D U and V. For U and V, determine those neighbors of S on the convex hull of $N(S)$ that represent the biggest progress toward U and V; call them U' and V', respectively. The set of next hop nodes is then all the nodes in the convex hull of $N(V)$ between and including U' and V'. The convex hull is used to ensure maximum progress; it also can be efficiently constructed.

Tessellating the plane

Apart from locally computed Voronoi diagrams, other, perhaps simpler, tessellations of the plane can also be considered. The biggest simplification would be to use a fixed tessellation into regions where each point in space is uniquely mapped to one region.

CHANG et al. [135] propose one such scheme. They use a fixed tessellation of the plane into hexagons where each hexagon either has a "manager" in charge of it or is classified as an obstacle to be rooted around. They describe rules on how to back out of dead ends in this simplified geometric structure.

Similarly, LIAO [499] describe the GeoGRID protocol. Here, the plane is divided into square grids where each grid has an elected gateway in charge of it. Only those gateway nodes propagate packets among different grids, resulting in a need to control the size of such a grid.

Mesh-based geocasting

Geocast Adaptive Mesh Environment for Routing (GAMER), a mesh-based protocol for geocasting, has been proposed by CAMP and LIU [118]. It improves upon other mesh-based geocasting protocols by adapting the density of the created mesh according to the mobility of the nodes in the network. Since mobility, however, is not the core focus of the present discussion, the reader is referred to reference [118] for details and for references to other mesh-based geocast approaches.

Geocasting using a unicast protocol – GeoTORA

As a last example of standard geocast, let us consider how to modify a unicast protocol to obtain a geocast protocol. The starting point is the Temporally Ordered Routing Algorithm (TORA) unicast ad hoc routing protocol [619]. The intuition behind TORA is to conceive of the graph as a "landscape" where different nodes have different heights above ground. If the destination of a unicast routing protocol is the lowest point in this landscape (e.g. at height zero) and if there are no local minima, then the forwarding process is trivial: simply pass on the packet downward. Formally, this intuition is captured by imposing a Directed Acyclic Graph (DAG) onto the original graph by orienting its edges. This DAG only has a single sink (a node without outgoing edges), which is the destination node. The essence of the TORA protocol is then in ensuring that this DAG structure is maintained despite link failures or node mobility.

On this basis, KO and VAIDYA [424] first show how to modify TORA to support anycasting (sending a packet to any arbitrary member of a given group). This can be achieved by simply assigning height 0 to all nodes in this anycast group. The DAG can still be constructed, using essentially the same rules as in TORA (the devil is, of course, in the details).

Once anycasting is in place, the extension to geocasting is also relatively simple: any node in the destination region joins the anycast group and, in addition, locally floods a received packet within the destination region, similar to other flooding protocols. It is also necessary to handle the case of an empty geocast region or of an empty geocast region with a node moving into it. These cases are described in reference [424].

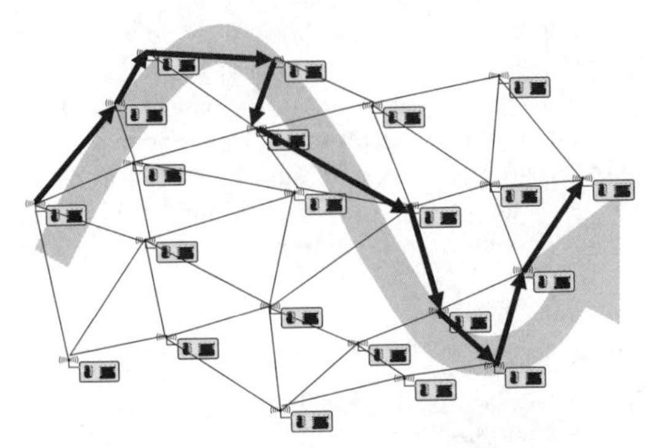

Figure 11.17 Trajectory-based forwarding

Trajectory-based forwarding (TBF)

In the previous approaches, the destination region was – intuitively – conceived of as a more or less convex region somewhere "far away". But this is not the only possible interpretation of geocasting as the somewhat different approach of Trajectory-Based Forwarding (TBF) [587, 600] shows (Figure 11.17). Instead of trying to send a packet to some region far away, the region of interest can actually be a path in the network. This path, or trajectory of the packet, can be embedded into the packet as a parametric description of the curve that the packet is supposed to follow; the parameter could be time or, preferably, the length of the path that the packet has followed. In this sense, trajectory-based routing combines aspects from source routing (as the trajectory is defined by the source) and geocasting. Different forms of such trajectories can be useful for different purposes, for example, a tree form for broadcasting or a "boomerang" (where the packet visits all nodes in the network and returns to the source node) for management of a network.

Given such a parametric description of a trajectory, the forwarding of a packet can follow different rules. For example, a node could forward a packet to its neighbor that minimizes the distance from the prescribed trajectory or to the one that results in the most advance on the trajectory (without deviating too much from it). Selecting the best forwarding policy depends on the actual application requirements.

11.5.3 Further reading on geographic routing

Impact of localization errors In a real system, it is unrealistic to expect that all nodes know their correct positions. Rather, they will only have approximately correct positions about their own and their neighbors' position. Several papers investigate the impact of such localization errors on, among other aspects, geographic routing protocols. He et al. [339] show that, for one exemplary routing protocol, delivery ratio and path length overhead are acceptable as long as the localization error is within about 40 % of the communication range. For face routing protocols, however, Seada et al. [748] report a higher susceptibility to imprecise localization information. They claim that the construction of a planar graph (e.g. a Gabriel graph) that is necessary for face routing is often to blame and suggest to allow edge removal between two nodes only if both these nodes intend to remove the edge because of the same witness. Kim et al. [419] provide further results.

Location services A location service maps from node identifier to (likely or last known) location of the node. This service is important for ad hoc or Internet-based geographic information but rarely needed in WSNs. Such "position databases" or "location tables" can be organized centrally or the information can be kept distributed in structures akin to routing tables.

BASAGNI et al. [61] suggest, in the DREAM protocol, to spread location updates information less frequently to distant nodes, exploiting the effect that the same absolute position change maps to a smaller angular change the further away an observer is (distance effect). Moreover, nodes trigger updates based on their own mobility. In a follow-up paper, BASAGNI et al. [60] point out that this mechanism results in relatively high accuracy of the distributed position tables.

LI et al. [483] describe a scalable, distributed position database. A node only updates a few dedicated location servers (which are implemented by normal nodes) with new information about its location; in queries these servers are used as well. The decision of which nodes uses which other nodes as location servers is based on so-called consistent hashing.

Finally, GROSSGLAUSER and VETTERLI [306] describe a scheme where location information is updated by the mobility of the nodes themselves. This removes any additional communication overhead and relies on implicit diffusion of information.

Location-Aided Routing (LAR) This protocol [423] uses location information to assist in the flooding phases of standard ad hoc routing protocols; the protocol is similar in many respects to the LBM described above. It assumes that a node, which tries to find a route to the destination node, has some notion of where the destination is likely going to be (the "expected zone"). Route requests are then addressed to the expected zone and intermediate nodes form – explicitly or implicitly – a "request zone" where only nodes in this request zone forward request packets. Several shapes of request zones are possible; the simplest one is perhaps the smallest rectangle that encompasses both the source node and the expected zone.

Making geocasting energy aware The previously discussed geocasting schemes paid no heed to the energy reserves of the wireless nodes. Geographic and Energy Aware Routing (GEAR) [919] is a geocasting scheme that introduces load-splitting among neighbors when forwarding toward the target region, trying to equalize the energy consumption of all nodes. As a result, extended network lifetime and better connectivity after partition are claimed.

Geographic routing without geographic coordinates As the results about imprecise information has shown, geographic routing does not depend on the actual location to work correctly, but some sufficiently close representation of the network layout will do. The question is how close this information really has to be. RAO et al. [681] introduce an interesting approach where the coordinates used for geographic routing are purely virtual ones and are constructed without actually recurring to the physical location of nodes at all (quite different in detail from the virtual polar coordinates described above).

The simplest case of virtual coordinates is when nodes at the edge of the network know their actual coordinates. Then, nodes iteratively try to estimate their real coordinates by computing their virtual coordinates as the average of the coordinates of their neighbors; initially, all nonperimeter nodes set their virtual coordinates to the center of the network. While this iterative relaxation does not produce good estimates of the real coordinates, it is rather surprising to note that the information in these virtual coordinates suffices to be used in geographic routing protocols, achieving good accuracy and path length. RAO et al. [681] also describe schemes where perimeter nodes do not know their location and show that, even then, virtual coordinates are still useful for geographic routing protocols.

Link asymmetry ZHOU et al. [934] show that geographic routing fails miserably in the presence of link asymmetry.

11.6 Mobile nodes

As discussed in Section 3.1.4, there are three essential sources of mobility in a WSN: mobility of the sensor nodes, mobility of data sinks, and mobility of the observed event. All of these three types of mobility are to a smaller or larger degree treated by the mechanisms already discussed and, even more so, by the mechanisms to be discussed in the chapter on data-centric networking. In addition, handling mobility is the core focus of mobile ad hoc networks research in the first place. Nevertheless, a few pointers to some specific approaches deserve some brief mention here.

11.6.1 Mobile sinks

The first case with special requirements is that of mobile sinks. One possible approach – two-tiered networking where data sources use a geographic mesh to broadcast their data and sinks subscribe to the data at their nearest mesh point [911] – has already been described in Section 11.4.4.

Another option [416] is based on explicit construction of a multicast tree. A mobile sink associates itself with a fixed sensor node, which will join on its behalf a multicast tree from the data source and acts as its proxy in this tree (if there are multiple sources, there are separate trees for each source); the tree will, in general, contain one such proxy node for every sink. This multicast tree can be constructed using any of the Steiner tree approximations described in Section 11.4.2; the authors themselves suggest the rate-based heuristic also described in that section.

Once this multicast tree has been constructed, it has to be maintained when a sink moves around and leaves the range of its current proxy, associating with a new proxy. There are two options: The sink moves in such a way that the previous proxy node is no longer a good choice in the multicast tree. In this case, the new proxy joins the tree and the old one is relieved of the duty to serve this node (it might still remain in the tree as it might have to supply other nodes with data). In the second case, the old proxy is still acceptable but now a multihop (unicast) connection between the old proxy and the sensor node currently closest to the mobile sink has to be set up; the old proxy uses it to forward data to the sink. Such a multihop extension of the tree must not exceed a given threshold; otherwise a new branch is constructed from scratch to the then-closest sensor node. These two cases are shown in Figure 11.18. This scheme trades off overhead in the tree maintenance against possibly longer paths due to the extension of the multicast tree by unicast branches.

Finally, the SINA architecture [758] also considers this problem. The idea here is to update the location of a mobile sink with a dedicated *resolver* node in the network, which is in charge of making sure that the query results are correctly delivered despite the mobile sink having moved away from the place where it issued the query.

11.6.2 Mobile data collectors

Sometimes, it is not possible or desirable to move the actual data sinks around but using multihop communication might also not be useful, for example, in rather sparse networks where communication distance between nodes and thus the energy required for communication are high. If the collection of data from sensor nodes is purely the objective and if this collection task can tolerate delays, then the concept of Mobile Ubiquitous LAN extensions (MULEs) [753] is applicable. A MULE is a mobile device, equipped with radio front ends to communicate with sensor nodes, that moves around between the sensor nodes, collects and buffers their data, and occasionally visits the actual data sink to off-load that data. Examples for MULEs can be autonomous robots but also animals or even humans are conceivable.

The interesting question pertaining to MULEs is the interdependence between movement pattern of the MULEs, time between visits to sensor nodes, data collection rate at the sensors, buffer space at sensors and MULEs, communication speed between MULE and sensor, and the resulting delay

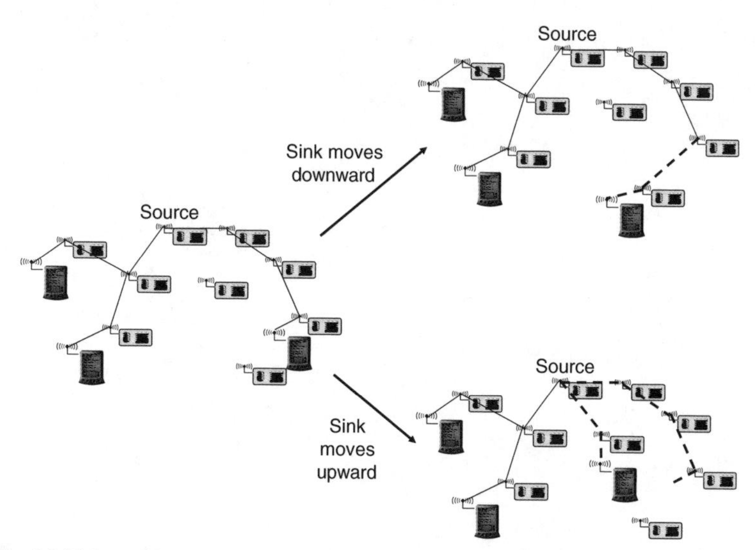

Figure 11.18 Multiple mobile sinks connected to a multicast tree (tree edges modified upon movement are dashed)

and data delivery rate at the actual data sink. These questions are analyzed in references [382, 753]. The authors claim that such MULEs are considerably more energy efficient than multihop communication and can increase lifetime of the network without impeding data collection quality too much.

11.6.3 Mobile regions

The geocast *destination* regions so far considered were static (the forwarding zones were sometimes dynamically adapted). For some applications like tracking mobile events, it would be useful to be able to specify a destination zone that changes its location (and possibly shape) over time. For such a moving zone, data should be delivered at time t to all nodes that are covered by the destination zone at time t. This service model is called **mobicast** [368, 369]. The challenge here is to come up with a protocol that ensures that data is delivered in time to the right nodes; the core technique is to send data to a forwarding zone that precedes the movement of the actual destination zone.

11.7 Conclusions

Supporting energy-efficient unicast and multicast communication in a wireless sensor network is a crucial optimization task and its solution draws upon insights from many different disciplines. Both the design of algorithms and their evaluation is a challenging task, requiring great care in selecting the proper assumptions and algorithmic principles, but they also pay off handsomely in extended capacity of lifetime of the network.

The mechanisms and schemes described in this chapter were mostly based on the assumption that nodes have a clearly defined address or at least location that could be used to designate the target of the communication. For wireless sensor networks, these mechanisms are important but they are complemented by mechanism that deal with the collection and dissemination of data directly. These concepts are described separately in Chapter 12 but cannot really be separated for a practical development of such a system.

12

Data-centric and content-based networking

Objectives of this Chapter

This chapter provides a different perspective on networking: Instead of addressing individual nodes as it is done in traditional routing, here *data* is the focus of attention. Abstractions and protocols that allow data to be routed as such are treated as is the combination with simple in-network processing techniques, namely aggregation. Moreover, storing data in the network itself is treated.

Chapter Outline

12.1 Introduction

12.1.1 The publish/subscribe interaction paradigm

The basic ideas of data-centric networking have been described already in Section 3.3.4. To recapitulate briefly: In a wireless sensor network, it is often the case that a node is interested in some information (for example, about the presence or absence of certain events) but does not care about the source of this information. This indifference to the information source is strengthened by the interchangeability of redundantly deployed sensor nodes, all observing the same or similar areas – it matters not which of these essentially identical nodes observes an event and reports it while the other nodes are sleeping (or are kept sleeping in the sense of intentional sampling [758]). Vice

Protocols and Architectures for Wireless Sensor Networks H. Karl and A. Willig

versa, a node reporting an event also does not care about the identity or even the number of sinks for the data that it provides. Moreover, both data source and data sink themselves can be switched off to save energy and are not immediately able to learn about the data request once it is made or about the data once it is available. And lastly, all these nodes can be, at any time, engaged in other activities so that data should be sent and made available in an asynchronous fashion.

In this view, the interaction in the network is **data-centric** as the identities of the nodes are irrelevant and as the other degrees of separation are required as well. To quote reference [123]:

> Flow of information – from sender to receiver – is determined by the specific interests of the receiver rather than by an explicit address assigned by the sender. With this communication pattern, receivers subscribe to information that is of interest to them without regard to any specific source (unless that is one of the selection criteria), while senders simply publish information without address it to any specific destination.

There are many options to realize such an interaction pattern. As the previous quotation already indicates, the most suitable and naturally matched paradigm here is **publish/subscribe**. The conceptual idea is essentially very simple: All nodes are connected to a "software bus" [251]. On this bus, data is made publicly available via a **publish** action; those nodes that have previously announced their interest in that particular kind of data by an appropriate **subscribe** action are then **notified** about the availability of this data. Figure 12.1 illustrates this concept; note how several publishers can publish data of the same kind and how notifications can be delivered to various subscribers.

This concept of publication and subscription matches the requirement for a data-centric wireless sensor network remarkably well. The publish/subscribe interaction pattern provides three essential properties concerning the relationship between providers and subscribers of information [251]:

Decoupling in space Publishers and subscribers need not to be aware of each other, they can be oblivious of their mutual identities and numbers.

Decoupling in time Publishing and notification of data can happen at different times; the "software bus" provides intermediate storage.

Decoupling in flows Interactions with the software bus can happen asynchronously without blocking.

Hence, precisely the properties required for data-centric networking in WSNs are fulfilled.

12.1.2 Addressing data

When subscribing to data or when publishing it, the question is how to *refer* to this data. The simplest case is the so-called **topic-based** published/subscribe variants [251]. Here, a set of *keywords*

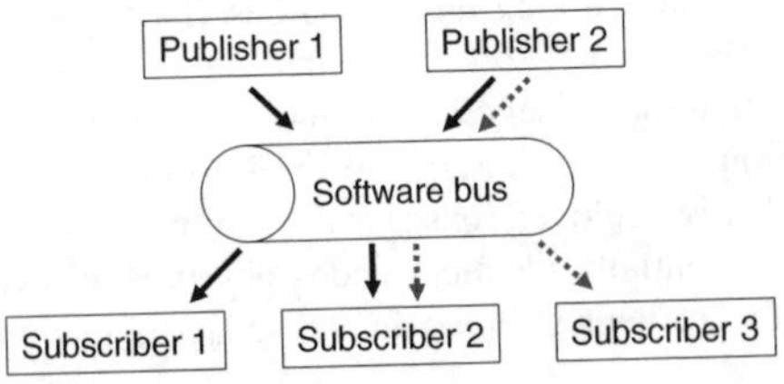

Figure 12.1 Software bus as abstraction of the logical interaction pattern of a publish/subscribe system (different arrows shapes indicate different publications/subscriptions)

exist into which the set of all data is grouped and publications and subscriptions happen using a chosen keyword. A typical example for topics are names of stocks traded at a stock exchange; when the price of a given stock changes, a notification for the corresponding topic is generated. As an extension, topics can be arranged hierarchically, akin to topic hierarchies found in newsgroup systems.

While the topic abstraction is simple, it lacks flexibility. A more expressive concept is to evaluate arbitrary predicates on the *content* of the entire data as such [123]. An example for such a predicate would be "Is temperature reported in the events larger than 25 °C?". Primitive predicates can be combined into more complex ones with standard logical operators (and, or, not) with the usual semantics; they can also be scoped in time ("only provide notifications within a certain time") or in space ("only provide notifications originating from a certain room"). Consequently, the abstraction used here is **named data**. In the parlance of the software bus, these predicates constitute filters describing which events/notifications shall be delivered to each individual subscriber. In the terminology of reference [343], the resulting names are *external* to the network topology (as they reflect data and not topological relationship) and immediately *relevant* to the application; they also remove the need for mapping between different naming abstractions.

To evaluate such a predicate, the data contained in the events need to follow a certain format (called "datagram model" in reference [123]). HEIDEMANN et al. [343] introduce an attribute-based naming scheme to this end. For the discussion at hand, the precise structure of this datagram model is not important; it can be free text or some predefined format.

Thus, there is no need for explicit keyword selection during publication and subscribers can flexibly describe their interests in particular subsets of all possibly published data. The content-based address of a node is then this very predicate (or its equivalent set of matching data), resulting in the notion of **content-based networking**. Section 7.5 has described content-based addresses already in more detail.

Because of the expressive power and the flexibility of content-based networking, it will be the main focus of this chapter. It is the most important variant of publish/subscribe to capture the notion of data-centric networking.

12.1.3 Implementation options

The most trivial implementation option for a software bus (regardless of the particular variant of publish/subscribe to be used) is a centralized solution: All subscriptions and publications are sent to a central node that evaluates the data-centric addresses (e.g. the content-based predicates) to decide to which node which publication has to be sent. Clearly, this is inadequate – but what other options exist?

Topic-based publish/subscribe is closely related to group communication. One idea is thus to form one topic per group and to use (energy-efficient) multicasting to distribute publications to their subscribers in the respective group.

While it might be conceivable to accordingly construct multicast groups even for a content-based networking scheme, this is fraught with numerous problems [123]: (i) The number of multicast groups would be highly dynamic and even difficult to determine as the content-determining predicates change. (ii) Using a smaller number of "summarizing" groups to distribute notifications would result in needless network traffic. (iii) Using, on the other hand, a large number of predetermined groups, reflecting the finest granularity of data that the predicates could resolve, would necessitate publishing to many groups; moreover, managing such a large number of groups would result in large administrative overhead.

Therefore, content-based networking cannot easily be mapped to the usual, identity-based multicast groups. Instead, **content-based forwarding and routing** are required. Each node has to store,

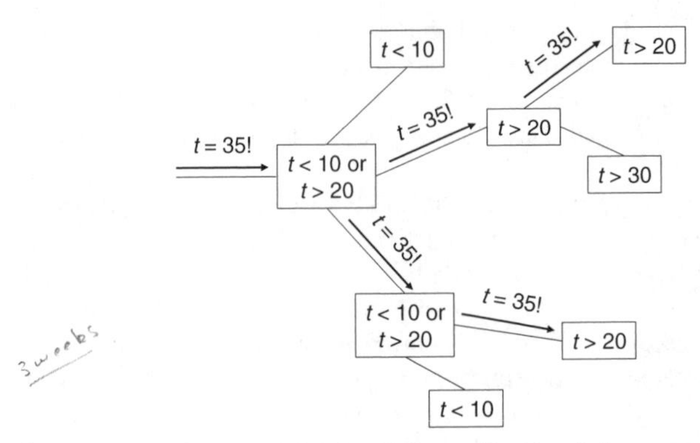

Figure 12.2 Content-based networking and forwarding, with subscriptions to temperature values as example

for each of its neighbors, a predicate describing the set of predicates that nodes in that subnetwork have subscribed to. For an incoming datagram containing a notification, a node then simply forwards this datagram to all of its neighbors where this predicate matches (Figure 12.2).

The challenge is then to keep this content-based forwarding table up to date with little overhead, to keep it small (as illustrated by merging the two subscriptions to $t > 20$ and $t > 30$ to a single subscription $t > 20$), and to organize it such that predicates can be easily and quickly evaluated. In some form or another, many of the approaches described in the remainder of this chapter address this problem, often without explicitly casting their contribution in this particular framework. MÜHL et al. [576] give an overview of various approaches, for example,flooding the subscriptions or exploiting information contained in the content-based filters to limit propagation of messages [121, 122, 575].

12.1.4 Distribution versus gathering of data – In-network processing

As publish/subscribe decouples publishers and subscribers in space, their respective numbers are no longer important. Hence, it would make little difference in principle whether the *distribution* of data from a small set of data sources to many or all nodes in the network is considered or vice versa the *gathering* of data from many or all sources to a single or a few sinks – performing a so-called **convergecast** – is considered. The notion of convergecast is shown in Figure 12.3 – note the correlation between temperature readings with one outlier. Evidently, a convergecast tree is closely related to a multicast tree. This figure also highlights one of the major problems of data gathering: there is a lot of traffic converging at the root node of the tree, yielding the **implosion** problem [374]; this effect is particularly annoying if redundant data is reported to the sink via multiple roots.

With respect to performance and to actual protocol implementations, there can be, however, different trade-offs involved depending on which of the two cases is considered. Thus, a loose categorization of protocols along these two cases is possible and different protocols for these

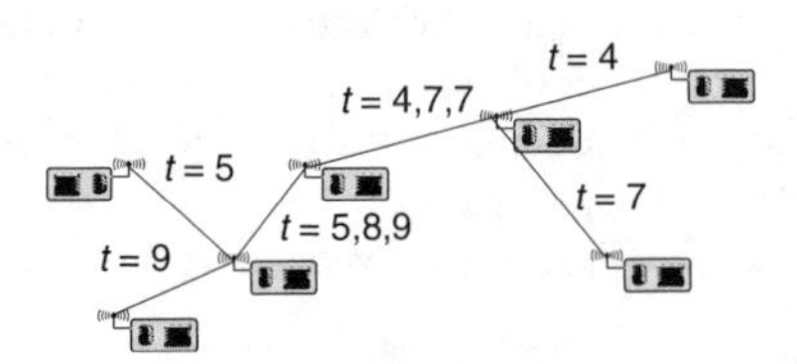

Figure 12.3 Using convergecast to collect temperature data from the sensor network

special cases can be developed, all implementing the same service interface but optimized for different cases [341].

Moreover, gathering data admits an additional optimization. Often, it is not necessary that all data from *all* nodes arrives at the sink. Frequently, an **aggregate** of the data – maximum, average, or minimum, for example, – is to be computed anyway. In such a case, performing these aggregation operations within the network is a viable option to reduce the amount of data that has to be transported. This option is obvious in the convergecast example of Figure 12.3 when, for example, the highest temperature reading is the relevant information to be extracted from the network. Data aggregation is the most prominent but not the only example of **in-network processing**. In some situations, in-network processing can also be profitably applied to data distribution tasks.

Strictly speaking, data aggregation and in-network processing in general are not tied to data-centric networking. A traditional convergecast can be formulated in a similar fashion. Nonetheless, data-centric networking and data aggregation are a good match in that they both emphasize the dependence of a wireless sensor network on data. For this reason, simple approaches to in-network processing – namely, aggregation – are treated in this chapter. More advanced techniques are summarized later. A further variant of in-network processing is **storing of data in the network**. It is also briefly summarized in an own section here.

12.2 Data-centric routing

Concrete protocol solutions to data-centric networking can be roughly categorized by the intended frequency of interaction. One group contains protocols that target repeated interactions – periodically reading a set of values from a sensor network, for example. Another group is so-called **one-shot queries**, where only a single request for data is to be answered by the network. Naturally, the first class of interactions admits solutions that invest certain effort in setting up a routing structure that typically cannot be amortized for one-short queries. Section 12.2.2 describes several solutions for repeated interactions; Section 12.2.1 deals with one-shot queries. Moreover, Section 12.2 in its entirety does not handle aggregation or other in-network processing aspects; these are treated in Section 12.3.

12.2.1 One-shot interactions

Disseminating big data sets via SPIN

One of the first data-centric dissemination protocols for wireless networks is Sensor Protocol for Information via Negotiation (SPIN) [345] (Figure 12.4). The target scenario is a network where one, several, or possibly all nodes have data that should be disseminated to the entire network. Moreover, the data per node is relatively large such that a unique name for each piece of data that a node holds can be easily created and is of small size relative to the data itself.

When applying, for example, simple flooding to such a scenario, the network will suffer from implosion and from overlap – the same area is observed by different nodes, each independently and needlessly reporting that data. Moreover, simple flooding is unaware of resource limitations in different nodes. To overcome these problems, HEINZELMAN et al. [345] suggest to use the names of the data to negotiate which nodes should forward which data.

This negotiation replaces the simple sending of data in a flooding protocol by a three-step process. First, a node that has obtained new data – either by local measurements or from some other node – *advertises* the name of this data to its neighbors. The receiver of an advertisement can compare it with its local knowledge and, if the advertised data is as yet unknown, the receiver can *request* the actual data. If the advertisements describe already known data (for example, because it has been received via another path or another node has already reported data about the same area), the advertisement is simply ignored. Only once a request for data is received, the actual data is transmitted.

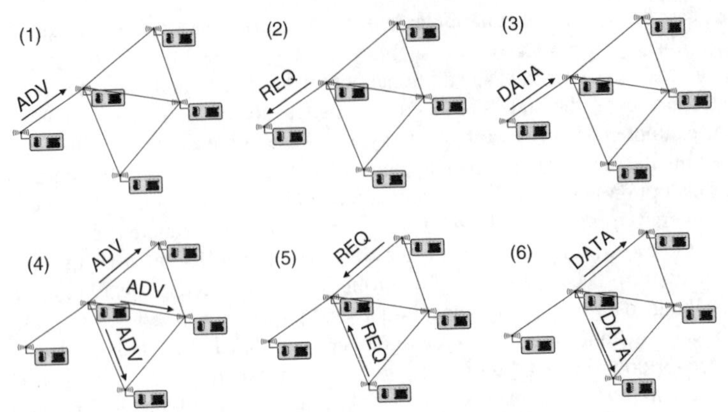

Figure 12.4 The basic operation of the SPIN protocol

The savings in this approach rest on the small size of the data description compared to the data itself. Once description of the data becomes comparable to data itself, it is not useful to first announce the data instead of simply sending it. The advantage is a relatively simple rule how to constrain, on the basis of the actual data, the flooding of data within the network. It is actually more powerful than the broadcasting constructions undertaken in Chapters 10 and 11 since it can take into account which *data* is actually missing and not only which node has not yet reported (possibly redundant) data.

This basic protocol idea is made more concrete by developing protocols for point-to-point networks and networks that enjoy a wireless multicast advantage (depending on the MAC, a WSN can be regarded as either of these two cases). Moreover, variants of the protocols that adapt node behavior to the remaining battery capacity are presented; nodes with low reserves reduce their participation in the protocol. While details differ, these protocols are claimed to be able to transmit 60 and 80 % more data for a given amount of energy than conventional protocols.

Active query forwarding

While SPIN is concerned with disseminating data to the entire network and is essentially a data-centric version of flooding, the ACtive QUery forwarding In sensoR nEtworks (ACQUIRE) mechanism [712] targets the collection of data from a network and is more comparable to gossiping and rumor routing (Section 11.2).

The target of the paper is the support of one-shot queries (which are likely not going to be repeated) that are complex (i.e. consist of several subqueries) and that pertain to replicated data where several nodes have sufficient information to answer the query (a typical example is the occurrence of an event that has been observed by multiple sensor nodes).

ACQUIRE leverages the ideas of gossiping and mobile code. A query is sent into the network, is partially resolved as far as possible at an intermediate node, and then forwarded onward (along with accumulated intermediate results) as long as the query has not yet been fully answered. Once this is the case, the query has turned itself into a response, which is then routed back to the node that has issued the query.

To assist in the (partial) resolution of the query at intermediate nodes, the node currently working on the query is allowed to draw upon data from its local vicinity, for example, the nodes at most d hops away (where d is a parameter). This local information can be updated if it is outdated. Moreover, such information can be used to guide the forwarding of the query to the next node; alternatively, the query could also be simply handed on to a randomly selected neighbor.

While ACQUIRE has a number of parameters that have to be selected – d, the time until local information is considered outdated – the authors show that it can achieve considerable savings.

12.2.2 Repeated interactions

The optimization space for one-shot queries or one-shot dissemination of data is somewhat limited, in particular, when no aggregation of data should be used. Hence, let us focus more on the case of repeated interactions where the exploration of the network topology might pay off.

Directed diffusion – Two-phase pull

The first example protocol to be discussed here is *directed diffusion* [378]. It is one possible realization of publish/subscribe for a wireless sensor network; it is mostly concerned with scalability issues and tries to find solutions that do not depend upon network-wide properties like globally unique node identifiers; rather, the goal is to find solutions that purely rest on **local interactions**. The most prevalent – albeit not the only – service pattern is subscription to data sources that will publish data at a selectable rate over a selectable duration.

Directed diffusion is actually more a design philosophy than a concrete protocol and there are a number of protocol variants that are optimized for different situations. We start here with the original and basic variant (the "two phase pull"), postponing issues like aggregation, geographical support, and so on, to later sections.

In this scheme, data distribution starts by nodes announcing their **interests** in certain kinds of named *data*, specifying their interests by a set of attribute-value pairs (see Section 7.5 and references [343, 378] for details); in the publish/subscribe parlance, this corresponds to a subscription to data. These interest messages are distributed through the network; in the simplest case they are flooded.

Given such an interest flood, it would be trivial to set up a convergecast tree with each node remembering the node from which it has first received the interest message from a given sink (reference [933] discusses alternative choices for the parent node in the tree); interests to different data and/or from different sinks would result in separate trees being constructed. But such a simple tree construction is faced with a serious impediment: In the absence of globally unique node identifiers, a node in the network cannot distinguish whether different interest messages originated at different data sinks and would thus require the construction of separate convergecast trees to inform all sinks of published data or whether these packets are owing to the same sink and have simply traveled via different paths. This predicament is highlighted by Figure 12.5.

For a node X in the example of Figure 12.5, there is, at first, only a single option – remember all neighbors from which an interest message has been received to, later on once data has been published, forward the actual data to all these neighbors. In the directed diffusion terminology, this is the setup of a **gradient** toward the sender of an interest. Each node stores, for each type of data received in an interest, in a **gradient cache** a separate set of gradients, potentially one for

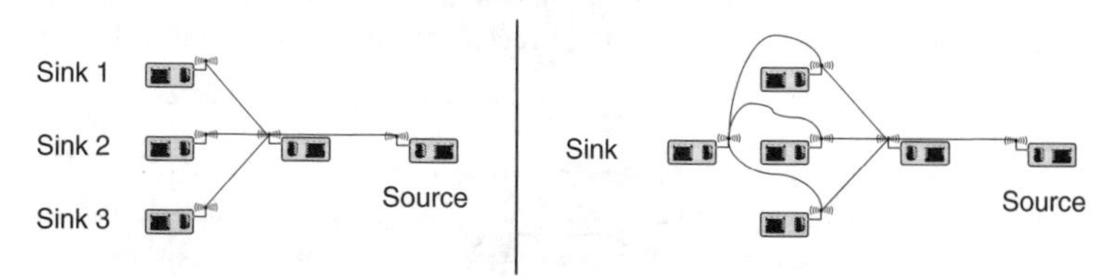

Figure 12.5 Inability of network node X to distinguish interest messages from a single or multiple sinks

each neighbor. Unlike the simple parent–child relationship in a tree, gradients will often be set up bidirectionally between two neighbors as both neighbors forward interest messages. In addition, a gradient is not simply a direction, but it also contains a *value*. This value represents, in a sense, the importance or usefulness of a given link. It can constitute different semantics depending on the concrete application that directed diffusion is supporting; a typical example is the *rate* with which data is transmitted over a given link (recall that directed diffusion is geared toward the support of periodic publications of data). Initially, these gradient values are the same for each neighbor; they are modified in the course of the protocol execution. Also, these gradients are initialized to low values, which are used to *explore* the network.

Once the gradients are set up, even with only preliminary values, data can be propagated. A node that can contribute actual data from local measurements becomes a source and starts to send data. It uses the highest rate of all its outgoing gradients to sample and send data. An intermediate node, in the simplest case, would forward all incoming data messages over all its outgoing gradients, potentially suppressing some of the data messages to adapt to the rate of each gradient. This simple scheme, however, results in unnecessary overhead in networks like the one shown in Figure 12.6, where data messages are needlessly repeated (due to the presence of loops in the gradient graph). Just checking the originator of these data messages is again not feasible because of the lack of globally unique identifiers. Hence, the **data cache** is introduced: Each node stores, for each known interest, the recently received data messages. If the same message comes in again – irrespective of from the same or different originators – it is silently discarded.

The data cache notwithstanding, Figure 12.6 also shows that two copies of the same data message would be delivered to the sink, constituting nonnegligible overhead. The gradient values, or more specifically the rates associated with the gradients, provide a lever to solve this problem. One idea is to try to limit redundancy in the received data. A neighboring node that contributes new data messages (which cannot be found in the data cache) should be preferred over neighbors that only provide stale copies, or rarely provide new data, or appear to have high error rates, or are otherwise unattractive. This "preference" of a neighbor can simply be mapped onto the rate of a gradient. A node can **reinforce** a neighbor by simply sending a new interest message to that neighbor asking for a higher rate of data transmission. If this new, required rate is higher that the data rate that an intermediate node is currently receiving, it in turn can reinforce its best neighbor with this higher rate. In the end, the reinforcement will percolate to the source(s) of the data messages. The nonreinforced gradients can be maintained as backups, they can be actively suppressed, or they can be left to die out in the sense of soft state information.

These two phases – first flooding the interest messages to explore the network and then again having information flow from the sink toward the sources during reinforcement – along with the fact that the sinks initiate the "pulling" of data explain this variant's classification as a "two-phase pull" procedure.

These mechanisms of interests, gradients, and reinforcements constitute the pivotal mechanisms in directed diffusion. It is worthwhile to reiterate that all of them are indeed strictly local, dispensing with the need for globally unique identifiers. Reference [378] contains further details how these mechanisms result in loop-free operation and how paths can be maintained in the presence of node or link failure (essentially, the reinforcement mechanism automatically adapts to the new topology). It should also be emphasized that, in principle, directed diffusion in the form described here can

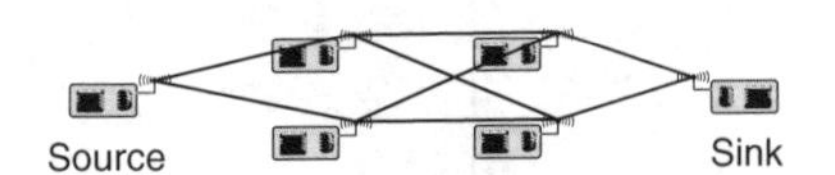

Figure 12.6 Multiple intersecting paths necessity a data cache in directed diffusion

handle both multiple sources and multiple sinks of data; the local rules result in a correct but not necessarily optimal flow of data messages.

Publish/subscribe assisted by geographic scoping

For many wireless service network interactions, specifying the region of interest is a natural desire. Hence, interests should be able to express this information and data-centric routing protocols should take it into account.

Geographic scoping in the original directed diffusion

In the original directed diffusion proposal [378], interest messages may contain information about the geographical region from where information is requested. When an interest is thus scoped, it is evidently superfluous to distribute it to nodes that are outside the given region of interest. The interest flooding can then be replaced by a geocasting of the interest message.[1] One specific option to do so is the GEAR protocol [341, 919], a geocasting protocol with support for routing around holes.

Content-based multicast

This approach [935] explicitly incorporates knowledge about the movement direction and speed of devices and environmental entities into the protocol. One example is fire fighters in a large chemical plant. Suppose that sensor nodes are capable not only of detecting the presence of dangerous vapors but can also measure wind speed and direction and thus predict the mostly likely movement of such dangers. On the basis of this information about movement of persons and dangers, persons should be warned if they are going to move into a dangerous area.

The challenge here is that the set of receivers for a given warning does not depend on the identities but rather on the movement data; it is in this sense data-centric. Zhou and Singh [935] propose a mixed "sensor push/receiver pull" approach to solve the problem. Sensor nodes geocast their warnings into a computed region of the network; receivers can pull such information looking ahead in their movement path.

To this end, the area in question is divided into "blocks" or "regions" with one lead node each (in this sense, it predates the later TTDD protocol [911] that uses somewhat similar ideas). This leader node collects all pushed warning messages; it also answers to pulled inquiries. Data exchange is supported by a simple geographic routing protocol.

Push diffusion – supporting few senders and many receivers

As directed diffusion represents both an interface/naming concept [343] and a concrete routing implementation (the one described above), it stands to reason that different routing protocols supporting the same interface have been developed. One such alternative routing protocol is *push diffusion* [341], which is intended for many receivers and only a few senders. A typical example is an application where sensor nodes cross-subscribe to each other to be informed about local events but where the amount of actual events is quite low. In such a situation, two-phase pull would perform purely as the sinks would generate a lot of traffic trying to set up (exploratory) gradients.

This problem is solved by reversing the roles: Instead of the sinks sending out interests, sources send out exploratory data (i.e. flood it since no gradients exist yet). Once data arrives at interested sinks, they will reinforce these gradients, and then, data at higher rate will only follow these reinforced paths. The flooding overhead is justified since the event detection rate of sources is quite small to begin with.

[1] Intanagonwiwat et al. [378] also mention that semantical information and "historical evidence" can be used for such more efficient distribution of the interest floods but this is not explored in detail.

One-phase pull – supporting many senders and few receivers

Similar to the above-described push diffusion, *pull diffusion* [341] is a specific routing protocol for the directed diffusion interface; this one is geared toward many senders and a small number of receivers.

As the name indicates, one-phase pull eliminates one of the flooding phases of two-phase pull, which constitute its major overhead. More precisely, interest messages are still flooded in the network (in the absence of geocasting options) but the interest messages set up direct parent–child relationships in the network between a node and the node from which it first receives an interest message – in effect, a tree is formed in the network. This is only possible using (e.g. randomized) flow identifiers in the interest messages, which is feasible only for a small number of messages. Moreover, one-phase pull more strongly depends on link symmetry than does two-phase pull.

Directed diffusion assisted by topology control

Reducing the flooding overhead inherent in two-phase pull is a promising means for improvement. The clustering discussion in Chapter 10 has introduced a variety of techniques for efficient broadcasting of information. In particular, passive clustering fits well with directed diffusion. HANDZISKI et al. [331] show how this combination works in detail. In particular, the passive clustering structure is constructed on the fly with the distribution of interest floods. This results not only in better energy efficiency but, in particular, the percentage of actually delivered events is considerably improved, mostly because of easing the contention on the MAC layer. In this sense, this work highlights the need for a careful adjustment of at least three different protocol layers – MAC, topology control, and data-centric routing – for an efficient wireless sensor network.

Increasing robustness by multiple forwarding paths

The previous approaches all attempted to confine forwarding to a single, hopefully optimal path from a given source to a given sink. Faced with node failures and unreliable links, it can be appropriate to increase the data forwarding's robustness by using multiple paths.

Several such schemes exist and many of them leverage multipath techniques as described in Section 11.3.4. One such approach, for example, is GRAdient Broadcast (GRAB) [912]. It does use a notion of gradients that differs from directed diffusion and is actually more similar to gradients in the sensor of ad hoc routing protocols like TORA [619], where distance (or rather, energy cost) to a given sink is expressed as the "height" of a node. Packets are given a credit and flow down these height gradients toward nodes with smaller cost than the credit remaining in the packet.

In such a context, robustness can be relatively easily achieved by giving a packet a larger credit, enabling it to also travel via additional paths. The larger the additional credit given by the source, the more (less optimal) paths become available, resulting in a wider mesh of paths. As a consequence, energy efficiency is traded off against additional robustness.

12.2.3 Further reading

There are a few additional aspects and concepts that deserve a brief mentioning:

Scheduling The question on *when* to send data within a convergecast tree is considered by CENTINTERNEL et al. [125]. They describe a scheme where a root node sends out a scheduling control packet that is passed on in the tree. Time slots are then assigned at each layer of the tree to make sure that data can flow through the tree with low delay due to MAC contention. Moreover, there is a scheme for speculatively assigning time slots for nodes that do not always have data to send.

Hierarchical dissemination The TTDD scheme [911] described in Section 11.4.4 can be regarded as an example of a hierarchical data-centric routing scheme.

Further rules for choosing neighbors in directed diffusion SCHURGERS and SRIVASTAVA [736] describe various rules how to select the next hop in directed diffusion, in particular, when the gradient information is ambiguous. Possible choices include a random selection or a selection based upon residual energy or to push away traffic from other flows.

Variations on Steiner tree approximations A relatively straightforward variation of the Takahashi Matsuyama heuristic (see Section 11.4.2) and its application to building up trees in data-centric routing is explored by KIM et al. [418]. This heuristic also incorporates some rate adaptation ideas similar to those described in reference [416].

Disseminating information non-uniformly TILAK et al. [818] introduce a notion of nonuniform data dissemination. The essential idea is that, for some applications, the farther away information is disseminated, the less accurate is has to be.

12.3 Data aggregation

12.3.1 Overview

When looking at data-centric networking in isolation, all messages still have to be delivered to all sinks. The real power of concentrating on data lies in the ability to *operate* on the data while it is transported in the network. The simplest example of such in-network processing is **aggregation** of data – computing a smaller representation of a number of messages that is equivalent (or at least suitably represents) in its content to all the individual messages – and only forwarding such aggregates through the network. Computing a mean or the maximum of the measured values of all sensors is a typical case in point. More advanced examples of aggregation might include approximating contours of regions of equivalent values (measured from the environment or node parameters like remaining battery capacity) or approximating lines or polygons that separate different regions [95].

The actual benefits of such aggregation depend on the location of the data sources, relative to the data sink. Intuitively, when all data sources are spread out, the paths to the sinks do not intersect, and there is little if any opportunity to aggregate data at some intermediate nodes. If, on the other hand, the data sources are all nearby – for example, when they all observe an event at a certain place – and they are located far away from the sink and their paths to the sink merge early on, the expected benefits of aggregation are large (Figure 12.7). This is in fact often the case [626] and the intuition about resulting benefits is confirmed by results [439].

The principal mechanics of data aggregation are thus relatively straightforward: Data flows from sources to a sink along a tree. Intermediate nodes in the tree apply some form of aggregation function to data they have collected from some or all of their children. This aggregated value, possibly along with additional administrative values (for example, the number of nodes that have contributed to a

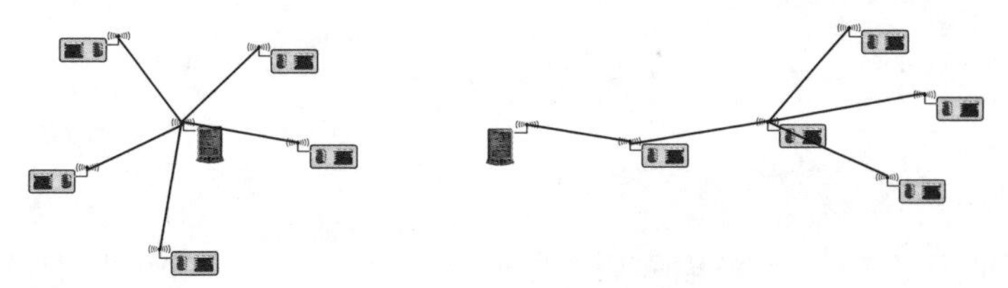

Figure 12.7 Different cases where data aggregation is pointless or promising

mean value) is then forwarded. Apart from the tree formulation, data aggregation can also be used in the context of gossiping data throughout the network; examples are discussed below.

The efficacy of data aggregation can be judged using different metrics (see, for example, reference [933]).

Accuracy Perhaps the most important metric is **accuracy** [95] – the difference between the resulting value at the sink and the true value – since not all data is delivered to the sink any longer; accuracy can be expressed as differences, ratios, statistics, or other values depending on the particular case.

Completeness Potentially an operational approximation of accuracy is **completeness** [314], the percentage of all readings that are included in the computation of the final aggregate at the sink.

Latency Aggregation can also increase the **latency** of reporting as intermediate nodes might have to wait for data.

Message overhead The main advantage of aggregation lies, of course, in the reduced **message overhead**, which should result in an improved energy efficiency and network lifetime. Aggregation protocols can often deliberately trade off between accuracy, message overhead, and latency and only provide estimates of the actual aggregated value.

The main open questions are thus:

- Which aggregation functions can be used, what categories exist?
- How can the tree be formed, where (and how) should aggregation points be placed?
- How long should a node wait for data from its children/neighbors?
- How should an interface look like that allows to easily express aggregation actions?

These questions will be answered in the remainder of this section, along with example protocols.

12.3.2 A database interface to describe aggregation operations

To cast the issues for aggregation protocols in a concrete context, it appears best to start with a specific interface description. One popular example for an interface that can express aggregation needs is inspired by database query languages, specifically, SQL. Madden et al. [531] describe how an SQL-like syntax ("Tiny Aggregation" or TAG for short) is suitable for wireless sensor networks. They imagine the WSN as a whole to represent a (virtual) relational database table called `sensors` against which queries can be executed. The syntax of such a query is defined in Listing 12.1.

Listing 12.1: Syntax of an SQL query to aggregate data from a sensor network [531]

```
SELECT {agg(expr), attributes} FROM sensors
WHERE {selectionPredicates}
GROUP BY {attributes}
HAVING {havingPredicates}
EPOCH DURATION i
```

In such a query, the phrase `agg(expr)` denotes the aggregation function, applied to a given expression; an example would be `AVG(temperature)` denoting that the average of all temperature readings is to be determined. The **`WHERE`** clause acts as a filter on the measured values before they enter the aggregation process; usually, these predicates are intended to be locally evaluated by each node (`WHERE` predicates requiring distributed evaluation are rarely considered and constitute an at least partially unsolved problem). The `GROUP BY` clause partitions the data into subsets and the `HAVING` clause further filters these groups. An example would be to compute average temperature values (`SELECT AVG(temperature)`) separately for each floor in a building (`GROUP BY floor`) but only from the fifth floor upward (`HAVING floor > 5`); the floor number for each temperature average can be obtained by `SELECT AVG(temperature), floor`.

The `EPOCH DURATION` indicates repeated interactions. Nodes periodically measure, transmit, and aggregate information and the epoch duration marks the period for these repetitions. Only data that belongs to the same epoch can be justifiably aggregated. The result is a periodic stream of data, representing the biggest difference to the usual SQL semantics.

Other database-inspired queries models for wireless sensor networks exist (e.g. COUGAR [269]) and have different models for interactions with the sensor network. The database abstraction for wireless sensor networks has attracted considerable research effort; some additional references are [88, 202, 253, 269, 347, 513, 527, 528, 529, 530, 711, 711, 713, 908, 909]. Execution plans are treated in references [30, 313, 712, 887].

12.3.3 Categories of aggregation operations

A basic differentiation of aggregation operations is between one-shot aggregations, where only a single request is to be answered, and continuous or periodic requests. This differentiation is akin to one-shot or repeated interactions discussed above and the resulting trade-offs are similar.

Moreover, aggregation operations can be distinguished according to the representation of intermediate results and according to the properties of the actual aggregation function.

Representation of intermediate results

When computing aggregates in an intermediate node, it is, in general, insufficient to only communicate the result of the actual aggregation function between nodes. An evident example is the computation of an average: A node that receives two averages from its children (assuming a tree structure) has no way of knowing how these two values should be processed unless it also knows how many readings have contributed to each average – these two numbers are necessary to properly weigh the individual averages. Hence, to compute an average, a tuple $<$ average, count $>$ or, briefly, $< a, c >$ should be exchanged between nodes; MADDEN et al. [531], for example, call these tuples **partial state records**. A new partial state record can be computed in an intermediate node as

$$< a_{\text{new}}, c_{\text{new}} > = < a_1 c_1 + a_2 c_2)/(c_1 + c_2), c_1 + c_2 >.$$

The computation becomes even simpler if only sum s and count c are exchanged between nodes; the update rule is then simply $< s, c > = < s_1 + s_2, c_1 + c_2 >$. The actual average is then only computed at the ultimate sink. In either case, a – usually trivial – function is necessary to extract the actual aggregate out of the partial state once it has been fully computed at the sink.

Such multivalued partial state records are, however, not always necessary. Aggregation functions like minimum or maximum can easily use a single value – the sofar determined minimum or maximum – to represent the partial state at each aggregation step. The nature of the aggregation function thus determines the precise form of the intermediate state record.

Aggregation functions

Given two partial state records $< x >$ and $< y >$, either received from a neighboring node or locally measured, an aggregation function f computes a new state record $< z > = f(< x >, < y >)$.

Which properties should f have to be usable as an aggregation function? MADDEN et al. [531] provide a taxonomy, partially based on previous work on data cubes and extended here by the composability definition of reference [314].

Duplicate sensitive Is the aggregation result changed if the measured value of a particular device (or some intermediate aggregate) is used in the computation more than once? If yes, the aggregation structure should be acyclic; otherwise, a wider range of topologies can be profitably used.

Examples for duplicate-sensitive aggregations are the sum of measured values (`SUM`), counting the number of certain instances (e.g. number of sensors that have raised an alarm, `COUNT` for short), as are the average (`AVG`), the median of a set of values (`MEDIAN`), and computing the histogram of values (`HISTOGRAM`). Minimum and maximum (`MIN` and `MAX`), on the other hand, are not sensitive to duplicates.

Summary or exemplary An exemplary aggregate is a single, in some sense representative, value out of a set of values. A summary aggregate is a function of the entire set and, typically, does not strongly depend on individual values.

`MAX` and `MEDIAN` are typical exemplary aggregates; `SUM` is, as expected, a summary aggregate.

Composable [314] An aggregation function f is said to be *composable* if the result of f applied to a set W of measurements can be computed by applying f to some partition of $W = W_1 \cup W_2$ (usually with $W_1 \cap W_2 = \emptyset$), using a known helper function g. Formally,

$$f(W) = g(f(W_1), f(W_2)).$$

Using this definition alone, practically any function is composable as no restriction is made on the partial state records. For example, even the median would be composable (contradicting intuition) by setting $f(W) =< \text{median}(W), W >$ and $g(< \text{median}(W_1), W_1 >, < \text{median}(W_2), W_2 >) =< \text{median}(W_1 \cup W_2), W_1 \cup W_2 >$. Evidently, this makes little sense and composability has to be considered in junction with the behavior and size of the partial state records. The following definition classifies this behavior.

Behavior of partial state records The partial state necessary to compute the aggregated value varies for different aggregation functions. The most important cases are:

Distributive The partial state *is* the aggregated value of a set of partial measurements. No additional function has to be applied at the sink. The partial state is of constant size. `MIN` is a typical distributive aggregate.

Algebraic In this case, the partial state is of constant size, often with the actual aggregate a part of it or a simple function. `AVG` is a typical example.

Content-sensitive For content-sensitive aggregates, the size and structure of the partial state depends on the values that have been actually measured. An example is the computation of a histogram of the measured values.

Holistic For holistic aggregation functions, the partial state needs to reflect all measured values, like in the median example above.

Unique These functions are similar to holistic ones except that here the partial state size is only proportional to the number of *distinct* values that have been observed (rather than to all of the observed values).

In a practical sense, aggregation functions with distributive and algebraic partial state are well amenable to in-network aggregation; content-sensitive functions may or may not be. Holistic aggregation functions, however, cannot practically be aggregated; for unique functions, aggregation might be applicable depending on the behavior of the values to be observed.

Monotonic Monotonic aggregation is coupled with the notion of the "magnitude" of a partial state record s, given by a function $m(s)$. An aggregation function is monotonic if it only increases the magnitude of the partial states it operates upon (or, equivalently, decreases). Formally, f is monotonic if and if $\forall\ s_1, s_2 : m(f(s_1, s_2)) \geq \max\{m(s_1), m(s_2)\}$ (or, alternatively, $\forall\ s_1, s_2 : m(f(s_1, s_2)) \leq \min\{m(s_1), m(s_2)\}$). Monotonic aggregation functions are relevant in the context of optimized execution of `HAVING` clauses; see below for details.

Timing aspects A further, orthogonal aspect concerns the timing aspects of aggregation. Is it done early or late, is in a one-short aggregation or does it concern periodic interactions? Is it aggregation in time (regarding, e.g. correlated data) or is it aggregation in space?

In most of the protocols considered in the literature, simple aggregation functions like minimum/maximum or summing and averaging are considered. In rarer cases, more complicated application examples like finding iso-barometric lines in a plane [531] are considered.

12.3.4 Placement of aggregation points

When collecting data toward a sink along a tree or along a routing structure such as the one resulting from directed diffusion, the aggregation points have to be well placed for maximum benefit. Intuitively (compare Figure 12.7), aggregation should happen close to the sources and many sinks should be aggregated as early as possible – the tree should have, figuratively, long trunks and bushy leaves. Directed diffusion does not necessarily result in a tree, but it is well suited to aggregation and, similar to the tree case, aggregation should happen as early as possible.

If the routing structure is grown without regard to the later aggregation, the resulting structure is not necessarily optimal (in fact, this is again a Steiner tree problem in disguise). The aggregation is, in a sense, opportunistic. INTANAGONWIWAT et al. [376, 377] consider how to influence the directed diffusion routing structure so as to optimize the aggregation benefits.

To ensure that aggregation points are placed near the sources, a simple variation of the Takahashi–Matsuyama heuristic is used. Initially, an energy-efficient path between a source and a sink is constructed, using the usual exploration and reinforcement rules of directed diffusion. Additional sources join this tree by searching for the shortest path to this tree. This scheme is implemented using local interactions.

Implicitly constructing good aggregation trees is also considered by ZHOU and KRISHNAMACHARI [933]. They look at the consequences of different rules concerning which node is chosen as a parent in the convergecast tree out of the set of neighboring nodes that have issued invitations to join the tree. The simplest rule is to use the first such node from which an invitation has been received (resulting in a breadth-first-search-like tree), to randomly pick one, the nearest node first, or a weighted randomization. As the tree construction immediately determines the placement of the aggregation points, these rules can make a considerable difference. Indeed, the authors show that none of these rules simultaneously achieves good network reliability, latency, and data aggregation ability and that a compromise has to be struck.

12.3.5 When to stop waiting for more data

When aggregating data, an intermediate node, as well as the sink, has to decide how long to wait for data from each of its children in a convergecast tree. In the simplest case, a node knows which of its neighbors are its children (by means of an acknowledgment of the invitation messages during tree formation) and waits for answers from all of them. This can, however, take a long time because of fluctuations in the radio channel with ensuing high error rates, temporary node failures, or simply because of a very imbalanced tree. Waiting a long time will result in more data entering the computation of the aggregate and thus to higher accuracy but it will also increase delay and, potentially, energy consumption because of the required idling of the radio receiver. A compromise has to be found.

A relatively simple scheme, where the times for each hop are essentially regarded as a constant, is described in reference [920]. Here, rules to set timer values based on a maximum waiting time of the source are described.

More challenging is the case when the time it takes a node to deliver its local measurement or its own aggregate to its parent in the tree is a random variable. When there is also some cost involved in waiting, this problem becomes an instance of the more general problem of **optimal stopping rules**. It is investigated in detail by BRODER and MITZENMACHER [100]. While they consider an Internet context, the mathematical treatment is independent of the motivating example and applicable to WSNs as it is.

Formally, an aggregating node sends requests for data to its n children at time 0. Later, at time t it will decide to return an aggregate of the k answers received up to time t to collect a *reward* $R_k(t)$. This reward is, in general, a function increasing in k and decreasing in t – one example would be exponential decay: $R_k(t) = k\mathrm{e}^{-\gamma t}$ for some constant γ. More generally, $R_k(t) = r_k(1 - Z(t))$ where the constants r_k are the "undiscounted" rewards for k collected answers (which would be incurred at $t = 0$) and $1 - Z(t)$ is the discount factor that approaches 0 as $t \to \infty$. The return times for the n children are assumed to be identically distributed, independent random variables with a distribution F that is known to the aggregating node. The goal is then to develop simple rules for the aggregating node to decide when to forward the sofar collected values to the parent node.

The key result of reference [100] is that optimal behavior of an aggregating node depends on the "shape" of the answer time distribution F and on the reward functions. More specifically, the **failure rate** of the distribution function F and of the discount function Z (assuming that Z is given by a distribution-like function) are the crucial properties.[2] Distributions of random variables like F and distribution-like functions like Z can be categorized as having *increasing* (or *decreasing*) failure rate if $r(t)$ is increasing (decreasing).

On this basis, BRODER and MITZENMACHER [100] show that for certain combinations of failure rates of F and Z, one of two possible strategies applies.

- In case 1, F has increasing failure rate, Z has decreasing failure rate, and the sum of the failure rates of F_0 and Z is decreasing. Here, the aggregating node should, upon receiving the kth response, either propagate the result to the parent node immediately or wait for the next child's results, no matter how long it takes.
- In case 2, F has decreasing failure rate, Z has increasing failure rate, and the sum of the failure rates of F_0 and Z is increasing. In this case, with k responses, the aggregating node should wait for a fixed time t_k before propagating the aggregated value to the parent.

In either case, answers arriving after propagation to the parent are ignored.

While this is not an all-encompassing answer as not all possible combinations are covered, it is an important guideline for practical system design.

[2] The failure rate $r(t)$ of a (nonnegative) random variable X with distribution function F with density f is defined as $r(t) = f(t)/(1 - F(t))$. The name failure rate comes from reliability theory; it describes the probability that an event is about to occur at time t provided it has not occurred already $\left(r(t) = \lim_{\Delta t \to 0} \frac{\Pr(t < X \leq t + \Delta t | X > t)}{\Delta t}\right)$.

12.3.6 Aggregation as an optimization problem

Before looking at some concrete protocol examples, it is instructive to look at data aggregation as an optimization protocol. As is commonly the case, the optimization formulation is not directly applicable as a protocol but gives upper bounds and general insights into the structure of the problem.

KALPAKIS et al. [395] considered the benefits of data aggregation by using a linear programming formulation. As a starting point, a sensor network with known position and energy supplies of all nodes is given, along with the cost to communicate between any two nodes. The network operates in rounds; in each round a node generates a data packet of unit size. The goal is to construct a "schedule" – an assignment of transmit/receive pairs for each round – such that the time until the first node runs out of energy is maximized.

This problem is translated into one of finding a flow network. Without aggregation, this is relatively straightforward. The integer program contains the flows from all nodes to the base station as decision variables and it can be solved by a linear relaxation. With data aggregation, the flow network has a similar integer program but it has a more complicated backward translation into a schedule.

The interesting result is the comparison of resulting network lifetimes with and without aggregation. Using of aggregation can improve network lifetime by factors of about 19–22, depending on network density. This large improvement rests on strong assumptions about network-wide knowledge about node positions that is employed in the optimization problem.

Moreover, the formulation as an integer program with a linear relaxation has a large computational overhead (even though it is still polynomial). In a follow-up paper, DASGUPTA et al. [193] consider a heuristic with reduced overhead to arrive at a practical but still centralized solution. The idea is simply to first cluster nodes together and then to execute the data gathering algorithm only for the clusters. The data aggregation within a cluster is then solved separately. While this divide et impera approach sacrifices optimality, the authors show that for practical cases the result is still within about 10 % of the (nearly) optimal solution. Reference [396] provides some more details.

This work is further extended in reference [194] by including coverage constraints. Nodes can act as either sensors or relays and a distributed protocol is proposed to decide this role assignment. The reader is referred to the paper for details.

12.3.7 Broadcasting an aggregated value

So far, attention was implicitly focused on using aggregation along a convergecast tree. Aggregation can also profitably be applied to broadcasting an aggregated value of all sensor readings within the entire WSN – for example, to inform all nodes of the currently highest measured temperature. Two examples are described here.

A gossiping-based solution

GUPTA et al. [314] discuss the case of providing an estimate of a composable, algebraic aggregation function to all nodes in the network. The task is complicated by potential failures of both links and nodes.

A naive fully distributed solution – every node sends its measurement to every other node in the network – is robust but does not scale. The core idea to realize scalability with acceptable message overhead is to introduce a hierarchy. The nodes are partitioned into groups with K members on average, where K is a known constant. Then, K of these groups are again collected into a group of the next hierarchy level and so on until only a single group remains. In each group at each hierarchy level, a leader is elected. Using such a hierarchy, the aggregate is computed bottom-up by group members sending their measured values/their computed estimates to their respective group leader.

At the end, the leader of the top-most group knows the aggregated value (because the aggregation function is assumed to be composable) and can redistribute it downward.

Building such a hierarchy in a sensor network should exploit geographic or, even better, radio proximity. In effect, this is a multilevel clustering problem. Alternative solutions in reference [314] include the use of hash functions; ideally, topology-aware hash functions combine these two aspects.

This hierarchy can also be used for a gossip-based approach, circumventing the inherent reliability problems of a leader-election-based solution. This gossiping approach proceeds in $\log_K N$ rounds, which is the height of the hierarchy. In round $i = 1 \dots \log_K N$, a node randomly selects other nodes that are in the same group as itself at hierarchy level i. To each selected node, it will send its own node identifier and its current best estimate (obtained from local measurement at the beginning or as the result of the previous round). This gossiping happens $K \log N$ times per round. At the end of a round, a node applies the aggregation function to the received values and proceeds to the next round (details are slightly more involved, please refer to reference [314]). At the end of the last round, each node has an estimate of the global aggregated value. The interesting aspect is that this algorithm is only poly-logarithmically suboptimal; its application to WSNs, however, is challenging because of the need for multihop gossiping.

Another interesting aspect of this paper is the relationship between broadcasting aggregate values and the consensus problem. As the consensus problem is not solvable under the present error model, neither is the aggregation broadcast task. The authors relate it to the Byzantine agreement and randomized consensus protocols.

Continuous, exemplary aggregation with adaptive accuracy

In many respects similar to the previous approach is a scheme proposed by Boulis et al. [95]. It also targets the distribution of an estimated aggregated value to all nodes in a WSN, it is intended for doing so continuously, and it is essentially based on an intelligent form of gossiping. Unlike the previous one, however, it is suitable only for exemplary aggregation functions (like minimum or maximum) and it does not need the construction of some form of hierarchical grouping.

In this distributed estimation algorithm, each node not only maintains an estimate of the current aggregated value, it also stores an estimate of the *confidence* a node puts into this value (even better, probability density functions are used, but this is usually impractical). This confidence for an aggregated scalar value A could be the variance of A.

The intelligence in this scheme lies in the fusing of the local estimate with new information either from local measurement or received from a neighboring node; in addition, rules are described to regulate when such a new local estimate is to be sent to neighbors. Since different fusing rules are required for the local and remote case, three different rules are required (Figure 12.8).

Fusing local estimate with local measurement Suppose a sensor node has some local estimate of the global aggregation function and it performs some local measurement, contributing to this aggregated value. A heuristic is needed to update the local estimate, taking into account the concrete aggregation function to be computed.

Both the local estimate and the local measurement suffer from uncertainty, expressed by the variance of these values. It is one option to approximate the true, but in general unknown, distribution function by a Gaussian distribution, to compare these distributions, and to derive a new estimated value and a new variance from this comparison. The details are slightly involved, but essentially a new distribution function is determined that is a representation of the new information.

Fusing local estimate with remote information Fusing a local and a remotely obtained estimate can be performed using the theory of covariance interaction (a generalization of Kalman

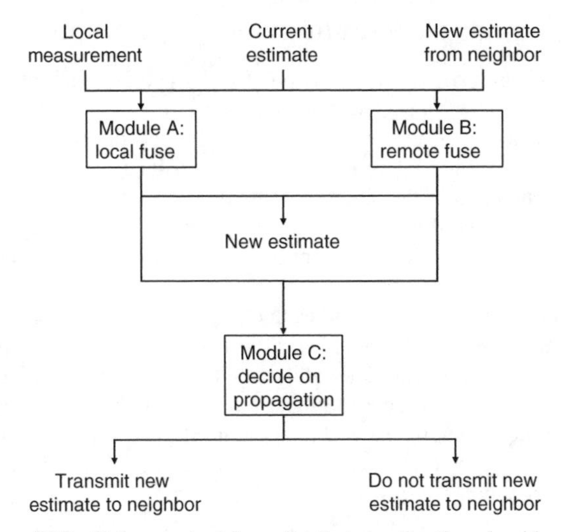

Figure 12.8 Rules required for a distributed estimation algorithm [95]

filters). Reference [95] discusses the formulas for the general case; for the maximum aggregation, the new variance is the smaller of the two old variances and the new estimated value is the one that corresponded to that particular variance.

Deciding whether to transmit local estimate after update Once a node has updated its local estimate, the question arises whether to distribute this new estimate to its neighbors. The intuition is to only distribute estimates if they would bring about "significant" changes in the neighbors' estimates in turn.

To decide this, every node X keeps track of the last estimates it has heard from each of its neighbors. Node X combines its own, new estimate with each of these remembered neighbor estimates, calculating the new value that the neighboring node would obtain if it were informed about the update at node X. Node X can determine the results that the other nodes would compute since all nodes use the same fusing algorithm.

Once the new result at a given neighbor Y is determined, node X computes the difference between node Y's current and its prospective, new estimate. Only if this difference, either in value or in confidence, is larger than a predefined threshold (e.g. in percent) will node X transmit its new estimate to node Y. The justification here is that these speculative calculations on behalf of the neighboring nodes consume much less energy than simply transmitting the packet.

This percentage threshold ultimately determines the best possible accuracy that the algorithm is able to achieve; it trades off energy consumption (fewer messages) against accuracy.

Reference [95] describes a further optimization when two-hop neighborhood information is available.

The resulting interaction is fully localized and does not need a routing structure (like a convergecast tree) in the network. It does, however, assume that indeed all nodes require an estimate of the global aggregate. In practice, the resulting accuracy is even better than the worst-case one determined by the propagation threshold and the energy consumption is quite good.

12.3.8 Information-directed routing and aggregation

A different take on the aggregation problem is possible by coming back to the ideas of gossiping and rumor routing. A query that is injected into the network travels around and collects information as it is forwarded from one node to the next. The pivotal question is to which node to forward a query. Ideally, that node should be chosen that can contribute most *information* to answer the query or that lies in the direction of the largest amount of information. Such an **information-driven routing** (also called *information-directed routing*), including information aggregation, is described in references [167, 513]. It is, in a sense, the most thorough incarnation of the idea of data-centric networking.

The application under consideration is target tracking; the goal is to provide an estimate of an event source as accurately as possible with as little energy as possible. Think of sensor nodes as microphones, for example, that can detect the sounds created by a moving object. Ideally, the query would be send to nodes that have the most information about the target – but the obvious predicament is that in order to know that and to avoid needless communication a node would have to communicate with its neighbors to find out which node has information. This riddle is solved by using methods of estimation theory.

Some background on estimation theory

Some background on the notions and notations of estimation theory as used here appear in order; the presentation follows, in a simplified manner, reference [167].

In target tracking and similar applications, the position $x(t)$ of a – possibly moving – object shall be determined by a set of N sensors (let us assume for the moment that all sensors know the position x_i of all other sensors). It is futile, however, to attempt to obtain the precise position; at best, estimates can be obtained. Such an estimate is represented by the **belief**, an *a posteriori* probability distribution of the (conjectured) target position x, given the measurements $z_1, \dots, z_N$ from N sensors: $\Pr(x|z_1, \dots, z_N)$. The actual estimate as such is then the expected value of this probability distribution, $\overline{x} = \int x \Pr(x|\dot{z}_1, \dots, z_N)\, \mathrm{d}x$.

In addition to the estimate, the belief, as a probability distribution, also expresses the **uncertainty** of the currently available information. It can be approximated by the covariance of the distribution,

$$\Sigma = \int (x - \overline{x})(x - \overline{x})^{\mathrm{T}} \Pr(x|z_1, \dots, z_N)\, \mathrm{d}x.$$

Intuitively, the belief is less uncertain if the corresponding distribution is tightly centered. This intuition can be formalized by introducing different **information utility measures** ψ, mapping the class of probability distributions to the real numbers. By convention, a large value $\psi(f_X)$, where $f_X(\cdot)$ is the density of the random variable X, indicates high certainty.

Such information utility measures can be based, for example, on the covariance Σ of a given distribution. The determinant $\det(\Sigma)$ of the covariance is proportional to the volume of the covariance ellipsoid;[3] the trace $\mathrm{trace}(\psi)$ is proportional to its circumference. Covariance-based information utility measures are suitable for distributions that can be well approximated by a Gaussian distribution (and are, in particular, unimodal). For multimodal, non-Gaussian distribution, the Shannon entropy $h(X)$ of a random variable X with support in S can serve as a measure, $h(X) = -\int_S f_X(x) \log f_X(x)\, \mathrm{d}x$. To make sure that large values correspond to high certainty, take the negative of these functions. Reference [167] discusses some additional measures, in particular, the geometric measure $\psi(f_X) = (x_j - \overline{x})^{\mathrm{T}} \Sigma^{-1} (x_j - \overline{x})$, that uses the so-called Mahalanobis

[3] An ellipsoid centered around the mean of the multivariate distribution such that there is a predefined random mass contained in it.

distance, is suited for target tracking using sensors that measure amplitudes of signals emanating from the tracked target; it expresses the conjectured reduction of uncertainty that sensor j can contribute, based on the current estimate $\overline{x}$ and its covariance matrix Σ.

Once such an information utility measure has been selected, it can be used to concisely compare distribution functions. It is clear that adding more observations will improve the estimate and its information utility. Thus, the information utility can be used to select the next sensor from which measurements should be incorporated. This selection is based, on the one hand, on the current estimate and its uncertainty, and on the other hand, on a candidate sensor's position and probability distribution of its measured value. On the basis of these facts, the probability distribution of the information utility measure *after* incorporating the measurements from a particular sensor node can be computed *without* having to know the concrete measured value. To reemphasize: not the actual information utility, *only* its probability distribution can be determined!

Nonetheless, the new information utility's probability distribution suffices. Computing it for various candidate sensors allows to select the next sensor according to one of several rules. Options include:

Best average case Choose that candidate sensor that has the highest *expected* information utility measure after its measured value has been incorporated.

Maximize the worst case Suppose each candidate sensor has measured a value that contributes the least information to the estimate. Pick then that sensor that still maximizes the obtained information utility.

Maximize best case Suppose, to the contrary, that each sensor has the most valuable possible measured value. Pick then that sensor that maximizes the obtained information utility.

A concrete routing algorithm has then two different metrics at its disposal. One metric is the classical energy cost required to communicate with a neighboring node. The other is the information utility that would – likely – result if that node would be asked to contribute its measured value to the query. How a concrete routing algorithm balances these two sources of information is the topic of the following paragraphs.

Information-Driven Sensor Query

Assume a clustered sensor network where the clusterhead initiates queries and only asks nodes within its own cluster for information. The Information-Driven Sensor Querying (IDSQ) algorithm deals with this case. A normal cluster member simply waits for a request and delivers its local measurement when asked to do so. The clusterhead successively polls cluster members until the estimate is good enough (i.e. the uncertainty small enough).

The selection regarding which sensor node to poll for information is based upon the current belief and the position of the neighboring nodes. On the basis of any rule discussed above, the clusterhead selects that node that promises the biggest improvement in uncertainty. Once the answer from that node has been received, the clusterhead updates its belief.

Note that this scheme is purely based upon information; energy efficiency considerations play no role here.

Constrained Anisotropic Diffusion Routing

The IDSQ scheme is relatively restricted, yet simple. More interesting is the case when a query cannot be answered from a set of nodes within a cluster. This case is solved by the Constrained Anisotropic Diffusion Routing (CADR) scheme. Here, a query is assumed to float through the network, akin to active query schemes described in Section 12.2.1. When choosing the next node to forward the query to, however, the prospective improvement of the information utility measure as well as energy cost to communicate with a neighbor are taken into account. The simplest way for balancing it is a fixed factor between 0 and 1, computing a weighted sum of energy and information impact. The node that maximizes this weighted, composite objective function will be the one that receives the query. Reference [167] discusses the consequences of different choices of this parameter.

Moreover, a node, upon receiving a query, can update it with its own measured values, in this sense aggregating information into the query. The selection of the next node then takes place based upon the updated estimate and the updated uncertainty estimate.

So far, global knowledge about sensor node positions has been assumed. It is also possible to work without this assumption. Nodes then route either only on the basis of position of their immediate neighbors, along the gradient of the weighted objective function composed of energy and information components, or biased toward the estimated current position of the target (to avoid slowly iterating along the gradient).

Avoiding dead ends

The CADR scheme is greedy; as a consequence, it can be stuck in dead ends (local maxima). Reference [513] repairs this shortcoming and also generalizes the set of supported scenarios (routing from a query starting point to a prespecified exit node). It is instructive to note that the problematic cases are similar to those that are encountered in geographic routing but solutions like GPSR are not necessarily applicable. The idea is to introduce a look-ahead mechanism, extracting information from several (possibly all) M hop paths before deciding which one of them to use. The details are beyond the scope of the presentation here; the reader is referred to reference [513].

Overall, these information-driven routing schemes manage to combine the notion of data-centric routing – send the query where the action is, using information observed by the network itself – with concepts of energy-efficient routing – weigh possible information gain against the cost to obtain it. The results achieved in tracking accuracy and energy efficiency are rather impressive.

12.3.9 Some further examples

Tiny Aggregation (TAG)

The Tiny Aggregation (TAG) scheme [530, 531] is currently perhaps the most popular aggregation scheme for wireless sensor networks; it popularized the use of SQL as an interface abstraction. Its current acclaim is in no small part due to the fact that a TAG implementation is available for TinyOS.

The basic operation is based on a convergecast tree where nodes send data upstream toward sink(s) according to the `EPOCH DURATION` specification in the query. Some consideration is given to the scheduling of data flows in this tree to enable long sleeping periods or to use pipelining [530]; however, much of these considerations depend on the concrete MAC layer in use.

TAG also includes support for grouping predicates. In the simplest case, each node maintains separate partial state records for each group (dynamically allocating them if a reply from a child node with a hitherto unheard-of group arrives); incoming replies are only aggregated with state records of their own group. If many groups exist, memory can become scarce and nodes can evict partial state records from their local memory by sending them to their parent, akin to caching strategies. This strategy is called *partial preaggregation*.

What is more, TAG also supports the in-network evaluation of `HAVING` predicates to reduce network traffic. One such applicable `HAVING` predicate would be `MAX(temperature) < 100`.

First of all, it is a monotonic predicate: Once the maximum temperature exceeds a given value, it will not be reduced by aggregating temperate values from other nodes. Moreover, it can be locally decided: Once a node sees a temperature value in a given group that violates the `HAVING` predicate, it knows that this predicate cannot possibly be repaired. It can then inform other nodes in the network of this violation for a given group or at least suppress it locally. Note that a predicate like `MAX(temperature)>100` is not useful for in-network curtailing of messages despite its being monotonic. A node cannot conclude, just because its own local value/aggregate is smaller than 100, that other nodes might not still report values that would make the group fulfill this `HAVING` predicate.

Reference [531] discusses some further optimizations like taking advantage of a shared channel, that is, snooping on the messages other nodes are transmitting and *hypothesis testing*. This latter technique is best explained for monotonic and exemplary aggregation functions. The idea is that along with a query a hypothesis of a possible result is communicated – when searching for the minimum sensor reading, the initiating data sink could include a hypothesis obtained from its local neighborhood in the query. This way, many nodes know that their value is not going to influence the result of the query and abstain from providing answers. The concrete savings in message overhead for this example depend on the network topology and on the spatial correlation of the observed values.

It comes as no surprise that the message savings realized by TAG largely depend on the particular aggregation function. For noncomposable aggregations like median, there is no benefit at all; for count and minimum only about 10 % of the bytes transmitted by a simple, centralized aggregation scheme are required.

Data funneling and coding by ordering

The scenario considered in reference [640] is that of a data sink requesting all sensors in a given geographical region to periodically send their measured values. The data sink is interested in obtaining *all* values of each, uniquely identified node. Under such circumstances, the aggregation techniques described above are not immediately applicable.

Two ideas are brought forward to tackle this scenario. The first idea is *data funneling*. After the request (or interest) has been flooded (in principle, using any geocasting method), the nodes in the region of interest can send their data to a node on the border of that very region. This node will act as an aggregation point and forward the readings to the data sink. To protect this border node from exhausting its energy resources, its role is rotated occasionally among all border nodes that face the data sink. To achieve this load balancing, all interior nodes must know when to send their data to which border node. This is possible by having all the border nodes flood the interest packet, along with their own identifiers, into the region. Then, all interior nodes know about all eligible border nodes and can apply a selection function to this record of information. Since all interior nodes apply the same function to the same data (assuming no border node announcements are lost in the region), at a given time all data will be sent to the same border node.

It is, however, not immediately obvious how a border node can actually perform data aggregation – after all, all readings have to be sent to the data sink. The simplest improvement can be made by concatenating all values into a single long packet, saving on the medium access overhead for many small packets. In addition, reference [640] describes a cleverer method called *coding by ordering*. This method leverages the fact that the *order* in which individual measurements are put into a larger packet is irrelevant and can be arbitrarily chosen. Thus, this choice can actually convey information! For example, if a value can only assume one of six values, the 3! orderings of three readings can encode this particular value and thus convey a fourth reading.

PEGASIS – Energy/delay metric

A further dimension is added to the data aggregation problem when looking not only at the consumed energy but also at the resulting delay before data is available at the sink. One approach that explicitly addresses these two aspects is Power-Efficient GAthering in Sensor Information Systems (PEGASIS) [507] (actually, a slight misnomer since the main focus is on being energy efficient, not so much on power efficient). The considered scenario is a homogeneous network of sensor nodes where all the nodes have to transmit their local measurement to a known sink, once per given round (rounds are somehow synchronized, for example, by high-powered beacon signals from the data sink). Measurements can be aggregated in intermediate nodes using any algebraic aggregation function, all nodes have global knowledge about sensor positions, all nodes have power control with in principle arbitrary range, and nodes may or may not have CDMA-capable radio transceivers.

The goal is to find a convergecast structure that has good energy consumption and delay behavior and that balances energy consumption among the sensor nodes. The proposed figure of merit is the product of the energy times the delay needed per round of data gathering.

An interesting aspect about PEGASIS is its convergecast structure. Unlike in most cases, it is not a general tree but a chain. This chain is constructed starting from the node farthest away from the data sink. The chain grows from one end only and the next node to be added is the as-yet unselected node closest to the current end node. One node in the chain is elected as leader; it will transmit the aggregated data to the data sink. This leader role is shifted one position in the chain with each round – in effect, the leader node can be arbitrarily far away from the data sink and potentially has to use high transmission power to deliver data to the data sink.

The actual data collection also takes place along the chain; various alternatives are possible here. The simplest option is to have the current leader node send a token out into the chain, have it propagate until the end, and return the data, aggregating it along the way. Once data arrives at the leader, the token is sent into the other part of the chain and the process repeats. Once data from both halves has arrived, it is forwarded to the data sink. Clearly, this approach leads to high delay as it does not exploit possible parallelism of transmissions in the network.

This parallelism is naturally achieved by a tree (and the help of a MAC protocol) but it is also possible on a chain. Assume that all nodes have CDMA transceivers such that parallel transmissions can go on without interfering with each other at all. Then a simple parallel transmission strategy is to first have direct neighbors send data to each other, for example, odd-numbered nodes send to even-numbered nodes. Then, in a second step, only those nodes that were receiving data (and have aggregated it) in the previous step remain active and one half of them sends their aggregated data to their neighbors. These steps repeat until, after $O(\log n)$ steps, the aggregated data has arrived, as a single transmission, at the current leader. In effect, a tree has been overlaid over the chain. After the leader has moved, the tree organization has to move also, but this is straightforward to compute – perhaps the biggest advantage of the chain construction.

When CDMA is not available, it is not possible to have arbitrarily close transmissions in parallel. The suggestion is then to only use three levels of such an aggregation hierarchy as opposed to $O(\log n)$ levels. The n nodes are divided into G groups (sequentially along the chain); within each group, a simple sequential aggregation takes place. Then, the leaders of these G groups form two subsets and aggregate within. Eventually, data is transmitted from one subset leader to the other, which transmits to the data sink. Of course, the starts of these groups and their respective leaders have to be rotated appropriately.

Both these latter schemes are compared to direct communication and to LEACH, using the product of energy and delay. The authors show a considerable improvement, particularly for the CDMA-based scheme.

12.3.10 Further reading on data aggregation

Clustering and aggregation Several of the clustering schemes discussed in Chapter 10 lend themselves easily to aggregating data in the clusterhead. LEACH, for example, uses the clusterheads to collect data from the cluster members and then forwards the aggregated values directly to the data sink (see Sections 5.4.1 and 10.4.5 for details on LEACH).

Application-independent aggregation The aggregation functions discussed so far all relied on *manipulating* data to some extent, needing semantic information on what the actual aggregation function is (a minimum requires different operations than a mean). In this sense, these are all application-dependent aggregation operations. The Application-Independent Data Aggregation (AIDA) approach [338] takes a different perspective: What can be gained by aggregation *without* manipulating data, without application knowledge? The set of possible operations to aggregate is thus small; essentially, it boils down to concatenating packets to be forwarded, exploiting queuing delays in a node to find opportunities for such operations. Nonetheless, reference [338] shows that considerable benefits can be obtained even using this simple operation.

Impact of link imperfections ZHAO et al. [927] point out the impact of real-world communication artifacts on aggregation structures. They show errors up to 50 % when using arbitrary links and recommend to discard links with high loss rate and asymmetric links from the construction of the aggregation structure. It deserves particular emphasis that their results are backed by real experiments.

Mobile devices Data dissemination between mobile devices, not all of which are connected to the Internet, is considered in reference [614].

Statistical accuracy OKINO and CORR [606] look at the question of how to collect data such that the results are statistically accurate.

Aggregation and game theory A game-theoretic approach to data collection is proposed in reference [397]. The set of routing paths is considered as the outcome of a game and the optimal solution is shown to be the Nash equilibrium.

Aggregation and security Security aspects of data aggregation are discussed, for example, in reference [662].

12.4 Data-centric storage

Accompanying the data-centric networking development are some considerations on *storing* data within the network. Data-centric storage comes back to the basic question of which entity needs to know which data. If nothing is known about this question, data could be distributed in the entire network (perhaps with the help of aggregation to make it more efficient) or nodes that produce data by measuring could locally store it and wait for a query to arrive. But the querier has to resort to flooding the network if it has no information where the data is (which would allow geographically scoping the flood) or which specific node has the data (so that traditional routing schemes could be applied). Neither of these options is convincing in the end.

The idea of Dynamic Code Scaling (DCS) [684] is to let the data itself describe where it is stored. More specifically, the name of the data (which is necessary for any data-centric approach anyway) is used to represent a key under which the data can be looked up. Under this key (under the node that is identified by the key), the actual data is then stored, much like in peer-to-peer networking

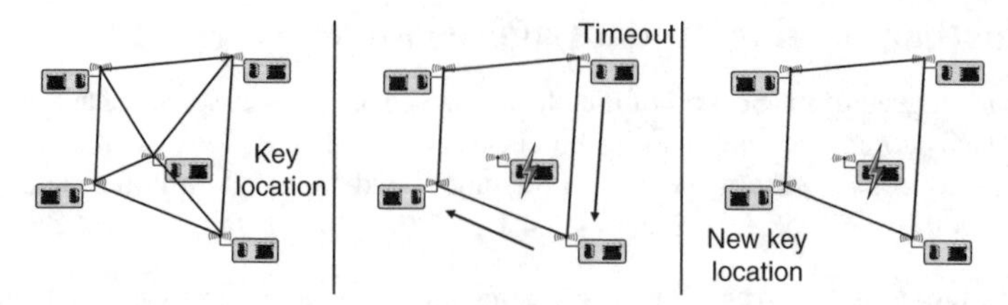

Figure 12.9 Operation of geographic hash tables: Selecting a new key location once a node storing some keys fails

and distributed hash tables. A query can then be routed directly to the node corresponding to the name of the data and directly retrieved. This concept saves on cost to distribute either data or queries in the entire network, replacing a flood by a unicast communication.

The precise role of a key can be defined in several ways. A natural match with wireless sensor networks and the one proposed in reference [684] is to use a geographic interpretation of the key: The name of data corresponds to a particular location and the mapping rule is known to all nodes. Then, both producer and querier of data can easily compute the storage location of a given data item and route their packets toward it, using some geographic routing protocol. This geographic key interpretation is the essence of a Geographic Hash Table (GHT) (Figure 12.9).

If for every so-computed location there were a node available at precisely this position, if nodes did not fail, and if nodes had infinite amount of storage, then there would be nothing more to add. Handling these imperfections, however, requires some further considerations.

Nodes not available at the hashed location When no node is available at precisely the location to which the key matches, it will not be possible to deliver the packet to such a node. Instead, the closest node to the computed position should represent an acceptable proxy. How this proxy or "home node" can be determined depends on the geographic routing protocol; reference [684] describes one scheme based on GPSR (see Section 11.5.1). In GPSR, once a packet has reached a node from which there is no other node closer to the destination, the packet will enter perimeter routing mode, traveling around the "hole" in the network where the hoped-for node at the destination does not exist. Eventually, the packet will return to the node where it started perimeter mode – this state can be detected and this node can be declared the home node for the given key value. Fortunately, it works for both putting data into storage and for queries.

Handling failing and new nodes When a node fails, all the key-value pairs stored in it would be lost. To protect against such data loss, the data should be replicated – ideally, to neighbor nodes. Here again, the perimeter mechanism lends itself to a simple solution. All nodes on the perimeter of a node become replicas to store the data. To detect a failed node and to keep the replicas consistent, all nodes occasionally send refresh packets around the perimeter, distributing new data and checking for the presence of a home node. If none is found, the node then closest to the key's virtual location assumes the role of the home node.

The same mechanism can be used to integrate new nodes. A new node would be integrated by GPSR and learn, via the refresh packets, of existing key-value pairs. It could also assume the home node role for appropriate locations. Alternatively, new nodes can be explicitly integrated without having to wait for refresh timeouts.

Handling limited storage per node It can easily be the case that too much data ends up at a single node, either from a single key or from different keys. Reference [684] proposed the use of mirror nodes in the vicinity of the home node to store such data, trading off storage space against some additional local communication.

These characteristics make GHT suitable for scenarios with stable and static nodes. One quite appealing property of GHT is that it not only reduces the average number of messages but also ameliorates hotspots in the network (details depend on particular variants of GHT). The most benefits are reaped in large networks with many detected events, but where only some of these events are actually queried for.

The ideas of data-centric storage are relatively new and are still explored. One interesting direction is, for example, the graph embedding procedure GEM [593] described in Section 11.5.1, which lends itself to the support of data-centric storage.

Distributed data storage is a relatively new part of WSN research. A few pointers to further relevant literature are [197, 274, 275, 305, 493, 713, 760, 782]. Most of these references also have a tight connection to a database abstraction.

12.5 Conclusions

Putting the data to be collected or disseminated in the focus of communication protocol design considerably changes the design paradigms. It is no longer possible or sensible to try to construct some form of routing structure in the network that is based on the identifiers of the nodes. Rather, the data as such has to guide the interaction of separate nodes in the network. This chapter has shown a couple of possible approaches. Many of them rest, in some form or another, on the notion of publishing named *data* and subscribing to certain names. This paradigm, in combination with data aggregation or other, advanced forms of in-network processing, admits crucial optimizations for wireless sensor networks.

13

Transport layer and quality of service

Objectives of this Chapter

The service obtained from "classical" networks like the Internet and from sensor networks differ. The Internet is supposed to transport independent byte streams, and intermediate nodes do not know more. In a sensor network, the nodes collaborate and interact with the environment; the nodes know the data they carry.

A key requirement is reliability. In sensor networks, reliability refers not only to the eventual delivery of data packets (transport reliability) but also to the ability to detect physical phenomena in the first place. The coverage of a sensor network is thus an important consideration.

This chapter discusses protocols and approaches to deal with reliability in a sensor network; we refer to these somewhat sloppily as transport protocols. These protocols are not "cleanly" placed on top of some network layer protocol. Instead, the unique constraints of sensor networks call for careful cross-layer design.

Chapter Outline

13.1 The transport layer and QoS in wireless sensor networks

In the Internet, a large number of independent users runs a multitude of applications and each user judges its QoS individually. The Internet is largely seen as a vehicle for moving streams of

Protocols and Architectures for Wireless Sensor Networks H. Karl and A. Willig

bytes over multiple hops from one place to another, and from the users perspective it is transport protocols (TCP, UDP) which provide this service. Consequently, the measures used for judging the quality of this service are related to the protocols, not to any application. Typical measures include delay, jitter, throughput/goodput, packet loss rate, and many more.

Things are different in sensor networks. A sensor network is *not* seen as a mere infrastructure for transporting data, but the nodes are tasked with collaboratively monitoring and controlling the physical environment. They have to process each other's data locally or while in transit toward some sink nodes and thus have to *know* the data they forward. Thus, users expect another service than they get from the Internet and the networks ability to support these application-dependent tasks is an important cornerstone of sensor network QoS besides the traditional network-oriented quality measures.

This chapter focuses on reliability as one of the key QoS measures. Reliability has many facets and encompasses more than reliable packet delivery. These different facets are described in Section 13.1.1.

In the Internet, the network layer and the transport layer play an important role in achieving data transport reliability. The network layer offers a best-effort service and the transport layer is responsible for achieving reliable and in-sequence delivery, congestion and flow control, and other things. This principle of pushing everything into the transport-layer protocols at the end nodes is coherent with the Internet's end-to-end principle [719]. Anything below the network layer is taken as a black box.

In sensor networks, this perspective is not optimal. In fact, there are good reasons to *not* think of reliability as something that is only provided by a protocol on top of a networking layer. One of the reasons is the unique constraints in sensor networks regarding energy, memory, and computational power. A second reason is provided by the observation that a sensor network as a whole has a specific task, and thus it is possible – and also advisable, given the constraints – to design all protocols *jointly*, allowing one protocol to explicitly make assumptions about the behavior of other protocols or even control them.

Accordingly, transport protocols in wireless sensor networks are conceived in this chapter as collections of mechanisms to provide certain services in an end-to-end fashion, without requiring these mechanisms to run entirely on top of a networking layer.

13.1.1 Quality of service/reliability

One of the most important qualities is reliability. In sensor networks, the notion of reliability has several facets:

- In the problem of **detection reliability**, the main question is whether the events the network is supposed to detect actually *can be* be detected. A necessary prerequisite is that possible event locations are *covered* by the sensing ranges of a number of sensors. The required node density depends on the sensor's sensing ranges, the shape of the sensing regions, and environmental conditions like, for example, the presence of obstacles. The **coverage problem** is discussed in Section 13.2.
- For cheap sensors, single sensor readings can be inaccurate. The user, however, wants to have credible information he can count on. Thus, to achieve **information accuracy**, multiple readings – obtained either over time or over space – should be combined to smooth out noise and detect outliers. On the other hand, too many similar sensor readings are a waste of energy. This issue, however, is closely tied to the application domain and to the specific sensor technology and is beyond the scope of this chapter.
- Given that an event has indeed been detected, this information must be reported reliably from the event location (or the sensors around it) to sink nodes, which are often several hops away. Conversely, applications like distribution of application code to the sensors require reliable data

delivery from sink nodes to individual or groups of sensors. We thus have the **reliable data transport** problem, discussed in more detail in Section 13.3.

Some applications require not only reliable but also timely delivery of data. This refers to time bounds between the time of occurrence or detection of an event and the time where the sink or the user knows about the event.

Most of the available literature is focused on reliability; the issue of timely delivery is only rarely addressed. And from the first sight, there seems to be an "impedance mismatch" between the goals of timely and quick delivery on the one hand, and some of the pertinent characteristics of sensor networks like small energy budgets, low duty cycles, spending time in sleep states, node failures, and multihop communication on the other hand. In this chapter, we concentrate on reliability and point only briefly to some of the work concerning delays in Section 13.5.4.

13.1.2 Transport protocols

Let us first examine the different tasks commonly attributed to transport protocols:

- **Reliable data transport**: This task requires the ability to detect and repair losses of packets in a multihop wireless network; appropriate mechanisms working on different layers are discussed in more detail in Section 13.3.
- **Flow control**: The receiver of a data stream might temporarily be unable to process incoming packets because of lack of memory or processor power. Flow control has so far not been a research issue in sensor networks; some of the reasons have been discussed in Section 6.1 in the context of the link layer.
- **Congestion control**: Congestion occurs when more packets are created than the network can carry and the network starts to drop packets. Dropping packets is a waste of energy and counteracts any efforts to achieve reliability or information accuracy. Congestion-control schemes try either to *avoid* this situation or to *react* to it in a reasonable manner. One important way to avoid congestion is to control the rate at which sensor nodes generate packets. Congestion control and rate control are discussed in Section 13.6.[1]
- **Network abstraction**: The transport layer offers a programming interface to applications, shielding the latter from the many complexities and vagaries of data transport. Since there is yet no standard transport protocol in sensor networks, there is no consensus on such an interface.

As stated in the introduction and will be exemplified in the course of this chapter, it is beneficial to implement reliability, flow control, and so on not in a single protocol running on top of a network layer, but to combine several mechanisms working on different layers.

Transport protocols in sensor networks are faced with other kinds of environments than "traditional" transport layers like the ubiquitous TCP are designed for. Some of the particular challenges for transport protocols in wireless sensor networks are the following:

- Wireless sensor networks are multihop wireless networks of homogeneous nodes. This is not an easy environment as the different problems of TCP over wireless channels illustrate (see Section 13.5.3).
- Any transport protocol must comply with the stringent energy constraints, memory constraints, or computational constraints of sensor nodes. Significant engineering efforts would be required to run heavyweight protocols like TCP on such nodes.
- Transport protocols are faced with variable topologies.

[1] Discussing congestion control in the context of transport protocols instead of the networking layer is somewhat artificial, but is at least inspired by an important role model, namely the Internet approach of keeping IP simple and putting everything into TCP.

13.2 Coverage and deployment

Many wireless sensor networks are tasked with surveillance of certain geographical areas, for example, to detect intruders, wildfires, or rare animals in a habitat. Putting all communication aspects aside, such an event can only be detected if there are sensors close enough that can actually sense the event. Two important questions arise:

- We are given a sensor deployment, that is a particular placement of sensors over a certain geographical area. Which points of this area are close enough to sensors such that an event taking place at this point can be sensed. Asked differently: which points are *covered*? **Coverage** is thus an important aspect of QoS in sensor networks.
- Given an area to be observed and some coverage requirements, what number of sensors is needed and where should they be placed? This question, henceforth labeled as the **deployment** problem, can be posed under several interesting constraints, for example, cost constraints, presence of obstacles, availability of different types of sensors, and so forth.

Coverage and deployment have a second important implication besides QoS. If there is some overprovision of sensors, it might be possible to switch some sensors into sleep mode without compromising coverage. This allows to save energy and to prolong the lifetime of the overall network.

The larger part of this section is devoted to the coverage problem, which is the more important one in many practical situations. For example, in applications like wildfire detection or habitat monitoring, large areas have to be observed with large numbers of sensors, and in such a situation, sensor deployment will often only be a loosely controlled process. Most of the presentation (and of the available literature) concentrates on coverage and deployment in two dimensions but many concepts carry easily over to the three-dimensional case. References to protocols dealing with three-dimensional coverage problems are given where appropriate.

After introducing some simple models for the sensing ability of sensors in the next Section 13.2.1, we discuss different measures for coverage in Section 13.2.2. An often-used theoretical deployment model, the "independent deployment", which is in fact a Poisson point process, is explained in Section 13.2.3. After this, we discuss coverage issues for independent deployments under two sensing models: For the Boolean sensing model, in Section 13.2.4 and for a more general sensing model in Section 13.2.5. The question how certain coverage measures can be determined a posteriori in a sensor field with unknown deployment characteristics is treated in Section 13.2.6. Some brief comments about coverage measures in sensor grids are made in Section 13.2.7. The issue of deployment is not discussed in detail, but some references to start with are given.

Many of the algorithms described in this section require knowledge of the number of sensors and their geographical positions. This knowledge can be obtained from running locationing algorithms, see Chapter 9.

13.2.1 Sensing models

A sensor transforms environmental stimuli into electrical signals. The quality (signal strength, noise) of the resulting signal depends, among other factors, on the distance between the sensor and the actual event. For example, the amplitude of sound waves decreases quadratically (more general: according to a power law) with increasing distance to the location of the sound-producing event. When an acoustic sensor node has a very large distance to the sound event, the sensor readings become indistinguishable from the case of no sound event occurring at all.

A second aspect of sensing quality is directionality. In an idealized scenario, a sensor has the same sensitivity in all directions; however, in practice, often certain directions are preferred. This can be either by construction (for example, video cameras) or as a result of sensor deployment, when, for example, a node's acoustic sensor is obstructed by other node components.

A third aspect is constituted by the possibility that the same sensor can generate different outputs for the same environmental stimulus at different times, for example, due to temperature variations the sensing circuitry is exposed to. Generalizing this observation, the signal delivered by a sensor for an external event at a certain distance is not a fixed value, but a distance-dependent random variable. One simple assumption here would be to model such a random variable as a constant plus some zero-mean Gaussian noise with either constant or distance-dependent variance (compare for example, [924, Sec. 2]).

Most **sensing models** used in the literature focus on the first aspect and assume omnidirectional sensing and no random variations (for example, [496, 510, 547]). In [510] two sensing models are introduced:

- In the **Boolean sensing model**, all sensors of the same sensor modality (temperature, humidity, ...) have a common sensing range r. Events within this sensing range are detected reliably, and events outside this range are not detected at all. Accordingly, the sensor output signal for a sensor node at position $\mathbf{p}$ observing an event at position $\mathbf{q}$ has strength:

$$S(\mathbf{p}, \mathbf{q}) = \begin{cases} \alpha & : \quad \|\mathbf{p} - \mathbf{q}\|_2 \le r \\ 0 & : \quad \text{otherwise} \end{cases} \tag{13.1}$$

  where $\|\cdot\|_2$ is the Euclidean distance between points and α is the constant sensor value.[2] Tian and Georganas [816] even assume that sensors can have different sensing ranges, for example, depending on their residual energy.
- In the **general sensing model**, the sensor also possesses a certain maximal sensing range but within this range the sensor output obeys a power law instead of being uniform:

$$S(\mathbf{p}, \mathbf{q}) = \begin{cases} \dfrac{\alpha}{\|\mathbf{p} - \mathbf{q}\|_2^{\beta}} & : \quad r_0 \le \|\mathbf{p} - \mathbf{q}\|_2 \le r \\ 0 & : \quad \text{otherwise} \end{cases} \tag{13.2}$$

  where r_0 is a certain minimum distance (to avoid division by zero) and β is a positive real number depending on the sensing modality and sensor technology. Example: For acoustic signals, the relationship between the source signal power and the sensed signal power can be modeled with $\beta = 2$ ([924, Sec. 1.2.2]).

It is pointed out by Meguerdichian et al. [547] that the sensing quality may also well depend on the time that the sensor is exposed to the external event. An example for this is a film for a photo camera. It is also worth noting that under the general sensing model knowledge of the decay exponent β and of $S(\mathbf{p}, \mathbf{q})$ allows node $\mathbf{p}$ to estimate its distance to $\mathbf{q}$. This is, for example, interesting in detection and event localization applications (compare Section 14.3) or in acoustic ranging (Chapter 9).

There is another important type of sensors to which coverage considerations do not directly apply. These **point sensors** detect a phenomenon only upon having direct contact with it. Savvides et al. [727] name chemical sensors as an example: These can sense toxics only by direct measurements. When the phenomenon of interest – say, a toxic plum – is known to have reasonably smooth boundaries and minimum extension, then it is possible to derive conditions for the sensor deployment ensuring detection of the plum with sufficiently high probability.

[2] The event to be sensed, for example, a sound event, is assumed to have "unit loudness". Otherwise $S(\cdot, \cdot)$ would have to be normalized with the loudness of the sound event.

13.2.2 Coverage measures

The notion of coverage has different meanings in the literature. In general, coverage measures refer to a sensor network deployed to monitor some specified terrain A having area $|A|$. Most often, this terrain is assumed to be two dimensional. Some of the introduced measures are the following:

- The **area coverage** f_a specifies the percentage of $|A|$ being covered [510]. If $f_a = 1$, we say that **full area coverage** is achieved. In the general case for a point $\mathbf{q} \in A$ to be covered, we require

$$C(\mathbf{q}) = \sum_{s \in \mathcal{S}} S(\mathbf{p}_s, \mathbf{q}) \geq \theta \tag{13.3}$$

  where $\mathcal{S}$ is the set of all sensors, $\mathbf{p}_s$ is the position of sensor $s \in \mathcal{S}$, and θ is a certain application-dependent threshold. The quantity $C(\mathbf{q})$ is also called **sensor field intensity** [547]. To express the requirement that under the Boolean sensing model at least one sensor covers $\mathbf{q}$, we can simply choose $\theta = \alpha$. Sometimes it is for reasons of fault tolerance required that an event is sensed by $k > 1$ sensors. Under the Boolean sensing model, this requirement is also denoted as k-coverage [364, 365] and it can be expressed by choosing $\theta = 3\alpha$. Clearly, such a formulation carries over to the general sensing model.
- The **node coverage** f_n describes the percentage of nodes whose sensing range can be fully covered by the sensing ranges of other nodes. When the overlapping neighbors are awake, such a node can be safely switched into sleep mode without reducing the area coverage.

The next set of coverage-related measures considers paths between two chosen points $\mathbf{q}_0$ (the source point) and $\mathbf{q}_1$ (the destination point), which may be inside or outside of A. Clearly, the latter case is only interesting when the paths under consideration cross the terrain A. Before defining the next measures, it is useful to introduce the notion of the distance between a point and a point set: Be V a set of points, for example, the locations of a sensor network, that is $V = \{\mathbf{p}_s : s \in \mathcal{S}\}$, and be p another point.[3] Then:

$$\text{dist}(\mathbf{p}, V) = \inf\left\{\|\mathbf{p} - \mathbf{q}\|_2 : \mathbf{q} \in V\right\}$$

Using this notion, the following coverage-related measures can be defined:

- The **detectability** P_d represents the probability that an object (for example, an intruder) moving from $\mathbf{q}_0$ to $\mathbf{q}_1$ is sensed [510]. When the intruder knows the network topology - that is, all nodes and their positions – and is always able to choose optimal paths, the detectability gives the probability that there exists no path between $\mathbf{q}_0$ and $\mathbf{q}_1$ such that for all points $\mathbf{p}$ of this path $\text{dist}(\mathbf{p}, \mathcal{S}) > r$ holds. To detect the intruder, there must be at least one point with $\text{dist}(\mathbf{p}, \mathcal{S}) \leq r$.[4] When the network topology is unknown to the intruder it can choose *some* path (for example, a straight line) and the intruder remains undetected when for all points $\mathbf{p}$ on this path $\text{dist}(\mathbf{p}, \mathcal{S}) > r$ holds.
- The **maximum breach path** or **worst-case coverage** between $\mathbf{q}_0$ and $\mathbf{q}_1$ is defined as follows [496, 546]: be $\Pi(\mathbf{q}_0, \mathbf{q}_1)$ the set of all paths between $\mathbf{q}_0$ and $\mathbf{q}_1$; the maximum breach distance is then given by:

$$\max_{\pi \in \Pi(\mathbf{q}_0, \mathbf{q}_1)} \min_{\mathbf{p} \in \pi} \text{dist}(\mathbf{p}, \mathcal{S}).$$

[3] It is assumed throughout this chapter that location information is available to the sensor nodes, that is each node knows at least its own location and the location of its neighbors. How this information can be obtained is the subject of Chapter 9.

[4] An alternative formulation is that for all points $\mathbf{p}$ on such a path $C(\mathbf{p}) < \theta$ holds.

Intuitively, the maximum breach path is a path through a sensor network having the largest minimum distance to any sensor node. Such a path would be chosen by an intruder wishing to keep an as-large-as-possible distance between himself and the sensors when moving between two points.

- The **maximal support path** or **best-case coverage path** between $\mathbf{q}_0$ and $\mathbf{q}_1$ is defined as [496, 546]:

$$\min_{\pi \in \Pi(\mathbf{q}_0, \mathbf{q}_1)} \max_{\mathbf{p} \in \pi} \operatorname{dist}(\mathbf{p}, \mathcal{S}).$$

Intuitively, the maximal support path is the path having the smallest maximal distance to the sensor set. This would be a path preferred by someone wishing to stay under best possible observation.

- Consider an object traveling during time interval $[t_1, t_2]$ on a certain path from $\mathbf{q}_0$ to $\mathbf{q}_1$ such that the position of the object at time t is given by $\mathbf{p}(t)$. The **exposure** [547] for this object is the integral of the sensor field intensity $C(\mathbf{p}(t))$ over the path $\mathbf{p}(t)$. Assuming that $\mathbf{p}(t)$ is continuously differentiable, the exposure can be represented as:

$$E(\mathbf{p}(t), t_1, t_2) = \int_{t_1}^{t_2} C(\mathbf{p}(t)) \left\| \frac{\mathrm{d}\mathbf{p}(t)}{\mathrm{d}t} \right\|_2 \mathrm{d}t$$

The exposure can be regarded as the average observability of the object. A variation is to not integrate over $C(\mathbf{p}(t))$ but to always consider only the closest sensor, that is, integrating over $C^*(\mathbf{p}(t)) = \sup_{s \in \mathcal{S}} S(\mathbf{p}_s, \mathbf{p}(t))$. The exposure problem is discussed in more detail in [547], [840], and [175].

13.2.3 Uniform random deployments: Poisson point processes

Some of the coverage measures have been investigated for random deployments in several references, for example, [509, 510]. The most common assumption for a random deployment is that of a **Poisson point process**, defined as follows (see [404, Sec. 1.3] and [405, Chap. 16]): Be $U \subset \mathbb{R}^n$ a subset of the n-dimensional space and be $\mathcal{A}$ a nonempty family of subsets of U. Each element $A \in \mathcal{A}$ is assumed to have a volume $\mu(A)$. Furthermore, let us assume that a number of "points" are scattered over U. We are essentially interested in counting the number of those points belonging to some $A \in \mathcal{A}$; this quantity is denoted as $N(A)$. In a Poisson point process with intensity $\lambda > 0$, the following properties hold:

- For each $A \in \mathcal{A}$, the number $N(A)$ is a random variable having a Poisson distribution with parameter $\lambda \cdot \mu(A)$, that is,

$$\Pr[N(A) = k] = \mathrm{e}^{-\lambda \cdot \mu(A)} \cdot \frac{(\lambda \cdot \mu(A))^k}{k!} \qquad k \in \mathbb{N}_0.$$

- When $A_1, \ldots, A_n \in \mathcal{A}$ are disjoint, the random variables $N(A_1), \ldots, N(A_n)$ are independent.

Poisson point processes are popular, for example, for modeling the number of stars in a certain space area or the number of bacteria cultures on a Petri dish. The striking feature of such a Poisson point process is that it matches the intuitive notion most people have of "random deployments": It is shown in [405, Chap. 16] for Poisson point processes that under the assumptions $\mu(U) > 0$ and $N(U) = k$ the k points are independent and uniformly distributed in U.[5]

[5] In the given reference, the Poisson point process is introduced in a more axiomatic way than we did here.

We can now start to answer questions regarding certain coverage measures for sensor networks under such a random deployment. The presentation in the Sections 13.2.4 and 13.2.5 is largely based on [509, 510].

13.2.4 Coverage of random deployments: Boolean sensing model

We first discuss the case of an infinite sensor network in the two-dimensional plane (i.e. $U = \mathbb{R}^2$) to avoid any boundary effects.

It is straightforward to find the area coverage f_a for a Poisson point process of intensity $\lambda > 0$ under the Boolean sensing model.[6] Be $\mathbf{q}$ a randomly chosen point in the sensor field. What we are asking for is the probability that there is at least one sensor s with $\|\mathbf{p}_s - \mathbf{q}\|_2$ being smaller than the common sensing radius r. Consider the situation shown in Figure 13.1, where a number of sensors and a selected point $\mathbf{q}$ are shown. This point is covered when there is at least one sensor present in the circle A of radius r around $\mathbf{q}$. This circle has area πr^2 and the probability to find at least one sensor in it is:

$$f_a = \Pr[N(A) \geq 1] = 1 - \Pr[N(A) = 0] = 1 - \mathrm{e}^{-\lambda \pi r^2}$$

To satisfy a prescribed area coverage f_a, this equation can be solved to determine the required intensity λ of the Poisson point process:

$$\lambda(f_a) = -\frac{\log(1 - f_a)}{\pi r^2}$$

As a numerical example, let us assume that $r = 1$ m and the desired coverage is $f_a = 0.99$. In this case, a sensor intensity of $\lambda \approx 1.47$ sensors per m^2 is needed. To achieve an even better coverage of $f_a = 0.999$, this number grows to $\lambda \approx 2.2$.

The node coverage f_n has been obtained in [510] through simulation. In Figure 13.2, area coverage and node coverage are compared for a large sensor network with varying Poisson process intensity λ. The important point to observe here is that area coverage increases faster than node coverage; from the figure, to achieve a node coverage of 50 %, the area coverage must have reached a value of ≈95 %; another data point is that with an area coverage of 47 % a node coverage of only ≈1 % is reached. An important conclusion is that simply achieving a desired area coverage is not sufficient when energy and lifetime constraints come into play, but the network must be significantly overprovisioned to achieve a useful node coverage.

The detectability along a path through the sensor network depends on the knowledge that the intruder has about the sensor network. When the intruder does not know the sensor locations and

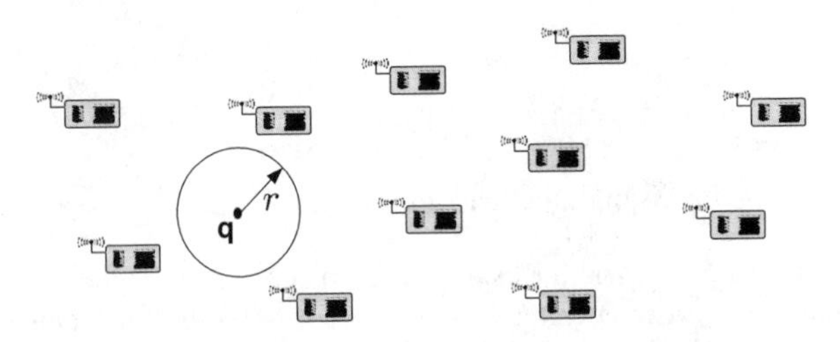

Figure 13.1 Determining area coverage f_a for Poisson point process

[6] Here is a small subtlety, since the area coverage can be understood in two ways. The first way is: We are given a fixed realization of a Poisson point process and pick a random point whose coverage probability we are interested in. The second way is: Given a fixed point in the area, in which fraction of the realizations of a Poisson point process is this point covered?

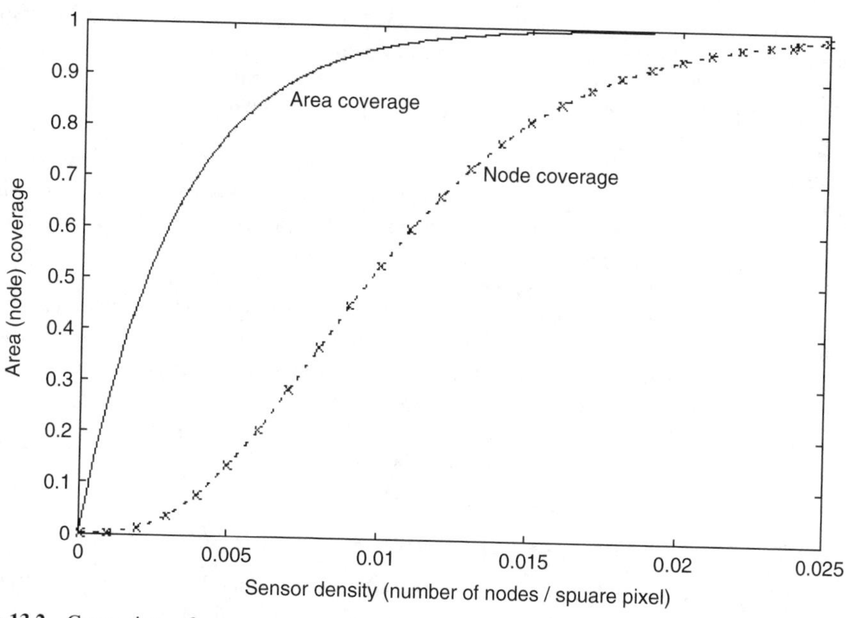

Figure 13.2 Comparison of area coverage and node coverage for varying intensity λ. Reproduced from [510, Fig. 1] by permission of IEEE

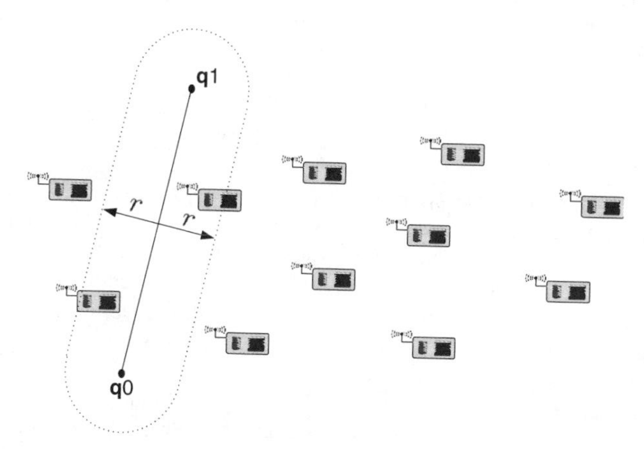

Figure 13.3 Minimum detectability path through a network for unknown node locations

chooses some arbitrary path between $\mathbf{q}_0$ and $\mathbf{q}_1$, it can be detected when there is at least one sensor in the tube of half width r around the path; compare Figure 13.3. The area of this tube can be minimized by choosing the most direct path between $\mathbf{q}_0$ and $\mathbf{q}_1$, that is, a straight line. Since the area of this tube is given by $\mu(A) = 2r\,\|\mathbf{q}_0 - \mathbf{q}_1\|_2 + \pi r^2$, the probability of finding at least one sensor in this area, and thus to detect the intruder, can be computed as:

$$P_d = \Pr[N(A) \geq 1] = 1 - \Pr[N(A) = 0] = 1 - \mathrm{e}^{-\lambda 2r\|\mathbf{q}_0 - \mathbf{q}_1\|_2 + \pi r^2}$$

Clearly, the larger the distance between $\mathbf{q}_0$ and $\mathbf{q}_1$, the more likely a detection becomes.

The second interesting case is when the intruder has perfect knowledge of the network - specifically, the positions of all sensor nodes – and can choose optimal paths. It is shown in [510] that for asymptotically large distances between $\mathbf{q}_0$ and $\mathbf{q}_1$ there exists a critical intensity λ_c such that for $\lambda < \lambda_c$ there exists almost surely a path which the intruder can use without being detected, whereas for $\lambda \geq \lambda_c$ the intruder is almost always detected.[7] This behavior can be attributed to the creation of **clusters**. Here, a cluster is a maximal set of sensors whose coverage regions form a connected area in the plane. It can be shown that for $\lambda < \lambda_c$ almost surely all clusters are of finite size, whereas for $\lambda \geq \lambda_c$ there exists almost surely at least one cluster of infinite size, which an intruder cannot cross without being detected. This critical intensity depends on the sensing range of the sensors.

LIU and TOWSLEY [510] discuss these three coverage measures also for the case of a strip-shaped sensor field having infinite extension in one direction but a finite width h. The most surprising result is that an intruder knowing the sensor positions can with probability one pass undetected from one side of the strip to the other. The reason for this is as follows: when the node positions within the strip are projected onto one of the strips side lines, the resulting node placement process is one-dimensional Poisson point process of finite density. With probability one, there are gaps between the projected sensing ranges of the nodes, which the intruder can use.

13.2.5 Coverage of random deployments: general sensing model

LIU and TOWSLEY [510] also discuss coverage measures for the generalized sensing model, that is, where the sensor output depends according to a power law on the distance between intruder and sensor (compare Equation 13.2).

Consider the situation where the sensors are placed according to a Poisson point process and the intruder has chosen some point $\mathbf{q}$. The sensor field intensity at $\mathbf{q}$ is given by (compare Equation 13.3):

$$C(\mathbf{q}) = \sum_{s \in \mathcal{S}} S(\mathbf{p}_s, \mathbf{q}) = \sum_{s \in \mathcal{S}} \frac{\alpha}{\|\mathbf{p} - \mathbf{q}\|_2^{\beta}},$$

not considering any bound r in the moment. $C(\mathbf{q})$ is actually a *random variable*, which depends on the sensor locations. A random variable of this kind is also known as **Poisson shot noise** (see, for example, [615, Sec. 10.2] and [518]) and the question of area coverage is closely related to the probability distribution function $F(\cdot)$ of $C(\mathbf{q})$.[8] In fact, since we denote a point $\mathbf{q}$ as covered if $C(\mathbf{q}) \geq \theta$ holds for some given threshold θ, $\mathbf{q}$ is covered with probability $1 - F(\theta)$. Closed-form expressions for the distribution function for the case of one and two dimensions are presented in [518] and [510], respectively. For the special case, of $\alpha = 1$ and $\beta = 4$ and for two different values of λ, the area coverage is plotted against the threshold value in Figure 13.4. Clearly, given a fixed threshold value θ, the higher area coverage can be achieved with a higher node density λ.

It is a somewhat surprising finding that under the power-law sensing model the node coverage is zero, that is it is not possible to remove or turn off a sensor without reducing the area coverage.[9] With respect to detectability, it is demonstrated in [510] that there exists a threshold density λ_c, such that an intruder wishing to move from $\mathbf{q}_0$ to $\mathbf{q}_1$ is almost surely detected when the distance between $\mathbf{q}_0$ and $\mathbf{q}_1$ is large and $\lambda > \lambda_c$ holds. This is shown by turning the general sensing model into a Boolean one, such that the coverage area of the Boolean sensing model is a subset of the

[7] Such critical densities or critical thresholds occur in many places in wireless networks, see for example, [443], [442].

[8] This distribution function is the same for all $\mathbf{q}$ because of the homogeneity of the Poisson point process.

[9] Essentially, it can be shown that there exist points $\mathbf{q}$ in the network with $C(\mathbf{q}) < \theta$, and when there are other points with $C(\mathbf{q}) > \theta$, there must be, according to the mean value theorem for continuous functions (and $C(\cdot)$ is easily seen to be continuous for a fixed deployment), points with $C(\mathbf{q}) = \theta$. Removing a positive term from $C(\cdot)$ by switching off a sensor turns a (covered) point $\mathbf{q}$ with $C(\mathbf{q}) = \theta$ into an uncovered point with $C(\mathbf{q}) < \theta$.

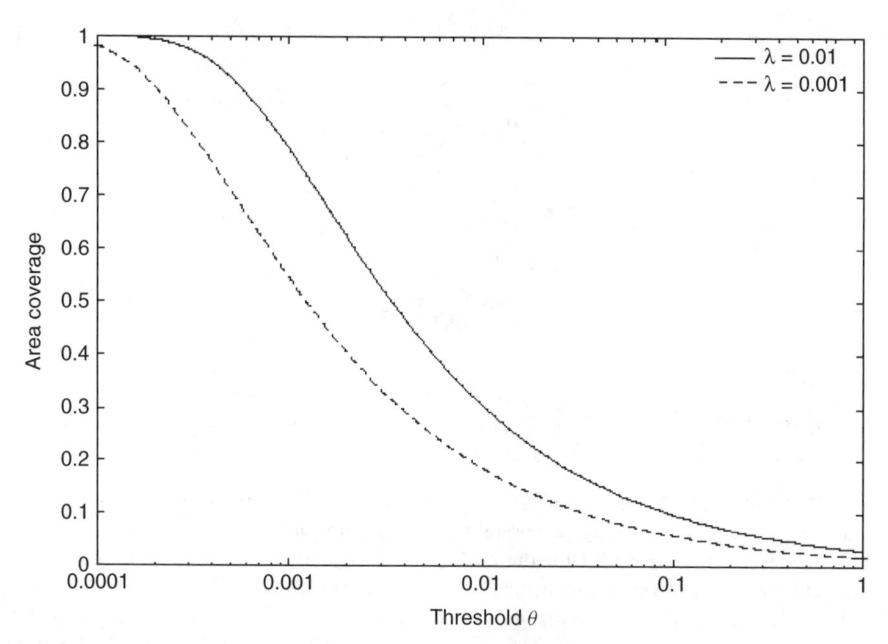

Figure 13.4 Area coverage versus threshold value θ with the general sensing model for two different λ, $\alpha = 1$ and $\beta = 4$ (from: [510, Fig. 6])

area covered under the general sensing model. Since there exists a threshold density λ_c under the Boolean sensing model such that for $\lambda > \lambda_c$ detection is almost sure, this must carry over to the general sensing model with its even larger area covered. It is not yet clear, however, what happens under the general sensing model when λ is below the threshold λ_c.

13.2.6 Coverage determination

The results discussed in the previous sections were theoretical in that coverage measures have been derived under given knowledge of the stochastic process governing node deployment.

In many practical situations, however, the actual deployment's underlying stochastic process as well as its parameters are unknown.[10] Instead, one is often faced with an arbitrary deployment and it is an important task now to judge the actual coverage measures. If protocols and algorithms are available to accomplish this, areas with reduced coverage can be identified and additional sensors can be deployed there.

Determining *k*-coverage

In [364, 365], a distributed protocol is proposed that checks whether a certain area is k-covered under the Boolean sensing model. Using the terminology introduced in Sections 13.2.1 and 13.2.2, this amounts to the question whether the sensor field intensity at all points is at least k times the fixed value α from Equation 13.1. The proposed protocol works for both uniform sensing

[10] It will often be a purely practical problem to achieve a deployment conforming to a Poisson point process with some prescribed density λ when the area to be covered is large. Just imagine sitting in a plane with a pile of sensors to be distributed over a forest: which flight path shall the pilot follow and when do you have to spread out sensors?

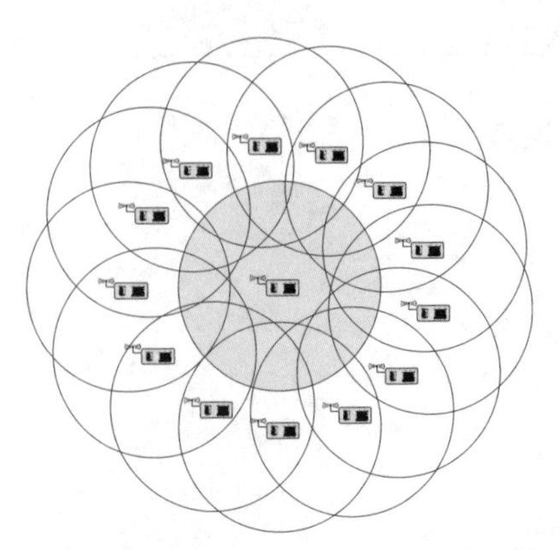

Figure 13.5 Example setup for perimeter coverage. Reproduced by permission of IEEE

ranges – all sensors have the same sensing range – as well as for nonuniform ones. Its key feature is that it avoids to check k-coverage separately for each point of the observed area.

We first discuss the general idea for the uniform case. Let us assume that a number of sensors $s_1, \ldots, s_n$ are spread over a two-dimensional area A. All sensors have the common sensing range r. Consider a fixed sensor s_i. A point $\mathbf{q}_i$ on the perimeter of s_i's sensing range is said to be **perimeter-covered** by s_j if $\mathbf{q}_i$ is within s_j's sensing range. The sensor s_i is said to be k-perimeter-covered when all points $\mathbf{q}_i$ on the perimeter of s_i are in the range of at least k other sensors, not including s_i itself. Consider for example the setup shown in Figure 13.5. Sensors are marked by black bullets, their sensing ranges as circles around them. One can see that all points on the perimeter of the highlighted sensor s_i in the center are covered by at least two other sensors. It is also important to note that there are points in the interior of s_i's sensing range that are only covered by s_i itself and which would not be covered at all when s_i fails. The key observation is that it can be shown that under mild assumptions the whole area is k-covered if and only if all sensors are k-perimeter-covered.[11] To turn this observation into a protocol, each sensor s_i has to check the coverage of its perimeter, which can be done locally. The word "locally" is defined here with respect to the sensing radius: it suffices for a node s_i to consider all neighbors s_j for which the distance $\left\|\mathbf{p}_{s_i} - \mathbf{p}_{s_j}\right\|_2$ is at most $2r$.[12]

Node s_i can determine the perimeter segment covered by another node s_j (more specifically: the angles $\phi_{j,1}$, $\phi_{j,2}$ in radians, which bound the covered perimeter segment) with the help of a simple geometric computation requiring only the node's positions; compare Figure 13.6. By collecting the perimeter angles covered by all neighbors, the node gets a picture of overall perimeter coverage, visualized in Figure 13.7. A simple way to determine whether node s_i is k-perimeter-covered is to go through this list, starting at an angle of 0 and ending with angle 2π. While traversing this list, a counter variable is incremented each time a segment starts and decremented when a segment ends. Clearly, it suffices to restrict this procedure to the angles 0, 2π, and all the particular angles $\phi_{j,1}$ and $\phi_{j,2}$ obtained from the previous step. When the counter always has a value of at least k, then indeed node s_i is k-perimeter-covered. For a static network, this computation has to be executed only occasionally to accommodate new nodes or dying nodes.

When a user at a sink node wants to check for k-coverage in a sensor network, an appropriate request can be flooded into the network, causing each node s_i to determine its perimeter coverage.

[11] For sensors, at the boundary of the target area it is impossible to have *all* points on their perimeter be covered by other sensors. It is reasonable to reduce to perimeter points *inside* the target area.

[12] Clearly, if r is in the same order of magnitude as the overall network diameter, the notion "local" loses a bit of its meaning.

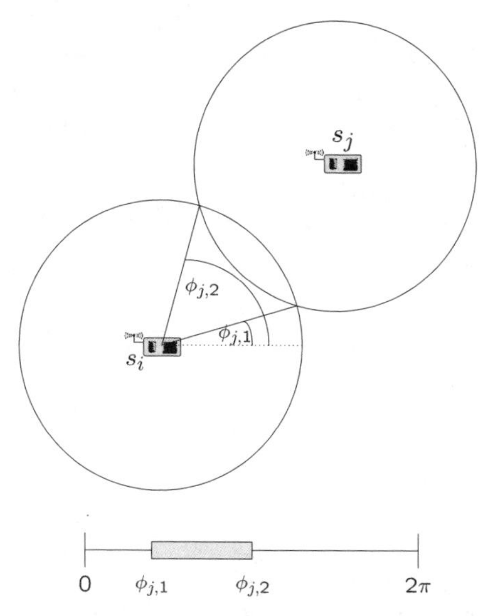

Figure 13.6 Example perimeter segment (adapted from: [364, Fig. 2a])

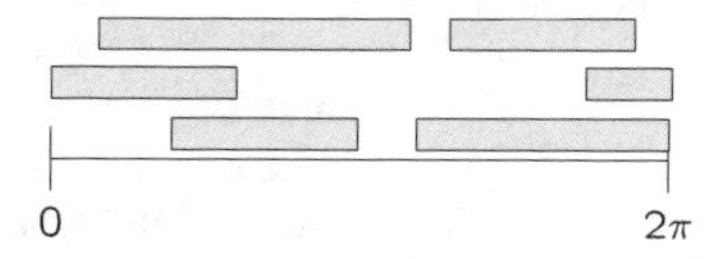

Figure 13.7 Overlapping multiple perimeter segments (adapted from: [364, Fig. 2b])

When s_i has perimeter coverage smaller than k, it can generate an appropriate report, for example, including details about the insufficiently covered segments. This information can guide placement of additional sensors.

It is shown in [364, 365] that this algorithm can also be used for the case of nonuniform or even noncircular sensing ranges; only the computation of node s_i's perimeter segment covered by another node s_j becomes slightly more complicated. A generalization to three dimensions is presented in [366].

Determining worst-case coverage

MEGUERDICHIAN et al. [546] investigate determination of the actual worst-case coverage for a random sensor field under the fairly general assumption that the sensing quality decreases with distance. Please remember that the worst-case coverage problem asks for a path through a network that an intruder would take to minimize risk of detection, that is, a path having the largest possible distance to the sensors. In [546], a centralized algorithm with polynomial time complexity is presented for this problem. The algorithm assumes availability of perfect location information. It is based on the construction of a Voronoi diagram (see Section 10.2.3, Section 11.5.2, and [29]) of the sensor network. In the two-dimensional case, such a diagram partitions the plane into a number of convex polygons such that: (i) exactly one sensor is contained in each polygon, and (ii) this sensor is the closest sensor to all other points lying truly within the respective polygon. The points on the

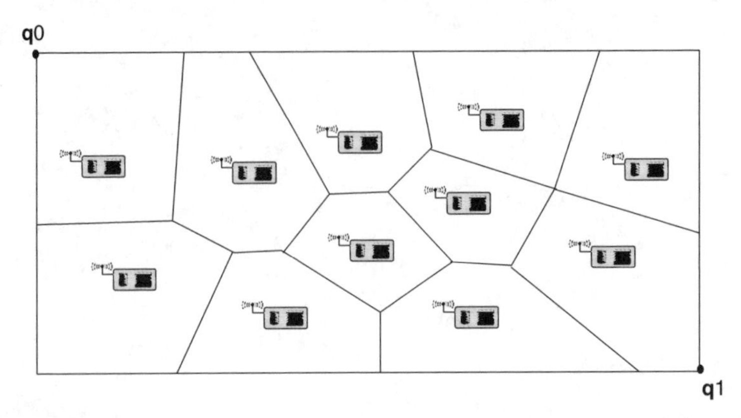

Figure 13.8 Example Voronoi diagram for the sensor network deployment of Figure 13.3

edges of a polygon have the same distance to the two neighboring sensors, except points on the boundary of the target region. Please refer to Figure 13.8 for an example of a Voronoi diagram. It is intuitively clear that an intruder wishing to minimize its visibility to sensors should choose a path exactly along the edges of the Voronoi polygons. In [546], an algorithm for finding a worst-case coverage path between two points $\mathbf{q}_0$ and $\mathbf{q}_1$ inside or outside the sensor field is devised, displayed in Listing 13.1. Essentially, this algorithm proceeds in three steps. The first is the construction of the (bounded) Voronoi diagram, that is, of a graph (U, L) whose vertex set U contains just the end points of the edges of the Voronoi polygons, and L is the set of edges. This graph is turned into a weighted graph (U, L'). The weight of each edge is the minimum distance of all points on this edge to the neighboring sensors. Be m the minimum weight and M is the maximum weight of the edges in L'. For a certain weight $w \in [m, M]$, a graph (U, L_w) is constructed containing only those edges from L' with weights larger than w. A breadth-first search is applied to (U, L_w) to check whether $\mathbf{q}_1$ can be reached from $\mathbf{q}_0$ using edges of weight larger than w. The maximum w for which such a path exists is obtained with a binary-search procedure. The path corresponding having this maximal value is the maximal breach path.

MEGUERDICHIAN et al. [546] investigate by simulation properties of the worst-case coverage/ maximum breach path for varying numbers of sensor nodes. Specifically, the following question is addressed: Given a certain sensor network one can determine the worst-case coverage w according to the previously discussed algorithm. Which improvement/reduction in worst-case coverage can be achieved by placing additional sensors at "good" locations? A good location would be an additional sensor along the edge of the found worst-case coverage path having the minimum weight. The reduction is determined by running the algorithm again after placing the additional sensor. In Figure 13.9, the average reduction after adding one, two, three, or four additional sensors is shown for varying initial sizes of the sensor network; each average is taken over 100 random deployments of the same initial number of nodes. It is an interesting result that already addition of only a single sensor can give significant improvements in the range between 10 and 25 %, whereas for further additional sensors the gains become smaller and smaller. In a second series of experiments, the worst-case coverage has been determined for random deployments of a variable number of sensors in unit area, but without opportunity to add further sensors at good positions. The results are shown in Figure 13.10. It can be seen that beyond a certain node density/number of nodes (here: around 100) any further increase in density gives less and less improvement in worst-case/breach coverage.

The determination of the best-case coverage path is based on Delaunay triangulations [496, 546], for which a distributed scheme has been presented in [496]. The reader should consult the given references for details.

Listing 13.1: Algorithm for finding the worst-case coverage path through a sensor network (paraphrased from [546])

```
// Initialize
Generate bounded Voronoi diagram V_0 = (U, L)
      // U = vertex set, L = edge set
      // this graph includes q_0 and q_1


// Create weighted graph
Initialize graph V_1 = (U, ∅)
foreach edge l in L {
   assign min distance of l to sensor field as weight of l
   add l to the edge set of V_1
}
m := min{l.weight | l ∈ edges of V_1}
M := max{l.weight | l ∈ edges of V_1}


while (M − m ≥ ε) {

   w := (M+m)/2

   Initialize graph V_w = (U, ∅)
   add all edges l from V_1 to V_w for which l.weight > w

   if breadth-first-search in V_w from q_0 to q_1 succeeds
   then
      m := w
   else
      M := w
   end
   return w
}
```

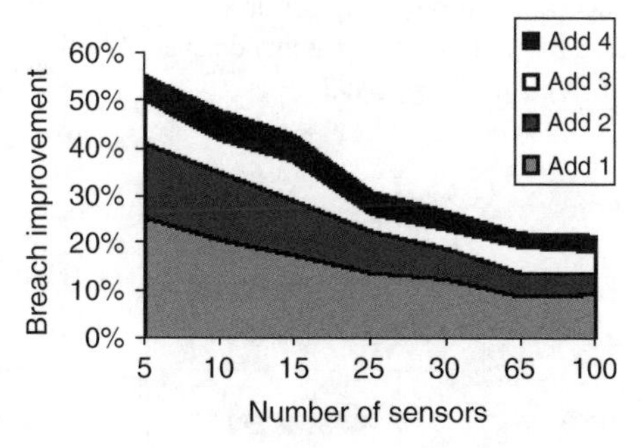

Figure 13.9 Effect of adding sensors on the worst-case coverage/maximal breach path. Reproduced from [546, Fig. 6] by permission of IEEE

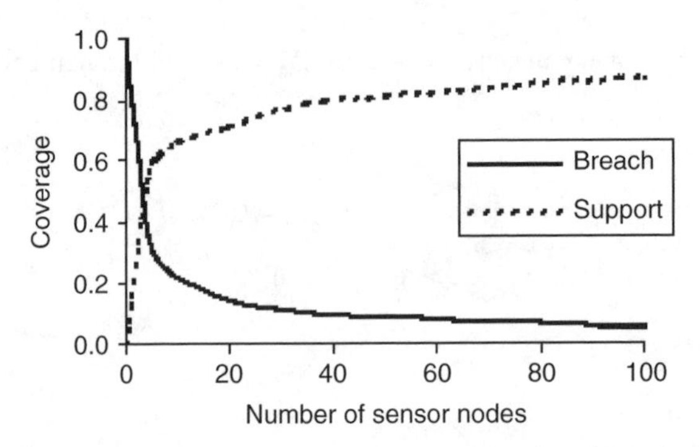

Figure 13.10 Worst-case coverage/maximal breach and best-coverage for random deployments in a unit area. Reproduced from [546, Fig. 8] by permission of the IEEE

13.2.7 Coverage of grid deployments

A few publications have also discussed coverage in grid deployments [509, 754]. In such a deployment, the whole sensor field is divided into an array of squares of side length D, and sensors are placed only at the centers of these squares. An example grid is shown in Figure 13.11.

LIU and TOWSLEY [509] consider the special case of sensors with homogeneous sensing range r under a Boolean sensing model. The sensor placement follows a spatial Bernoulli process, that is for each square center an independent Bernoulli experiment with success probability p tells whether this square is occupied by a sensor or not. Two cases have been considered in [509]; we start with the case $r < D/2$. In such a setup, the area coverage is clearly given by $f_a = \frac{p\pi r^2}{D^2}$, the node coverage is zero, and the probability of detection of an intelligent intruder is also zero, since he can always choose a path along edges of the square without ever being sensed. In the second case, it is assumed that $r = D/2$ holds. Here, area and node coverage remain the same, but the situation with respect to detectability in an infinite grid changes. Two squares are called *neighbors* when they are both occupied and have an edge in common. A group of occupied squares in which all squares are direct or indirect neighbors is called a *cluster*. It can now be shown that there exists a critical probability p_c, such that if $p > p_c$ there exists almost surely an unbounded cluster, which cannot be avoided by an intruder; accordingly, the detection probability is one. When the node density p is below p_c, all clusters are almost surely of finite size, and an intelligent intruder can find a way through the network, leading to a detection probability of zero. This "phase transition" behavior is similar to the case of random deployments discussed in Sections 13.2.4 and 13.2.5. The critical value p_c is given in [509] as $p_c = 0.5928$.

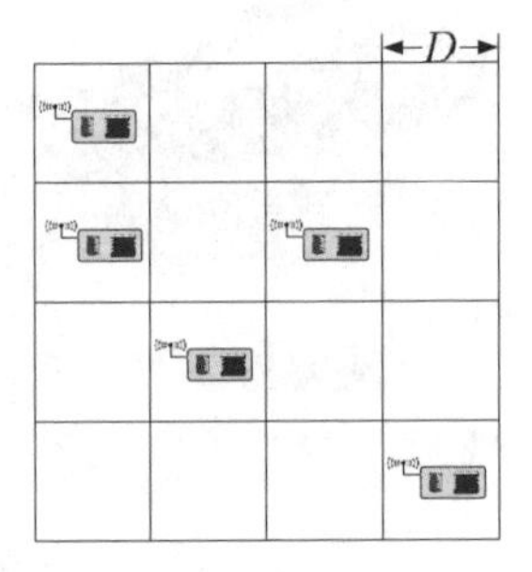

Figure 13.11 An example sensor grid

A different situation is considered in SHAKKOTAI et al. [754]. Here, the grid does not have infinite extension but is confined to a unit square. In this square, a number of n nodes are placed. Consequently, each square of the grid has a side length of $1/\sqrt{n}$. For simplicity, it is assumed that for each node the communication range and the sensing range (under the Boolean sensing model) are the same, being $r(n)$. Furthermore, all nodes have the same sensing range. Each node is active (i.e. not failed or sleeping) with probability $p(n)$, the nodes being independent of each other. Both full area coverage and connectivity are considered, with connectivity being defined as the ability of any active node to communicate directly or indirectly with any other active node. SHAKKOTAI et al. [754] have investigated asymptotic results, that is they consider large values for n. The following results are interesting:

- A necessary and sufficient condition for achieving both connectivity and full area coverage is that $p(n)$ and $r(n)$ obey

$$p(n) \cdot r^2(n) \sim \frac{\log n}{n}$$

  saying that for large n both a high degree of unreliability and small sensing radii can be used and still connectivity and coverage are maintained.[13]
- When $D(n)$ is the maximum number of hops required for communication between arbitrary active nodes, it can be shown that with probability one the following holds:

$$\sqrt{2} < r(n) \cdot D(n) < \frac{2}{1 - \frac{2}{\sqrt{\pi c}}},$$

  given that the relationship $p(n) \cdot r^2(n) \geq c \cdot \frac{\log n}{n}$ is satisfied for some $c > \frac{4}{\pi}$.
- It can be shown for small values of the success probability $p(n)$ that connectivity can be achieved without necessarily achieving full area coverage.

13.2.8 Further reading

- Closely related to the problem of coverage determination is the problem of local node density estimation. In [692], the **density inference protocol** (DIP) is presented, which estimates local density on the basis of observed collision rates at the MAC layer during dedicated time periods.
- The node coverage measure provides some indication of which fraction of nodes can be switched into sleep mode (thus saving energy) without compromising coverage. Knowing this number does not tell *which* sensors are good candidates for being switched off and how to coordinate multiple good candidates covering a given point so as not to switch off simultaneously. This requires local coordination. Some approaches and protocols for this are described in [816], [913], and [364].
- KOSKINEN [432] investigates the expected area coverage under a Boolean sensing model for a finite region, thus explicitly including boundary effects. Two scenarios are investigated. In the first one, all sensors as well as the point $\mathbf{q}$ of interest are confined to lie *inside* a disk. In the second scenario, the unit disk $\mathcal{D}$ is embedded into a larger region $\mathcal{D}'$ such that between each point in $\mathcal{D}$ and the boundary of $\mathcal{D}'$ there is a distance of at least r. The n sensors are spread in $\mathcal{D}'$ but the point of interest $\mathbf{q}$ is confined to $\mathcal{D}$. In the first scenario points $\mathbf{q}$ close to the disk boundary have a reduced coverage, as becomes apparent from Figure 13.12: For point $\mathbf{q}_1$, there is a full circle within which a sensor can detect the intruder at $\mathbf{q}_1$; however, at $\mathbf{q}_0$, the

[13] A function $f(\cdot)$ behaves asymptotically as a function $g(\cdot)$, written as $f(x) \sim g(x)$, if there exists some $c \in \mathbb{R}, c \neq 0$ such that $\lim_{x\to\infty} \frac{f(x)}{g(x)} = c$ holds (clearly, $g(x) \neq 0$ is required).

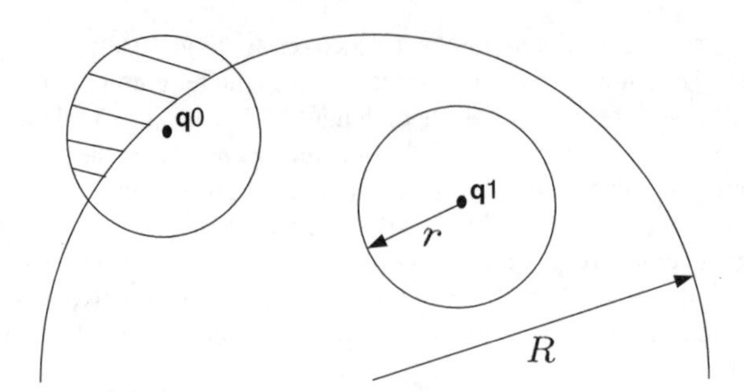

Figure 13.12 Boundary effects for area coverage

"detection area" is reduced. The expected area coverage can then be obtained by "averaging" over all possible $\mathbf{q}_0$, and in [432] a closed-form expression is given. One result presented in [432] is that in the second scenario slightly more nodes are needed to ensure an average area coverage of 99 %. Furthermore, KOSKINEN [432] investigates the required node density and/or sensing range r required to achieve full area coverage of some target domain.

- In PERILLO and HEINZELMAN [631], [632], the scheduling of node's sleep mode of nodes is jointly optimized with route selection between regions of interest and sink nodes to optimize sensor network lifetime while keeping a sufficient number of sensors awake to not compromise sensing quality. The problem is formulated as a generalized maximum flow problem with certain constraints, which in turn is solved as a linear programming problem in polynomial time.
- In a nutshell, the problem of deployment is concerned with finding the required number of sensors and their positions to fulfill some coverage goal, for example, to achieve full area coverage. This problem can be posed under several additional constraints like cost minimization, the need to cover some parts better than others, the availability of different types of sensors having different costs and sensing abilities, and so forth. Similar questions come up in the area of cellular network planning; see for example [565]. A related problem is known from the area of computational geometry; it is called the **art gallery problem** [609]. In this problem, an art gallery (for example, described by a polygon) has to be covered with the minimum number of sensors such that all points in the gallery are covered. This problem can be solved in two dimensions, but becomes NP-hard in three dimensions [365]. In CHAKRABARTY et al. [129], the deployment problem in three dimensions with two types of sensors (cheap and small range vs. costly and large range) is posed as a linear programming problem with the goal of minimizing overall costs. DHILLON et al. [204] consider a similar problem but assume imprecise sensors. With such sensors, an event at distance d from the sensor is not detected reliably, but only with a certain probability, which in general depends on d. They propose an algorithm for sensor placement that strives to achieve a given minimum detection probability for all points of a given area. Further references are [358], [945], and [173, 174].
- In BYERS and NASSER [112] it is proposed to schedule the activities (sensing, transmitting/receiving, aggregating) of a number of nodes such that a **utility function** is maximized. One part of this function captures the subset of sensors tasked with sensing and the achieved sensing quality. A utility function measures the utility to a user (for example, information quality) provided by different partitions of the set of sensors into the three different activities.

13.3 Reliable data transport

The problem of reliable data transport over wireless multihop networks like wireless sensor networks is not an easy one. There are three main sources of packet losses:

- The wireless channel can introduce (lots of) transmission errors, the packets of different nodes can collide, or nodes can lose packets because of other failures.
- Packets can be dropped in the network because of congestion, that is, overload of intermediate nodes.
- The receiver might drop packets because they arrive too quickly.

The focus of this section is on the first issue, that is, on error recovery. The second issue, congestion control, is discussed in Section 13.6. Solutions to the third problem are often called **flow control**; Section 6.1 explains why flow control is not of big importance in sensor networks.

In Section 13.3.1, we discuss the different reliability requirements encountered in the realm of sensor networks. Section 13.4 discusses the problem of reliable delivery of single packets, using techniques like acknowledgments, packet duplications, and so forth. Section 13.5 describes two block delivery protocols, namely PSFQ (Section 13.5.1) and RMST (Section 13.5.2). This section includes also a discussion of why or why not proven solutions from the IP world (most notably TCP) are useful in sensor networks (Section 13.5.3). Stream delivery is discussed in the context of congestion control (Section 13.6) because the main reliability-related control knob for stream delivery, the rate by which sensors generate packets, must be controlled to avoid congestion and adverse effects on the desired reliability.

Several physical layer mechanisms have been developed to increase transmission robustness, for example, FEC, choice of modulation schemes and transmit power, diversity mechanisms, and several more (see Chapter 4). With such mechanisms, a packet is more likely to make a single hop but they cannot eliminate all losses. The additional mechanisms needed to achieve end-to-end reliability are the focus of this chapter.

13.3.1 Reliability requirements in sensor networks

What are the requirements for reliable data transport in wireless sensor networks? A first glance toward this question can be gained by comparing sensor networks with other networks.

In traditional networks like the Internet, the transport protocols (TCP, UDP) and the underlying network layer protocols have essentially no clue which kind of data they transport. In fact, a key design requirement for these protocols is **data transparency**. Such a protocol must strive to deliver every single bit to the receiver(s), since nothing is known about the relative importance of the different data bits. On the other hand, sensor networks are *not* designed with the goal of transporting multiple independent data streams. Sensor networks are data-centric and rely on in-network processing. The reliability requirements are pretty much application specific and the protocols can take advantage of this; they know the data they carry.

Several data transport tasks for wireless sensor networks have been discussed in the literature. These can be roughly classified into the following orthogonal axes:

Single packet versus block versus stream delivery: The cases of delivering only a single packet on the one hand and of delivering a number or even an infinite stream of packets on the other hand differ substantially in the protocol mechanisms usable in either case. In the **single packet delivery** problem, a single packet must be reliably transported between two nodes. It may be argued that such a requirement will not occur in dense wireless sensor networks where many nodes observe the same phenomenon and report highly correlated data. However, there are arguments against this claim. The first one is that not all sensor networks will be dense. Secondly, data aggregation is an important strategy in wireless sensor networks to condense many redundant or correlated measurements into a small piece of data. Aggregation drastically reduces the amount of data that must be transmitted to distant sink nodes, but on the other hand, the reliability requirements for the summary packet are much higher than for any of the individual sensor readings. In the **block delivery** problem, a finite data block comprising multiple packets must be delivered to a sensor or a set of sensors. Some application examples for this are the retasking of a sensor network (i.e. the distribution of new application code to a set of sensor nodes) or the injection of user queries [849]. Finally, in the **stream delivery** problem, a theoretically unbounded number of packets has to be transported between two nodes. An example for this is periodic measurement reports.

Sink-to-sensors versus sensors-to-sink versus local sensor-to-sensor: It can be assumed that most communications in sensor networks are not between arbitrary peer nodes, but information flows either from sensor nodes toward a single or a few sink/gateway nodes or in the opposite direction, from sinks to sensors. In the latter case, the groups of sensors can be geographically specified ("all sensors in the conference room") or by other attributes ("all temperature sensors with less than 50 % battery capacity"). Other applications like, for example, target tracking (see Section 14.3.1 require reliable handover of target state (e.g. estimated position, speed) between neighboring nodes close to the target's trajectory.

Guaranteed versus stochastic delivery: In the case of **guaranteed delivery**, it is expected that *all* transmitted packets reach the destination; anything else is considered a failure. For example, when a block of application code is distributed to a set of sensors, losing any packet renders the code block useless. In general, guaranteed delivery is challenging and costly in terms of energy and bandwidth expenditure, specifically over links with sometimes high error rates like wireless ones. Furthermore, many applications can live with some losses, provided that there are not too many of them. The concept of **stochastic delivery** guarantees allows a limited amount of losses. There are several ways to specify stochastic guarantees. For example, one might specify that for periodic data delivery within every k subsequent packets at least m packets must reach the destination; any number below m is considered a failure. Such a specification has similarities to the concept of (m, k)-firm deadlines [328, 862]. It is applicable when subsequent sensor readings are highly correlated, for example, because a slowly varying physical process is monitored. A similar specification requires that at least m out of k sensors detecting an event must deliver a packet to the sink or to a local clusterhead for proper detection/aggregation. A third approach to deal with stochastic guarantees is to specify a **delivery probability** simply as the long-term fraction of arriving packets. In general, the higher the desired delivery probability, the higher are the energy costs needed to achieve this.

As the reliability requirements are application dependent, several works have focused on developing transport protocols for "single points" or "small point sets" in the "space" spanned by the previous three axes. Generally applicable transport protocols that are lightweight enough to run on constrained sensor nodes seem not to have appeared yet. Even in the Internet world, there is no single protocol: TCP is used for unicast applications, whereas for reliable multicast, other protocols like, for example, Scalable Reliable Multicast (SRM) [263] have been developed.

However, even for more specialized protocols, it is a design challenge to achieve a small footprint in terms of code size, size of runtime data, and computational complexity.

13.4 Single packet delivery

The problem is to deliver a single packet from a source to a sink node over multiple hops. As pointed out in Chapter 6 for the single-hop case, there is a trade-off between achievable reliability and energy costs. It is thus appropriate to consider stochastic guarantees and to measure reliability in this case as a **packet delivery probability**.

The following discussion is structured according to the number of assumed network paths between source and sink. In **multipath solutions**, the presence of multiple paths is taken for granted and can be exploited. In **single-path solutions**, this assumption is not made.

13.4.1 Using a single path

Taking all the single-hop physical-layer mechanisms like FEC or transmit power variation for granted, the prime mechanism to improve packet delivery probability are retransmissions and usage of multiple packets.

In a retransmission scheme, in general three issues have to be resolved: (i) Who detects losses and what are the indicators used? (ii) Who requests retransmissions? and (iii) Who carries out these retransmissions? In single packet delivery, the data packet can get lost. Only the transmitting node has a chance to detect this and the canonical way is to use timeouts for acknowledgment packets. It is also the transmitting node who requests and performs retransmissions. To convince the transmitter about successful packet delivery, the receiver has to send a **positive acknowledgment**, that is the receiver has to confirm that indeed a packet has been received.

There is more flexibility in case of block or stream delivery. For example, it is possible to let the receiver detect losses (e.g. by checking for holes in the received sequence numbers) and request retransmission of missing packets by using **negative acknowledgment** (NACK) packets. A NACK indicates packets that failed for some reason. If additionally NACKs carry **implicit acknowledgments** of other packets, then there is no necessity to send positive acknowledgments for every packet, thus saving lots of energy.

For single-hop delivery in wireless multihop networks, two standard approaches using positive acknowledgments are the following:

- MAC-layer retransmissions: when a node on the path forwards the data packet, it expects to receive a MAC-layer acknowledgment. Typically, the transmitter makes a bounded number of trials to successfully forward the packet and drops it after this number has been exhausted. However, for small data packets, the acknowledgments create significant overhead, which is invariably expended even on exceptionally good channels.
- End-to-end retransmissions: the source node needs to buffer the packet until an acknowledgment from the sink node arrives. Again, the number of trials made by the source node is typically bounded. End-to-end retransmissions can be combined with MAC-layer retransmissions.

Hence, one of the first questions is whether to rely entirely on end-to-end acknowledgments or to additionally use MAC-layer/link-layer acknowledgments and retransmissions. We discuss this choice with an example.

Example 13.1 (Is it efficient to use only end-to-end acknowledgments?) Let us consider the situation where a single data packet of length l_d bits has to be transmitted from a sensor node (source) over a fixed path to some sink node n hops away. The sink node always generates an acknowledgment packet of size l_a bits called *sink acknowledgment*, which travels back to the source. When the source does not receive the sink acknowledgment within a certain time, it retransmits the data packet. The number of retransmissions is not bounded, since in this example we are interested in the energy needed until the packet is eventually delivered.

We investigate two cases. In the first case, all packets (data or acknowledgment packets) are transmitted without any per-hop ARQ protocol, that is there are no MAC-layer acknowledgments. In the second case, the MAC layer or link layer performs a bounded number k of trials to deliver a packet. The required MAC-layer acknowledgments have a size of l_m bits. We do not consider the problem of correct timer settings in this example.

To keep things easy, the error behavior of all the wireless channels is modeled as a simple Binary Symmetric Channel (BSC) with bit-error probability p (see Section 4.2.4). For a BSC, the probability that a packet of l bits length can be transmitted successfully is given by $P(l) = (1 - p)^l$. The expected number of trials a packet needs to pass one hop is given by $\frac{1}{P(l)}$. Similar to the Example 6.3, we model the energy costs for transmitting/receiving a packet of size l bits as:

$$e_t(l) = e_{t,0} + e_t \cdot l \qquad e_r(l) = e_{r,0} + e_r \cdot l,$$

where $e_{t,0}$ and $e_{r,0}$ are fixed energy costs spent each time transmitting/receiving a packet. Parameters e_t, e_r are the energy spent on transmitting/receiving a single bit. We assume that the costs for receiving a packet are incurred independent of whether reception is successful or not.

In the pure end-to-end case without MAC-layer acknowledgments, a successful trial requires that the data packet reaches the sink (taking all n hops successfully) and the sink acknowledgment reaches the source afterward, again over n hops. The overall energy costs are composed of two components: a number of failed trials and a final successful trial. The probability of a successful trial is $P(l_d)^n \cdot P(l_a)^n$ and the associated energy costs are $n(e_t(l_d) + e_r(l_d)) + n(e_t(l_a) + e_r(l_a))$. The probabilities of the different failure outcomes and their associated energy costs are displayed in Table 13.1; it is straightforward to compute the average costs of a failed trial from this. Taking all this together, one immediately obtains the overall expected energy $E_{e,-}$ needed to eventually deliver the packet and to get the sink acknowledgment back to the source.

The analysis for the case where the MAC layer is allowed to make a number k of trials to deliver a packet is slightly more complex, since different per-trial and per-hop outcomes must be considered for an accurate determination of the average energy costs. For a single trial to transmit a packet of size l bits on the MAC layer, three cases can occur:

- The packet does not make it and consequently the receiver does not receive anything; the probability for this is $1 - P(l)$ and the costs are $e_t(l) + e_r(l)$;
- The packet is successfully received, the receiver generates a MAC-layer ack but this ack does not reach the original transmitter; the probability for this is $P(l) \cdot (1 - P(l_m))$ and the cost is $e_t(l) + e_r(l) + e_t(l_m) + e_r(l_m)$.

Table 13.1 Outcomes and energy costs for the pure end-to-end case

Outcome	Probability	Energy costs
data makes 0 hops	$1 - P(l_d)$	$e_t(l_d) + e_r(d)$
data: 1 hop	$P(l_d) \cdot (1 - P(l_d))$	$2(e_t(l_d) + e_r(l_d))$
...	...	...
data: $n - 1$ hops	$P(l_d)^{n-1} \cdot (1 - P(l_d))$	$n(e_t(l_d) + e_r(l_d))$
data: n hops, ack: 0 hops	$P(l_d)^n \cdot (1 - P(l_a))$	$n(e_t(l_d) + e_r(l_d)) + e_t(l_a) + e_r(l_a)$
data: n hops, ack: 1 hops	$P(l_d)^n \cdot P(l_a) \cdot (1 - P(l_a))$	$n(e_t(l_d) + e_r(l_d)) + 2(e_t(l_a) + e_r(l_a))$
...	...	...
data: n hops, ack: $n - 1$ hops	$P(l_d)^n \cdot P(l_a)^{n-1} \cdot (1 - P(l_a))$	$n(e_t(l_d) + e_r(l_d)) + n(e_t(l_a) + e_r(l_a))$
data: n hops, ack: n hops	$1 - \sum \text{other} = P(l_d)^n \cdot P(l_a)^n$	$n(e_t(l_d) + e_r(l_d)) + n(e_t(l_a) + e_r(l_a))$

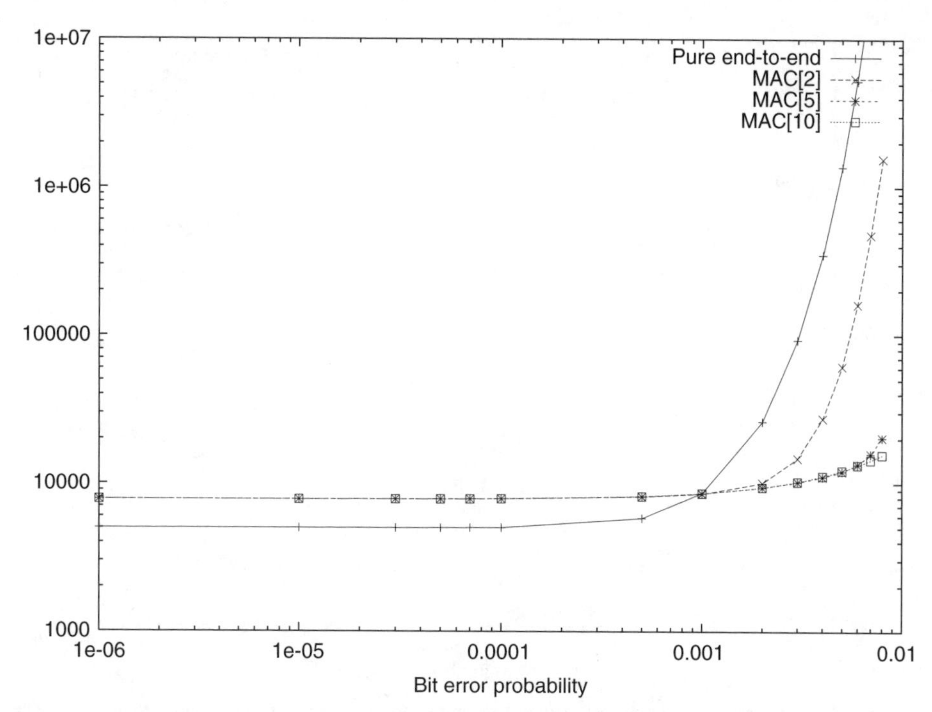

Figure 13.13 Comparing expected costs (in energy units) for pure end-to-end acknowledgments versus end-to-end acknowledgments plus k trials at the MAC layer for varying bit-error rate p and $n = 10$ hops

- Both the packet and MAC-layer acknowledgment are successfully transmitted; this happens with probability $1 - (1 - P(l)) - (P(l) \cdot (1 - P(l_m))) = P(l) \cdot P(l_m)$ and again the costs are $e_t(l) + e_r(l) + e_t(l_m) + e_r(l_m)$.

By extending this analysis to take k trials per hop into account, one can compute the overall expected energy $E_{e,k}$ needed to eventually deliver the data packet and get the sink acknowledgment back to the source node.

In Figure 13.13, we compare the average costs $E_{e,-}$ for the case of end-to-end acknowledgments without MAC-layer acks with the average costs $E_{e,k}$ for the case including k MAC-layer trials. The parameter varied is the bit-error rate p. The assumed parameter settings are $e_t = e_r = 1$, $e_{t,0} = e_{r,0} = 50$ of energy units, a data packet length of $l_d = 100$ bits, a sink acknowledgment length of $l_a = 50$ bits, and a MAC-layer acknowledgment length of $l_m = 20$ bits. The number of hops is $n = 10$ and k has been chosen as $k \in \{2, 5, 10\}$. There are two regimes: For low bit error rates, the pure end-to-end scheme is more energy efficient since most packets arrive successfully and MAC-layer acknowledgments are wasted. On the other hand, when the bit-error rate is higher than a certain threshold, MAC-layer acknowledgments can keep the energy costs within reasonable bounds whereas the costs for the case without MAC-layer acknowledgments explode. It is beneficial to allow more trials k as the BER increases.

It is also instructive to compare both approaches for varying number of hops n while all other parameters including the bit-error rate p are kept fixed. In Figure 13.14, the average energy costs are shown for $p = 0.001$. Again, there is a threshold below which MAC-layer

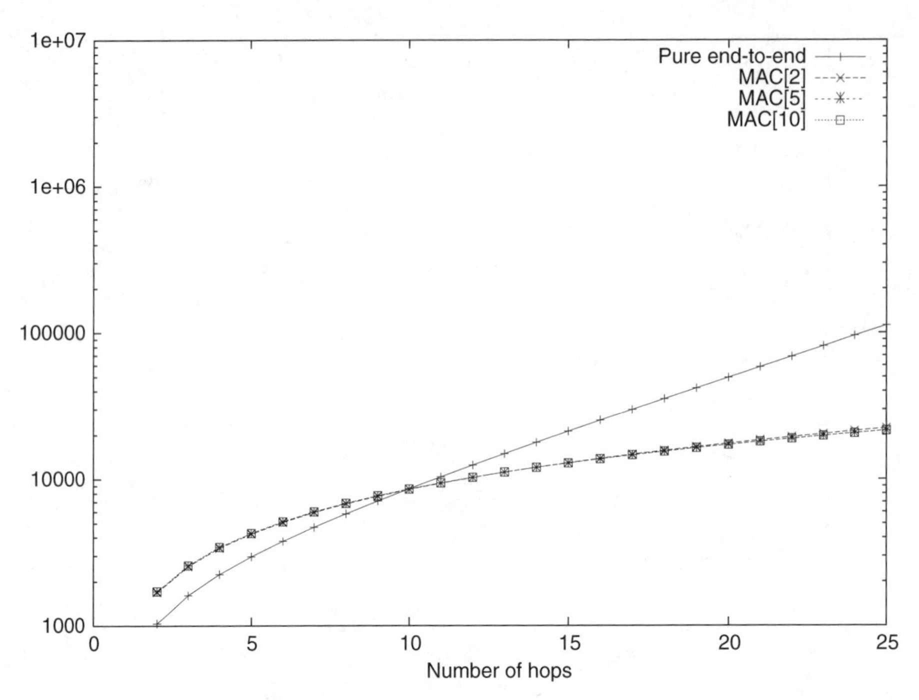

Figure 13.14 Comparing expected costs (in energy units) for pure end-to-end acknowledgments versus a combination of end-to-end and k MAC layer trials for fixed bit-error rate $p = 0.001$ and varying number n of hops

acknowledgments are a waste of energy, but for larger numbers of hops, they outperform the pure end-to-end case significantly. The energy costs for the pure end-to-end scheme increase *exponentially* with the number of hops. For bounded k, the scheme with MAC retransmissions has a similar behavior, but curves start to diverge at significantly higher BER values. It can be shown that for unbounded k the average energy costs are *linear* in the number of hops.

There is another advantage of using MAC-layer acknowledgments, not shown in the figures. Both approaches (with/without MAC) differ not only in the overall amount of energy spent but also in the distribution of energy expenditure over nodes. When MAC-layer acknowledgments without any restrictions regarding the number of retransmissions are used, each node spends on average the same amount of energy to forward the packet. In all other cases, the source node and likely the first hops are involved in *every* retransmission, even when the packet failed many hops away of them just before the sink. The distribution of energy expenditure over the nodes becomes more even as the number k of MAC trials increases.

One important issue has not been addressed in this example: the problem of setting timers. When positive acknowledgments are used, the transmitter has to start a timer for each packet. If the timer expires, the transmitter infers that the packet is lost and a retransmission is initiated. Setting timers for MAC-layer acknowledgments is fairly easy because the scope is only a single hop plus some small delays due to MAC and node processing. Setting timers for end-to-end transmissions over multiple hops is much more tricky. If no end-to-end round-trip time estimates are available from previous packets, the source node needs some global information to come up with a reasonable guess for the timeout value. At minimum, the source node needs to know the hop distance to the sink or some bound on the maximum number of hops in the overall network (the network diameter) as well as an estimate on the expected time needed to make one hop. However, these numbers are variable, although on different timescales. On longer timescales, the number of hops in a stationary network can change because of deployment of new nodes or death of existing nodes. On the other hand, the time needed to make one hop depends on the current state of the wireless channel, the allowed number of MAC retransmissions, the MAC protocol overheads, the node's buffer occupancy, the amount of local congestion, and so forth. In short, this delay can be highly variable. Another issue not to be underestimated in sensor networks is that with end-to-end acknowledgments the source node has to *buffer* all packets in transit for possible retransmission.

In DEB et al. [198], a method called Hop-by-Hop Reliability (HHR) has been proposed, which does not use MAC-layer acknowledgments but sends the same packet to the next forwarder (or **upstream node**) multiple times. The desired end-to-end delivery probability r is translated into a number of hop-by-hop delivery probabilities r_i, such that $\prod_{i=1}^{n} r_i = r$ holds. A node i knows its local packet error probability P_i toward the upstream neighbor j and chooses the number of packets N_i such that j receives at least one of the packets correctly with probability r_i. Under the assumption of a BSC, the probability that j receives at least one packet is given as $1 - P_i^{N_i}$. The number N_i can be computed from this by solving $r_i = 1 - P_i^{N_i}$ for N_i. A variation of the HHR approach is the Hop-by-Hop Reliability with Acknowledgments (HHRA) scheme, in which a node i sends up to N_i packets, but after each packet, it waits for an acknowledgment and aborts further transmission if an ack indeed arrives. For both protocols, the overhead, that is, the average overall number of packets needed to transmit a data packet over n hops, has been compared in [198] under the following assumptions: (i) acknowledgments are perfectly reliable and do not contribute to the packet count, and (ii) $r_i = r^{1/n}$, that is all hops require the same reliability. In Figure 13.15, the overhead is shown for $n = 10$, $r = 0.7$, and varying packet loss rate e common for all hops. There are some similarities to the results obtained in Example 13.1: for low error rates, MAC acknowledgments are wasted, and for higher error rates, they are beneficial.

It must be noted, though, that the idea of sending the same packet multiple times to the same upstream neighbor loses a bit of its charm when considering bursty channel errors. As long as a

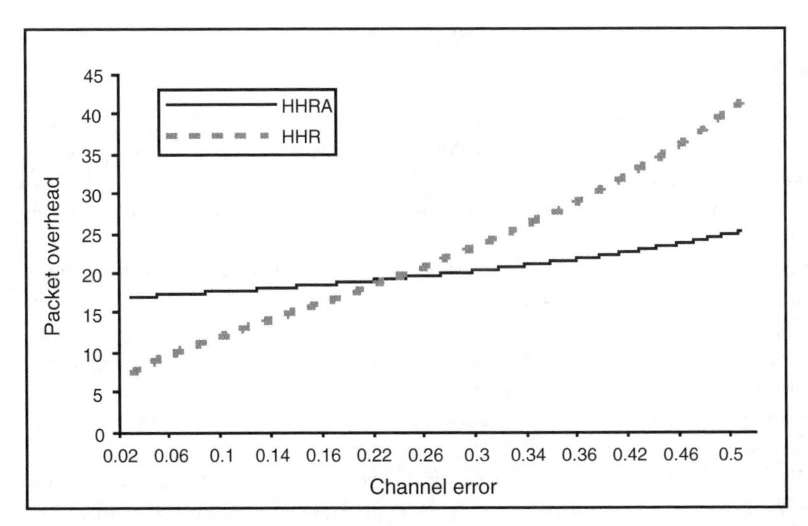

Figure 13.15 Overhead comparison of HHR and HHRA for $n = 10$, $r = 0.7$, and varying packet error rate e. Reproduced from [198, Fig. 3] by permission of ACM

channel is in a deep fade, all subsequent packets are likely erroneous and thus wasted. When delay is not an issue, postponing schemes can be interesting here (see Section 6.2.2).

13.4.2 Using multiple paths

The existence of multiple paths between a source and a sink node can be exploited in several ways.

Providing alternative routes

A first possibility is to set up multiple routes, to choose a preferred route from these, and to switch to another one when the preferred route fails [276]. These routes can be either pairwise node-disjoint and nonintersecting paths or the paths can be **braided paths**. In GANESAN et al. [276], localized protocols for setting up both types of paths are discussed and investigated for their **resilience** (the probability that there exists an alternative path when the current one fails) as well as their maintenance overhead. It is beneficial to choose "good" routes in the first place, that is, energetically feasible routes that have low per-hop error rates. The per-hop error rates can be estimated from physical-layer attributes like signal strength or from counting acknowledgments and retransmission from underlying MAC-layer and link-layer protocols [787]; see also Section 6.4.2. Instead of switching to another path in case of problems, it is also viable to *repair* a failed path by rerouting it locally around failed hops [817].

Instead of transmitting a single packet over *one* of these paths, it is also an option to send multiple packets over multiple paths, like, for example, the ReInForM approach discussed below. ReInForM sends multiple copies of the same packet. DULMANN et al. [214] present a variation of this approach that tries to reduce the overhead. They adopt the concept of parity packets as, for example, used in multimedia applications [119]. Specifically, the source node adds a number of redundancy bits to its data packet, and the resulting larger packet is fragmented into smaller ones. Each of these fragments is transmitted on a separate path. The coding scheme is chosen such that not all fragments must be present at the receiver to successfully decode the packet and some fragment losses are therefore tolerable.

ReInForM

The ReInForM approach developed by DEB et al. [199] is based on the idea of sending multiple copies of the same packet over multiple, randomly chosen routes. Packet duplication occurs not only at the source node but also on intermediate nodes. All nodes decide on the number of duplicates to be created on the basis of local error rates, the hop distance to the sink node, and the required reliability.

The protocol works as follows. Each node i knows its hop distance n_i to the sink and furthermore knows all its neighbors j and their respective hop distance n_j. Node i classifies its neighbors into three sets H_i^-, H_i^0 and H_i^+. The set H_i^- contains all neighbors j with $n_j = n_i - 1$, H_i^0 contains the neighbors j with $n_j = n_i$ and H_i^+ contains those for which $n_j = n_i + 1$ holds. Furthermore, node i possesses an estimate of its local packet error rate e_i.

The protocol operation starts at the source s and the goal is to deliver a packet with reliability r_s. To decide about the number of paths P, the source uses its locally estimated error rate e_s and the assumptions of a BSC and independent paths. The probability that a single packet fails over a path with n_s hops is given by $1 - (1 - e)^{n_s}$. If P packets are sent over P paths, the probability that none of these copies reaches the sink is $(1 - (1 - e)^{n_s})^P$ and thus the probability that at least one packet reaches the sink is $1 - (1 - (1 - e)^{n_s})^P$. This probability shall at minimum be r_s and the required number of paths P can be obtained as:

$$P = \frac{\log(1 - r_s)}{\log\left(1 - (1 - e)^{n_s}\right)}. \tag{13.4}$$

Which of s's neighbors are selected as starting points of these paths? First, the source selects a distinguished *next-hop* neighbor t from H_s^-; this is guaranteed to exist in a connected network. The next-hop neighbor forwards the packet unconditionally, all other neighbors do not. The probability that t really receives its copy and becomes a forwarder is just $1 - e_s$ and thus the number of paths to be distributed over the *remaining* neighbors of s is given as $P' = P - (1 - e_s)$. These paths are allocated to the nodes in H_s^-, H_s^0, and H_s^+ such that H_s^- has precedence over H_s^0, which in turn has precedence over H_s^+. The source computes the values P_s^-, P_s^0, and P_s^+ determining the number of paths that shall be created from *each* of s's neighbors in the sets H_s^-, H_s^0, and H_s^+, respectively. Finally, the source prepares a packet containing its hop distance n_s, its local error rate e_s, the next-hop neighbor t, and the three values P_s^-, P_s^0, and P_s^+. Finally, this packet is broadcast.

When the next-hop node t receives the packet, it accepts the role of a (new) source and behaves in almost exactly the same way the source has; t became a *forwarder*. Any other node u receiving the packet first determines the value $P_u \in \{P_s^-, P_s^-, P_s^-\}$ according to the class to which u belongs; for example, if $n_u = n_s - 1$, then $P_u = P_s^-$. Node u decides to become a forwarder when $P_u \geq 1$. If $P_u < 1$ node, u becomes a forwarder only with probability P_u. Before forwarding the packet, however, any forwarder u computes its local required reliability as

$$r_u = 1 - \left(1 - (1 - e_s)^{n_s - 1}\right)^{P_u}.$$

In Figure 13.16, the forwarding behavior of this algorithm in a random sensor network of 300 nodes spread out in a square area of 100 m side length and having a transmission range of 20 m is illustrated. The source is in the lower left corner, and the sink in the upper right corner of the field. The desired reliability r_s is 70 % and the packet error rate is uniformly 30 %. The upper half shows the paths taken for a *single* packet issued by the source, and the lower half displays the case of 10 packets. The following points are remarkable:

- The random choice of the next-hop neighbor provides some load balancing across nodes; different paths are chosen instead of always favoring a single path or a small number of paths.
- The algorithm has a tendency to create most of the duplicates close to the source; later on packet duplications are more rare. This is a direct result from Equation 13.4 because the number of hops n_i to consider becomes smaller.

In Figure 13.17, the achieved packet delivery probability (taken over 200 packets issued by the source) for ReInForM with 40 and 70 % reliability target is compared with two other schemes: (i) flooding, and (ii) sending only a single packet over a single path. The number of packets created in the network for the different schemes is displayed in Figure 13.18; both figures belong to the same experiment. The parameter varied is the (uniform) packet error rate. It can be observed that flooding is always a safe but expensive bet, the single packet approach breaks down quickly (as expected), and ReInForM is able to maintain the desired reliability target. It can also be seen that a higher reliability target or an increased packet error rate incurs a higher energy cost.

HHB and HHBA

In Deb et al. [198], two further protocols besides HHR and HHRA are presented, the Hop-by-Hop Broadcast (HHB) and Hop-by-Hop Broadcast with Acknowledgments (HHBA) protocols.

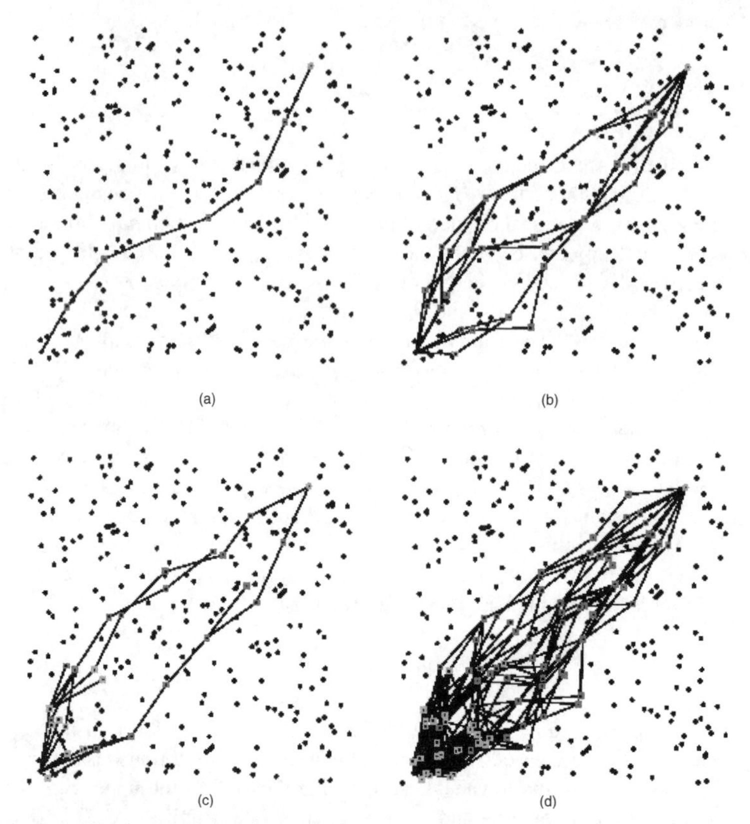

Figure 13.16 Illustration of ReInForM's forwarding behavior for target delivery probability of 70 %. Upper row: Single packet issued by the source with packet error rate of 0 % (left part) and 30 % (right part). Lower row: Source issues 10 packets . Reproduced from [199, Fig. 3] by permission of IEEE

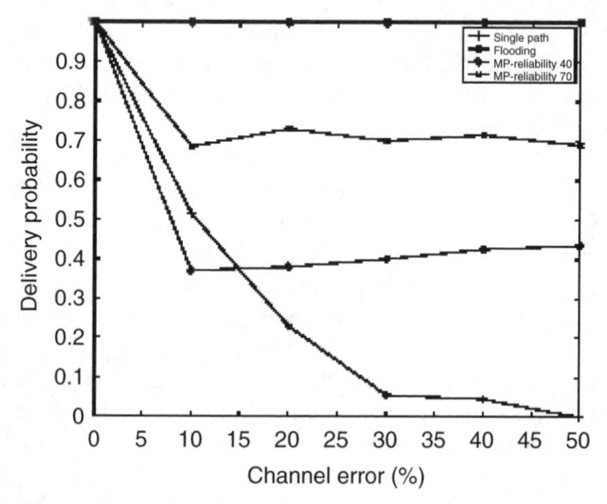

Figure 13.17 Achieved delivery probability for flooding, single packet delivery, and ReInForM with reliability levels of 40 and 70 % respectively for varying packet loss rate. Reproduced from [199] by permission of IEEE

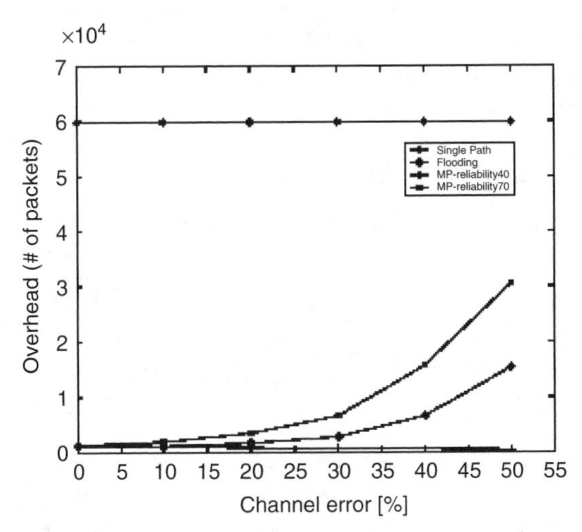

Figure 13.18 Required overhead for flooding, single packet delivery, and ReInForM with reliability levels of 40 and 70 % respectively for varying packet loss rate. Reproduced from [199, Fig. 5] by permission of IEEE

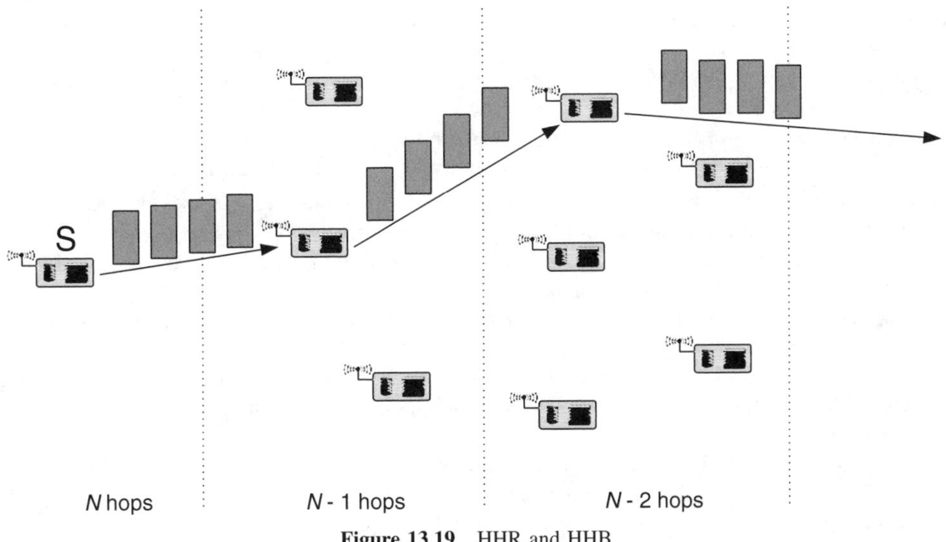

Figure 13.19 HHR and HHB

We first explain the difference between HHR and HHB; please refer also to Figure 13.19. In HHR, the source picks *one* neighbor and sends the same packet in unicast mode to this neighbor N times, hoping that at least one packet gets through. This is illustrated in Figure 13.19 by having $N = 4$ packets on each arc. If instead the source would *broadcast* the packets, it would suffice if *any* of the $k = 3$ neighbors having a $n - 1$-hop distance to the sink receives *any* of the packet copies correctly. By using the mediums broadcast property, one can reduce the number N of copies generated in each hop, sending only $N' < N$ packets. With a packet error rate of e_s, the delivery probability is given by $1 - e_s^{kN'}$. Given a required delivery probability r_s, this can be solved for N' and one gets $N' = N/k$. Hence, the source needs to send much less packets.

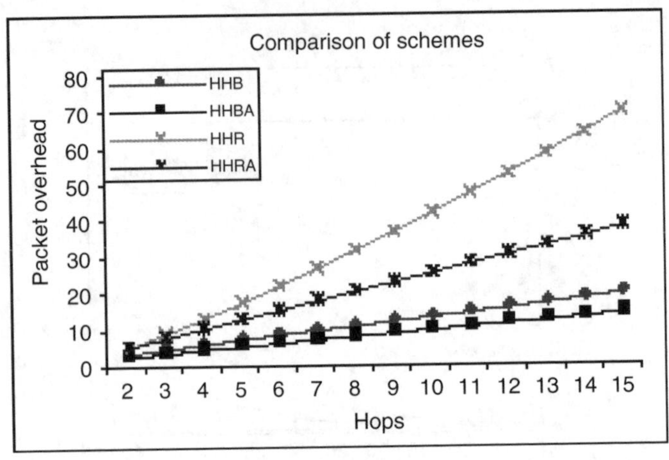

Figure 13.20 Comparison of the overhead induced by the HHR, HHRA, HHB and HHBA protocols for 50 % packet loss rate, 70 % desired delivery probability and varying number of hops. Reproduced from [198, Fig. 5] by permission of ACM

However, another problem occurs. The number k of $n-1$-hop neighbors that decide to forward the packet must be controlled; otherwise such a scheme is hardly distinguishable from flooding. A *deterministic control* like, for example, running a leader election protocol between these k nodes, involves signaling overhead, which may easily eat up all energy savings. Accordingly, a stochastic control is desirable that *on average* lets one of the k nodes forward the packet. The average number of $n-1$ hop neighbors receiving the packet is given by $k(1-e_s^{N'})$, and if any of these forwards with probability $\frac{1}{k(1-e_s^{N'})}$, then indeed there is, on average, one forwarder. However, this requires all these nodes to know the number k. This number is therefore included into the packet. Any intermediate node receiving the packet filters out duplicates (i.e. packets having a sequence number recently seen) and forwards the packet at most once. When the intermediate node has decided to forward the packet, it behaves pretty much as a new source node.

In the HHBA protocol, the source node also sends N' packets, but with a larger spacing than with HHB. The larger spacing is sufficient to accommodate acknowledgment packets. Specifically, only a node that has decided to become a forwarder sends back an acknowledgment. After receiving the acknowledgment, the source stops to transmit duplicates. Hence, it may happen that the source needs to send fewer than N' packets.

In Figure 13.20, the four schemes HHR, HHRA, HHB, and HHBA are compared for their overhead, that is, the overall number of generated packets required to deliver a packet with 70 % probability over a random sensor network with uniform packet loss rates of 50 % and varying number of hops. It can be seen that the overhead increases linearly with the number of hops and that the broadcast-based schemes have significant advantages.

13.4.3 Multiple receivers

The approaches discussed so far are all targeted to deliver a single packet to a single receiver. The task to transmit a single packet reliably to multiple receivers is more tricky and to our knowledge not yet addressed in full detail. One problem is that using positive acknowledgments would lead to the **ack implosion** problem.

An obvious solution is clearly flooding/reliable broadcast [811] or directed flooding, when geographic information is available and the nodes to be reached are confined to a geographic region.

When the set of nodes to be reached is a large fraction of the overall number of nodes, flooding can be a reasonable choice.

13.4.4 Summary

We summarize some of the main lessons that can be learned from the results and protocols discussed so far:

- Pure end-to-end recovery is only advantageous when the single hops show virtually no errors. For "real-life" error rates like those found on wireless channels, local error recovery based on MAC-layer acknowledgments is advisable. The MAC-layer behavior can also support selection of feasible routes [787]: when many trials are needed for a certain link it is likely a bad link and should be avoided.
- Higher desired delivery probability or a larger number of hops leads to increased energy consumption.
- Flooding is a simple scheme achieving excellent delivery probability, but at a high price.
- Positive acknowledgments are needed to ensure reliability levels. However, these have some problems: (i) they are even sent when the channel is good, and (ii) there is the ack implosion problem when a packet is to be delivered to multiple receivers.

13.5 Block delivery

The delivery of large data items is required, for example, when time series data has to be transported from sensors to a sink or when the sink sends out new application code or new user queries to retask the whole network. To keep the packet error rates reasonably low (see Section 6.3) and to comply with packet-size limitations of the transceiver, it is often mandatory to split up the data block into multiple packets/segments/fragments. One important feature of such a block transfer is that NACKs can be used. As explained in Section 13.4.1 this potentially reduces the number of acknowledgment packets.

A NACK can be regarded as a retransmission request issued by the receiver. When intermediate nodes cache the segments, they can serve such a request as well as the original source node could, but with the benefit that the NACK and the following retransmitted segment do not need to travel the whole distance between source and sink node. Such a node is also called a **recovery server** [617]. In an extreme case, all nodes in the network spend some buffer for caching.

We discuss two schemes incorporating these ideas.

13.5.1 PSFQ: block delivery in the sink-to-sensors case

The Pump Slowly Fetch Quickly (PSFQ) protocol presented by Wan et al. [849] addresses the case where an ordered block of packets is to be delivered from one sink node (for example, a user terminal) to a set of sensor nodes. Some major applications of such a protocol are the distribution of new application or protocol code for retasking the sensor network or the injection of complex queries. Clearly, a sensor node has to receive the entire code block before it can start to work with it; losses are not tolerable.[14]

There is a need to distribute packet blocks to individual sensors, groups of sensors, or even the whole sensor network. For the purposes of the present discussion, it is assumed that the underlying network layer offers appropriate services and addressing mechanisms.

The basic idea of the PSFQ protocol can be described as follows. The data source **pumps** the packets making up the code block one after another into the network, using a large period

[14] Another problem of such a retasking operation not considered here is how a single sensor decides when to *activate* the new code – in many cases, such an activation makes sense only when all other sensors are ready to run the new code, too.

and a broadcast or directed broadcast mechanism. Nodes receiving those packets store them into an internal buffer and, when they are received in-sequence, forward them to downstream nodes. An intermediate node receiving an out-of-sequence segment does not forward it immediately, but quickly requests the missing segments from the upstream neighbor. This operation is called a **fetch operation** and corresponds to a NACK. Between two pumps, multiple fetch trials can be made. Therefore: Pumping is slow and fetching is quick. As soon as the missing packets arrive, the intermediate node continues to pump the packets in-sequence into the network, again using a broadcast operation. The recovery is thus local, not end-to-end. The protocol assumes that losses are entirely due to channel errors and not due to congestion, and PSFQ contains no mechanisms to deal with congestion.[15]

How does the sink know that all packets are distributed and the new code block can be enabled? If the time between pumps is large enough to accommodate sufficient numbers of fetches/retransmissions and if furthermore the network diameter is known, it can be estimated when the packet stream will be delivered successfully.

In the next few sections, the protocol operation is sketched in some more detail.

Behavior of the data source

The data source splits the available code block or file into a series of packets or segments. Besides the code, each packet (also called **inject message**) contains four additional fields: (i) a *file id* identifies code blocks, (ii) a *file length* indicates the length of the code block, (iii) the **sequence number** identifies particular segments/packets within a code block, and (iv) a Time To Live (TTL) field allows to restrict the scope of the code distribution operation to those nodes that are at most k hops away from the data source.

The data source broadcasts the packets one after another with a spacing of $T_{\min}$ seconds. The choice of $T_{\min}$ is subject to a number of considerations. First, it must be large enough to accommodate some number of fetch operations; Wan et al. [849] suggest a number of five fetch operations. Secondly, by choosing a larger interval, the intermediate and end nodes have sufficient time to process the incoming segments and one has a simplistic kind of flow control. Third, there must be sufficient time for downstream nodes to repump the segments.

All intermediate nodes have a cache for incoming segments. To keep discussion simple, it is assumed that this cache is large enough to accommodate all program fragments. Please refer to Wan et al. [849] for a discussion of the protocol behavior in case of smaller cache sizes.

Handling a duplicate packet

Any intermediate node receiving a segment checks whether the segment has already been received in the past by looking it up in the cache, using file id and segment number as key. If the segment is found, it is silently dropped. This is a convenient way to prevent forwarding loops.

If the packet has not yet been received, its TTL field is decremented. If this field is zero, afterward the node stops forwarding the packet. Otherwise, it starts the forwarding process by looking at the packet's sequence number. Two cases can be distinguished.

Handling an in-sequence packet

A new packet is received in-sequence when all packets belonging to the same code block with lower sequence numbers have been received. In this case, the packet is scheduled for repumping. This operation is handled by the following rules:

[15] In standard TCP, just the opposite assumption is made, leading to serious performance problems of TCP over wireless links [41].

- The node picks a random time from the interval $[T_{\min}, T_{\max}]$ and starts an appropriate timer.
- During this waiting time, the node listens on the channel to check whether other nodes broadcast the same segment. If four other packets have been received, the node cancels the timer and abandons the repumping operation.[16]
- If the timer expires, the packet is rebroadcast/repumped. The random delay is useful in reducing the probability of hidden-terminal situations, since for broadcast packets no prevention measures on the MAC level (like for example, RTS/CTS dialogues) are applied.

Handling an out-of-order packet

The behavior in case of receiving an out-of-order packet is more complex. The general strategy followed by PSFQ is to *not* repump the packet, but instead the node tries to request (fetch) the missing segments as quickly as possible. When this is successful, the node continues pumping the packets in their correct order.

The decision to suppress the pump operation for the time being can best be explained by an example, shown in Figure 13.21. The loss of packet 3 triggers a fetch operation at node *A* as soon as *A* receives packet 4. When *A* would forward packet 4, nodes *B* and *C* would detect the same loss event and start fetch operations for packet 3 themselves. In this case, three nodes start fetch operations for a single loss event. Furthermore, the fetch operations of *B* and *C* are likely useless as long as *A* does not have the missing segment. Therefore, suppressing the pump operation prevents **loss propagation**.

Upon detecting the out-of-order packet, node *A* prepares a **NACK message** (negative acknowledgment) that includes the file id, the file length, as well as a list of missing segments (a node can have multiple outstanding segments, for example, due to bursty losses).[17] The node broadcasts such NACK messages every T_r seconds (with some randomization to avoid collisions with other node's NACKs) as long as there are missing segments and the maximum number of trials is not exhausted. The value of T_r is smaller than $T_{\max}$; actually, their ratio defines how many fetch operations a node can execute before the next pump.

When *A* receives NACK messages from other nodes requesting the same segments, it suppresses its own NACK for a while. In the meantime, *A* listens on the medium whether the answer packet with the right segment can be picked up. When *A* does not receive an answer (for example, because the answering node is out of *A*'s range), node *A* resumes transmission of NACK packets.

The NACK packets are broadcast. In dense sensor networks, they may reach several neighbors and possibly different neighbors have different portions of the missing segments in their cache. To avoid collisions of all these answer packets, the following procedure is applied. Suppose that node *A*'s NACK indicates segments 3, 6, 7, and 9 as missing and the NACK is heard by other nodes *B* and *C*. Node *B* has only the segments up to and including 6, but node *C* has all segments. Both nodes *B* and *C* – which cannot necessarily hear each other – start a timer with a random value between 0 and T_r. When this timer expires, either node sends a single packet containing segment

[16] The rationale for this is the following [849]: The goal of node *A*'s repump operation is to reach neighbors of *A* not having the packet yet. If, however, *A* has already received the segment *k* times from other neighbors, significant fractions of *A*'s neighborhood are already covered by these "foreign" broadcasts. It is shown in Ni et al. [596] for randomly deployed nodes that the additional coverage that can be gained by *A* rebroadcasting the packet diminishes as *k* increases. For example, for $k \geq 4$, the expected additional coverage is $\leq 0.05\,\%$. The decision to avoid rebroadcasting for $k \geq 4$ has the advantage of avoiding collisions. Furthermore, if receiving is significantly cheaper than transmitting, there is also an energy advantage gained by not sending the packet.

[17] The choice of negative acknowledgments over positive ones has two advantages. First, positive acknowledgments are always transmitted, in good as well as bad channel states. In contrast, negative acknowledgments occur only during bad channel states; during good times there is no overhead. Secondly, there is the problem of acknowledgment implosion: Just imagine a node has 10 neighbors and each one would send a positive ack. This involves not only lots of packets, but also requires the sender to keep track of which of its neighbors has *not* sent an ack and thus needs a retransmission.

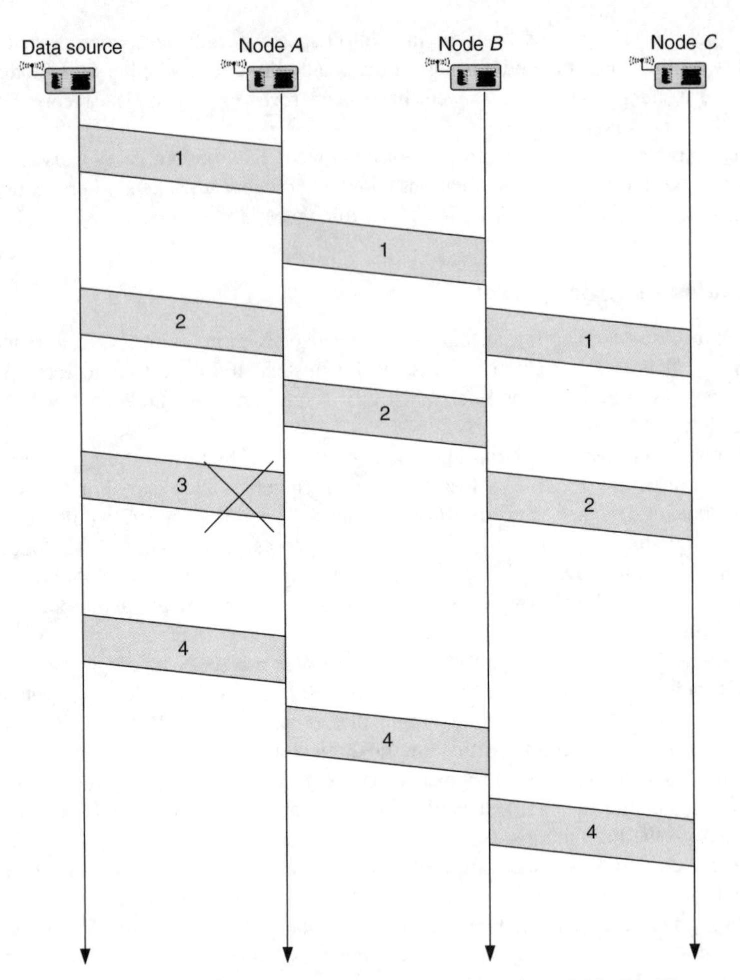

Figure 13.21 Propagation of loss events (adapted from [849, Fig. 3])

3. If B receives a packet with segment 3 before its timer expires, the timer is canceled and B keeps quiet (clearly, C behaves the same way). After T_r seconds have passed, the same procedure starts over, but now for the packet with segment number 6. When $2 \cdot T_r$ seconds have passed, node B ends its activities, since it has no segments beyond 6 in its cache. However, node C continues in the same manner as before. This way, there is only one eligible segment number in each time interval of T_r seconds. The randomization strives to reduce collisions.

The PSFQ protocol contains provisions for upstream nodes to propagate A's NACK packets further toward the sink, when they have been received multiple times and the requested segments are not present. However, this behavior is assumed to be the rare exception and thus the recovery is only local.

Proactive fetch

The recovery method described so far works only when a node A has a chance to detect loss of segment n by successfully receiving any segment with a higher sequence number than n. It fails when the last segment or the k last segments of a file are missing. However, given that a node has received at least one of the segments, it can derive the overall number of segments comprising the code block (from the file length parameter) as well as a coarse estimate of the time when it *should* have received the missing segments. This estimate is based on the knowledge of T_{max}.

A node A sets a timer to the value T_{pro} each time it receives a new packet. Upon expiration of this timer node, A enters the fetch mode and requests all the missing segments from its upstream neighbors; this is called **proactive fetch**. Choosing T_{pro} too small might cause premature fetch operations, choosing it too large induces longer file delivery delays. Another consideration is the assumed cache size: when T_{pro} is too large, the missing segments may already have disappeared from the caches of upstream nodes and the NACK messages must be propagated further. The choice proposed in [849] makes T_{pro} dependent on the difference between the last received segment number and the maximum segment number. The rationale is to fetch a single missing segment quickly, but to be patient when more segments are missing because of a persistently bad channel. Specifically, it is proposed $T_{pro} = \alpha \cdot (S_{max} - S_{last}) \cdot T_{max}$ for some $\alpha \geq 1$ describing the "eagerness" of the proactive fetch operation.

Report operation

The PSFQ protocol also specifies a report facility, which allows the data source to assess how many nodes have already received the complete code block and can thus switch to the new software. The sink node requests reporting by setting a reserved bit in the TTL field of an inject message. Essentially, report messages are generated by the most distant nodes (those that receive packets with a TTL of one) and travel back to the data source. All intermediate nodes piggyback their own data onto these packets. If this would exceed the maximum packet size, the incoming report packet is simply forwarded and the intermediate node creates a new one.

The data generated by an end node or an intermediate node contains the node's own address and a summary of the already received segments. When a node receives a report packet having a record with its own node address (for example, due to a routing loop), the packet is dropped silently. Intermediate nodes receiving packets with the report request bit set, but which do not receive any actual report packets for some time, start generation of report packets by their own.

Performance results

The performance of PSFQ has been investigated by simulation and in an experimental test bed. PSFQ is compared with a reliable multicast protocol from the IP world, namely an idealized version of the Scalable Reliable Multicast (SRM) protocol presented by FLOYD et al. [263] within the context of a distributed whiteboard implementation. Reliable multicast protocols offer services similar to PSFQ. SRM also uses the concept of local recovery, but there are some differences compared to PSFQ:

- in-sequence delivery is not enforced, and
- each node transmits its last received sequence number periodically to the multicast group to detect losses of the last segment of a code block.

Therefore, SRM involves more signaling traffic than PSFQ, which sends NACK packets only upon a (proactive) fetch operation.

The investigated SRM version is somewhat idealized, since all operations concerning construction and maintenance of the multicast tree are not taken into account and the resulting tree is an ideal shortest-path tree to all receiving nodes.

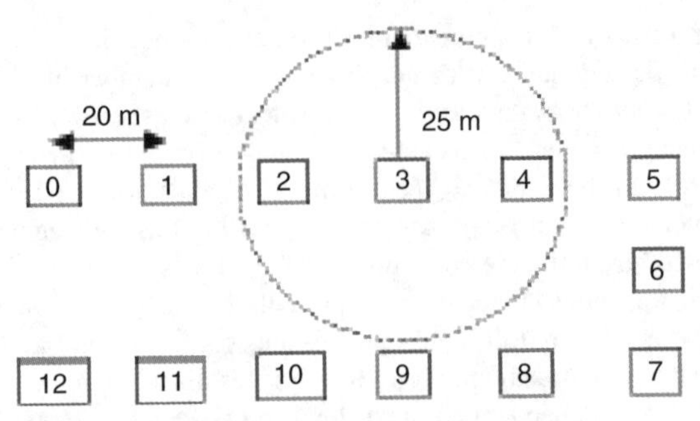

Figure 13.22 Network deployment scenario for comparison of PSFQ and SRM. Reproduced from [849, Fig. 4] by permission of ACM

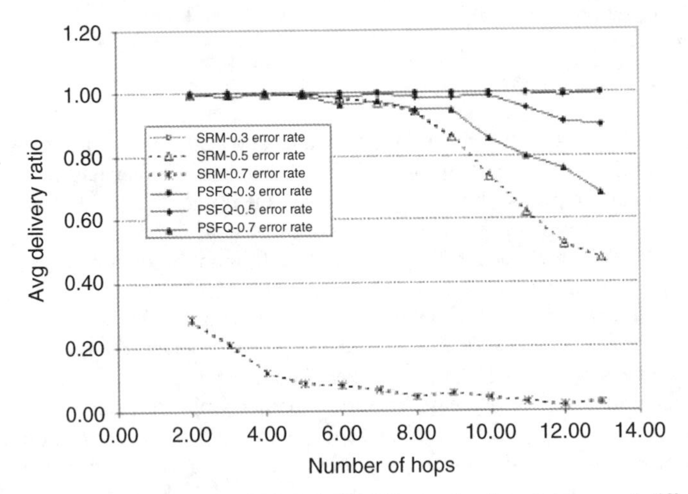

Figure 13.23 Average delivery ratio of PSFQ and SRM for varying hop distance and different packet loss rates. Reproduced from [849, Fig. 5] by permission of ACM

The investigated network scenario is depicted in Figure 13.22. It consists of 13 nodes with a spacing of 20 m between nodes and a transmission radius of 25 m. The nodes use simple MAC-layer broadcasts for PSFQ, whereas for SRM a CSMA-CA variant with RTS/CTS handshake and up to four link-layer retransmissions is used. Hence, SRM should suffer less from collisions than does PSFQ. The channel loses packets according to a BSC model with some packet loss probability. The data source (node 0) wishes to deploy a code block of 2.5 kB, segmented into 50 packets carrying 50 bytes of user data each. The chosen parameters for PSFQ are $T_{\min} = 50\,\text{ms}$, $T_{\max} = 100\,\text{ms}$ and $T_r = 20\,\text{ms}$.

In Figure 13.23, the average delivery ratio at the different sensor nodes (taken over 10 replications) is shown for different values of the packet loss rate and for varying maximal hop distance. The average delivery ratio simply counts how many of the packets issued by the data source have reached the respective node 100 s after the data source has sent the last packet. It can be seen that the delivery ratio decreases with increasing packet loss rate and with increasing number of hops. Furthermore, both schemes have problems in delivering the file to all nodes when the packet loss rate is 50 % or more. However, SRM has much bigger problems than does PSFQ in delivering the file over very lossy channels.

Further results presented by WAN et al. [849] indicate that PSFQ is not only more reliable but delivers code blocks faster in a regime of high packet loss rate (≥50 %). On the other hand, SRM is faster for packet loss rates below 40 %. These results are obtained in a scenario covering only the first three hops where both schemes are able to deliver all packets even when faced with a high packet loss rate.

A third interesting performance measure is the communication cost, defined as the ratio of the number of data packets delivered at nodes to the total number of packets involved in the whole code block transfer. These costs have also been investigated for a three-hop scenario and it turns out that PSFQ consistently needs about half the energy of SRM, even after disregarding all of SRM's costs involved at the MAC and link layer (RTS, CTS packets, MAC-layer acknowledgments).

To summarize, PSFQ is well adapted to the problem of reliably distributing code blocks from a data source to a number of sensor nodes. An implementation of PSFQ under TinyOS requires 2 kB of code, which is acceptable for sensor nodes. One problem of PSFQ is that it requires nodes to be awake constantly and to overhear the channel during the file transfer. However, these costs are acceptable when code downloads are a rare operation.

13.5.2 RMST: block delivery in the sensors-to-sink case

The Reliable Multisegment Transport (RMST) protocol described by STANN and HEIDEMANN [787] is designed toward guaranteed delivery of large blocks of data from sensors to sinks. It is tightly integrated with directed diffusion [378], described in Section 12.2.2. The large data block is fragmented by source nodes into a number of fragments and these are transmitted. On the other end, the sink node collects all incoming fragments and delivers the whole block as soon as all fragments have arrived. RMST is not designed to include explicit congestion control, to guarantee in-sequence delivery of fragments at the sink, or to obey any time bounds.

Design of RMST mechanisms

The design of RMST combines repair mechanisms on different layers:

- It exploits MAC-layer retransmissions to increase the chance of data packets to make it over a single hop. Specifically, the IEEE 802.11 MAC layer is taken as a basis. For unicast frames like data frames, the whole machinery of MAC-layer acknowledgments and the virtual carrier-sense operation (involving RTS/CTS frames) is used. For broadcast operations like, for example, in interest dissemination, neither of these mechanisms is applied. STANN and HEIDEMANN [787] suggest to use at least three trials on the MAC layer.
- In RMSTs **cached mode**, all nodes in the network and specifically those on a reinforced path between source and sink cache the fragments and check for missing ones. When directed diffusion constructs the *reinforced path* from a source to a subscribing sink, the nodes on this path learn also about the identity of their upstream neighbors, that is, those neighbors on the reinforced path that are closer to the source. Accordingly, a **back channel** is associated with a reinforced path, and is just the same path but in reverse direction, from the sink to the sensor(s). Intermediate nodes in cached mode are required to check periodically for missing fragments, which can be either holes in their local fragment list or the fragment list is truncated at the end of the block. If there are any missing fragments, a **NACK packet** indicating the missing fragments is sent along the back channel as MAC-layer unicast packet. When an upstream node has the missing fragments in its cache, it retransmits these along the reinforced path. Otherwise it forwards the NACK packet further on the back channel. However, NACK forwarding should happen rarely when nodes have enough memory and the path does not change in the meantime. It is important to note that the fragments as well as the NACK packets are represented as attributes on the diffusion layer.

Furthermore, the NACK mechanism works independent of the behavior of underlying MAC protocols. Hence, this mechanism can be regarded as a transport-layer mechanism.
- There is also a **noncached mode** in which intermediate nodes maintain no caches, only the subscribing sink has one to collect fragments. In this mode, it is entirely up to the sink to detect losses and issue NACK packets.
- On the application layer, the source regularly sends out all fragments comprising the block and does so until the sink explicitly unsubscribes.
- The diffusion routing mechanism keeps track of node failures and of using good routes: the source node send out exploratory messages regularly, leading to establishment of a new reinforced path (along with its back channel).

These mechanisms have been investigated in different combinations, with the goal to minimize the costs associated with repair actions.

A set of additional attributes is used to integrate RMST into directed diffusion. A data block is uniquely identified by an attribute RmstNo, a fragment within a data block is identified by its FragNo attribute, and the total number of fragments making up the data block is given by the MaxFrag attribute. The NACK packet is also defined by a special attribute.

Evaluation

We discuss some of the results presented in Stann and Heidemann [787]. The investigated scenario consists of 21 nodes arranged in a 3×7 grid such that each node reaches only its immediate neighbors. There is a single source and a single sink node, both placed at opposite places in the grid. Packets from the source to the sink have to take at least six hops. The MAC layer of the nodes is IEEE 802.11. The block size is 5 kB, fragmented into 50 fragments of 100 bytes size. Three variants are considered for the behavior of the MAC layer:

- In the *no-ARQ* variant, all packets use the broadcast-MAC address and consequently no RTS, CTS, or acknowledgment packets are used.
- In the *all-ARQ* variant, all MAC packets are targeted toward a single node and include all the RTS-/CTS-ACK machinery. Broadcasts are implemented by sending a separate unicast packet to every neighbor.
- In the *selective-ARQ* variant, data and NACK packets use the unicast machinery, and other types of packets like interest packets are disseminated by MAC-layer broadcasts.

The wireless channels follow a simple BSC error model with varying packet error rates.

The evaluation investigates the repair costs of different combinations of the above mechanisms. These costs are measured by counting the overall number of bytes transmitted in the network to deliver the data block, including any RTS, CTS, and MAC-layer acknowledgment packets. This number is normalized to the byte count needed by a baseline scheme, which runs over a perfect channel and uses no-ARQ on the MAC layer. Included in all byte counts are the costs associated with directed diffusion, like, for example, the costs of interest dissemination. In Figure 13.24, a number of different schemes (X, Y) are compared: Y denotes the MAC-layer mode (no-ARQ, all-ARQ, selective-ARQ), whereas X denotes the scheme used on the application and transport layer:

- A hyphen indicates that no transport layer is used at all; instead the source transmits all blocks periodically until the sink unsubscribes.
- An N indicates the noncached mode, that is only the sink issues NACK packets.
- A C indicates the cached mode, that is all nodes have caches and the ability to issue NACK packets.

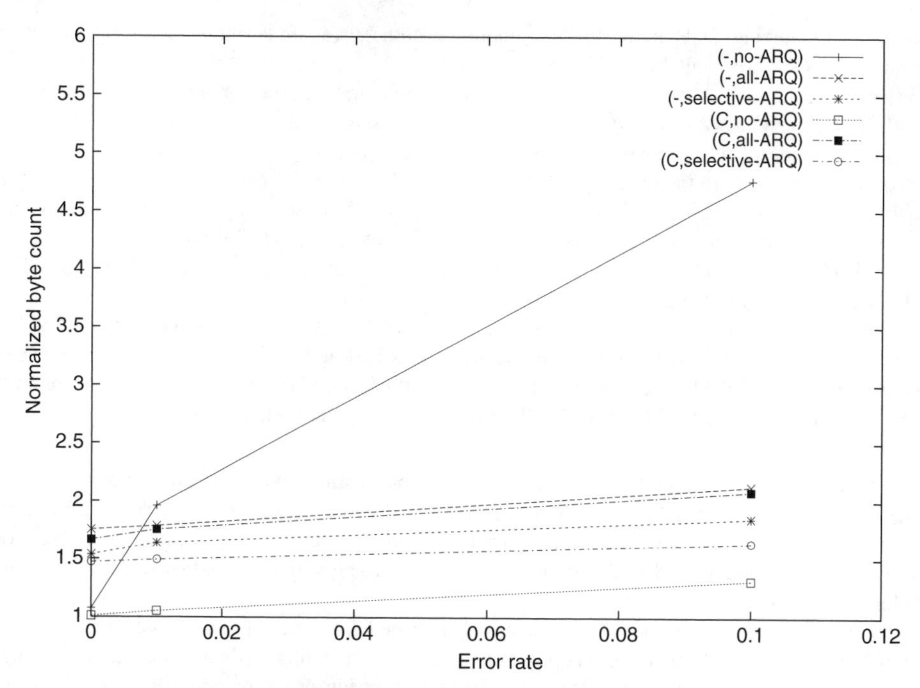

Figure 13.24 Normalized overhead of different combinations of mechanisms for varying packet loss rate (Figure based on data from [787])

In Figure 13.24, the results for no transport layer at all versus the cached mode are compared. The following points are important:

- For zero errors, the schemes employing no-ARQ on the MAC layer are the cheapest ones, as could be expected. For the other error rates (1 to 10 %), using either MAC-layer or transport-layer repairs gives significant energy savings.
- The selective-ARQ schemes are always cheaper than their all-ARQ counterparts because of their lower overhead for broadcasts.
- The difference between (-, all-ARQ) and (*C*, all-ARQ) is rather small; the same holds true for the difference between (-, selective-ARQ) and (*C*, selective-ARQ). Hence, combining transport- and MAC-layer recovery brings little additional benefit.
- Interestingly, hop-by-hop recovery on the transport layer (i.e. the cached mode) with no-ARQ is cheaper than any other scheme. Reliability is provided here by the transport layer without incurring the various MAC-layer costs.

Furthermore, it is shown that for all-ARQ and selective-ARQ, the two transport-layer schemes (cached, noncached) exhibit almost the same performance, whereas for no-ARQ, the noncached scheme did not manage to deliver the data block within the simulation time. A major reason is loss of the sinks NACK packets.

13.5.3 What about TCP?

When starting to think about reliability over multihop networks, immediately TCP comes to mind [790, 893]. And this is not without reason: TCP is mature, pretty well understood, and tons of publications watching TCP under any imaginable circumstances or designing more or less useful variations of TCP are available.

When the goal is to connect a sensor network through some gateway or sink node to the outside world, it is clearly an obvious choice to run a TCP stack and maybe an HTTP server on top in the gateway node. This way, arbitrary Internet hosts can access the data collected in the sensor network by connecting to the gateway. This is even possible when the gateway node is also resource constrained, since several lightweight TCP implementations exist, some with full functionalities and others with reduced functionalities; see for example, DUNKELS [217], RAMAKRISHNAN [675], and LIN et al. [506]. In RIIHIJÄRVI et al. [693], even an FPGA implementation of subsets of HTTP, TCP, and IPv6 within 10 kgates is described.[18]

Another option would be to use TCP *within* the network as a means to provide reliable end-to-end communication, for example, between sensor nodes and sink/gateway nodes or between peer sensor nodes. Although there are some efforts toward this goal [215, 218, 219], there seems to be some "impedance mismatch" between TCP and many sensor network concepts:

- TCP relies on the concept of individually addressable stations with globally unique addresses or networkwide unique addresses. Wireless sensor networks, however, are data-centric and the individual nodes are unimportant. As we have seen in Chapter 7, this paradigm is served more efficiently by using only locally unique node addresses, geographic addresses or data-centric addressing schemes.
- TCP connects two distant nodes and treats all intermediate nodes (routers) as "dumb" entities, which merely forward blocks of bits having no particular meaning to them. In contrast, in sensor networks, intermediate nodes perform in-network processing or aggregation of data.
- TCP carries lots of per-segment overhead, each TCP segment has a minimum header of 20 bytes for port numbers, sequence numbers, checksum, window size, and more. Additionally, address fields are needed, which in the Internet are part of a 20 bytes IP header [790]. On the other hand, most data packets in a sensor network will be small, just a few bits of sensor data. It is possible to reduce header sizes by header compression techniques like RObust Header Compression (ROHC) [92], [380]. However, these techniques require a preliminary setup of contexts between nodes.
- TCP also has some runtime overhead. Most notably, TCP requires separate checksums for each TCP segment.
- TCP strives for perfect reliability and accepts no losses. In sensor networks, this attitude is not only costly in terms of energy, but also not required in many applications. For example:
  - Given that most data in sensor networks flows from the sensors toward a single sink or a few sinks, the TCP flows will be essentially unidirectional. The TCP receiver (here: the sink) has the habit of sending extra acknowledgment segments when there is no data flowing into the reverse direction to which it can piggyback the acknowledgment. Therefore, lots of extra packets carrying TCP acknowledgments are created and travel to the source nodes.
  - TCP is application blind and cannot know that a sensor measurement missing from a particular node can be replaced by a correlated measurement from a nearby node. So, retransmissions are requested even when the data is essentially available.
- TCP does not match well the load patterns of many sensor network applications. Consider, for example, a large sensor network tasked with wildfire detection. For long times nothing happens, but once a fire starts, many sensors wake up and start to periodically transmit data. Setting up a TCP connection requires a three-way-handshake between transmitter and receiver. So, at the beginning of a wildfire many nodes initiate three-way-handshakes at the same time, leading to

[18] Just to give an idea of the code complexity of a full TCP implementation: The TCP code in a recent Linux distribution (Debian unstable, third quarter of 2004) is well beyond 200 kB of C code with well over 7000 lines of code, not counting any code belonging to the IP or lower layers.

significant congestion at the few sink nodes and delaying connection setup. Setting up all the connections in advance, say, at network configuration time, would require that the TCP contexts in the sink have to exist for very long times and the sink must not have any reboot during that time.

An additional problem is that TCP is not exactly famous for squeezing the best throughput performance out of wireless links [41, 361] or to work particularly well in wireless multihop environments [267, 289, 800]. This is in parts due to TCP's congestion-control algorithm, which reduces the transmission rate upon packet losses. TCP's standard assumption is that losses are due to network congestion – in which case it is wise to reduce the transmission rate – and not due to link errors.

13.5.4 Further reading

- We point very briefly to some work on keeping end-to-end delay bounds. This is a very challenging problem [786] and is not only a transport but also a routing issue [340]. In Okino and Corr [606], a concept of statistical delay bounds is introduced, called *(α, β)-currency*. Specifically, a node i in the network transmitting packets to a sink is said to be (α, β)-current if for the packet delays $\delta(i)$, taken as a random variable, the following holds: $\Pr[\delta(i) \leq \beta] \geq \alpha$. The whole network is said to be (α, β)-current if all of its nodes are so. Okino and Corr [606] discuss an algorithm for determining the fractions of time that a node transmits packets and the time that a node receives packets for forwarding purposes. Currentness results for grid deployments are derived analytically and in a test bed. Lu et al. [519] devise the RAP protocol stack composed of a prioritized MAC protocol, a packet scheduling policy, geographic forwarding, and a transport-layer protocol called *location-addressed protocol*.
- Park and Sivakumar [617] propose an intermediary solution between letting all intermediate nodes cache segments and letting only the end nodes cache segments. They propose to use dedicated **recovery servers** to which all retransmission requests are forwarded. When there are enough servers, retransmission requests and answers travel only a small number of hops. This research is part of the GARUDA project[19] at Georgia Tech. This project considers the case of reliable block delivery from sink to sensors.
- Reason and Rabaey [688] present experimental results of radio energy consumption and reliability in a network of 25 PicoRadio nodes [667] arranged in a grid. On the MAC layer they investigate CSMA with preamble sampling ([227, 228]; see also Section 5.2.5). Reliability is targeted only at the link layer by using per-hop acknowledgments and retransmissions; there is no end-to-end scheme. When looking at the achieved packet loss rates for every single node, it shows up that (unsurprisingly) nodes having a higher hop distance to the central node have higher loss rates. Furthermore, with lower radio duty cycles of the proposed on-demand spatial TDMA technique, more collisions are provoked, which have an adverse effect on reliability, because any collision eats up one trial on the MAC layer and fewer trials remain to combat channel errors.
- Agrawal et al. [12] propose the Simple Wireless Sensor Protocol (SWSP), which is inspired by Transmission Control Protocol (TCP) but with some differences: (i) There is no congestion-control component, (ii) the window size is always small, (iii) a node supports only a single connection, for example, to the sink node, and (iv) the sink node can use a single acknowledgment packet to issue acknowledgment information to *all* connections at the same time. Nodes are supposed to set up their connections early and keepalive messages are used to maintain the connection.

[19] `http://www.ece.gatech.edu/research/GNAN/work/garuda.html`

- The problem of reliable group communications/multicast has received some attention in the context of ad hoc networks, but mostly without considering energy aspects. Some references to start with are Luo et al. [524], Wu and Bonnet [896], and Tang et al. [812].

13.6 Congestion control and rate control

Congestion occurs when over a prolonged period of time more packets are generated than the network (as a whole or locally) can actually carry. Usually, nodes have some buffer space available, which can handle transient overloads. Any packet in excess of the available buffer space is dropped, wasting all the energy spent on this packet so far. Clearly, the larger this buffer space is, the more overload can be carried and packet dropping occurs later. On the other hand, longer queues impose longer end-to-end delays and the protocols need longer time to react on congestion states.

13.6.1 Congestion situations in sensor networks

In sensor networks, there are some typical situations that are amenable to congestion. Consider, for example, applications where sensors are quiet most of the time but start periodic packet generation upon some external event. When many sensors recognize this event simultaneously, a traffic hot spot around the event location is created. On the other hand, when the network has a structure in which many sensors report to a few sink nodes, the area around the sinks can become a hot spot even if the traffic around the single event locations is bearable. Clearly, in such hot spots, we have not only packet drops due to buffer overflows, but in case CSMA-type protocols or ALOHA-type MAC protocols are used, we also have an increased collision rate and longer access delays, costing energy at the MAC layer (see Chapter 5) and causing locally generated packets to pile up.

Some of the implications of congestion in sensor networks have been investigated by Tilak et al. [819] using simulations. They considered among others a scenario where a varying number of nodes is randomly deployed over an $800 \times 800\,\text{m}^2$ area. A physical phenomenon moves through this area toward a randomly chosen destination at a randomly chosen speed between 1 and 2 meters per second. Each node has a sensing range of 200 m and measures its distance to the phenomenon; the accuracy of these measurements is within 5 % of the sensed distance. As soon as a sensor detects the phenomenon, it starts to transmit packets at a given reporting period; when the phenomenon moves out of the sensing range, the sensor stops. The sensors report their readings to a sink node. The sink node works with slotted time and produces a position estimate for every time-slot, based on the packets received during that slot. The sensor nodes run an IEEE 802.11 MAC protocol and DSR as routing protocol; each node is able to buffer five packets.[20]

In Figure 13.25, the achieved goodput is displayed for three different numbers of nodes and varying reporting rate. The goodput is defined here as the ratio of packets reaching the sink to the overall number of packets transmitted throughout the network. The following points are noteworthy:

- For all numbers of nodes, the goodput decreases as the reporting rate increases. Furthermore, the goodput decays rather quickly.
- The larger the number of nodes, the stronger the decrease in goodput. Hence, for a given reporting rate, increasing the number of nodes in the network/the node density actually decreases goodput, wasting more energy. It is even shown that on average individual sensors run out of energy more quickly as the node density increases. Therefore: *Increasing node density shortens node/network lifetime* as long as no additional mechanisms are devised.

[20] The choice of IEEE 802.11 MAC here and in other sections of this chapter has to do with the fact that the ubiquitous ns-2 simulation tool (see `http://www.isi.edu/nsnam/ns/`) has models for this and for many other protocols (DSR, directed diffusion, ...) used in the domains of ad hoc and sensor networking.

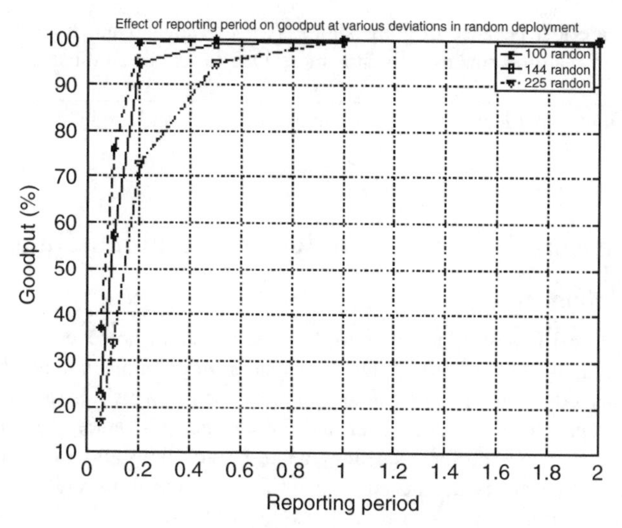

Figure 13.25 Goodput versus reporting period for three different numbers of nodes. Reproduced from [819, Fig. 6] by permission of ACM

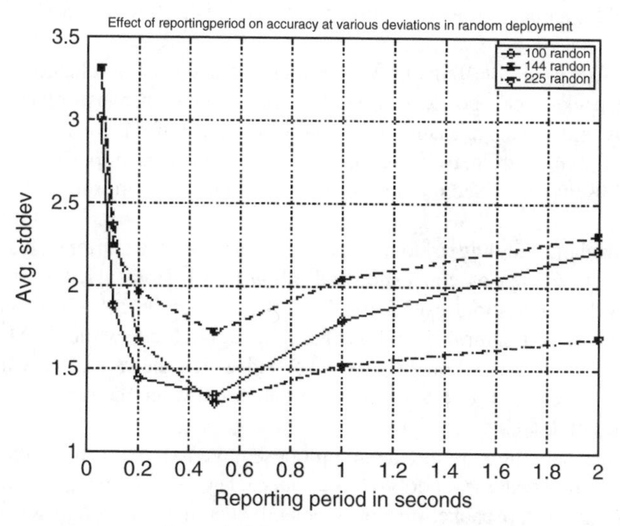

Figure 13.26 Accuracy versus reporting period for three different numbers of nodes. Reproduced from [819, Fig. 9] by permission of ACM

- Similar results hold true for a grid deployment.
- It is also shown that sink nodes should be close to phenomenon; larger distances provoke congestion as more packets are created

The achieved accuracy is shown for the same setup in Figure 13.26. The accuracy is defined as the average (taken over all time-slots) standard deviation between estimated and true position of the node phenomenon. It can be observed that for all numbers of sensors there is a reporting period giving the best accuracy. For smaller reporting periods, accuracy decreases quickly because more and more packets are lost because of congestion.[21] On the other hand, for lower reporting rates, the fewer number of readings makes estimates more inaccurate. To summarize:

> Congestion can decrease network/node lifetime and reduce information accuracy.

13.6.2 Mechanisms for congestion detection and handling

Congestion detection

Sensor nodes can use different approaches to detect congestion; however, these methods are all local in the sense that a node can only judge the situation in its immediate neighborhood.

The detection methods proposed so far in the context of sensor networks rely on two fundamental indicators: The occupancy of a nodes buffer and the channel utilization. The simplest method is to compare the instantaneous buffer occupancy against some threshold value. If the threshold is exceeded, a congestion state is diagnosed. However, when the threshold makes up for a large fraction of the total buffer size, congestion states are detected (too) late. An improved method, like, for example, used in ESRT (see Section 13.6.3), takes the "growth trend" into account: The buffer occupation is sampled regularly and congestion is diagnosed when the instantaneous buffer level is above some threshold and additionally the buffer size has grown in the immediate past. An above-threshold occupancy level combined with negative growth is a sign that congestion is resolved.

It is shown by WAN et al. [850] that buffer occupancy alone is not a reliable congestion indicator, specifically when packets can get lost already *on the channel* because of collisions or hidden-terminal situations and have no chance to actually reach a buffer. Only the situations of a full buffer and an empty buffer are reasonable indicators of (non)congestion. Consequently, the CODA framework [850] discussed in Section 13.6.4 uses a second congestion indicator, namely channel sampling.

The goal of channel sampling is to obtain an estimate of the current channel utilization U. This estimate is in turn used as congestion indication [850]. However, the relationship between the utilization and the congestion level depends on the MAC protocol: For example, with TDMA the channel can be almost saturated without harming throughput, while CSMA variants have a certain maximum channel utilization $U_{\max} < 1$ beyond which the rate of collisions increases and the goodput actually decreases [68]. Congestion is diagnosed when the channel utilization is within some neighborhood of $U_{\max}$.

How to do channel sampling? In the method proposed by WAN et al. [850], the channel sampling algorithm is triggered when the node's packet queue becomes nonempty. This is the moment where the node wants to start packet transmission and when the question of whether the channel is congested or not becomes relevant. The time after starting the sampling operation is subdivided into **sampling epochs**, with the duration of one epoch spanning multiple packets. Within an epoch, the channel is periodically sampled, say N times. Between the sampling instants, the node can switch off its transceiver to conserve energy. When in epoch n a number M out of N samples indicate a busy channel, this epoch's utilization is $\Theta_n = M/N$. The estimates for K consecutive epochs can be combined, for example, with exponential weighting. The weighting factor and the number K can be used to tune the estimator.

[21] The packets are dropped indiscriminately without taking into consideration their sensor data precision, which directly depends on the distance between sensor and phenomenon. It would be interesting to investigate the resulting accuracy when the dropping policy tries to keep more accurate readings and drop the more imprecise ones.

Congestion handling

Some of the mechanisms proposed in the context of wireless sensor networks to *avoid* congestion in the first place or to *react* on it are the following:

- **Rate control**: The rate by which sensor nodes transmit their own sensor readings can be controlled. Alternatively, when this rate is fixed for certain applications, the number of nodes generating at this rate can be controlled. This control can be executed in an end-to-end fashion or locally. In the end-to-end case, the ultimate receiver, for example, a sink node, causes the transmitting nodes to reduce their rate. The sink uses some feedback mechanism like, for example, acknowledgments or dedicated signaling packets. In the local case, a node A might be signaled by its next-hop forwarder B that B's buffers are full. Node A can reduce its rate or propagate the overflow signal further backward [850]. The actual desired target rate depends on the accuracy requirements of the user and the general trade-off is that relaxing these requirements tends to reduce the frequency and duration of congestion states. Examples for rate control approaches are discussed in Sections 13.6.3 and 13.6.4.
- **Packet dropping**: When a forwarding node having full buffers receives a new packet, it clearly must drop the new packet or an old one from the buffer.[22] A node can make a better-informed dropping decision when it has some information at hand about the importance of the packet. One option is to label each packet with an explicit priority value and to let the forwarder drop the packet of lowest priority, either new or already buffered.
- In-network processing and aggregation: Since a sensor network is deployed toward specific applications, forwarders know the data they forward and can compress, drop, or aggregate it accordingly. This is the option truly distinguishing sensor networks from other types of networks.

Clearly, other types of networks allow further mechanisms. For example, in ATM networks [196], a combination of admission control, careful route selection, and resource reservation in intermediate nodes is used to avoid congestion in the first place. However, such a mechanism requires a reasonably stable network, per-connection state in intermediate nodes, and significant signaling and, in general, seems too heavyweight for wireless sensor networks.

13.6.3 Protocols with rate control

ESRT

The Event-to-Sink Reliable Transport (ESRT) protocol developed by SANKARASUBRAMANIAM et al. [720] considers the situation where a number of sensors observing an event report periodically to a single sink node. The sink node does not care which nodes send packets; it is only interested in receiving a sufficient number of packets to present reliable and credible information to the user. This is called **event-to-sink reliability**. ESRT works by carefully adjusting the sensor's reporting rate to achieve two goals:

- a sufficient number of packets is received, and
- not many more packets than needed are received to avoid congestion and save energy.

[22] In the Internet, often an alternative strategy called *random* early drop [262] is followed, explicitly designed for TCP: Routers start to drop packets randomly already before their buffer is full. A TCP connection hit by such a loss automatically reduces its rate, hopefully avoiding full buffers in routers. Given that really something similar to TCP runs in a sensor network, this strategy must be accompanied by using highly reliable link layers to avoid losses through channel errors and to keep TCP's assumption that losses are a sign of congestion working.

Most of the protocol operation is carried out at the sink. The protocol is round based. A single round is called a **decision period** and has a fixed duration τ. The sink counts the number of packets r_i received during round i and compares this with the nominal/desired number of packets R, which is needed to achieve sufficient information fidelity. The ratio $\eta_i = \frac{r_i}{R}$ is the normalized measure of reliability, and the goal of the protocol is to keep η_i in the interval $[1 - \varepsilon, 1 + \varepsilon]$ with ε as usual being arbitrary, positive, and small. This ε-corridor allows for stable protocol behavior despite small variations. The main control knob of the protocol is the sensor node's reporting frequency f. The frequency f_{i+1} to be used in round $i + 1$ is computed by the sink at the end of round i and broadcast to all sensors. It is assumed that the sink is not power constrained and can transmit sufficiently strong signals to reach all nodes. The sensors adopt the new reporting frequency immediately.

The precise operation of the protocol can be motivated with the help of Figure 13.27, which displays the observed reliability η for varying reporting frequency f in a random deployment (see Section 13.2.3). In this deployment, 200 nodes are spread over a field of $100 \times 100\,\text{m}^2$. The nodes have a radio range of 40 m and $n = 81$ nodes sense a given event. One of the nodes is the sink node. The curve is representative in the sense that similar shapes are obtained for other numbers of reporting sensors, n, too. The following things are noteworthy:

- In the first phase, the achieved normalized reliability increases linearly (observe the logarithmic axis scale!) with increasing reporting frequency until a maximal normalized reliability is achieved at $f = f_{\max}$. Below and up to $f_{\max}$, the network shows no signs of congestion.
- After this point, the reliability drops and starts to fluctuate. This can be attributed to the increasing influence of congestion, since too many packets are generated and nodes start to drop packets. The fluctuations are random effects.

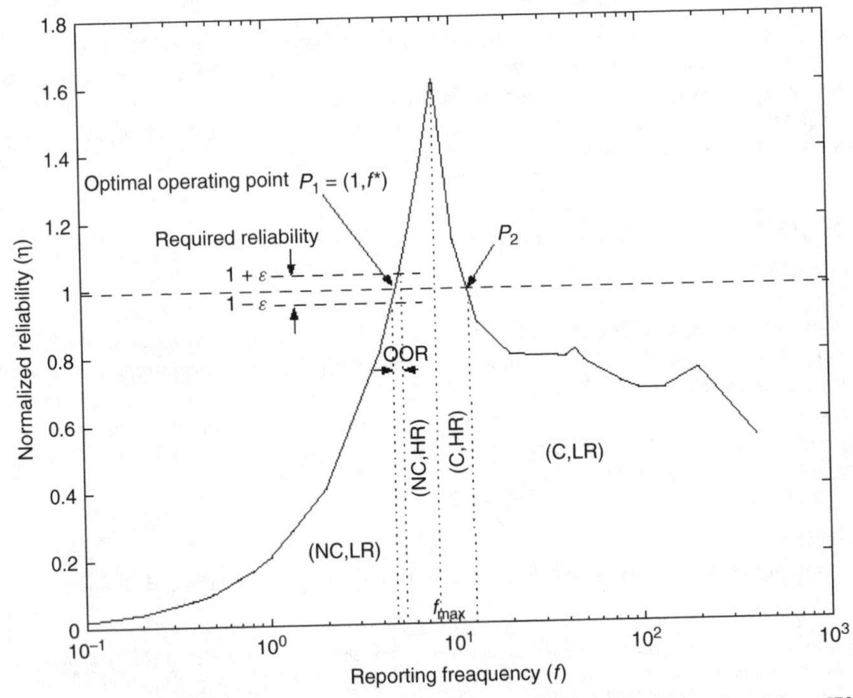

Figure 13.27 Achieved normalized reliability η versus reporting frequency f. Reproduced from [720, Fig. 4] by permission of ACM

Please note that the curve hits the desired reliability $\eta = 1$ two times: from below (point P_1) and from above (point P_2).

In round i, the protocol uses the frequency f_i. At the end of this round, the sink determines the reliability η_i and knows also whether congestion has occurred in round i. This information is used to determine frequency f_{i+1}, which is then broadcast. Congestion is detected by the sensor nodes by inspecting their local packet buffers. Briefly, when the estimated rate of buffer growth and knowledge of the current buffer contents indicate that the buffer will overflow in the next round, a node sets a congestion indication bit in outgoing packets. When the sink receives such a packet, it infers a congestion indication.

The protocol distinguishes five different regimes, indicated in Figure 13.27.

1. In the leftmost part, the reliability is below $1 - \varepsilon$ and there is no congestion. This part is called (NC, LR) for No Congestion, Low Reliability. The protocol increases the reporting frequency f_{i+1} for the next round as $f_{i+1} = \frac{f_i}{\eta_i}$, motivated by the linearity of the curve in this part.
2. For a small frequency window, the reliability is in $[1 - \varepsilon, 1 + \varepsilon]$, there are no signs of congestion, and the protocol operates close to P_1. This is the **optimal operating region** (OOR) and the protocol chooses $f_{i+1} = f_i$ to stay in this region.
3. For frequencies between the OOR and $f_{\max}$, there is more reliability than needed but still no congestion. A reduction of reliability is feasible and also useful to save energy. ESRT performs a moderate reduction of the reporting frequency:

$$f_{i+1} = \frac{f_i}{2}\left(1 + \frac{1}{\eta_i}\right).$$

4. For frequencies between $f_{\max}$ and the frequency at P_2, there is congestion and high reliability (C, HR). Again, it is a good idea to reduce the frequency, but now a bit more aggressively: $f_{i+1} = \frac{f_i}{\eta_i}$.
5. Beyond the frequency at P_2, there is congestion and low reliability (C, LR). This region is to be left even quicker, with $f_{i+1} = f_i^{\eta_i/k}$, where k is the number of times that the network was in (C, LR) consecutively.

It becomes clear from this behavior description that the protocol tries to move to the optimal operating region. It can be shown analytically that the protocol indeed converges to the optimal operating point when in the noncongested regime, indeed, the reliability increases linearly with the frequency. However, the speed of convergence depends on ε, with smaller values of ε having a longer convergence time. The convergence behavior of ESRT is illustrated in Figure 13.28, where for the above described setup with $n = 81$ transmitting nodes both the convergence toward the optimal operating point and the associated energy consumption in every round of duration $\tau = 10\,\mathrm{s}$ is displayed.

This protocol has some advantages: The sink does not need to know global network properties like the number of nodes, and most of the protocol complexity is in the sink. The sensor nodes just need to be able to capture the sink's commands to set new reporting frequencies. A disadvantage of this protocol is that the reporting rate is the only control knob of the protocol and that there is no way to accommodate different node densities. If in a sparsely populated area the rate is reduced, then the information about this area becomes more noisy. On the other hand, if the rate is increased in an area with high node density, local congestion can occur, reducing the rate of packets arriving at the sink and causing the latter to increase packet generation rates further. A second disadvantage is the assumption that sink broadcasts are heard everywhere, which may often not be true.

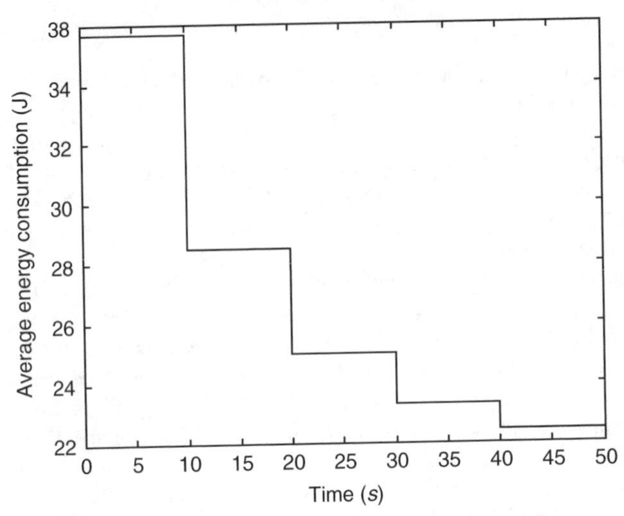

Figure 13.28 Average energy consumption versus time for ESRT protocol, starting in state (NC, HR). Reproduced from [720, Fig. 13] by permission of ACM

An algorithm based on the Gur game

The algorithm proposed in IYER and KLEINROCK [381] also relies on networkwide broadcasts; however, it does not control the reporting frequency but lets sensors choose randomly and independently between sleeping and being awake for the next round (thus avoiding any local coordination among sensors). A sensor being awake sends a report, but a sleeping sensor sends nothing. At the end of a round, the process starts over.

The sink node controls the probability by which individual sensors make their decision. It follows the goal to receive a desired number k^* of packets in every round. Using probabilities as control knob to achieve a desired number of incoming packets N_P would be easy, if the sink knows the number N of living nodes in the network. It can just send the probability N_P/N to the sensors and each sensor carries out an independent Bernoulli experiment to decide whether to sleep or not. However, N is hard to obtain and also time variable because of node death or deployment. Another option would be to use an algorithm similar in spirit to ESRT: the sink could start with some initial guess p_0 and adjust this probability on the basis of the number of number of packets received.

The algorithm presented by IYER and KLEINROCK [381] is another option. It is based on the **Gur game**: There are a number of independent players not aware of each other, and a referee. In each round, a player decides among two given alternatives, say: "yes" and "no". The referee counts the number k of yes-votes, determines a reward probability $r = r(k)$ from this, and announces this to all players. Each player makes an independent random experiment: The player is rewarded with probability r; with probability $1 - r$ the player is penalized. The reward function $r(k)$ is fixed and assumes a maximum for the value k^* of desired "yes" votes. Now the question is: How should the players behave to maximize their rewards, given that they have no knowledge about the number of other players, about their behavior, the reward function $r(k)$, the optimal value k^*, and the values k_i observed by the referee in round i?

The surprising answer is that it is possible to find a finite-state machine, which runs independently in every player and which changes state every time a reward or penalty is incurred. There are $2M$ states, from $-M$ to -1 and further from 1 to M. The states and their possible transitions along with their associated probabilities are displayed in Figure 13.29. The parameter M is also called *memory* parameter. State changes happen when the referee announces r; when the automaton is in a

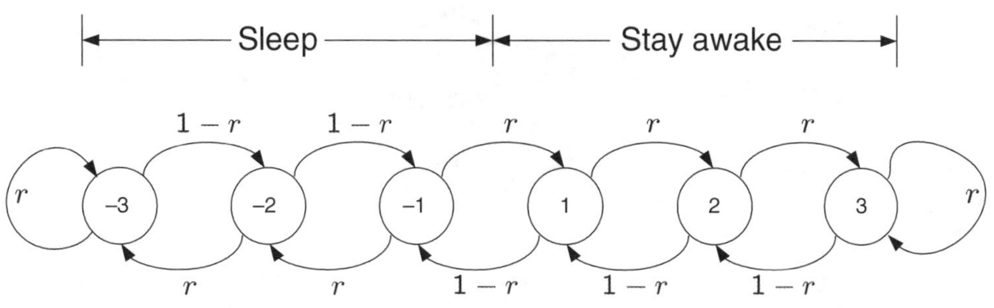

Figure 13.29 Example automaton with memory $M = 3$

negative state $-i$, it moves with probability r to state $-i-1$ (in case $i = -M$ the automaton stays there) and with probability $1-r$ it moves one step to the right. For positive states, the behavior is symmetric: upon rewards the state is increased (state i becomes $\min\{M, i+1\}$) and upon penalties the state is decreased. Hence, when the automaton experiences subsequent penalties, it moves to the center, and with subsequent rewards it moves to the fringes. When the player is in a positive state it votes with "yes"; otherwise it votes with "no". The important thing is: It can be shown that under this construction the number of yes-votes k_i received by the referee in round i converges to k^*, except for random fluctuations. However, as discussed in FROLIK [266], convergence depends on the start states of the players. An example with $M = 3$ is provided in which the network fails to converge when *all* nodes start in state 3.

This algorithm can be adapted for the situation in a sensor network. Each sensor node runs an instance of this automaton, all sensors and the sink are time synchronized, and all sensors are ready to receive the sink's feedback r at the end of round i, whether or not they were sleeping during the round. When a sensor's automaton is in a positive state, it remains awake in the next round $i+1$ and transmits a packet; otherwise, the sensor goes to sleep mode. The sink node can obtain k_i simply by counting the packets arriving in round i. A key design decision is the sink node's reward function; IYER and KLEINROCK [381] have the particular shape $r(k_i) \sim a + b \cdot e^{-c(k_i - k^*)^2}$ with $a, b \in (0, 1)$ and $a + b = 1$, c can be any positive number. This function has its maximum at the desired value k^*, and each sensor is rewarded when $k_i = k^*$, since $r(k^*) = 1$ holds. Once this has been reached, all automatons move more and more to the fringe and no sensor changes its behavior (sleeping, keeping awake) anymore.

Distortions to this behavior can occur when packets can get lost or new nodes are added/old ones die. Further distortions are introduced by packet delays, that is, the times needed for packets to travel from sensors to the sink. These times can be random and furthermore they can be larger than the duration of a round (IYER and KLEINROCK [381] have used a round of 1 s length). One example evolution of k_i over time is shown in Figure 13.30 for a scenario with initially 100 sensors, a memory size of $M = 1$, and random initial states for the sensors, a desired rate of $k^* = 35$ packets per second, $a = 0.2$, $b = 0.8$ and $c = -0.002$. The duration of a round is 1 s and the packet delays are randomly chosen between 0 and 5 s. New sensors are created every 100 s in the mean and they live 101 s on average. Under these random distortions, the k_i vary significantly but always come back to the target value k^*.

This scheme is especially interesting for cluster-based sensor networks (like, for example, LEACH [344]), where the clusterhead assumes the role of the sink node [266]. This would relax two problems of this scheme when applied to a whole multihop sensor network. One problem is that sensors need not only be awake for sensing but often also for purposes of forwarding without having an explicit sensing task. Another problem is that this algorithm is insensitive against variations in node density. Areas with low density are treated the same way as areas with high

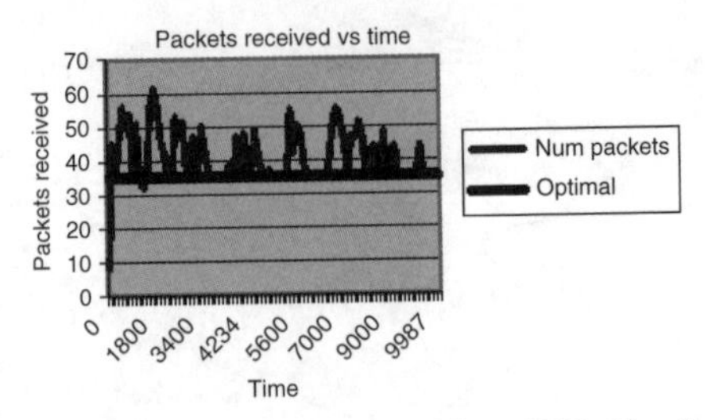

Figure 13.30 Evolution of k_i over time. Reproduced from [381, Fig. 4] by permission of IEEE

density, but any sensor missing in a low density area is more valuable than a sensor in a high density area.

13.6.4 The CODA congestion-control framework

The COngestion Detection and Avoidance (CODA) approach described by Wan et al. [850] combines a congestion-detection mechanism with two congestion-control mechanisms working on different timescales. These mechanisms are targeted toward different congestion scenarios. An open-loop hop-by-hop backpressure mechanism aims to resolve transient congestion situations like, for example, hot spots around the sensor nodes first observing an external event. Secondly, the acknowledgment-based closed-loop mechanism addresses resolution of persistent congestion states through the use of acknowledgments and a self-clocking mechanism.

Open-loop hop-by-hop backpressure mechanism

The basic idea is simple: When a node detects a congestion situation, it regularly broadcasts **backpressure messages** as long as the congestion situation persists. Additionally, it invokes some application-dependent policy to deal with the congestion situation: The node might decide to drop some of the packets it is supposed to forward, it can reduce the rate of its own measurement reports, or it can simply stop forwarding for a while and continue later on.

A node B receiving a backpressure message from another node A also has several options. For example, B can also start to drop packets or reduce its rate, since A is not willing to accept packets anyway. Node B can also decide to forward the backpressure message further toward the data source.[23] If node B is itself congested, it can increment a counter in the backpressure message before forwarding it. As soon as the message reaches a noncongested node, the size of the congested area can be roughly inferred from the counter value. This can be used on longer timescales as a hint for routing protocols to avoid the whole area.

CODA's backpressure messages can also indicate a **chosen node**. This node is allowed to continue its transmissions upon hearing a backpressure message; all other nodes have to apply their backpressure policy.

This method is open-loop because the node issuing the backpressure message receives no direct feedback.

Closed-loop regulation mechanism

A sink node might receive data from multiple sources. This may potentially create a persistent hot spot close to the sink or in other regions of the network. This situation requires a mechanism

[23] How the next upstream neighbor toward the data source is identified is outside the scope of CODA.

to reduce the packet generation rates of the source nodes persistently. Relying entirely on the previously discussed backpressure mechanism is not sufficient: when nodes receive no backpressure messages anymore, they continue to forward packets and to send at their regular rates, this way quickly creating the next congestion state. Secondly, these backpressure messages would have to travel several hops from the sink to the source nodes, creating additional network load.

The closed-loop regulation mechanism is requested by the sources when their packet generation rate exceeds a given fraction r of the locally available channel capacity. Specifically, the sources set a specific bit in their data packets, which triggers the generation of acknowledgment packets in the sink. The rate of acknowledgment packets is application specific; the sink can, for example, generate one acknowledgment per 100 data packets it receives from the source. How the acknowledgments are disseminated to the source nodes is not within the scope of CODA, but is requested from other protocols like directed diffusion. This opens the possibility to restrict the acknowledgments only to sources observing a certain phenomenon while other sources remain unregulated.

A source requesting acknowledgments expects to receive a minimum number of acks over a certain time period. If less acknowledgments are received, the source slows down its own local packet generation according to some policy.

A regulating effect can be achieved in several ways:

- The sink can decrease its acknowledgment rate or even stop sending acknowledgments when it diagnoses its own neighborhood as busy or if the desired minimum number of packets needed to achieve a certain information accuracy is not reached because of congestion losses.
- The sink issues acknowledgments at their nominal rate and trusts that these get lost in the hot spots, slowing down the sources "beyond" these hot spots.

The sources stop to request further regulation as soon as their generation rate drops below the threshold value r.

Some results

Wan et al. [850] have investigated the performance of CODA both in a test bed (including some initial parameter tuning) and by simulation. The simulation model considers both the open-loop and closed-loop congestion-control methods. The source nodes implement a simple reduction policy, applied whenever a backpressure message is received or when the number of acknowledgments is insufficient: They simply halve their generation rate. Intermediate nodes stop any transmission for some random number of packet times before continuing, except when they are the chosen nodes. An IEEE 802.11 MAC is used, however, without MAC-layer acknowledgments or RTS/CTS exchanges. The radio range of a node is 40 m. Suppression messages are generated when the estimated channel utilization is beyond 80 % of $U_{\max}$. The sink node generates acknowledgment packets in the closed-loop mechanism every 100 packets. Directed diffusion is employed for routing purposes.

In one of the investigated scenarios, 30 nodes are deployed randomly in a rectangular area. Three of these nodes are sink nodes and another six are source nodes, the selection being made randomly. A sink requests data from two of the sources. The sources generate event data packets of 64 bytes size, but they have different packet rates of at maximum 20 packets per second. Four of the sources are switched on and off at random times between 10 and 20 s simulated time.

In Figure 13.31, the evolution over time of the number of received packets at the sink nodes is displayed. This number serves as an indicator for the achieved information accuracy. Three different schemes are compared: (i) A scheme without any congestion control (labeled noCC), (ii) a scheme with open-loop control only (OCC), and (iii) a scheme with combined open-loop and closed-loop schemes (CCC). In Figure 13.32, the evolution of the number of dropped packets over time is shown for the same setup. The following points are remarkable:

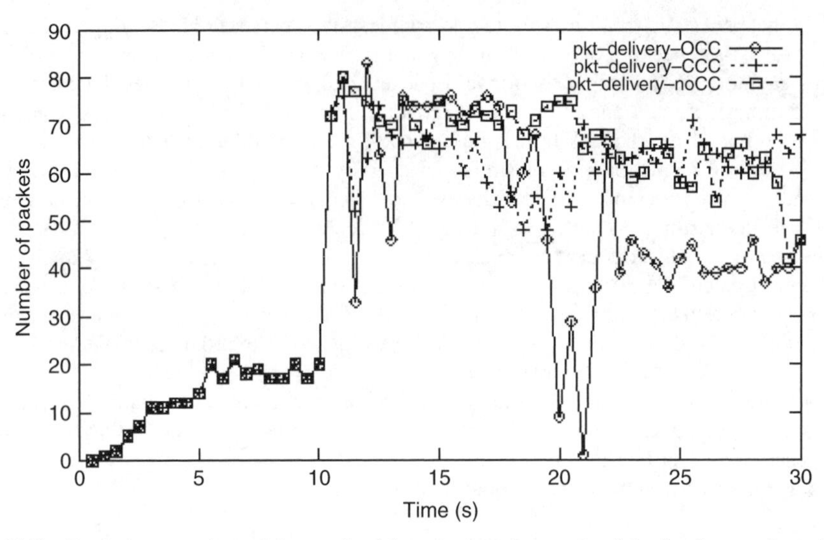

Figure 13.31 Evolution over time of the number of received packets at the sinks for three variants: Without congestion control (noCC), with open-loop congestion control only (OCC) and with combined open-loop and closed-loop congestion control (CCC). Reproduced from [850, Fig. 11a] by permission of ACM

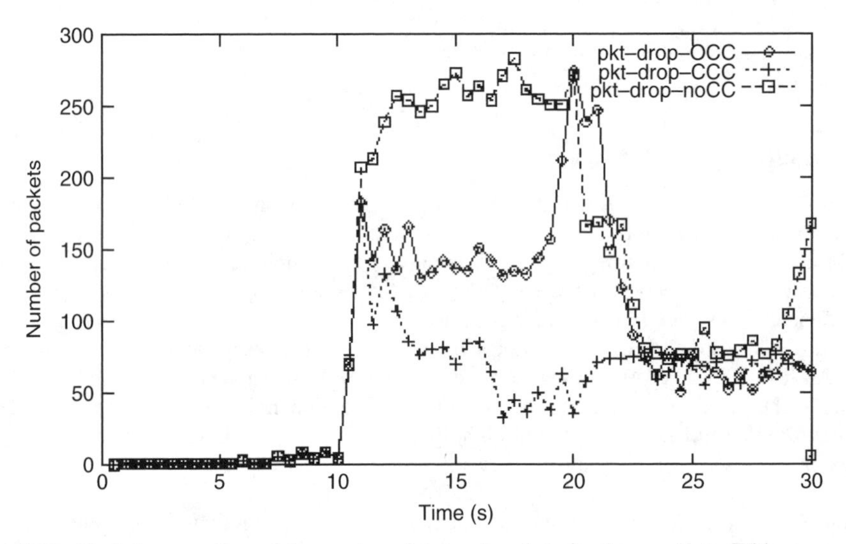

Figure 13.32 Evolution over time of the number of dropped packets for three variants: Without congestion control (noCC), with open-loop congestion control only (OCC) and with combined open-loop and closed-loop congestion control (CCC). Reproduced from [850, Fig. 11b] by permission of ACM

- There is a phase around 20 s where the number of packets delivered by OCC is much lower than for the other schemes. This is around the same time where the packet drop rate of OCC reaches the same order as the noCC scheme; in all other situations, OCC drops less packets.
- The combined scheme CCC achieves a good portion of noCC's delivery rates, but drops much less packets during the congested time period between 10 and 20 s.

A general trend observed in different scenarios is that the congestion-control algorithms tend to achieve not the highest possible delivery rate, but the small price paid here is compensated by significant reductions in the number of dropped packets.

13.6.5 Further reading

- In Frolik [266], the ,Gur game strategy is modified by making the reward probability $r(k)$ dependent on the lifetime of the network and the expected lifetime of the individual nodes. As lifetime increases, the reward probability decreases too, and the sensors tend to move away from the fringe states toward the middle states, thus changing their behavior (sleep, awake) frequently. Otherwise, once the desired number of packets has been reached, the automatons move to the fringe and stay there forever. This, however, means that a sensor deciding once to stay awake does so forever, while the other sensors sleep forever.
- In [477], an overload management policy for a specific distributed control system is derived, having optimal control system performance.
- In Woo and Culler [888], a CSMA-based MAC-layer mechanism is proposed by which a node regulates its local packet generation rate with the objective of distributing bandwidth fairly over nodes.
- Hull et al. [370] discuss a method of explicit **bandwidth allocation** for different data streams. This is achieved by mapping data attributes to traffic classes. Each node classifies packets and runs a scheduler, which selects packets for transmission according to their class. The scheduler also controls the outgoing rate per class. A second ingredient of the proposal is a hop-by-hop congestion-control scheme. In this scheme, nodes send overflow messages as soon as their buffer occupancy exceeds a threshold; upstream nodes reduce their rate. A third ingredient is to avoid hot spots around sink nodes by deploying more sink nodes and to let source nodes decide about good/nearby sink nodes.
- Congestion control has been already been considered for wireless multihop ad hoc networks, mostly in the context of TCP; see for example, [16] and [622].

14

Advanced application support

Objectives of this Chapter

The objective of this chapter is to complement the techniques introduced so far with advanced techniques to support applications. In particular, in-network processing in various forms is discussed, detailing techniques that are both clearly related to a specific application and generally applicable. Next, security questions for wireless sensor networks are highlighted. Finally, examples for some concrete applications are given and the techniques used for their realization are analyzed and presented.

Chapter Outline

14.1 Advanced in-network processing

14.1.1 Going beyond mere aggregation of data

The most energy-efficient communication is the one that is not required in the first place. Aggregation as a relatively simple scheme for avoiding transmission has already been discussed in Section 12.3. In that context, however, the aggregation functions have been, for the most part, relatively simple – averaging, maximum, or minimum, for example. In this section, more sophisticated means for reducing the number of messages or the length of messages that have to be send by WSN nodes are discussed. This section discusses basic algorithms and protocols; it is complemented by the Section 14.3, which goes in more detail of several exemplary applications.

The crucial consideration that can be exploited for many advanced in-network processing schemes is the following. In most networks, sensor nodes are relatively close by each other. When they observe a function of the physical environment (such as temperature, humidity, etc.), the changes of this function over a given distance are likely to be small in most cases. Hence, the measurements taken by nearby sensors are **correlated** in space. Moreover, since physical processes do not usually change abruptly, the measurements from a given sensor are also going to be correlated in time. This **correlation in time and space** is the main lever exploited by many in-network processing schemes. In the presence of correlation, some very simple things change: For example, the optimal tree for collecting all data from the network is no longer a shortest-path tree but rather tends toward a traveling-salesperson structure [183]. Most of the work focuses on using proper distributed coding schemes to exploit the correlation among sources; the Slepian–Wolf theorem [774] is a basic result here.

Such a correlation structure opens the door for sophisticated compression schemes, reducing the number of bits to be transmitted. On the other hand, if an expected correlation structure is actually not present – if a "surprising" event happens – a lot of information can be extracted [295] and has to be reported at high priority.

Another aspect is **distributed signal processing** (some papers also use the term Collaborative Signal and Information Processing (CSIP) to emphasize the collaborative aspect): Making use of the processing power of the sensor nodes to preprocess the measured data, obtaining a different, more compact form of representation of the observed data, or a reduction to relevant aspects. A typical example here is distributed tracking of vehicles [142, 925]: Each sensor can observe some partial information – for example, a distance estimate to a moving object – but only taken together does this information allow to compute and update a position estimate. The trivial solution to send all values to a central processing site is inefficient regarding energy spent for communication. Rather, preprocessing in the field is advisable. Depending on the type of processing to be done and on whether processing or communication constitutes the efficiency bottleneck, SOHRABI et al. [778] distinguish between **coherent** and **noncoherent cooperative functions**. A few remarks on this field are given in Section 14.1.2; it is treated in more depth for some examples in Section 14.3.

A counterpart to data collection, possibly using clever aggregation schemes, is the distribution of data (broadcasting). This issue has been extensively studied in Section 11.4, but there only simple store-and-forward functionality and independence among transmissions had been assumed. But proximity among sensors can also be used in aiding in the broadcasting of data by cleverly encoding which sensor transmits which data. This relatively new field of **network coding** is surveyed regarding basic ideas and fundamental results in Section 14.1.4.

There are some additional aspects that would fit to this topic. One important consideration is **sensor fusion**: How can measurements taken from different types of sensors be combined? Sensor fusion has been briefly discussed in the context of localization and positioning in Chapter 9; further examples can be found in references [102, 326, 854]. But as these topics are not a characteristic of wireless sensor networks, their treatment is left to specialized textbooks.

14.1.2 Distributed signal processing

Performing signal processing already within the network itself can be done for many different application examples. A generic discussion of these concepts is relatively difficult as the techniques are often closely tied to a specific application such as tracking or beamforming. Therefore, only a few ideas are highlighted here; more details and aspects can be found in Section 14.3.

Parallel algorithms in a WSN – FFT

A first idea to perform distributed signal processing in a WSN is to use a standard algorithm used for signal processing that is amenable to a parallel implementation. For such an algorithm, it stands to reason to investigate whether and how it can be used in a WSN and what the consequences for the energy efficiency of the WSN are, in particular, how communication overhead trades off against computation complexity (owing to, e.g., the need to perform computations redundantly).

One algorithm that meets these conditions is Fast Fourier Transform. Its distributed execution in a WSN context has been investigated by CHIASSERINI and RAO [152]. They use the standard "butterfly" structure of a parallel FFT computation and map it onto WSN nodes. In addition, a careful inspection of the algorithmic structure shows that some redundant computations can be removed by introducing asymmetric computations (one node uses intermediate results of another node instead of raw data) without increasing the amount of data that has to be communicated. The paper characterizes how normalized energy consumption and computation time relate to each other, depending on the number of sensors used for the distributed computation.

Applying MPEG encoding to sensor networks

GOEL and IMIELINSKI [295] point out a fetching analogy between encoding a movie and optimizing data transmission in a WSN. Suppose a set of sensors is placed in the field, each sensor with its own (x, y) coordinates and periodically reporting its measured value. Equate the set of readings from all sensors per one measurement period with a single frame of a movie – each senor contributes one "pixel" to this virtual image, and the measured value corresponds to the brightness or color of a pixel. Successive measurement periods correspond to successive frames in the movie.

Depending on the particular application scenario of the WSN, this analogy is more or less close. For a WSN monitoring moving objects, for example, there is a spatiotemporal correlation between successive measurement periods that is very similar to that of a movie and, hence, some of the basic coding and compression ideas of MPEG coding should be applicable.

MPEG, in particular, exploits **predictions** of future pixel values to reduce the data rate that has to be transmitted. Such **prediction-based monitoring** is also proposed by GOEL and IMIELINSKI [295]. They assume a clustered organization of the sensor network. Initially, each sensor node reports all its measurements to its clusterhead. The clusterhead computes, using MPEG-inspired algorithms, predictions for the likely future measurements of each sensor and returns a prediction model to each sensor. The sensor, upon receiving this model, need only transmit data to the clusterhead in the future if it is considerably different from the predicted value ("It is not news if we can predict it" [295]). As it is unlikely to be able to *exactly* predict future values, certain error margins have to be allowed for. If, on the other hand, a measured value does not match the predicted value, the sensor transmits to the clusterhead to correct this prediction error; the clusterhead can then also adapt its prediction model for such a sensor.

Computing confidence in mobile agent results

The mobile agent concept has been described in Section 11.2: An "agent" wanders through the network, collecting information from each node. Such a simple collection task can be extended by more sophisticated in-network processing when nodes are not necessarily fully confident that their measured results are correct or when nodes provide only interval measurements (because of faulty equipment, e.g.). In the first case, a node might announce a **confidence interval** describing the measured yes/no reading to a mobile agent ("I am between 40 and 70 % certain that I just saw a truck go by"). This scenario is investigated by QI et al. [666].

The core idea is to associate a random distribution of confidence with the boundaries announced by a node – Gaussian, uniform, or some other distribution. A mobile agent arriving at a node collects this distribution and continues to the next node. There, the already-collected description of the confidence distribution has to be merged with the local confidence distribution.

The merging algorithm rests on previous work by PRASAD et al. [655]. In that paper, a simple interval description for the values that a single sensor considers plausible is used. These intervals are overlapped, counting how many sensors consider a certain value plausible. Then, this overlapped function is analyzed at different levels of granularity (resolution) to derive a final interval, describing the belief of all sensors.

This algorithm is modified and extended by QI et al. [666]. In particular, a measure of "intermediate accuracy" is introduced (based on the width and height of the confidence intervals – if the probability mass is well centered, the accuracy of this interval description is likely high) that allows the mobile agent to determine when it has computed a result of sufficient accuracy. Then, the mobile agent can return to the base station and deliver its (estimated) result. In some aspects, this scheme is similar to a localization scheme described in Section 9.5.4; it also bears some resemblance with a data broadcasting scheme described in Section 12.3.7.

14.1.3 Distributed source coding

In a sensor network, a common situation is that multiple sensors make periodic measurements and transmit them to a central point where data is collected. In order to reduce energy consumption, the *rate* with which these data transmissions should take place should be as low as possible but high enough to allow the data collector to learn about the measured data with high probability.

In information-theoretic terms, the sensors are distributed information sources and the question is which rates should be used to encode their data. Since the sources are distributed, a **distributed source coding** problem has to be solved. The problem is also closely related to **compression** of data before transmission.

When the measurements of different sensors are statistically independent, there is little that can be done to reduce the required minimum data rates. In practice, however, the likely presence of correlation between readings (owing to similarity of measurements of a physical process in close proximity) admits some surprising solutions. These are described here – but first, some background on source coding appears necessary.

Source coding

Suppose an information source is given that periodically sends one of A symbols of the alphabet $\mathcal{A} = \{1, \ldots, A\}$; let X be the random variable describing which symbol is chosen by the source and let $P(X = x) = p_X(x), x \in \mathcal{A}$ denote the random variables for the probability that the source picks symbol x.[1] Let $\mathbf{X} = (X_1, \ldots, X_n)$ denote a sequence of n symbols successively generated by the source; the individual realizations X_i are assumed to be independent of each other and identically distributed. The probability of a particular sequence $\mathbf{x} = (x_1, \ldots, x_n)$ occurring is given by

$$P_{\mathbf{X}}(\mathbf{x}) = \prod_{i=1}^{n} p_X(x_i),$$

owing to the independence assumption.

A sequence $(x_1, \ldots, x_n)$ can be regarded as a block of n successive characters. Which kinds of blocks do appear? For sufficiently large n, it can be expected that the character "1" appears $np_X(1)$ times, character "2" appears $np_X(2)$ times, and so on. Put the other way around, blocks where there are noticeably different numbers of characters present are highly unlikely and blocks essentially only differ in the order in which characters are sent. This approximation only holds, however, if n is sufficiently large, that is, if the block length is long enough. The encoding and decoding of entire blocks – **block coding** – is thus a technique that cannot work on single characters produced by a source but only on entire sequences of such characters.

[1] The presentation here follows [774]. Introductions to this topic are available via any good textbook on information theory [180].

One such block T of length n appears with probability

$$\begin{aligned} p_T &= p_X(1)^{np_X(1)} \cdots p_X(A)^{np_X(A)} \\ &= \mathrm{e}^{\ln p_X(1)^{np_X(1)}} \cdots \mathrm{e}^{\ln p_X(A)^{np_X(A)}} \\ &= \mathrm{e}^{np_X(1)\ln p_X(1)} \cdots \mathrm{e}^{np_X(A)\ln p_X(A)} \\ &= \mathrm{e}^{\sum_{i=0}^{n} np_X(i)\ln p_X(i)} \\ &= \mathrm{e}^{n\sum_{i=0}^{n} p_X(i)\ln p_X(i)} \\ &= \mathrm{e}^{-nH(X)} \end{aligned}$$

where

$$H(X) = -\sum_{i=0}^{n} p_X(i) \ln p_X(i)$$

is called the **entropy** of the random variable X (or of the source described by X).

As each ordering of characters in such a block is equally likely (because of independence between repetitions), their probabilities p_T are all the same. Thus, there are $N_T = 1/p_T = \mathrm{e}^{nH(X)}$ different blocks than can occur with nonnegligible probability. To represent N_T (encoding which block actually has occurred), $O(\log N_T)$ bits are necessary. Since n characters should be encoded in the given block, effectively

$$R = 1/n \log N_T = 1/n \log \mathrm{e}^{nH(X)} = H(X)$$

bits have to be transmitted per character generated by the source. R is called the **rate** of the source (or of the random variable). This means that, for some $\eta > 0$, an encoder/decoder pair can be found that works at block length n and rate $H(X) + \eta$ and that transmits the source characters at arbitrarily small (but not negligible) decoding error. Such a pair does not exist at any rate $H(X) - \eta$ (for any pair working at this rate, the decoding error cannot be made arbitrarily small).

Correlated sources – the Slepian–Wolf theorem

What happens when more than a single source is present? Consider two sources described by their random variables X and Y with entropy $H(X)$ and $H(Y)$, respectively, and both transmitting to a common sink. Obviously, the information of both sources can be transmitted at a total rate $H(X) + H(Y)$. When the sources are independent (and unaware of each other), there is little that can be done at a source coding level to reduce this rate.

There is, however, a possibility for correlated sources. Suppose X and Y are two correlated sources, that is they have a joint probability density $p_{X,Y}(x, y)$ that is *not* the product of the marginal densities p_X and p_Y. It is easy to imagine that two encoders would be able to exploit this correlation if both encoders had access to both sources (Figure 14.1). The argument uses the conditional probability $p_{X|Y}(x|y) = p_{XY}(x, y)/p_Y(y)$ and its associated entropy

$$H(X|Y) = -\sum_y p_Y(y) \sum_x p_{X|Y}(x|y) \log p_{X|Y}(x|y);$$

the intuition is that $H(X|Y)$ expresses the remaining uncertainty in X once Y is known. The procedure is then to encode source Y with rate $H(Y) + \varepsilon_Y$ for some small, positive ε_Y. At this

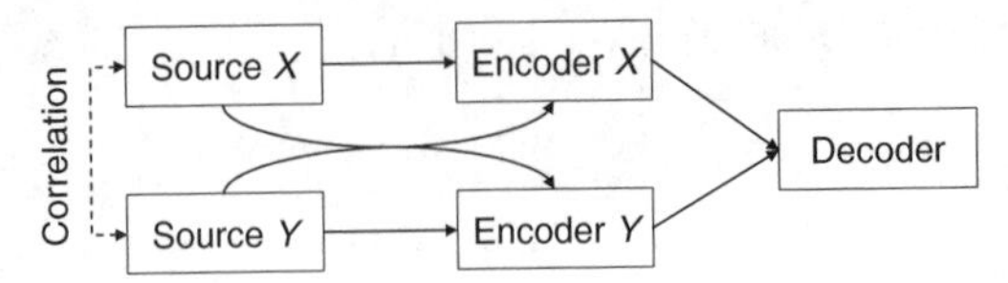

Figure 14.1 Coding two correlated sources with knowledge about the other source

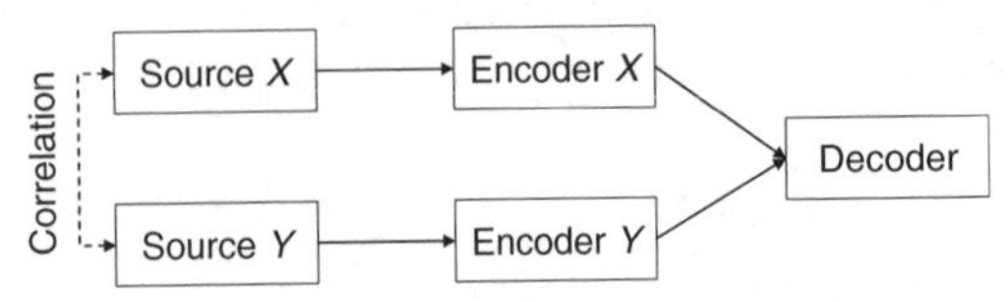

Figure 14.2 Separately coding two correlated sources

rate, the information from Y can be decoded with small decoding error, giving the decoder access to Y's data. It then suffices to encode source X with rate $H(X|Y) + \varepsilon_X$ (for some small, positive ε_X). Hence, the total rate to encode X and Y is only $H(Y) + H(X|Y) + \varepsilon_X + \varepsilon_Y$. Obviously, there is asymmetry between X and Y as one source is the "primary" one but they can switch roles arbitrarily.

But granting X's encoder access to Y's data would require, in a WSN, additional traffic to transport Y's data to X – but reducing the amount of data traffic is the very objective here. The realistic scenario is thus illustrated by Figure 14.2: Two correlated, but separate sources with two separate encoders. The question is whether it is possible, in such a situation, to reduce the total rate compared to the obvious solution $H(X) + H(Y)$ by exploiting correlation between X and Y.

The answer is, somewhat surprisingly, yes! In fact, the very same rates can be achieved as if X and Y were mutually known. This is the main theorem proven by Slepian and Wolf [774] and published in 1973. The nonintuitive part is that it suffices for the encoder of X to know the joint distribution of X and Y but it does not need to have access to the actual realizations of Y. The proof of this theorem is somewhat involved and, moreover, nonconstructive: It does not allow the immediate construction of such encoders or decoders. It is also an asymptotic argument that holds strictly only if the block length goes to infinity. The reader is referred to reference [774] for details.

Nonetheless, it is a seminal result, opening the door for a lot of research with many variations. Some of the early work includes references [898, 899, 900], dealing, for example, with lossy source coding and the availability of side information. The extension to continuous sources (as opposed to the discrete sources described here) has also been undertaken. It took some time, however, before a practical construction for such codes appeared. Before describing the first such scheme next, let us examine a short example.

Example 14.1 (Correlation via Hamming distance [651, 652]) Suppose two sources X and Y are given, both of which are 7-bit, binary random variables (i.e. they can assume values $0, \dots, 127$). They are correlated in such a way that the realizations of X and Y will always have a Hamming distance of at most 1 (i.e. at most a single bit will differ).

Here, the rate per source is $H(X) = H(Y) = 7$ bits. The rate to transmit both sources is $H(X, Y) = H(Y) + H(X|Y) = 10$ bits, which is smaller than the simple sum of $H(X) + H(Y) = 14$ bits. The Slepian–Wolf theorem guarantees the existence of encoder and decoder as long as the rates are chosen such that $R_X > 3$, $R_Y > 3$, and $R_X + R_Y > 10$. Hence, distributed source coding does promise a considerable gain.

Syndrome coding – DISCUS

PRADHAN and RAMCHANDRAN [651] present the first approach for construction of actual codes that achieve the Slepian–Wolf performance (see also references [653, 901] as surveys/tutorials). The idea is perhaps best explained using the example just discussed; for simplicity of illustration, let us assume only 3-bit-long sources X and Y.

The Slepian–Wolf theorem suggests (as one option) that source Y is completely transmitted to the decoder. How can X be encoded such that the decoder can deduce its original value, using the known value of Y? When a certain value of Y is given, the correlation structure assumed here restricts the possible values of X. For example, when $y = 000$, x can only be 000, 001, 010, or 100; all other combinations are not possible (Figure 14.3). Symmetrically, when x is 000, the encoder of X *knows* that y can be only one of these values. It also knows that the decoder, upon receiving any of these values from Y, can limit the set of possible values of X to all those bit combinations that have Hamming distance of at most 1 from this set of four values. Specifically, the only possible values that the decoder could speculate about are 000, 001, 010, 100, 011, 101, and 110 – 111, on the other hand, is *not possible* and the encoder at X knows that the decoder will know that as well!

Therefore, it is obvious to the encoder at X that the decoder can use the value of Y to distinguish between 000 and 111. The encoder therefore does not need to waste bandwidth to communicate this distinction to the decoder.

An analogous argument shows that the encoder of X does not have to bother with distinguishing between the two elements of the following pairs: $\{001, 110\}$, $\{010, 101\}$, and $\{100, 011\}$. These pairs are called **cosets**. Hence, X's encoder only has to transmit enough information to allow the decoder to distinguish between these four cosets, requiring *two* instead of three bits!

Example 14.2 (Coset-based distributed source coding) Consider the following example: $x = 101$, $y = 100$. Upon receiving y, the decoder knows that the only possible values of x are 100, 000, 110, and 101 owing to the correlation structure. The encoder of X tells the decoder that x belongs to the coset $\{010, 101\}$, using two bits to do so. Since only 101 is consistent with both these facts, the decoder knows that $x = 101$.

The terminology of "cosets" hints at some channel-coding background of this scheme. With each coset, a unique **syndrome** is associated via a channel code's parity matrix (the details are beyond the scope of this presentation; please refer to reference [651] or some literature on channel coding). This syndrome can be communicated to the decoder. Thus, derives the name Distributed Source Coding Using Syndromes (DISCUS) for this scheme.

Reference [651] goes beyond this presentation. It also discusses how to apply these basic ideas to encoding with a fidelity criterion, admitting continuous-valued random variables X and Y. The

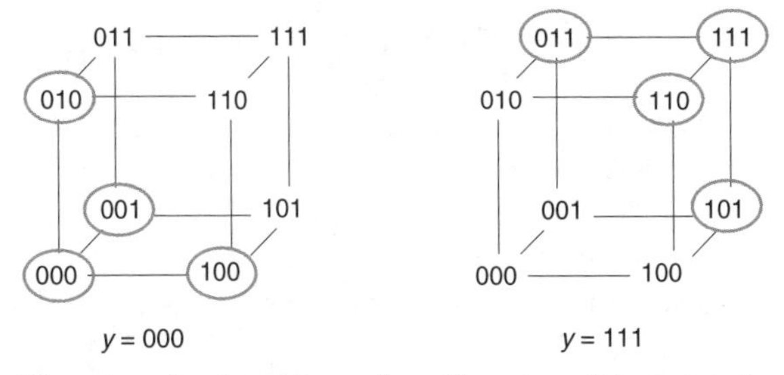

Figure 14.3 Possible values of x for a given y for a Hamming–distance-based correlation structure

correlation in this case is expressed by $Y_i = X_i + N_i$, where the N_i are a sequence of independent and identically distributed random variables independent of X_i.

Some extensions

In this scheme, there is again an asymmetry between the two sources: One of them has to send at the full rate, and the other at reduced rate. While it is possible to switch roles between sources periodically to smooth out this asymmetry, it might be desirable to find a scheme that allows both sources to have the same data rate at the same total rate (this is admissible by the Slepian–Wolf theorem). Such a symmetric version of the DISCUS scheme is presented in reference [652].

References [164, 165] describes a recursive codebook construction scheme and appropriate encoder and decoder mechanisms.

Additional references on the topic are [1, 37, 282, 459, 732, 928].

14.1.4 Network coding

A rather innovative way of using network resources is opened up by **network coding** [15, 492]. In a traditional network, a router essentially only receives packets on incoming links and redistributes them to its outgoing links. In the spirit of in-network processing, what happens when a router is allowed to also modify data and redistribute the result of coding over already received packets to its outgoing links? This is the essential question of network coding and it does provide some surprising insights.

To illustrate the concept, consider the example shown in Figure 14.4: Node S is trying to multicast information to nodes T and U, each intermediate link has a capacity of 1 unit of data per unit of time, and each node can send and/or receive different data units at the same time over different links. Using simple store-and-forward networking, node S can transmit one data unit per unit time to T and U; the additional links via nodes C and D are of no avail here.[2] At best, node S could transmit data x to node A and data y to node B. These nodes broadcast their received data; node C forwards, say, x to D and from there to U. It is not possible, however, to transport two units of data to *both* receivers in one unit of time.

This result changes, however, when node C is allowed to process the received data and to forward the results of such processing. In particular, let node C not just forward one of the received data units, but the result of computing the logical XOR of x and y (Figure 14.5). Node D then broadcasts this data to both T and U who can then use x or y to compute their missing data unit. Thus, two units of data can be transmitted in one unit of time.

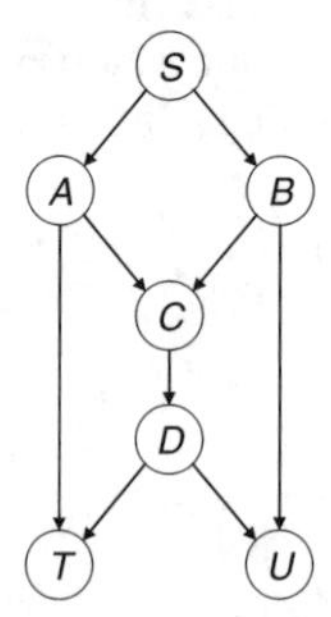

Figure 14.4 Example network for network coding [15]

[2] Delays are ignored here or pipelining is assumed.

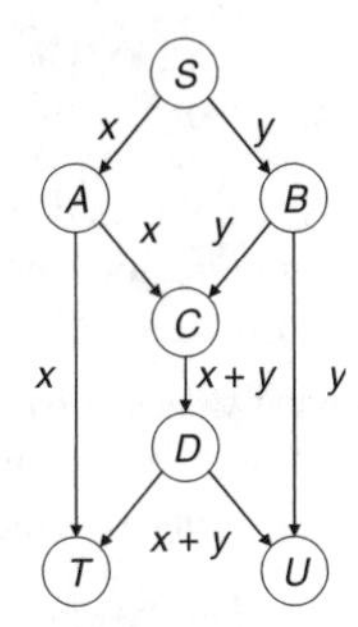

Figure 14.5 Example network when actually using network coding [15]

This fundamental idea can be generalized to the theory of network coding – applying ideas from coding theory to the processing and forwarding of data in a network. As a fundamental result, AHLSWEDE et al. [15] have established a theorem that corresponds to the max-flow/min-cut theorem. They consider a graph G with limited edge capacities, a single source s and several destinations $t_1, \ldots, t_n$ to which the data from the source should be transmitted at rate h. They show that this data distribution task is solvable if and only if for each destination t_i a maximum flow exists that is larger than h. The proof is somewhat complicated and requires the explicit constructions of codes that realize the in-network processing task.

In another basic paper on network coding, LI et al. [492] have shown that *linear coding* is a simple coding scheme that nevertheless achieves the optimal performance as formulated by the max-flow theorem. In linear coding, the only allowed transformations that an intermediate node may perform are linear transformations on blocks of data, perceived as vectors in a suitably defined vector space. KOETTER and MEDARD [428] extend this work by considering how coding can be used to make a network robust against permanent errors (where usually, rerouting had to be employed).

In these papers, transmissions are not restricted by any costs incurred by the nodes but only limited by the link capacities. In this sense, these papers fit better to a wired network. Two references that approach the application of network coding to wireless networks are [522, 897]. LUN et al. [522] cast network coding for wireless networks as a linear optimization problem. The interesting (and surprising) consequence is that with network coding not only is the performance of the network improved beyond that of traditional multicast routing but the optimization problem also becomes simpler than the (NP-complete) Steiner tree problem usually used to model this case. WU et al. [897] explicitly take the wireless multicast advantage into account when using network coding in a cross-layer approach to improve multicasting in wireless networks.

Overall, the field of network coding is, at the time of this writing, still relatively young and full of vigor. It is also, sadly, still relatively little known in the networking research community. Interesting results from a cross-pollination of wireless sensor network research and the more information-theoretic network coding research are still to be expected.

14.1.5 Further issues

A few more research issues and approaches should be briefly mentioned:

Cleaning "noisy" sensor readings ELNAHRAWY and NATH [232] present a plausible approach to the problem of noisy sensor readings. It is based on Bayesian estimation theory.

Beamforming Beamforming – trying to track a mobile source – is a prime example of in-network processing and described in more detail in the following section on specific applications. Two further references are [142, 907]. Reference [457] also considers heterogeneous networks (i.e. networks with more powerful nodes added) for such an application.

Compression and routing Reference [728] combines the concepts of network coding and routing. They characterize conditions for the rate/distortion function to be applied so that all nodes can broadcast their readings in the entire network, given a prescribed quantization error.

In a somewhat similar context, PATTEM et al. [623] look at the performance of routing in combination with compression ad aggregation in presence of spatial correlation of the observed data. They claim a static clustering scheme achieves near-optimal performance.

Transport capacity and measurement accuracy MARCO et al. [538] address the principal question how much data needs to be extracted from a measurement field to attain a given measurement error and how this number relates to the measurement accuracy of the sensor network. Moreover, they study how the number of sensors influences these results.

14.2 Security

Network security [729, 731, 785] is one of the most pressing concerns in all wireless networks, including wireless sensor networks. In this section, we briefly introduce the security problem and explain some of the specifics of wireless sensor networks. The discussion is in parts based on SCHÄFER [729, 730].

14.2.1 Fundamentals

Network designers have to be aware of and decide about suitable mechanisms to implement one or more of the following general **security goals** [729, Sec. 1.2]:

Confidentiality Information should only be revealed to authorized entities; any other entity should not be able to discover the information from eavesdropping or from reading memories.

Data integrity The receiver of information wants to be sure that it is not modified in transit, either intentionally or by accident. To distinguish unmodified "wanted" information from unmodified bogus information, the originator must be identifiable uniquely.

Accountability The entity requesting a service, triggering an action, or sending a packet must be uniquely identifiable.

Availability Legitimate entities should be able to access a certain service/information and to enjoy proper operation.

Controlled access A service or information access should only be granted to authorized entities.

Any security analysis must start with stating the desired security goals, followed by an assessment of the possible risks or security threats posed by an attacker. Some common threats are eavesdropping, masquerading (i.e. pretending to have another entity's identity), authorization violation (using services without being allowed to use them), provoking loss or modification of information, forgery (i.e. creating new information), repudiation, and sabotage.

When considering networking, some of the common attacks are eavesdropping as a purely passive attack, and insertion, deletion, or replaying of packets as an active attack. Attacks can be placed on all the layers of a given protocol stack.

Many countermeasures have been developed against these threats. These mechanisms frequently rely on symmetric or asymmetric **cryptographic algorithms** [731], [548]. These algorithms can be used to encrypt data packets, to sign these with almost unique hash/cryptographic check values, or to create certificates. Cryptographic algorithms essentially work by applying certain operations on combinations of the user data and specific **key values**, which optimally are only known to the sender and the receiver of a packet. Distributing these keys to the users and taking care of their lifecycle are essential parts of **key management** protocols. In practice, key management turns out to be the most complex part of security protocols; the raw encryption and decryption procedures are small but important building blocks.

14.2.2 Security considerations in wireless sensor networks

Can security measures and cryptographic protocols in wireless sensor networks be considered in the same way as for other types of networks? There is some consensus that the answer seems to be "no", for the following reasons:

- The network infrastructure of a WSN is made up of small, cheap nodes spread over a possibly hostile area. Unlike other types of networks, it is often impossible to prevent the sensor nodes from being physically accessed by attackers. This is also referred to as **node capture**. It is reasonable to assume that an attacker can achieve full control over a captured node, that is he can read its memory or influence the operation of the node software. Special secure memory devices would be needed to prevent the attacker from reading the memory; however, these will only rarely be present in cheap sensor nodes.
- The constraints regarding memory and computational capabilities are a serious obstacle for implementing cryptographic algorithms. Especially asymmetric key cryptography is considered too heavyweight for small processors, let alone the key management involved. The usage of several cryptographic block ciphers in sensor networks has been investigated in reference [471].
- When in-network processing is to be performed, intermediate nodes need to access and modify the information contained in packets; hence, a larger number of parties is involved in end-to-end information transfers.
- The finite energy budget of sensor nodes opens up a particularly attractive line of attacks: to force victim sensor nodes to exhaust their energy budget quickly and to die.

An additional challenge pointed out by Schäfer [730] is that attackers can have much more energy at their disposal than the sensor nodes. All security measures carried out by a sensor node require extra energy and stressing the node by attacks can cause premature depletion. This amounts to one particular kind of a **denial-of-service** attack (DoS). In the following, some of these DoS attacks are briefly described.

14.2.3 Denial-of-service attacks

Wood and Stankovic [891] consider a number of different denial-of-service attacks in sensor networks, working at different levels. Denial-of-service attacks in general can try to [730] (i) disable services, or (ii) to deplete service providers, for example, by overusing the service. To disable a sensor network's service, an attacker might simply destroy nodes. Although sensor networks have some resilience to node failures, the attacker can distort the network by destroying a large number of nodes or by focusing on especially important nodes, for example, sensor nodes in the vicinity of sinks that are needed for forwarding. In the following, however, we discuss protocol-related attacks.

Physical-layer and link-layer attacks

With **physical-layer jamming**, an attacker simply distorts radio communication. One way to achieve this is to place attacker nodes somewhere into the network and let them continuously send radio signals in the sensor network's frequency band. Especially effective is such an attack when the attacker nodes are close to sink nodes, effectively reducing a user's ability to control the network or to acquire data from it. A single attacker node can distort many neighbors at once and, by strategical placement of a number of attacker nodes, the whole sensor network can be disabled.

One possible countermeasure is the use of modulation schemes with some robustness against interference, for example, frequency-hopping or direct-sequence spread-spectrum techniques ([293, 297, 557]; see also Section 4.2.5). A second possible countermeasure is that the uncompromised sensor nodes reduce their duty cycle upon detecting such an attack. If the attacker has itself only a finite energy budget, it can persevere only for a limited time. A third countermeasure can be taken by routing protocols: If the attacker jams only a limited area, packets may be routed around. In protocols like directed diffusion, frequent interest dissemination can find working routes. Finally, sensor nodes with different physical layers can switch between these (for example, between a radio and an infrared transceiver).

A cleverer attacker can take knowledge about the protocols into account to save energy, giving rise to **link-layer jamming**. Especially, the MAC protocol is a good candidate. Let us consider, for example, protocols based on exchange of RTS/CTS packets (see Section 5.1.2) like PAMAS (Section 5.3.2) or S-MAC (Section 5.2.2). Whenever an attacker node a receives an RTS packet issued by some node x, it can answer with a jamming signal, interfering with any CTS packet sent to x. As a consequence, x has no transmit opportunity, backs off and tries again later with another RTS packet. According to Wood and Stankovic [891] no effective countermeasure against such an attack exists. The attacker might exploit the MAC protocol further to save energy. For example, in S-MAC the attacker can adapt its activity periods to the schedules of its neighbors.

Another ugly attack exploits MAC protocols using immediate acknowledgments and retransmissions. Upon receiving a data frame from node x, the attacker node can jam the acknowledgement frame destined to x. This causes x to back off, retransmit the same packet and to waste energy. Another way of depleting a node x is to continuously send RTS packets to this node, causing him to answer with CTS packets.

Network-layer attacks

Several types of attacks can be executed on the network layer. First, attacker nodes can behave similar to normal nodes; specifically, they can participate in routing protocols or dissemination of interests with the goal of directing routes to itself and to drop packets later on. This attack is called **black hole** attack. For example, in distance-vector protocols, the attacker can pretend to have particularly good routes to the sink. Dropping of packets destroys information, and furthermore, the forged route advertisements attract lots of traffic around the attacker, causing increased congestion levels and contention.

In a similar kind of attack, so-called **misdirections**, the adversary creates wrong routes, for example, by sending wrong route advertisement packets or by falsely answering route request packets. A wrong route can, for example, contain a loop and cause waste of energy. Another possible effect is that traffic does not reach the intended sink nodes. Instead of creating wrong routes, an adversary can also cause creation of unnecessary routes, for example, by issuing route lookup requests. All nodes participating in route selection waste their energy.

Even without actively trying to be included as a forwarder into routes, an attacker node can drop other nodes' packets and forward only its own packets. Such an attack is called **neglect and greed**. The attacker node can drop packets in a random fashion or all of them. Routing or data dissemination protocols that cache routes (like DSR or directed diffusion) are vulnerable to this attack. The attacker node participates in route setup and distorts, later on, the forwarding of data packets. When this behavior has been detected, the network may set up alternate routes or a source node can send multiple copies of a packet over node-disjoint routes from the beginning.

All these attacks have their source in adversary nodes participating in routing protocols. To prevent this, authentication and/or authorization mechanisms are needed to restrict routing protocols only to trustworthy nodes. Protocols for this purpose are beyond the scope of this chapter.

An attack called **homing** seeks to determine the geographic locations of certain important nodes in the network, for example clusterhead nodes. This information can be obtained from eavesdropping location-centric protocols. Once this information has been determined, the adversary can direct other attacks to these nodes. Clearly, a good way to prevent this attack is encryption of location information.

Transport layer and application attacks

If the transport layer uses explicit connections between identifiable nodes, either end of the connection needs to maintain some form of connection control block (CCB). Similar to TCP syn flood attacks, an attacker can issue a large number of connection setup requests and cause exhaustion of memory at the end nodes because of large numbers of unneeded CCBs.

Another kind of attack identified by WOOD and STANKOVIC [891] is **desynchronization**, which can be applied to transport protocols resting on sequence numbers. By issuing forged packets with wrong sequence numbers, the attacker can cause wasteful retransmissions or even cause the participants to end the connection.

In sensor networks deployed to detect certain environmental events, an attacker node can generate sensor data indicating this event, causing nodes in the vicinity or even the whole network to wake up and to start various activities. Possible countermeasures can be developed starting from outlier detection techniques.

14.2.4 Further reading

- Reference [730] is a survey article on sensor network security. It discusses some key management protocols in further detail. Key management issues in sensor networks are also discussed by ESCHENAUER and GLIGOR [244] and ZHU et al. [937].
- KARLOF and WAGNER [406] consider secure routing and network-layer attacks in sensor networks in some more detail and discuss also a number of countermeasures.
- PERRIG et al. [638] consider among others the problem of secure broadcast; see also [637].
- The *ACM Conference on Computer and Communication Security (CCS)* and the *IEEE Symposium on Security and Privacy* regularly present papers on sensor network security issues.

14.3 Application-specific support

In this section, we briefly describe three different tasks that sensor networks might be tasked with and which likely are important building blocks of sensor network applications. The first one is detection and tracking of (mobile) targets, for example, intruders into some site, the second one is detection of edges or of contours/isolines in the level of a continuous physical phenomenon, and the third is to obtain an estimate of a physical field. We also sketch some ideas of how these tasks can be approached with sensor networks.

All these applications are good showcases for geographic forwarding and geographic addressing since they make heavy use of position information and incorporate this into forwarding and routing decisions; this is especially true for tracking.

14.3.1 Target detection and tracking

The goal is to detect targets entering the area observed by a sensor network (for example, an elephant entering a habitat), to estimate their initial position and to update/"track" the position estimate as the target moves (**tracking**). Sometimes it is also required to *classify* the target, that is, to assign it to one of a finite number of possible target classes (for example, elephant versus giraffe versus ice bears).

Tracking a *single target* is already a challenging task, and tracking *multiple targets* has additional complexities. Consider, as an example, that the sensor network has detected two elephants entering the network at different places A and B. The elephants meet, spend some time together, and later on either elephant goes its own way, leaving the sensor field at locations C and D. Now, which elephant has left the field at location C? The one coming in from A or the other one? In a similar scenario but with one elephant replaced by a giraffe, classification can help with deciding this question. In general, however, it is necessary to *associate* sensor readings or position estimates to tracks. For example, a position estimate $\mathbf{q}$ can be associated to the single track whose estimated current position is closest to $\mathbf{q}$.

Networking Requirements

Let us have a brief look at the networking and signal processing requirements of the different tasks (compare Li et al. [482] and Brooks et al. [101]):

Detection Typically, a node can autonomously detect the target's presence by comparing its own sensor readings against some application-specific threshold, possibly after averaging multiple readings over a certain time window.

Target localization Depending on the sensing model (compare Section 13.2.1), the readings of a certain minimum number of nodes have to be combined. For example, to detect an acoustic event under the general sensing model, three noncollinear sensors are needed when sound intensity at the source is known, whereas four sensors are needed when this intensity is not known.

Classification An individual node can classify a target, for example, by looking at its spectral properties. Such classification algorithms work typically on time series of sensor readings. The classification result (or **decision**) is then communicated to other nodes. In a collaborative classification approach, some central node can either collect time series or decisions from individual nodes. Time series have a larger volume and thus require larger bandwidth, but in general provide more information to the central node. On the other hand, communication of decisions requires less bandwidth but information is lost. The trade-offs depend on the application and the accuracy/failure rate targets. Brooks et al. [101] explores these trade-offs further.

Tracking tracking involves several nodes, specifically those close to the target's trajectory. Tracking is discussed in more detail next.

A tracking approach

Let us consider the situation shown in Figure 14.6. There are at least two options:

Centralized processing All nodes sensing the target report their (timestamped) readings to the sink node, which combines them to obtain the desired estimates. For example, in the directed diffusion framework, the sink can issue corresponding interests.

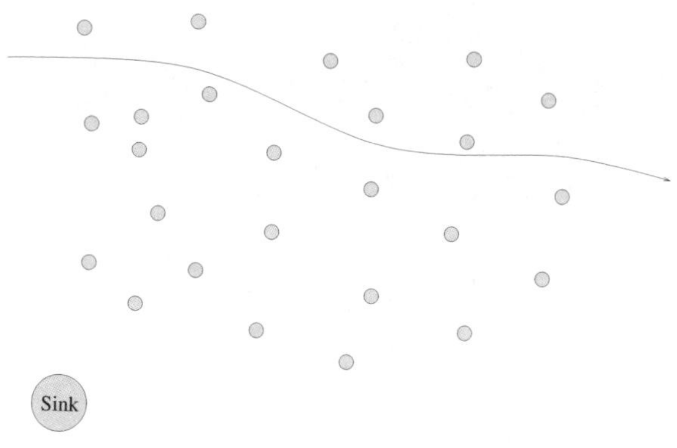

Figure 14.6 Example tracking scenario

Localized processing A **manager node** in the vicinity of the target collects local sensor readings and estimates the target position. By considering the current position as well as past positions and their timestamps, the manager node can also estimate the speed and heading of the target. As the target moves away, the manager hands off the whole target state (history of positions including the current one, speed, and heading estimate) to a new manager node. This new node is chosen such that its location is close to the projected target position at the time of handoff; the choice is thus *geography based*. This way, the estimated state *follows* the target in space and time. Additionally, manager nodes send target position reports to the sink from time to time.

The first approach imposes significant communication overhead when the target is far away from the sink. Another drawback is that in the presence of multiple targets the sink has to associate every position estimate with a target track, which requires significant computational efforts [101, 326]. On the other hand, in the second approach, the sensor nodes are mostly faced to only a single or a few targets simultaneously and can thus keep the amount of computation for association of position estimates to targets at reasonable levels. The second approach can also be combined with sleeping, since only nodes in the neighborhood of the target need to be awake and sensors can go back to sleep mode as soon as the target moves away. Different schemes for selecting the next manager node have been proposed.

In the work presented by Brooks et al. [101], the sensor network is subdivided dynamically into smaller units, called **spatial cells** (Figure 14.7). Cells can be in three different states:

Alerted cell An alerted cell has not yet detected the target, but expects it to enter the cell soon. One approach to become an alerted cell is to put cells from the network fringe a priori into the alerted state. A second way to enter this state is when the cell receives a target state packet from a neighbor cell's manager node; see below.

Active cell An active cell has detected the target and possesses an estimate of its state. This estimate is freshly created (for example, when a target enters the observed region) or is an updated version of the state received from a neighbor cell. The manager node predicts future positions of the target and alerts neighboring cells accordingly by transmitting the updated target state into that cell. As illustrated in Figure 14.7, it might well happen that a manager node alerts two or more neighboring cells, for example, cell 2 alerts the two cells 3a and 3b.

Sleeping cell A sleeping cell is neither alerted nor active, but ready to accept target state packets and to enter the alerted state.

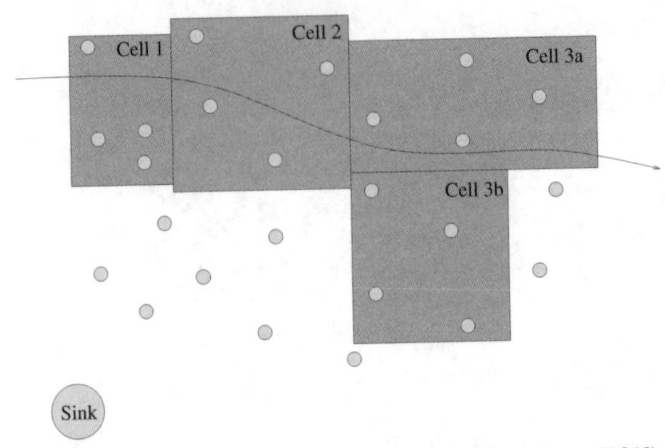

Figure 14.7 Example tracking scenario (adapted from: [101])

It is important to note that in this framework all decisions are purely local and involve only local communications. However, some important issues have to be resolved (i) How are cells constituted? (ii) How should packets be routed from cell to cell? and (iii) How are manager nodes within a cell determined? For the routing part, a reactive location-centric routing scheme called UW-routing is used. Route requests (RREQ) are directed toward cells and the first node in the target cell receiving an RREQ answers with a route reply packet. For computing the cell size, the current manager node takes, for example, the target speed and the sensing capabilities into account. For finding the next manager nodes, two promising techniques have been investigated in BROOKS et al. [101]:

- The manager node predicts the track from an extended Kalman filter and hands the target state over to nodes close to this track.
- In the **lateral inhibition** scheme, a potential new manager node A sends a packet indicating its interest to continue the track when it has detected the target. A delays the packet according to its estimated "goodness of fit" derived from its own sensor readings. For example, node A can take a signal amplitude as an estimate of its distance to the target. The shorter the distance (i.e. the better its goodness of fit), the shorter the waiting time. If another potential manager node B announces its interest earlier than A, then node A defers.

Further references

The IDSQ approach for selecting the next manager node is presented by ZHAO et al. [926] as well as by ZHAO and GUIBAS [924] (it is also discussed in more detail in Section 12.3.8). In a nutshell, in this *information-driven* approach, a manager node requests sensing values according to their anticipated information-utility, that is, in their projected ability to reduce the uncertainty in the position estimate. The next manager node is chosen according to a similar criterion.

A specific sensor network application for tracking the position of zebras is presented in reference [388]. Here, a box with a GPS receiver and some memory is attached to a zebra. The box records the zebra positions with a certain sampling rate. Since typical zebras cannot be expected to appear regularly at some data exchange point, a different approach has been chosen. Whenever two zebras meet, they exchange parts of their trajectories. This way, any zebra carries also information about the trajectory of those zebras it has met so far. Scientists can now drive through the habitat and try to meet as many zebras as possible, downloading their data.

Two further references dealing with detection, localization, or tracking algorithms are [310] and [925].

14.3.2 Contour/edge detection

Some sensor network applications require detection of contours or edges. Consider, as an example, a large field of chemical sensors. In case of an accident, it is important to get an idea of the position, extent, and shape of a toxic plum. We discuss the problem and some solutions in more detail, mostly for the case of static sensor fields.

Problem description

A slightly more abstract formulation of such a task assumes that the sensor network has to observe some scalar field and the user is interested in the isolines of this field.[3]

An even more abstract setting is to assume that each sensor is able to evaluate a Boolean **event predicate** [156] and the goal of an edge detection algorithm is to find the boundaries between areas where the predicate, evaluated by perfect sensors without any measurement errors, evaluates to true or false, respectively. The **interior** of a phenomenon is then defined as the area $U \subset \mathbb{R}^2$ for which the event predicate yields true, and the **exterior** is the complement of the interior. In calculus, the edge would correspond to the boundary of the interior, that is within every ε-neighborhood of an edge point (x, y), points from both the exterior and interior are contained. With this definition, with probability one no single sensor of a set of randomly deployed sensors would lie directly on an edge. As a more practical definition, a sensor is considered to be on the edge if (i) the sensor is in the interior and (ii) it has a distance smaller than some prescribed r to at least one point on the edge. The number r is called **tolerance radius**.

The difference between contour and edge detection is characterized by Chintalapudi and Govindan [156]. In edge detection, there is an explicit notion of interior and exterior points; in contour detection, this is not the case.

We can also distinguish different tasks in edge/contour detection:

- A single sensor wants to determine whether it is an interior, exterior, or edge sensor but there is no immediate need to communicate this result further to any other node. For example, an exterior sensor might choose longer sleep periods than an interior sensor.
- A user wants obtain an explicit geometric description of the edge/contour. Accordingly, this shape must be determined and communicated to the user. The complexity of a shape description relates directly to the communication overhead, as the following examples illustrate:
  - If the network designer assumes beforehand that all contours have circular shape, three parameters suffice to describe a contour in the plane – the center point (x and y coordinates) and the radius of the circle determined by the protocol. Accordingly, the whole description can be encapsulated into a single small packet.
  - If the contour is described by a polygon with n points, a number of $2n$ values must be transported to the sink node. The number n depends on the number of sensor nodes in the vicinity of the contour and on the number of individual points each node contributes.

The edge/contour detection problem has some similarities to edge or contour detection in the computer vision/image processing field [265]. However, there are also important differences:

- Image processing algorithms work on pixels that are nicely arranged in a grid. They can therefore rely on techniques that require this regularity, for example, Fourier transform techniques. It is,

[3] An isoline in a scalar field is an open or closed path such that all points on this path have (approximately) the same amplitude. In a continuously differentiable scalar field with nonvanishing gradients, the isolines are also locally continuous [264, Chap. 8].

however, reasonable to assume that such a grid placement is not the dominant case in sensor networks. Instead, the edges/contours have to be estimated from irregularly placed points.

- The sensor readings can be noisy.

Localized edge detection

Chintalapudi and Govindan [156] discuss three different schemes by which a node can locally decide whether it is within some radius $r > 0$ of an edge or not. All these schemes base their decision on locally obtained information. Specifically, a node s collects from all other nodes t within some radius $R > r$ of s their respective positions (x_t, y_t) and the value e_t of t's event predicate. This information is combined with s's own event predicate. The radius R is called **probing radius** and is a measure of the size of the neighborhood s considers and thus a measure of the communication overhead, which scales roughly as $O(R^2)$ for a sensor field of homogeneous density. The three schemes are as follows.

- A statistical scheme: node s collects the neighbors's positions and event predicates and computes some statistics of these values. As an example, s could count the number N_T of neighbors (including s) for which the event predicate evaluates to true and the number N_F of neighbors with value false. Node s decides to be an edge node if

$$1 - \frac{|N_\mathrm{T} - N_\mathrm{F}|}{N_\mathrm{T} + N_\mathrm{F}} \geq \gamma_0$$

  holds for some threshold value $\gamma_0 \in [0, 1]$. The choice of this threshold value depends on the sensor density, on the ratio R/r, on the probability that an event predicate is erroneous, and on the desired maximum rate of bogus edge decisions.
- A scheme inspired by image processing: a modified high-pass filter with weights considering the arbitrary node placement is used and the resulting spatial variation in both x and y is compared against a threshold value.
- A classifier-based approach: it is assumed that the edge/contour is "large" compared to R and that it can locally be well approximated by a straight line. The classifier running in node s tries to find a line such that the number of sensors having the same event predicate value on either side of the line is maximized. If the resulting line has a point with distance $<r$ to node s, then s assumes to be on the edge; otherwise s is an interior or exterior node.

Some important performance parameters for these schemes are the percentage of missed detections (i.e. where an edge sensor fails to detect this), the percentage of false detections (i.e. where a sensor believes to be an edge node while it is in fact not), the communication overhead, and the "thickness" of the determined edge (i.e. the average distance of edge nodes to the true edge). False detections can result from two sources. The first one is "really false" induced by noise in the event predicates, and the second source (called *unwanted detections*) is failure to distinguish edges passing within distance r from s and edges passing at a distance between r and R. Some important observations made by Chintalapudi and Govindan [156] are the following:

- There is an energy-accuracy trade-off. Increasing the probing radius R increases the accuracy (increased detection probability and thinner edges for nonincreasing false detection rates) but the communication overhead increases as well, since node s collects $O(R^2)$ event predicates.
- For $R/r \geq 2$ and linear edges, all schemes perform comparably well; for elliptical edges, the classifier-based scheme has an inferior detection capability.

- The first two schemes have an increasing amount of false positives as the ratio R/r increases, since the number of unwanted detections increases. On the other hand, for the classifier-based scheme, the amount of false detections *decreases* with increasing R/r since unwanted detections are almost ruled out a priori and really false detections become less likely with increasing number of neighbors considered.

A global contour detection scheme

The algorithm described by NOWAK and MITRA [601] considers a sensor field with randomly deployed nodes. An edge partitions the field into two parts such that in each part the sensor readings are equivalent. For ease of explanation, the field is assumed to be quadratic (Figure 14.8).

The algorithm essentially constructs a pruned quadtree in a bottom-up manner. In the first step, the sensor field is subdivided into squares or clusters at the finest resolution, for example, by a recursive dyadic partition algorithm. Within each square (i, j). a clusterhead collects the sensor readings/event predicates from the other nodes within its cluster and computes the average value $\theta_{i,j}$ and the sum of squared errors $R_{i,j}$ between the sensor readings and the average value. In the following step, the algorithm considers four neighboring clusters (arranged as a square of larger size) and checks whether these should be combined into a single larger cluster, represented by a new clusterhead. This check, based on a **complexity penalized estimator**, rests on two different terms. First, the new clusterhead computes a new average value θ taking all sensor readings of the new cluster into account. Then, the sum of squared errors R between θ and all the sensor readings of the large cluster is computed.[4] The second parameter relates to the number of leafs in the pruned tree. A smaller number of leaves leads to a more compact representation of the overall edge and thus less data must be sent to the user at the sink node. Ultimately, the decision of whether or not the four clusters are combined into a single one balances the decrease (if any) of the sum of squared errors R with respect to the sum of the $R_{i,j}$ of the component clusters on the one hand and the reduction in communication overhead on the other hand. This approach of combining (or not) four neighboring clusters into a new one is repeated toward higher and higher levels of aggregation.

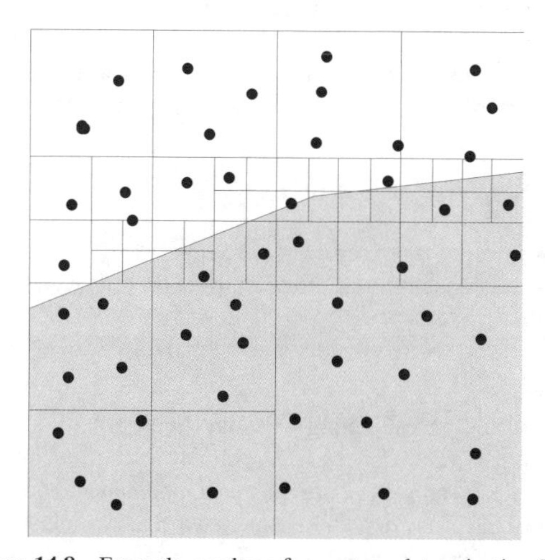

Figure 14.8 Example quadtree for contour determination [601]

[4] It is *not* necessary to pass all the single sensor readings from the children clusterheads to the new clusterhead; it suffices to send certain summary statistics.

It is useful to think of the pruned tree as an approximation of the sensor field by a two-dimensional step function, having a constant value in each of the surviving squares. After a certain number of iterations, the tree does not change anymore and the clusterheads of all surviving clusters can send their geographic coordinates and cluster sizes to the user who can derive the boundary from this information. It is shown that under certain assumptions the resulting tree has on average a number $O(\sqrt{n})$ of clusters.

NOWAK and MITRA [601] characterize the mean-square error of this scheme. The mean-square error is the sum over all sensors between the true sensor reading and the value of the step function according to the pruned quadtree, divided by the number of sensors. The mean-square error depends, among others, on the number of sensors and the "smoothness" of the boundary. For Lipschitz-continuous boundaries, the mean-square error behaves asymptotically as $\sim\sqrt{\frac{\log n}{n}}$, that is it goes to zero as n increases. By making even stronger assumptions about the boundary (for example, by requiring it to be a straight line), the speed of convergence can be made even faster.

Summarizing, by increasing the number of nodes, the mean-square error decreases, but on the other hand, the communication costs for constructing the pruned quadtree and for returning the final information to the user increase.

Further reading

In SAVVIDES et al. [727], an interesting scheme using mobile sensors is proposed. Once such a sensor has detected an event of interest (e.g. a plum of toxic gases), the node tries to move along the plum's perimeter and determine its boundary. Each node decides locally about its movements according to network connectivity requirements, the evolution of the plum perimeter, the terrain in question, and the presence of other nodes. This task can be seen as an instantiation of search problems for multiple mobile robots (see, for example, HAYES [336]).

In the setting considered by DANTU and SUKHATME [190], a static and randomly deployed sensor field is assumed. Furthermore, there is a single mobile sensor that desires to drive along the isolines of a scalar field. The mobile sensor communicates with neighboring static sensors, obtains the highest and lowest amplitude, and determines a new direction from this. To make this work, the scalar field is assumed to have a particularly simple shape, namely, concentric contours around a single source, the height of which decreases with distance from the source.

LIU et al. [516] use geometric dual-spaces for edge detection problems.

14.3.3 Field sampling

Consider a situation where a set of point sensors (see Section 13.2.1) like chemical sensors is tasked with drawing a "map" of the concentration of a certain chemical over the sensor field.

Problem description

Such a task can be formulated as estimating a **scalar field**, where a scalar field in two dimensions is simply a mapping $f : D \subset \mathbb{R}^2 \mapsto \mathbb{R}$. A similar definition holds for three-dimensional space. In general, scalar fields coming from physical phenomena are time variable. However, like most of the work on field sampling in sensor networks, we concentrate on a *snapshot*, that is, the state of the field at a certain time. The contribution of a single sensor to field estimation is a scalar measurement value associated with the sensors' geographical position, which must be known. This measurement value can be noisy. Furthermore, the measurement value provided by the sensor is a *quantized version* of the true scalar value, creating additional quantization noise.

What quality can such a reconstruction have and what is the associated communication overhead? Disregarding noise and quantization in the sensor readings for the moment, the quality depends on the **smoothness** or regularity of the sensor field. Some theoretical considerations are the following:

- A *perfect* reconstruction is only possible when the n sampled points delivered by the sensors are sufficient to determine the parameters of simple functions like polynomials. When an infinite number of sensors are available that additionally are evenly spaced, then according to Nyquist-type theorems [180], a broader class of functions can be perfectly reconstructed, the band-limited functions.[5]
- If the scalar field cannot be confined to simple functions, the n samples can only be used to derive an *approximation* to the true field. Different measures can be used to judge the quality of the approximation $\widehat{f}(\cdot)$ to the scalar field $f(\cdot)$, for example, the maximal difference $\|\widehat{f}(\cdot) - f(\cdot)\|_\infty$ or the L_2 distance $\|\widehat{f}(\cdot) - f(\cdot)\|_2 = \int(\widehat{f}(x) - f(x))^2\,\mathrm{d}x$.

Given that sensor readings are noisy and quantized, it is appropriate to turn to other quality measures, for example, the Mean Squared Error (MSE)

$$\frac{1}{|F|}\int_F E\left[(\widehat{f}(\cdot) - f(\cdot))^2\right]\mathrm{d}\mu$$

between the true field and the estimated field over the area F. Under particular assumptions on the smoothness of the field, for certain types of approximating fields and for certain assumptions on the stochastic process governing the measurement noise it is possible to find bounds for the MSE. In general, these MSE bounds depend on the number n of sensor nodes such that increasing n decreases the MSE. However, as pointed out by NOWAK et al. [602], a lower bound for the MSE is given by $O(n^{-1})$, which is the decay rate achievable for the simplest possible field estimation, a constant value.

An important problem for field estimation is the communication overhead. One obvious approach is to let every sensor send its reading toward the sink node and, for quickly varying scalar fields, one cannot hope to get away with significantly less transmissions when reasonable error bounds are desired. On the other hand, for smooth sensor fields neighboring sensors will have correlated/very similar sensor readings. In such a case, it is sufficient to locally compute some summary statistics and to send the result to the sink without compromising MSE targets too much. Even if there is additional local communication overhead, such a scheme will often pay out, since many sensor-to-sink packets will be saved. Similar ideas are explored in concepts for **distributed source coding** or **distributed compression** [34, 210, 539, 578, 653, 654, 809, 810].

A scheme using pruned trees of squares and platelets

NOWAK et al. [602] propose to adapt the concept of pruned quadtrees (as already used for contour detection) to field estimation as well, trying to find a good compromise between approximation quality (as measured by the MSE) and the communication overhead, which relates directly to the size of the tree. Besides, this approach preserves the possibility to find boundaries between different regions of the field, this way being able to treat noncontinuous or bandwidth-unlimited fields.

The n point sensors are arranged in a $\sqrt{n} \times \sqrt{n}$ regular grid in the square $[0, 1]^2$. For convenience, we assume that n is a power of two. Again, an initial dyadic partition of $[0, 1]^2$ consisting of n elementary squares hosting a single sensor is created, and each sensor makes a noisy measurement

[5] Briefly, for a function $f(\cdot)$ to be band-limited, it is required that the spectrum of $f(\cdot)$ (given by its Fourier transform) has bounded support.

of the scalar field in its square. The noise is assumed to consist of iid zero-mean Gaussian random variables with common variance for all sensors. NOWAK et al. [602] propose two different methods of estimating the field within a square:

- In the first method, the square is simply represented by a constant value having the smallest least squares error with respect to the sensor readings within this square. This estimator is referred to as the **Haar estimator**.
- The second method uses **platelets**. Specifically, the square is subdivided by a line that connects two points on the square boundary. This line partitions the square and for each partition the least squares error is determined. The line is chosen so as to minimize the sum of these errors. The set of possible lines is discretized by restricting the possible boundary points to a finite set. A platelet can be described by six points. One problem with platelets occurs when trying to combine four squares represented by platelets into a larger one; without further considerations the four child clusterheads would have to pass all the single sensor readings to the higher-level clusterhead. NOWAK et al. [602] propose a variation that comes reasonable close to the true platelet scheme and that passes only summary statistics of constant size; this scheme is referred to as constant-overhead platelet scheme in the following.

NOWAK et al. [602] demonstrate that for such pruned trees it is possible to find MSE bounds under certain regularity assumptions posed to the scalar field and for particular noise models. Specifically, they investigate the case of Hölder-α regular regions with Hölder-α regular[6] boundaries, assuming values of $\alpha \in \{1, 2\}$. The Hölder-2 regular fields are smoother than the Hölder-1 regular fields and it indeed turns out that lower MSE values can be achieved. NOWAK et al. [602] show for complexity-penalized estimators using the Haar or platelet approaches that the following holds:

- For Hölder-1 regular regions with Hölder-1 regular boundaries, the best possible estimator according to either the Haar or the platelet approach has an MSE performance bounded as follows:

$$O\left(n^{-1/2}\right) \leq \text{MSE} \leq O\left(\left(\frac{\log n}{n}\right)^{1/2}\right).$$

- For Hölder-2 regular regions with Hölder-2 regular boundaries and using the platelet approach, these bounds improve:

$$O\left(n^{-2/3}\right) \leq \text{MSE} \leq O\left(\left(\frac{\log n}{n}\right)^{2/3}\right).$$

- The scheme with the Haar estimator and the constant-overhead platelet scheme have similar communication costs when making the assumption that the costs of transmitting a packet are linear with the distance:
  - The in-network costs occur because of local communications of the clusterheads in the pruning process. To combine four elementary clusters, the packets of the clusterheads have to travel to an immediate neighbor. To combine four level-two clusters, the clusterheads packets have to travel twice this distance and so forth. It is shown that these in-network costs on average are given by:

$$O(\sqrt{n}).$$

[6] A function $f : D \subset \mathbb{R}^2 \mapsto \mathbb{C}$ is Hölder-α continuous with respect to a distance measure $\|\cdot\|$ if there exists a constant $c > 0$ such that for all $\mathbf{x}, \mathbf{y} \in D$ it is true that $|f(\mathbf{x}) - f(\mathbf{y})| \leq c \cdot \|\mathbf{x} - \mathbf{y}\|^\alpha$ holds; according to `http://cnx.rice.edu/content/m11172/latest/` the function is Hölder-regular of order $\gamma = \alpha + q$ when the q-th derivative of $f(\cdot)$ exists and is Hölder-α continuous.

– The external communication costs refer to the transmission of the final pruned quadtree to a sink node. Again, these costs behave as:

$$O(\sqrt{n}).$$

In summary, the scheme with pruned quadtrees has expected communication costs of $O(\sqrt{n})$.

Further Reading

Marco et al. [539] investigate field estimation from an information-theoretic perspective, trading off the number of nodes n on the one hand and the number of bits b_n making up a single quantized sensor reading on the other hand, given that the field is to be reconstructed at a sink with some prescribed MSE target. It is shown among others for stationary random fields with nonconstant autocorrelation that $b_n \to 0$ as $n \to \infty$. This is possible since, as the number of sensors increases for fixed size of the region to be covered, the sensor readings become more and more correlated. This can be exploited by sensors for reducing the number of bits used for representing values even when sensors have to make their measurements independently and without direct knowledge of neighboring sensors values. This is a consequence of Slepian–Wolf-type theorems [653, 774]. However, Marco et al. [539] show also that independent of the particular coding method $b_n \cdot n \to \infty$ as $n \to \infty$, that is the reduction in the number of bits needed by a single sensors decays not as quickly as the number of sensors increases.

Kumar et al. [455, 456] consider sampling of band-limited as well as non-band-limited functions with rapidly decaying spectra in one-dimensional sensor fields, exploring the trade-offs between sensor quality (i.e. number of bits k used by a sensor to encode its reading) and sensor density under a distortion criterion measuring the maximum difference $\|\cdot\|_\infty$ between true and estimated field. Dong et al. [208] consider reconstruction of another type of scalar field, namely a Gauss Markov field governed by a certain linear stochastic differential equation.

Bibliography

[1] A. Aaron and B. Girod. Compression with Side Information Using Turbo Codes. In *Proceedings of the IEEE Data Compression Conference*, pages 252–261, Snowbird, UT, April 2002.

[2] T. Abdelzaher, B. Blum, Q. Cao, Y. Chen, D. Evans, J. George, S. George, L. Gu, T. He, S. Krishnamurthy, L. Luo, S. Son, J. A. Stankovic, R. Stoleru, and A. Wood. EnviroTrack: Towards an Environmental Computing Paradigm for Distributed Sensor Networks. In *Proceedings of the IEEE International Conference on Distributed Computing Systems (ICDCS)*, Tokyo, Japan, March 2004.

[3] A. A. Abidi, G. J. Pottie, and W. J. Kaiser. Power-Conscious Design of Wireless Circuits and Systems. *Proceedings of the IEEE*, 88(10): 1528–1545, 2000.

[4] H. Abrach, J. Carlson, H. Dai, J. Rose, A. Sheth, B. Shucker, and R. Han. MANTIS: System Support for MultimodAL NETworks of In-situ Sensors. In *Proceedings of 2nd ACM International Workshop on Wireless Sensor Networks and Applications (WSNA)*, San Diego, CA, September 2003.

[5] N. Abramson. Development of the ALOHANET. *IEEE Transactions on Information Theory*, 31(2): 119–123, 1985.

[6] N. Abramson, editor. *Multiple Access Communications – Foundations for Emerging Technologies*. IEEE Press, New York, 1993.

[7] N. Abramson. Multiple Access in Wireless Digital Networks. *Proceedings of the IEEE*, 82(9): 1360–1370, 1994.

[8] ACPI – Advanced Configuration & Power Interface. `http://www.acpi.info/`, August 2003.

[9] S. Adireddy and L. Tong. Medium Access Control with Channel State Information for Large Sensor Networks. In *Proceedings of the 2002 IEEE International Workshop on Multimedia Signal Processing*, St. Thomas, Virgin Islands, December 2002.

[10] W. Adjie-Winoto, E. Schwartz, H. Balakrishnan, and J. Lilley. The Design and Implementation of an Intentional Naming System. In *Proceedings of the 17th ACM Symposium on Operating Systems Principles (SOSP'99)*, pages 186–201, Kiawah Island, SC, December 1999.

[11] S. Agarwal, S. V. Krishnamurthy, R. H. Katz, and S. K. Dao. Distributed Power Control in Ad-hoc Wireless Networks. In *Proceedings of the Personal Indoor Mobile Radio Conference (PIMRC)*, San Diego, CA, 2001.

[12] P. Agrawal, T. S. Teck, and A. L. Ananda. A Lightweight Protocol for Wireless Sensor Networks. In *Proceedings of the 2003 IEEE Wireless Communications and Networking (WCNC 2003)*, pages 1280–1285, New Orleans, LA, March 2003.

[13] D. Aguayo, J. Bicket, S. Biswas, G. Judd, and R. Morris. Link-Level Measurements from an 802.11b Mesh Network. In *Proceedings of ACM SIGCOMM'2004 Conference*, Portland, Oregon, DC, August 2004.

[14] L. Ahling and J. Zander. *Principles of Wireless Communication*. Studentlitteratur, 1997.

Protocols and Architectures for Wireless Sensor Networks H. Karl and A. Willig

[15] R. Ahlswede, N. Cai, S.-Y. R. Li, and R. W. Yeung. Network Information Flow. *IEEE Transaction on Information Theory*, 46(4): 1204–1216, 2000.

[16] G.-S. Ahn, A. T. Campbell, A. Veres, and L.-H. Sun. SWAN: Service Differentiation in Stateless Wireless Ad Hoc Networks. In *Proceedings of IEEE INFOCOM 2002*, New York, June 2002.

[17] I. F. Akyildiz, W. Su, Y. Sankasubramaniam, and E. Cayirci. Wireless Sensor Networks: A Survey. *Computer Networks*, 38: 393–422, 2002.

[18] I. F. Akyildiz, J. McNair, L.Carrasco, and R. Puigjaner. Medium Access Control Protocols for Multimedia Traffic in Wireless Networks. *IEEE Network Magazine*, 13(4): 39–47, 1999.

[19] J. Alonso, A. Dunkels, and T. Voigt. Bounds on the Energy Consumption of Routings in Wireless Sensor Networks. In *Proceeding of the 2nd International. Workshop on Modeling and Optimization in Mobile, Ad Hoc and Wireless Networks*, pages 62–70, Cambridge, UK, March 2004.

[20] K. Altinkemer, F. S. Salman, and P. Bellur. Solving the Minimum Energy Broadcasting Problem in Ad Hoc Wireless Networks by Integer Programming. In *Proceedings of the 2nd International Workshop on Modeling and Optimization in Mobile, Ad Hoc and Wireless Networks*, pages 48–54, Cambridge, UK, March 2004.

[21] A. D. Amis and R. Prakash. Load-balancing clusters in wireless ad hoc networks. In *Proceedings of the 3rd IEEE Symposium on Application-Specific Systems and Software Engineering Technology*, pages 25–32, Los Alamitos, CA, March 2000.

[22] A. D. Amis, R. Prakash, T. H. P. Vuong, and D. T. Huynh. Max-Min D-Cluster Formation in Wireless Ad Hoc Networks. In *Proceedings of INFOCOM*, New York, NY, March 1999.

[23] G. Anastasi, L. Lenzini, E. Mingozzi, A. Hettich, and A. Krämling. MAC Protocols for Wideband Wireless Local Access: Evolution Towards Wireless ATM. *IEEE Personal Communications*, 5(5): 53–64, 1998.

[24] E. Anceaume and I. Puaut. A Taxonomy of Clock Synchronization Algorithms. IRISA Research Report No. PI 1103, IRISA, 1997.

[25] A. Annamalai and V. K. Bhargava. Analysis and Optimization of Adaptive Multicopy Transmission ARQ Protocols for Time-Varying Channels. *IEEE Transactions on Communications*, 46(10): 1356–1368, 1998.

[26] G. Asada, M. Dong, T. S. Lin, F. Newberg, G. Pottie, and W. J. Kaiser. Wireless Integrated Network Sensors: Low Power Systems on a chip. In *Proceedings of the 1998 European Solid State Circuits Conference*, The Hague, Netherlands, 1998.

[27] R. B. Ash. *Information Theory*. Dover Publications, New York, 1990.

[28] ATmega 128(L) Preliminary Complete. ATmel product documentation, 2004.

[29] F. Aurenhammer. Voronoi Diagrams - A Survey of a Fundamental Geometric Data Structure. *ACM Computing Surveys*, 23(3): 345–405, 1991.

[30] R. Avnur and J. M. Hellerstein. Eddies: Continuously Adaptive Query Processing. In *Proceedings of the 2000 ACM SIGMOD International Conference on Management of Data*, pages 261–272, Dallas, TX, May 2000.

[31] B. Awerbuch. Optimal Distributed Algorithms for Minimum Weight Spanning tree, Counting, Leader Election and Related Problems. In *Proceedings of the 19th Annual ACM Symposium on Theory of Computing*, New York, May 1987.

[32] E. Ayanoglu, I. Chih-Lin, R. D. Gitlin, and J. E. Mazo. Diversity Coding for Tansparent Self-healing and Fault-tolerant Communication Networks. *IEEE Transactions on Communications*, 41(11): 1377–1386, 1993.

[33] E. Ayanoglu, S. Paul, T. F. LaPorta, K. K. Sabnani, and R. D. Gitlin. AIRMAIL: A Link-Layer Protocol for Wireless Networks. *Wireless Networks*, 1(1): 47–60, 1995.

[34] S. Jun Baek, G. D Veciana, and X. Su. Minimizing Energy Consumption in Large-Scale Sensor Networks Through Distributed Data Compression and Hierarchical Aggregation. *IEEE Journal on Selected Areas in Communications*, 22(6): 1130–1140, 2004.

[35] P. Bahl and V. N. Padmanabhan. RADAR: An In-Building RF-Based User Location and Tracking System. In *Proceedings of the IEEE INFOCOM*, pages 775–784, Tel-Aviv, Israel, April 2000.

[36] H. Bai and M. Atiquzzaman. Error Modeling Schemes for Fading Channels in Wireless Communications: A Survey. *IEEE Communications Surveys and Tutorials*, 5(2), 2003. http://www.comsoc.org/livepubs/surveys.

[37] J. Bajcsy and P. Mitran. Coding for the Slepian-Wolf Problem with Turbo-Codes. In *Proceedings of the IEEE Global Communications Conference (GLOBECOM)*, volume 2, pages 1400–1404, San Antonio, TX, November 2001.

[38] D. J. Baker and A. Ephremides. The Architectural Organization of a Mobile Radio Network via a Distributed Algorithm. *IEEE Transactions on Communications*, COM-29(11): 1694–1701, 1981.

[39] D. J. Baker and A. Ephremides. A Distributed Algorithm for Organizing Mobile Radio Telecommunication Networks. In *Proceedings of the 2nd IEEE International Conference on Distributed Computing Systems*, pages 476–483, Paris, France, April 1981.

[40] D. J. Baker, A. Ephremides, and J. A. Flynn. The Design and Simulation of a Mobile Radio Network with Distributed Control. *IEEE Journal on Selected Areas in Communications*, SAC-2(1): 226–237, 1984.

[41] H. Balakrishnan, V. Padmanabhan, S. Seshan, and R. H. Katz. A Comparison of Mechanisms for Improving TCP Performance over Wireless Links. *IEEE/ACM Transactions on Networking*, 5(6): 756ff, 1997.

[42] H. Balakrishnan, V. N. Padmanabhan, S. Seshan, M. Stemm, E. Amir, and R. H. Katz. TCP Improvements for Heterogeneous Networks: The Daedalus Approach. In *Proceedings of the 35th Annual Allerton Conference on Communication, Control, and Computing*, Urbana, IL, October 1997.

[43] H. Balakrishnan, C. L. Barrett, V. S. A. Kumar, M. V. Marathe, and S. Thite. The Distance-2 Matching Problem and its Relationship to the MAC-Layer Capacity of Ad Hoc Wireless Networks. *IEEE Journal on Selected Areas in Communications*, 22(6): 1069–1079, 2004.

[44] P. Baldi, L. De Nardis, and M.-G. Di Benedetto. Modeling and Optimization of UWB Communication Networks Through a Flexible Cost Function. *IEEE Journal on Selected Areas in Communications*, 20(9): 1733–1744, 2002.

[45] H. Baldus, K. Klabunde, and G. Muesch. Reliable Set-up of Medical Body-Sensor Networks. In *Proceedings of the Wireless Sensor Networks, First European Workshop (EWSN 2004)*, Berlin, Germany, January 2004.

[46] P. G. M. Baltus and R. Dekker. Optimizing RF Front Ends for Low Power. *Proceedings of the IEEE*, 88(10): 1546–1559, 2000.

[47] N. Bambos. Toward Power-Sensitive Network Architectures in Wireless Communications: Concepts, Issues, and Design Aspects. *IEEE Personal Communications*, 5: 50–59, 1998.

[48] N. Bambos and S. Kandukuri. Power Controlled Multiple Access (PCMA) in Wireless Communication Networks. In *Proceedings of the IEEE INFOCOM 2000*, Tel-Aviv, Israel, March 2000.

[49] S. Bandyopadhyay and E. J. Coyle. An Energy Efficient Hierarchical Clustering Algorithm for Wireless Sensor Networks. In *Proceedings of the IEEE INFOCOM*, San Francisco, CA, March 2003.

[50] A. Banerjea. Simulation Study of the Capacity Effects of Dispersity Routing for Fault Tolerant Real-Time Channels. *ACM SIGCOMM Computer Communication Review*, 26: 194–205, 1996.

[51] S. Banerjee and A. Misra. Minimum Energy Paths for Reliable communication in Multi-hop Wireless Networks. In *Proceedings of the 3rd ACM International Symposium on Mobile Ad Hoc Networking and Computing (MobiHoc)*, Lausanne, Switzerland, 2002.

[52] S. Bannerjee and S. Khuller. A Clustering Scheme for Hierarchical Control in Wireless Networks. In *Proceedings of the IEEE INFOCOM 2001*, Anchorage, AK, April 2001.

[53] L. Bao and J. J. Garcia-Luna-Aceves. Topology Management in Ad Hoc Networks. In *Proceedings of the 4th ACM International Symposium on Mobile Ad Hoc Networking and Computing (MobiHoc)*, Annapolis, MD, 2003.

[54] L. Bao and J. J. Garcia-Luna-Aceves. A New Approach to Channel Access Scheduling for Ad Hoc Networks. In *Proceedings of the Seventh Annual International Conference on Mobile Computing and Networking 2001 (MobiCom)*, Rome, Italy, July 2001.

[55] D. Barkai. *Peer-to-Peer Computing: Technologies for Sharing and Collaborating on the Net*. Intel Press, 2002.

[56] C.L. Barrett, S.J. Eidenbenz, and L. Kroc. Parametric Probabilistic Sensor Network Routing. In *Proceedings of the 2nd ACM International Workshop on Wireless Sensor Networks and Applications (WSNA)*, San Diego, CA, September 2003.

[57] K.A. Bartlett, R.A. Scantlebury, and P.T. Wilkinson. A Note on Reliable Full-Duplex Transmission over Half Duplex Lines. *Communications of the ACM*, 12(5): 260ff, 1969.

[58] S. Basagni. Distributed Clustering for Ad Hoc Networks. In A. Y. Zomaya, D. F. Hsu, O. Ibarra, S. Origuchi, D. Nassimi, and M. Palis, editors, *Proceedings of the International Symposium on Parallel Architectures, Algorithms, and Networks (I-SPAN)*, pages 310–315. IEEE Computer Society, Perth/Fremantle, Australia, June 1999.

[59] S. Basagni, I. Chlamtac, and A. Farago. A Generalized Clustering Algorithm for Peer-to-Peer Networks. In *Proceedings of the Workshop on Algorithmic Aspects of Communication (Satellite workshop of ICALP)*, Bologna, Italy, July 1997. Invited paper.

[60] S. Basagni, I. Chlamtac, and V. R. Syrotiuk. Geographic Messaging in Wireless Ad Hoc Networks. In *Proceedings of the 48th IEEE Vehicular Technology Conference*, pages 1957–1961, Houston, TX, May 1999.

[61] S. Basagni, I. Chlamtac, V. R. Syrotiuk, and B. A. Woodward. A Distance Routing Effect Algorithm for Mobility (DREAM). In *Proceedings of the 4th ACM/IEEE International Conference on Mobile Computing and Networking (MOBICOM)*, pages 76–84, Dallas, TX, October 1998.

[62] L. Benini, A. Bogliolo, and G. De Micheli. A Survey of Design Techniques for System-Level Dynamic Power Management. *IEEE Transactions on VLSI Systems*, 8(3): 299–316, 2000.

[63] L. Benini and G. De Micheli. *Dynamic Power Management Design Techniques and CAD Tools*. Kluwer, 1997.

[64] F. Bennett, D. Clarke, J. B. Evans, A. Hopper, A. Jones, and D. Leask. Piconet: Embedded Mobile Networking. *IEEE Personal Communications*, 4(5): 8–15, 1997.

[65] P. Bergamo, D. Maniezzo, A. Travasoni, A. Giovanardi, G. Mazzini, and M. Zorzi. Distributed Power Control for Energy Efficient Routing in Ad Hoc Networks. *Wireless Networks*, 10(1), 2004.

[66] C. Berrou. The Ten-Year-Old Turbo Codes are Entering into Service. *IEEE Communications Magazine*, 41(8): 110–116, 2003.

[67] C. Berrou and A. Glavieux. Near Optimum Error Correcting Coding and Decoding: Turbo-Codes. *IEEE Transactions on Communications*, 44(10): 1261–1271, 1996.

[68] D. Bertsekas and R. Gallager. *Data Networks*. Prentice Hall, Englewood Cliffs, NJ, 1987.

[69] C. Bettstetter. On the Minimum Node Degree and Connectivity of a Wireless Multihop Network. In *Proceedings of the 3rd ACM International Symposium on Mobile Ad Hoc Networking and Computing (MobiHoc)*, Lausanne, Switzerland, 2002.

[70] C. Bettstetter. Topology Properties of Ad Hoc Networks with Random Waypoint Mobility. In *Proceedings of the 4th ACM International Symposium on Mobile Ad Hoc Networking and Computing (MobiHoc)*, Annapolis, MD, 2003.

[71] C. Bettstetter and C. Hartmann. Connectivity of Wireless Multihop Networks in a Shadow Fading Environment. In *Proceedings of the 6th ACM International Workshop on Modeling, Analysis and Simulation of Wireless and Mobile Systmes (MSWiM)*, San Diego, CA, September 2003.

[72] J. Beutel, O. Kasten, F. Mattern, K. Römer, F. Siegemund, and L. Thiele. Prototyping Wireless Sensor Network Applications with BTNodes. In *Proceedings of the Wireless Sensor Networks, First European Workshop (EWSN 2004)*, Berlin, Germany, January 2004.

[73] P. Bhagwat, P. Bhattacharya, A. Krishna, and S. K. Tripathi. Using Channel State Dependent Packet Scheduling to Improve TCP Throughput Over Wireless LANs. *Wireless Networks*, 3(1): 91–102, 1997.

[74] Vijay K. Bhargava and Ivan J. Fair. Forward Error Correction Coding. In Jerry D. Gibson, editor, *The Communications Handbook*, pages 166–180. CRC Press/IEEE Press, Boca Raton, FL, 1996.

[75] V. Bharghavan. MACAW: A Media Access Protocol for Wireless LANs. In *Proceedings of ACM SIGCOMM'94 Conference*, London, UK, 1994.

[76] V. Bharghavan. A Dynamic Addressing Scheme for Wireless Media Access. In *Proceedings of IEEE ICC 95*, pages 756–760, Seattle, WA, June 1995.

[77] S. Bhattarcharya, H. Kim, S. Prabh, and T. Abdelzaher. Energy-Conserving Data Placement and Asynchronous Multicast in Wireless Sensor Networks. In *Proceedings of ACM/USENIX International Conference on Mobile Systems, Applications, and Services (MobiSys)*, pages 173–186, San Francisco, CA, May 2003.

[78] E. Biglieri. Digital Modulation Techniques. In Jerry D. Gibson, editor, *The Communications Handbook*, pages 273–287. CRC Press / IEEE Press, Boca Raton, FL, 1996.

[79] E. Biglieri, G. Caire, and G. Taricco. Coding and Modulation under Power Constraints. *IEEE Personal Communications*, 5(3): 32–38, 1998.

[80] E. Biglieri, J. Proakis, and S. Shamai. Fading Channels: Information-Theoretic and Communications Aspects. *IEEE Transactions on Information Theory*, 44(6): 2619–2692, 1998.

[81] L. Blazevic, S. Giordano, and J.-Y. Le Boudec. Self-Organize Terminodes Routing. *Journal of Cluster Computing*, 5(2): 205–218, 2002.

[82] N. Blefari-Melazzi, M.-G. Di Benedetto, M. Gerla, H. Luediger, M. Z. Win, and P. Withington, editors. Ultra-Wideband Radio Communication in Multiaccess Wireless Communications. *Journal on Selected Areas in Communications*, 20(9), 2002.

[83] D. Blough, M. Leoncini, G. Resta, and P. Santi. The K-Neigh Protocol for Symmetric Topology Control in Ad Hoc Networks. In *Proceedings of the 4th ACM International Symposium on Mobile Ad Hoc Networking and Computing (MobiHoc)*, Annapolis, MD, 2003.

[84] P. Blum, L. Meier, and L. Thiele. Improved Interval-Based Clock Synchronization in Sensor Networks. In *Proceedings of the Symposium on Information Processing in Sensor Networks (IPSN '04)*, Berkeley, CA, April 2004.

[85] A. Bogliolo, L. Benini, E. Lattanzi, and G. De Micheli. Specification and Analysis of Power-Managed Systems. *Proceedings of the IEEE*, 92(8): 1308–1346, 2004.

[86] E. Bonabeau, M. Doriga, and G. Theraulaz. *Swarm Intelligence: From Natural to Artifical Systems*. Oxford University Press, 1999.

[87] J. A. Bondy and U. S. R. Murty. *Graph Theory with Applications*. Elsevier North-Holland, 1976.

[88] P. Bonnet, J. E. Gehrke, and P. Seshadri. Querying the Physical World. *IEEE Personal Communications*, 7(5): 10–15, 2000. `http://lecs.cs.ucla.edu/Courses/CS213-Win02/Readings/PCM/Querying.pdf`.

[89] S. A. Borbash and M. J. McGlynn. Birthday Protocols for Low Energy Deployment and Flexible Neighbour Discovery in Ad Hoc Wireless Networks. In *Proceedings of the 2nd ACM International Symposium on Mobile Ad Hoc Networking and Computing (MobiHoc) 2001*, Long Beach, CA, 2001.

[90] C. Borcea, C. Intanagonwiwat, A. Saxena, and L. Iftode. Self-Routing in Pervasive Computing Environments Using Smart Messages. *Proceedings of the 1st IEEE International Conference on Pervasive Computing and Communications*, page 87. IEEE Computer Society, March 2003.

[91] G. Boriello and R. Want. Embedded Computation Meets the World Wide Web. *Communications of the ACM* , 43(5): 59–66, 2000.

[92] C. Bormann, editor, C. Burmeister, M. Degermark, H. Fukushima, H. Hannu, L-E. Jonsson, R. Hakenbeg, T. Koren, K. Le, Z. Liu, A. Martensson, A. Miyazaki, K. Svanbro, T. Wiebke, T. Yoshimura, and H. Zheng. Network Working Group. RObust Header Compression (ROHC): Framework and Four Profiles: RTP, UDP, ESP, and uncompressed. RFC 3095, 2001.

[93] P. Bose, P. Morin, I. Stojmenovic, and J. Urrutia. Routing with Guaranteed Delivery in Ad Hoc Wireless Networks. In *Proceedings of the 3rd International Workshop on Discrete Algorithms and Methods for Mobile Computing and Communications*, pages 48–55, Seattle, WA, 1999.

[94] A. Boukerche, X. Cheng, and J. Linus. Energy-Aware Data-Centric Routing in Microsensor Networks. In *Proceedings of the 6th ACM International Workshop on Modeling, Analysis and Simulation of Wireless and Mobile Systems (MSWiM)*, San Diego, CA, September 2003.

[95] A. Boulis, S. Ganeriwal, and M. B. Srivastava. Aggregation in Sensor Networks: An Energy Accuracy Trade-off. In *Proceedings of the 1st IEEE International Workshop on Sensor Network Protocols and Applications (SNPA)*, Anchorage, AK, May 2003.

[96] A. Boulis, C. C. Han, and M. B. Srivastava. Design and Implementation of a Framework for Programmable and Efficient Sensor Networks. In *Proceedings of the ACM/USENIX International Conference on Mobile Systems, Applications, and Services (MobiSys)*, San Francisco, CA, May 2003.

[97] R. Braden, T. Faber, and M. Handley. From Protocol Stack to Protocol Heap: Role-Based Architecture. *ACM SIGCOMM Computer Communication Review*, 33(1): 17–22, 2003.

[98] J. Bradshaw, editor. *Software Agents*. AAAI Press/MIT Press, Menlo Park, CA, 1996.

[99] D. Braginsky and D. Estrin. Rumour Routing Algorithm for Sensor Networks. In *Proceedings of the 1st Workshop on Sensor Networks and Applications*, Atlanta, GA, September 2002.

[100] A. Broder and M. Mitzenmacher. Optimal Plans for Aggregation. *Proceedings of the 21st Annual Symposium on Principles of Distributed Computing*, pages 144–152. ACM Press, 2002.

[101] R. R. Brooks, P. Ramanathan, and A. M. Sayeed. Distributed Target Classification and Tracking in Sensor Networks. *Proceedings of the IEEE*, 91(8): 1163–1171, 2003.

[102] S. Brooks and S. Iyengar. *Multi-Sensor Fusion*. Prentice-Hall, 1998.

[103] BTnodes A Distributed Environment for Prototyping Ad Hoc Networks, October 2004. Website `http://www.btnode.ethz.ch`.

[104] N. Bulusu, D. Estrin, L. Girod, and J. Heidemann. Scalable Coordination for Wireless Sensor Networks: Self-Configuring Localization Systems. In *Proceedings of the Sixth International Symposium on Communiation Theory and Applications*, Ambleside, Lake District, UK, July 2001. `http://www.isi.edu/scadds/papers/iscta-2001.ps`.

[105] N. Bulusu, J. Heidemann, V. Bychkovskiy, and D. Estrin. Density-Adaptive Beacon Placement Algorithms for Localization in Ad Hoc Wireless Networks. In *Proceedings of the INFOCOM*, New York, June 2002.

[106] N. Bulusu, J. Heidemann, and D. Estrin. GPS-Less Low Cost Outdoor Localization For Very Small Devices. *IEEE Personal Communications Magazine*, 7(5): 28–34, 2000.

[107] N. Bulusu, J. Heidemann, and D. Estrin. Adaptive Beacon Placement. In *Proceedings of the International Conference on Distributed Computing Systems (ICDCS)*, Mesa, AZ, 2001.

[108] N. Bulusu, J. Heidemann, D. Estrin, and T. Tran. Self-Configuring Localization Systems: Design and Experimental Evaluation. *ACM Transactions on Embedded Computing Systems*, 3(1): 24–60, 2004.

[109] T. Burd, A. Pering, A. Stratakos, and R. Brodersen. A Dynamic Voltage Scaled Microprocessor System. *IEEE Journal of Solid-State Circuits*, 35(11): 1571–1580, 2000.

[110] J. Burrell, T. Brooke, and R. Beckwith. Vineyard Computing: Sensor Networks in Agricultural Production. *IEEE Pervasive Computing*, 3(1): 38–45, 2004.

[111] S. F. Bush and A. B. Kulkarni. *Active Networks and Active Network Management: A Proactive Management Framework*. Plenum, 2001.

[112] J. Byers and G. Nasser. Utility-Based Decision-Making in Wireless Sensor Networks. In *Proceedings of the First Annual Workshop on Mobile and Ad Hoc Networking and Computing (MobiHOC'00)*, Boston, MA, August 2000.

[113] M. Cagalj, J.-P. Hubaux, and C. Enz. Minimum-Energy Broadcast in All-Wireless Networks: NP-Completeness and Distribution Issues. In *Proceedings of the 8th International Conference on Mobile Computing and Networking (ACM Mobicom)*, Atlanta, GA, September 2002.

[114] E. Callaway, P. Gorday, L. Hester, J. A. Gutierrez, M. Naeve, B. Heile, and V. Bahl. Home Networking with IEEE 802.15.4: A Developing Standard for Low-Rate Wireless Personal Area Networks. *IEEE Communications Magazine*, 40(8): 70–77, 2002.

[115] E. H. Callaway. *Wireless Sensor Networks – Architectures and Protocols*. Auerbach, Boca Raton, FL, 2003.

[116] R. Cam and C. Leung. Multiplexed ARQ for Time-Varying Channels – Part I: System Model and Throughput Analysis. *IEEE Transactions on Communications*, 46(1): 41–51, 1998.

[117] R. Cam and C. Leung. Multiplexed ARQ for Time-Varying Channels – Part II: Postponed Retransmission Modification and Numerical Results. *IEEE Transactions on Communications*, 46(3): 314–326, 1998.

[118] T. Camp and Y. Liu. An Adaptive Mesh-Based Protocol for Geocast Routing. *Journal of Parallel and Distributed Computing*, 62(2): 196–213, 2003.

[119] G. Carle and E. W. Biersack. Survey of Error Recovery Techniques for IP-Based Audio-Visual Multicast Applications. *IEEE Network Magazine*, 11(6): 24–36, 1997.

[120] J. Cartigny, D. Simplot, and I. Stojmenovic. Localized Minimum-Energy Broadcasting in Ad-Hoc Networks. In *Proceedings of the IEEE INFOCOM*, San Francisco, CA, March 2003.

[121] A. Carzaniga. *Architectures for an Event Notification Service Scalable to Wide-area Networks*. PhD thesis, Politecnico di Milano, Milano, Italy, December 1998.

[122] A. Carzaniga, D. S. Rosenblum, and A. L. Wolf. Design and Evaluation of a Wide-Area Event Notification Service. *ACM Transactions on Computer Systems (TOCS)*, 19(3): 332–383, 2001.

[123] A. Carzaniga and A. L. Wolf. Content-Based Networking: A New Communication Infrastructure. In *Proceedings of the NSF Workshop on an Infrastructure for Mobile and Wireless Systems*, Scottsdale, AZ, October 2001.

[124] J. K. Cavers. *Mobile Channel Characteristics*. Kluwer Academic Publishers, Boston, MA, 2000.

[125] U. Centinternel, A. Flinders, and Y. Sun. Power-Efficient Data Dissemination in Wireless Sensor Networks. In *Proceedings of the MobiDE*, San Diego, CA, September 2003.

[126] A. Cerpa, J. Elson, D. Estrin, L. Girod, M. Hamilton, and J. Zhao. Habitat Monitoring: Application Driver for Wireless Communications Technology. In *Proceedings of the ACM SIGCOMM Workshop on Data Communications in Latin America and the Caribbean*, San Jose, Costa Rica, 2001.

[127] A. Cerpa and D. Estrin. ASCENT: Adaptive Self-Configuring Sensor Networks Topologies. In *Proceedings of the INFOCOM*, New York, June 2002.

[128] K. Chakrabarty, S. S. Iyengar, H. Qi, and E. Cho. Coding Theory Framework for Target Location in Distributed Sensor Networks. In *Proceedings of the International Symposium on Information Technology: Coding and Computing*, pages 130–134, Las Vegas, NV, 2001.

[129] K. Chakrabarty, S. S. Iyengar, H. Qi, and E. Cho. Grid Coverage for Surveillance and Target Location in Distributed Sensor Networks. *IEEE Transactions on Computers*, 51(12): 1448–1453, 2002.

[130] H. Chan and A. Perrig. ACE: An Emergent Algorithm for Highly Uniform Cluster Formation. In H. Karl, A. Willig, and A. Wolisz, editors, *Proceedings of 1st European Workshop on Wireless Sensor Networks (EWSN)*, volume 2920 of *LNCS*, pages 154–171. Springer, Berlin, Germany, January 2004.

[131] R. Chandra, V. Ramasubramanian, and K. Birman. Anonymous Gossip: Improving Multicast Reliability in Mobile Ad-Hoc Networks. In *Proceedings of the International Conference on Distributed Computing Systems (ICDCS)*, pages 275–283, Mesa, AZ, 2001.

[132] A. Chandrakasan, R. Amirtharajah, C. S. Cho, and J. Goodman. Design Considerations for Distributed Microsensor Systems. In *Custom Integrated Circuits Conference*, pages 279–286, San Diego, CA, 1999.

[133] A. Chandrakasan, S. Sheng, and R. Brodersen. Low-Power CMOS Digital Design. *IEEE Journal of Solid-State Circuits*, 27(4): 473–484, 1992.

[134] A. P. Chandrakasan, R. Min, M. Bhardwaj, S.-H. Cho, and A. Wang. Power Aware Wireless Microsensor Systems. In *Proceedings of the ESSCIRC 2002*, Florence, Italy, September 2002.

[135] C.-Y. Chang, C.-T. Chang, and S.-C. Tu. Obstacle-Free Geocasting Protocols for Single/Multi-Destination Short Message Services in Ad Hoc Networks. *Wireless Networks*, 9(2): 143–155, 2001.

[136] J. H. Chang and L. Tassiulas. Routing for Maximum System Lifetime in Wireless Ad-Hoc Networks. In *Proceedings of the 39th Annual Allerton Conference on Communication, Control, and Computing*, Allerton, NY, October 1999.

[137] J.-H. Chang and L. Tassiulas. Energy Conserving Routing in Wireless Ad-Hoc Networks. In *Proceedings of the IEEE INFOCOM*, Tel-Aviv, Israel, March 2000.

[138] J. H. Chang and L. Tassiulas. Fast Approximate Algorithms for Maximum Lifetime Routing in Wireless Ad-Hoc Networks. In *Proceedings of the IFIP Networking*, May 2000.

[139] D. Charkraborty, A. Joshi, T. Finin, and Y. Yesha. GSD: A Novel Groupbased Service Discovery Protocol for MANETs. In *Proceedings of the 4th IEEE Conference on Mobile and Wireless Communication Networks*, Stockholm, Sweden, 2002.

[140] M. Chatterjee, S. Das, and D. Turgut. WCA: A Weighted Clustering Algorithm for Mobile Ad Hoc Networks. *Cluster Computing Journal*, 5: 193–204, 2002.

[141] B. Chen, K. Jamieson, H. Balakrishnan, and R. Morris. Span: An Energy-Efficient Coordination Algorithm for Topology Maintenance in Ad Hoc Wireless Networks. *Wireless Networks*, 8(5): 481–494, 2002.

[142] J. C. Chen, K. Yao, and R. E. Hudson. Source Localization and Beamforming. *IEEE Signal Processing Magazine*, 19(2): 30–39, 2002.

[143] J.-C. Chen, K. M. Sivalingam, P. Agrawal, and S. Kishore. A Comparison of MAC Protocols for Wireless Local Networks Based on Battery Power Consumption. In *Proceedings of the INFOCOM 1998*, San Francisco, CA, March 1998. .

[144] K. Chen and K. Nahrstedt. Effective Location-Guided Tree Construction Algorithms for Small Group Multicast in MANET. In *Proceedings of the INFOCOM*, New York, June 2002.

[145] R. Chen, K. C. Chua, B. T. Tan, and C. S. Ng. Adaptive Error Coding Using Channel Prediction. *Wireless Networks*, 5(1): 23–32, 1999.

[146] W. Chen and N. Huang. The Strongly Connecting Problem on Multihop Packet Radio Networks. *IEEE Transactions on Communications*, 37: 293–295, 1989.

[147] X. Chen and J. Wu. Chapter Multicasting Techniques in Mobile Ad Hoc Networks. *The Handbook of Ad Hoc Wireless Networks*, pages 2-1–2-16. CRC Press, 2003.

[148] Y. P. Chen and A. L. Liestman. Approximating Minimum Size Weakly-Connected Dominating Sets for Clustering Mobile Ad Hoc Networks. In *Proceedings of the 3rd ACM International Symposium on Mobile Ad Hoc Networking and Computing (MobiHoc)*, Lausanne, Switzerland, 2002.

[149] C.-C. Chiang, H. Wu, W. Liu, and M. Gerla. Routing in Clustered Multihop, Mobile Wireless Networks. In *Proceedings of the IEEE Singapore International Conference on Networks*, pages 197–211, Singapore, Malaysia, 1997.

[150] C.-C. Chiang, H.-K. Wu, W. Liu, and M. Gerla. Routing in Clustered Multihop, Mobile Wireless Networks with Fading Channel. In *Proceedings of the IEEE SICON*, pages 197–211, Singapore, 1997.

[151] C. Chiasserini, I. Chlamtac, P. Monti, and A. Nucci. Energy-Efficient Design of Wireless ad Hoc Networks. In *Proceedings of the IFIP Networking*, Pisa, Italy, 2002.

[152] C.-F. Chiasserini and R. Rao. On the Concept of Distributed Digital Signal Processing in Wireless Sensor Networks. In *Proceedings of the. IEEE Military Communication Conference (MILCOM)*, Anaheim, CA, October 2002.

[153] C. F. Chiasserini and R. R. Rao. Energy Efficient Battery Management. In *Proceedings of the IEEE INFOCOM*, Tel-Aviv, Israel, March 2000.

[154] C.-F. Chiasserini and R. R. Rao. Coexistence Mechanisms for Interference Mitigation in the 2.4-GHz ISM Band. *IEEE Transactions on Wireless Communications*, 2(5): 964–975, 2003.

[155] C. Chien, I. Elgorriaga, and C. McConaghy. Low-Power Direct-Sequence Spread-Spectrum Modem Architecture for Distributed Wireless Sensor Networks. In *Proceedings of the International Symposium on Low Power Electronics and Design (ISLPED)*, Huntington Beach, CA, August 2001.

[156] K. K. Chintalapudi and R. Govindan. Localized Edge Detection in Wireless Sensor Networks. In *Proceedings of the IEEE ICC Workshop on Sensor Network Protocols and Applications*, Anchorage, AK, April 2003.

[157] CC1000 Single Chip Very Low Power RF Transceiver. Chipcon Product Data Sheet. `http://www.chipcon.com/files/CC1000_Data_Sheet_2_1.pdf`.

[158] CC2420 2.4 GHz IEEE 802.15.4 / Zigbee RF Transceiver. Chipcon Product Data Sheet. `http://www.chipcon.com/files/CC2420_Data_Sheet_1_0.pdf`.

[159] I. Chlamtac and A. Farago. A New Approach to the Design and Analysis of Peer-to-Peer Mobile Networks. *Wireless Networks*, 5(3): 149–156, 1999.

[160] I. Chlamtac and A. Farago. Making Transmission Schedules Immune to Topology Changes in Multi-Hop Packet Radio Networks. *IEEE/ACM Transactions on Networking*, 2(1): 23–29, 1994.

[161] I. Chlamtac, A. Farago, and H. Zhang. Time-Spread Multiple-Access (TSMA) Protocols for Multihop Mobile Radio Networks. *IEEE/ACM Transactions on Networking*, 5(6): 804–812, 1997.

[162] I. Chlamtac, C. Petrioli, and J. Redi. Energy-Conserving Selective Repeat ARQ Protocols for Wireless Data Networks. In *Proceedings of the IEEE Personal, Indoor and Mobile Radio Conference (PIMRC '98)*, Boston, MA, September 1998.

[163] I. Chlamtac, C. Petrioli, and J. Redi. Energy-Conserving Access Protocols for Identification Networks. *IEEE/ACM Transactions on Networking*, 7(1): 51–59, 1999.

[164] J. Chou, D. Petrovic, and K. Ramchandran. Tracking and Exploiting Correlations in Dense Sensor Networks. In *Asilomar Conference on Signals, Systems, and Computers*, Pacific Grove, CA, November 2002.

[165] J. Chou, D. Petrovic, and K. Ramchandran. A Distributed and Adaptive Signal Processing Approach to Reducing Energy Consumption in Sensor Networks. In *Proceedings of the IEEE INFOCOM*, San Francisco, CA, March 2003.

[166] E. Christensen, F. Curbera, G. Meredith, and S. Weerawarana. Web Service Description Language (WSDL) 1.1. W2C Note, `http://www.w3.org/TR/wsdl`, March 2001.

[167] M. Chu, H. Haussecker, and F. Zhao. Scalable Information-Driven Sensor Querying and Routing for Ad Hoc Heterogeneous Sensor Networks. *International Journal of High Performance Computing Applications*, 16(3): 293–313, 2002.

[168] P. B. Chu, N. R. Lo, E. C. Berg, and K. S. J. Pister. Optical Communication Using Micro Corner Cube Reflectors. In *Proceedings of IEEE MEMS Workshop*, pages 350–355, Nagoya, Japan, 1997.

[169] I. Cidon and M. Sidi. Distributed Assignment Algorithms for Multihop Packet Radio Networks. *IEEE Transactions on Computers*, 38: 1353–1361, October 1989.

[170] D. D. Clark and D. L. Tennenhouse. Architectural Consideration for a New Generation of Protocols. In *Proceedings of the SIGCOMM '90*, pages 200–208, Philadelphia, PA, September 1990.

[171] A. E. F. Clementi, P. Crescenzi, P. Penna, G. Rossi, and P. Vocca. On the Complexity of Computing Minimum Energy Consumption Broadcast Subgraphs. In *Proceedings of the 18th Annual Symposium on Theoretical Aspects of Computer Science*, pages 121–131, Dresden, Germany, February 2001.

[172] A. E. F. Clementi, P. Penna, and R. Silvestri. Hardness Results for the Power Range Assignment Problem in Packet Radio Networks. In *Proceedings of the 2nd International Workshop on Approximation Algorithms for Combinatorial Optimization Problems (APPROX)*, pages 197–208, Berkeley, CA, 1999.

[173] T. Clouqueur, V. Phipatanasuphorn, P. Ramanathan, and K. K. Saluja. Sensor Deployment Strategy for Target Detection. In *Proceedings of the First ACM International Workshop on Wireless Sensor Networks and Applications (WSNA'02)*, pages 42–48, Atlanta, GA, 2002.

[174] T. Clouqueur, V. Phipatanasuphorn, P. Ramanathan, and K. K. Saluja. Sensor Deployment Strategy for Detection of Targets Traversing a Region. *MONET - Mobile Networks and Applications*, 8(4): 453–461, 2003.

[175] T. Clouqueur, P. Ramanathan, and K. K. Saluja. Exposure of Variable Speed Targets through a Sensor Field. In *Proceedings of the 6th Annual Conference on Information Fusion*, July 2003.

[176] W. R. Cockayne and M. Zyda. *Mobile Agents*. Prentice Hall, 1998.

[177] W. S. Conner, J. Chhabra, M. Yarvis, and L. Krishnamurthy. Experimental Evaluation of Synchronization and Topology Control for In-building Sensor Network Applications. In *Proceedings of the 2nd ACM International Workshop on Wireless Sensor Networks and Applications (WSNA)*, San Diego, CA, September 2003.

[178] D. J. Costello, J. Hagenauer, H. Imai, and S. B. Wicker. Applications of Error-Control Coding. *IEEE Transactions on Information Theory*, 44(6): 2531–2560, 1998.

[179] G. Coulouris, J. Dollimore, and T. Kindberg. *Distributed Systems – Concepts and Design*. Addison-Wesley, Harlow, England, third edition, 2001.

[180] T. M. Cover and J. A. Thomas. *Elements of Information Theory*. John Wiley & Sons, New York, 1991.

[181] J. M. Cramer, M. Z. Win, and R. A. Scholtz. Impulse Radio Multipath Characteristics and Diversity Reception. In *Proceedings of the IEEE International Conference on Communications (ICC)*, Atlanta, GA, 1998.

[182] P. Crescenzi and V. Kann. A Compendium of NP Optimization Problems. `http://www.nada.kth.se/~viggo/wwwcompendium/wwwcompendium.html`, February 2004.

[183] R. Cristescu and M. Vetterli. Power Efficient Gathering of Correlated Data: Optimization, NP-Completeness and Heuristics. In *Proceedings of the 4th ACM International Symposium on Mobile Ad Hoc Networking and Computing (MobiHoc)*, Annapolis, MD, 2003.

[184] F. Cristian. Probabilistic Clock Synchronization. *Distributed Computing*, 3: 146–158, 1989.

[185] R. L. Cruz and A. V. Santhanam. Optimal Routing, Link Scheduling and Power Control in Multi-hop Wireless Networks. In *Proceedings of the IEEE INFOCOM*, San Francisco, CA, March 2003.

[186] B. J. Culpepper, L. Dung, and M. Moh. Design and Analysis of Hybrid Indirect Transmission (HIT) for Data Gathering in Wireless Micro Sensor Networks. *ACM Mobile Computing and Communications Review*, 8(1): 61–83, 2004.

[187] F. Cuomo and C. Martello. MAC Principles for an Ultra Wide Band Wireless Access. In *Proceedings of the Global Telecommunications Conference (GLOBECOM)*, volume 6, pages 3548–3552, San Antonio, TX, 2001.

[188] F. Dai and J. Wu. Distributed Dominant Pruning in Ad Hoc Networks. In *Proceedings of the International . Conference on Communications (ICC)*, Anchorage, AK, May 2003.

[189] H. Dai and R. Han. TSync: A Lightweight Bidirectional Time Synchronization Service for Wireless Sensor Networks. *ACM SIGMOBILE Mobile Computing and Communications Review*, 8(1): 125–139, 2004.

[190] K. Dantu and G. S. Sukhatme. Poster Abstract: Contour Detection Using Actuated Sensor Networks. In *Proceedings of the ACM SenSys 03*, Los Angeles, CA, November 2003. Poster Abstract.

[191] A. K. Das, R. J. Marks, M. El-Sharkawi, P. Arabshabi, and A. Gray. Minimum Power Broadcast Trees for Wireless Networks: Integer Programming Formulations. In *Proceedings of the IEEE INFOCOM*, San Francisco, CA, March 2003.

[192] B. Das and V. Bharghavan. Routing in Ad-Hoc Networks Using Minimum Connected Dominating Sets. In *Proceedings of the International Conference on Communication (ICC)*, Montreal, Canada, June 1997.

[193] K. Dasgupta, K. Kalpakis, and P. Namjoshi. An Efficient Clustering-based Heuristic for Data Gathering and Aggregation in Sensor Networks. In *Proceedings of the IEEE Wireless Communications and Networking Conference (WCNC)*, New Orleans, LA, March 2003.

[194] K. Dasguptta, M. Kukreja, and K. Kalpakis. Topology-Aware Placement and Role Assignment for Energy-Efficient Information Gathering in Sensor Network. In *Proceedings of the 8th IEEE Symposium on Computers and Communications (ISCC)*, pages 341–348, Kemer, Turkey, July 2003.

[195] S. De, C. Qiao, and H. Wu. Meshed Multipath Routing: An Efficient Strategy in Sensor Networks. In *Proceedings of the IEEE Wireless Communications and Networking Conference (WCNC)*, New Orleans, LA, March 2003.

[196] M. de Prycker. *Asynchronous Transfer Mode – Solution for Broadband ISDN*. Prentice Hall, 1995.

[197] B. Deb, S. Bhatnagar, and B. Nath. Multi-Resolution State Retrieval in Sensor Networks. In *Proceedings of the 1st IEEE International Workshop on Sensor Network Protocols and Applications (SNPA)*, Anchorage, AK, May 2003.

[198] B. Deb, S. Bhatnagar, and B. Nath. Information Assurance in Sensor Networks. In *Proceedings of the 2nd ACM International Workshop on Wireless Sensor Networks and Applications (WSNA)*, San Diego, CA, September 2003.

[199] B. Deb, S. Bhatnagar, and B. Nath. ReInForM: Reliable Information Forwarding using Multiple Paths in Sensor Networks. In *Proceedings of the 28th Annual IEEE Conference on Local Computer Networks (LCN 2003)*, Bonn, Germany, October 2003.

[200] J.-D. Decotignie. Wireless Fieldbusses – A Survey of Issues and Solutions. In *Proceedings of the 15th IFAC World Congress on Automatic Control (IFAC 2002)*, Barcelona, Spain, 2002.

[201] A. Demers, D. Greene, C. Hauser, W. Irish, J. Larson, S. Shenker, H. Sturgis, D. Swinehart, and D. Terry. Epidemic Algorithms for Replicated Database Maintenance. In *Proceedings of the Annual ACM Symposium on Principles of Distributed Computing (PODC)*, pages 1–12, Vancouver, BC, Canada, 1987.

[202] M. Demirbas and H. Ferhatosmanoglu. Peer-to-Peer Spatial Queries in Sensor Networks. In *Proceedings of the International Conference on Peer-to-Peer Computing*, Linköping, Sweden, September 2003.

[203] J. Deng and Z. J. Haas. Dual Busy Tone Multiple Access (DBTMA): A New Medium Access Control for Packet Radio Networks. In *Proceedings of the IEEE ICUPC'98*, Florence, Italy, October 1998.

[204] S. S. Dhillon, K. Chakrabarty, and S. S. Iyengar. Sensor Placement for Grid Coverage under Imprecise Detections. In *Proceedings of the International Conference on Information Fusion (FUSION 2002)*, pages 1581–1587, Annapolis, MD, 2002.

[205] J. Diaz, M. D. Penrose, J. Petit, and M. Serna. Convergence Theorems for Some Layout Measures on Random Lattice and Random Geometric Graphs. *Combinatorics, Probability, and Computing*, 6: 489–511, 2000.

[206] L. Doherty, L. El Ghaoui, and K. S. J. Pister. Convex Position Estimation in Wireless Sensor Networks. In *Proceedings of the IEEE INFOCOM*, pages 1655–1663, Anchorage, AK, April 2001.

[207] M. Dong, L. Tong, and B. M. Sadler. Source Reconstruction via Mobile Agents in Sensor Networks: Throughput-Distortion Characteristics. In *Proceedings of the IEEE Military Communication Conference (Milcom)*, Boston, MA, October 2003.

[208] M. Dong, L. Tong, and B. M. Sadler. Optimal Reconstruction of Gauss Markov Field in Large Sensor Networks. In *Proceedings of First International Symposium on Control, Communications and Signal Processing*, Hammamet, Tunesia, March 2003.

[209] A. Doufexi, S. Armour, M. Butler, A. Nix, D. Bull, and J. McGeehan. A Comparison of the HIPERLAN/2 and IEEE 802.11a Wireless LAN Standards. *IEEE Communications Magazine*, 40(5): 172–180, 2002.

[210] S. C. Draper and G. W. Wornell. Side Information Aware Coding Strategies for Sensor Networks. *IEEE Journal on Selected Areas in Communications*, 22 (6): 966–976, 2004.

[211] R. Droms. *Dynamic Host Configuration Protocol*. RFC 1541, 1993.

[212] R. Dube, C. D. Rais, K.-Y. Wang, and S. K. Tripathi. Signal Stability Based Adaptive Routing (SSA) for Ad-Hoc Mobile Networks. *IEEE Personal Communications Magazine*, 4(1): 36–45, February 1997.

[213] D. Duchamp and N.F. Reynolds. Measured Performance of Wireless LAN. In *Proceedings of 17th Conference on Local Computer Networks, Minneapolis*, Minneapolis, MN, 1992.

[214] S. Dulmann, T. Nieberg, J. Wu, and P. Havinga. Trade-Off between Traffic Overhead and Reliability in Multipath Routing for Wireless Sensor Networks. In *Proceedings of the IEEE Wireless Communications and Networking Conference (WCNC)*, New Orleans, LA, March 2003.

[215] A. Dunkels, J. Alonso, and T. Voigt. Making TPC/IP Viable for Wireless Sensor Networks. In *Proceedings of the Work-in-Progress Session of the 1st European Workshop on Wireless Sensor Networks (EWSN)*, Technical Report TKN-04-001 of Technical University Berlin, Telecommunication Networks Group, Berlin, Germany, January 2004.

[216] A. Dunkels, D. Grönvall, and T. Voigt. Contiki – a Lightweight and Flexible Operating System for Tiny Networked Sensors. In *Proceedings of the First IEEE Workshop on Embedded Networked Sensors (EmNetS)*, Tampa, FL, November 2004.

[217] A. Dunkels. Full TCP/IP for 8-Bit Architectures. In *Proceedings of the First International Conference on Mobile Applications, Systems and Services (MOBISYS 2003)*, San Francisco, CA, May 2003.

[218] A. Dunkels, J. Alonso, T. Voigt, H. Ritter, and J. Schiller. Connecting Wireless Sensornets with TCP/IP Networks. In *Proceedings of the Second International Conference on Wired/Wireless Internet Communications (WWIC2004)*, Frankfurt, Germany, February 2004.

[219] A. Dunkels, T. Voigt, J. Alonso, and H. Ritter. Distributed TCP Caching for Wireless Sensor Networks. In *Proceedings of the Third Mediterranean ad Hoc Networking Conference (MedHocNet)*, June 2004.

[220] R. Eberhart and J. Kennedy. *Swarm Intelligence*. Morgan Kaufmann, 2001.

[221] J.-P. Ebert, B. Burns, and A. Wolisz. A Trace-Based Approach for Determining the Energy Consumption of a WLAN Network Interface. In *Proceedings of the European Wireless*, pages 230–236, Florence, Italy, February 2002.

[222] J.-P. Ebert and A. Wolisz. Combined Tuning of RF Power and Medium Access Control for WLANs. *MONET - Mobile Networks and Applications*, 6(5): 417–426, 2000.

[223] D. A. Eckhardt and P. Steenkiste. A Trace-Based Evaluation of Adaptive Error Correction for a Wireless Local Area Network. *MONET - Mobile Networks and Applications*, 4: 273–287, 1999.

[224] eCos. The eCos Operating System. `http://www.redhat.com/ecos`.

[225] K. Egevang and P. Francis. The IP Network Address Translator (NAT). RFC 1631, May 1994.

[226] A. El-Hoiydi, J.-D. Decotignie, C. Enz, and E. Le Roux. Poster Abstract: WiseMAC, an Ultra Low Power MAC Protocol for the WiseNET Wireless Sensor Network. In *Proceedings of the ACM SenSys 03*, Los Angeles, CA, November 2003. Poster Abstract.

[227] A. El-Hoiydi. ALOHA with Preamble Sampling for Sporadic Traffic an Ad Hoc Wireless Sensor Networks. In *Proceedings of the IEEE International Conference on Communications (ICC)*, New York, April 2002.

[228] A. El-Hoiydi. Spatial TDMA and CSMA with Preamble Sampling for Low Power Ad Hoc Wireless Sensor Networks. In *Proceedings of the International Symposium on Computation and Communication*, 2002.

[229] A. El-Rabbany. *Introduction to GPS: The Global Positioning System*. Artech House, 2002.

[230] M. Elaoud and P. Ramanathan. Adaptive Use of Error-Correcting Codes for Real-Time Communication in Wireless Networks. In *Proceedings of the INFOCOM 1998*, San Francisco, March 1998.

[231] E. O. Elliot. Estimates of Error Rates for Codes on Burst-Noise Channels. *Bell Systems Technical Journal*, 42: 1977–1997, 1963.

[232] E. Elnahrawy and B. Nath. Cleaning and Querying Noisy Sensors. In *Proceedings of the 2nd ACM International Workshop on Wireless Sensor Networks and Applications (WSNA)*, San Diego, CA, September 2003.

[233] J. Elson and D. Estrin. An Address-Free Architecture for Dynamic Sensor Networks. Technical Report 00-724, Computer Science Department USC, January 2000.

[234] J. Elson and D. Estrin. Random, Ephemeral Transaction Identifiers in Dynamic Sensor Networks. In *Proceedings of the 21st International Conference on Distributed Computing Systems (ICDCS-21)*, Phoenix, AZ, April 2001.

[235] J. Elson and D. Estrin. Time Synchronization for Wireless Sensor Networks. In *Proceedings of the 2001 International Parallel and Distributed Processing Symposium (IPDPS), Workshop on Parallel and Distributed Computing Issues in Wireless Networks and Mobile Computing*, pages 1965–1970, San Francisco, CA, April 2001.

[236] J. Elson, L. Girod, and D. Estrin. Fine-Grained Network Time Synchronization using Reference Broadcasts. In *Proceedings of the Fifth Symposium on Operating Systems Design and Implementation (OSDI 2002)*, Boston, MA, December 2002.

[237] J. Elson, L. Girod, and D. Estrin. Short Paper: A Wireless Time-Synchronized COTS Sensor Platform, Part I: System Architecture. In *Proceedings of the IEEE CAS Workshop on Wireless Communications and Networking*, Pasadena, CA, September 2002.

[238] J. Elson and K. Römer. Wireless Sensor Networks: A New Regime for Time Synchronization. In *Proceedings of the First Workshop on Hot Topics In Networks (HotNets-I)*, Princeton, NJ, October 2002.

[239] J. E. Elson. *Time Synchronization in Wireless Sensor Networks*. PhD dissertation, University of California, Los Angeles, CA, Department of Computer Science, 2003.

[240] R. F. Ember - Embedded. *Design of an IEEE 802.15.4 Compliant, EmberNet Ready and ZigBee Ready Communication Module using the EM2420 RF Transceiver*, 2004.

[241] Y. Ephraim and N. Merhav. Hidden Markov Processes. *IEEE Transactions on Information Theory*, 48(6): 1518–1569, 2002.

[242] D. Eppstein. Chapter Spanning trees and Spanners, *Handbook of Computational Geometry*, pages 425–461. Elsevier, Amsterdam, NL, 2000.

[243] A. H. Epstein. Milimeter-Scale, MEMS Gas Turbine Engines. In *Proceedings of the ASME Turbo Expo 2003 – Power for Land, Sea, and Air*, Atlanta, GA, 2003.

[244] L. Eschenauer and V. D. Gligor. A Key-Management Scheme for Distributed Sensor Networks. In *Proceedings of the 9th ACM conference on Computer and Communication Security, CCS'02*, Washington, DC, 2003.

[245] D. Estrin, L. Girod, G. Pottie, and M. Srivastava. Instrumenting the World with Wireless Sensor Networks. In *Proceedings of the International Conference on Acoustics, Speech and Signal Processing (ICASSP 2001)*, Salt Lake City, UT, May 2001.

[246] D. Estrin, R. Govindan, J. Heidemann, and S. Kumar. Next Century Challenges: Scalable Coordination in Sensor Networks. In *Proceedings of the Fifth Annual International Conference on Mobile Computing and Networks (MobiCom 1999)*, Seattle, Washington, DC, 1999.

[247] ETSI. *TR 101 683, HIPERLAN Type 2: System Overview*. ETSI, February 2000.

[248] ETSI. *TS 101 475, BRAN, HIPERLAN Type 2: Physical (PHY) Layer*. ETSI, March 2000.

[249] ETSI. *TS 101 761-1, BRAN, HIPERLAN Type 2: Data Link Control (DLC) Layer, Part 1: Basic Data Transport Function*. ETSI, March 2000.

[250] ETSI. *TS 101 761-2, BRAN, HIPERLAN Type 2: Data Link Control (DLC) Layer, Part 2: Radio Link Control Protocol Basic Functions*. ETSI, March 2000.

[251] P. Th. Eugster, P. A. Felber, R. Guerraoui, and A.-M. Kermarrec. The Many Faces of Publish/Subscribe. *ACM Computing Surveys (CSUR)*, 35(2): 114–131, 2003.

[252] K. Fall. A Delay-Tolerant Network Architecture for Challenged Internets. In *Proceedings of the ACM SIGCOMM*, pages 27–34, Karlsruhe, Germany, 2003.

[253] A. Faradjian, J. E. Gehrke, and P. Bonnet. GADT: A Probability Space ADT for Representing and Querying the Physical World. In *Proceedings of the 18th International Conference on Data Engineering (ICDE)*, San Jose, CA, February 2002.

[254] L. M. Feeney and M. Nilsson. Investigating the Energy Consumption of a Wireless Network Interface in an Ad Hoc Networking Environment. In *Proceedings of the IEEE INFOCOM 2001*, Anchorage, Alaska, AK, April 2001.

[255] W. Feller. *An Introduction to Probability Theory and Its Applications - Volume I*. John Wiley, New York, third edition, 1968.

[256] A. Ferreira and A. Jarry. Complexity of Minimum Spanning Tree in Evolving Graphs and the Minimum-Energy Broadcast Routing Problem. In *Proceedings of the 2nd International Workshop on Modeling and Optimization in Mobile, Ad Hoc and Wireless Networks*, pages 55–61, Cambridge, UK, March 2004.

[257] A. Festag. Optimization of Handover Performance by Link Layer Triggers in IP-Based Networks; Parameters, Protocol Extensions, and APIs for Implementation. Technical Report TKN-02-014, Telecommunication Networks Group, Technische Universität Berlin, July 2002.

[258] G. Finn. Routing and Addressing Problems in Large Metropolitan-Scale Internetworks. ISI Research Report ISI/RR-87-180, University of Southern California, March 1987.

[259] K. Flautner, S. Reinhardt, and T. Mudge. Automatic Performance-Setting for Dynamic Voltage Scaling. In *Proceedings of the 7th ACM Annual International Conference on Mobile Computing and Networking (Mobicom)*, pages 260–271, Rome, Italy, July 2001.

[260] J. Flinn, S. Y. Park, and M. Satyanarayanan. Balancing Performance, Energy, and Quality in Pervasive Computing. In *Proceedings of the IEEE 22nd International Conference on Distributed Computing Systems (ICDCS)*, pages 217–226, Vienna, Austria, July 2002.

[261] C. Florens and R. McEliece. Packets Distribution Algorithms for Sensor Networks. In *Proceedings of the IEEE INFOCOM*, San Francisco, CA, March 2003.

[262] S. Floyd and V. Jacobson. Random Early Detection Gateways for Congestion Avoidance. *IEEE/ACM Transactions on Networking*, 1(4): 397–413, 1993.

[263] S. Floyd, V. Jacobson, C.-G. Liu, S. McCanne, and L. Zhang. A Reliable Multicast Framework for Light-Weight Sessions and Application Level Framing. *IEEE/ACM Transactions on Networking*, 5(6): 784–803, 1997.

[264] O. Forster. *Analysis 2 – Differentialrechnung Im* $\mathbb{R}^n$*, Gewöhnliche Differentialgleichungen*. Vieweg, Braunschweig/Wiesbaden, 5th edition, 1977.

[265] D. A. Forsyth and J. Ponce. *Computer Vision: A Modern Approach*. Prentice Hall, Englewood Cliffs, NJ, 2003.

[266] J. Frolik. QoS Control for Random Access Wireless Sensor Networks. In *Proceedings of the 2004 Wireless Communication and Networking Conference (WCNC04)*, Atlanta, GA, March 2004.

[267] Z. Fu, P. Zerfos, H. Luo, S. Lu, L. Zhang, and M. Gerla. The Impact of Multihop Wireless Channel on TCP Throughput and Loss. In *Proceedings of the IEEE INFOCOM*, San Francisco, CA, March 2003.

[268] C. L. Fullmer and J. J. Garcia-Luna-Aceves. Solutions to Hidden Terminal Problems in Wireless Networks. In *Proceedings of ACM SIGCOMM'97 Conference*, pages 39–49, Cannes, France, September 1997.

[269] W. F. Fung, D. Sun, and J. Gehrke. COUGAR: The Network is the Database. *Proceedings of ACM SIGMOD International Conference on Management of Data*, pages 621–621. ACM Press, 2002.

[270] R. Gandhi, S. Parthasarathy, and A. Mishra. Minimizing Broadcast Latency and Redundancy in Ad Hoc Networks. In *Proceedings of the 4th ACM International Symposium on Mobile Ad Hoc Networking and Computing (MobiHoc)*, Annapolis, MD, 2003.

[271] S. Ganeriwal, R. Kumar, S. Adlakha, and M. Srivastava. Network-Wide Time Synchronization in Sensor Networks. Technical Report NESL 01-01-2003, Networked and Embedded Systems Lab (NESL), University of California, Los Angeles (UCLA), 2003.

[272] S. Ganeriwal, R. Kumar, and M. B. Srivastava. Timing-Sync Protocol for Sensor Networks. In *Proceedings of the 1st ACM International Conference on Embedded Networked Sensor Systems (SenSys)*, pages 138–149, Los Angeles, CA, November 2003.

[273] D. Ganesan, A. Cerpa, W. Ye, Y. Yu, J. Zhao, and D. Estrin. Networking Issues in Wireless Sensor Networks. *Journal of Parallel and Distributed Computing (JPDC)*, 64(7): 799–814, 2004, Special issue on Frontiers in Distributed Sensor Networks.

[274] D. Ganesan, D. Estrin, and J. Heidemann. DIMENSIONS: Why do we need a New Data Handling Architecture for Sensor Networks? *ACM SIGCOMM Computer Communication Review*, 33(1): 143–148, 2003.

[275] D. Ganesan, B. Greenstein, D. Perelyubskiy, D. Estrin, and J. Heideman. An Evaluation of Multi-resolution Storage for Sensor Networks. In *Proceedings of the 1st ACM International Conference on Embedded Networked Sensor Systems (SenSys)*, pages 89–102, Los Angeles, CA, November 2003.

[276] D. Ganesan, R. Govindan, S. Shenker, and D. Estrin. Highly-Resilient, Energy-Efficient Multipath Routing in Wireless Sensor Networks. *Mobile Computing and Communications Review (MC2R)*, 1(2): 28–36, 2002.

[277] D. Ganesan, B. Krishnamachari, A. Woo, D. Culler, D. Estrin, and S. Wicker. Complex Behavior at Scale: An Experimental Study of Low-Power Wireless Sensor Networks. Technical Report UCLA/CSD-TR 02-0013, Computer Science Department, University of California, Los Angeles (UCLA), CA, 2002.

[278] J. Gao, L. J. Guibas, J. Hershberger, L. Zhang, and A. Zhu. Discrete Mobile Centers. In *Proceedings of the 17th ACM Symposium on Computational Geometry (SoCG)*, pages 190–198, Medfords, MA, June 2001.

[279] J. Gao, L. J. Guibas, J. Hershberger, L. Zhang, and A. Zhu. Geometric Spanners for Routing in Mobile Networks. In *Proceedings of the 2nd ACM International Symposium on Mobile Ad Hoc Networking and Computing (MobiHoc)*, Long Beach, CA, 2001.

[280] R. X. Gao and P. Hünerberg. Design of a CDMA-Based Wireless Data Transmitter for Embedded Sensing. *IEEE Transactions on Instrumentation and Measurement*, 51(6): 1259–1265, 2002.

[281] R.X. Gao and P. Hunerberg. CDMA-Based Wireless Data Transmitter for Embedded Sensors. In *Proceedings of the 18th IEEE Instrumentation and Measurement Technology Conference*, volume 3, pages 1778 –1783, Budapest, Hungary, 2001.

[282] J. Garcia-Frias, W. Zhong, and Y. Zhao. Iterative Decoding Schemes for Source and Channel Coding of Correlated Sources. In *Proceedings of the 36th Asilomar Conference on Signals, Systems, and Computers*, Monterey, CA, November 2002.

[283] J. J. Garcia-Luna-Aceves and E. L. Madruga. The Core-Assisted Mesh Protocol. *IEEE Journal on Selected Areas in Communications*, 17(8), 1999.

[284] M. S. Gast. *802.11 Wireless Networks – The Definitive Guide*. O'Reilly, Sebastopol, CA, 2002.

[285] D. Gay, P. Levis, R. von Behren, M. Welsh, E. Brewer, and D. Culler. The nesC Language: A Holistic Approach to Networked Embedded Systems. *Proceedings of ACM SIGPLAN Conference on Programming Language Design and Implementation*, pages 1–11. ACM Press, 2003.

[286] C. N. Georghiades. Synchronization. In Jerry D. Gibson, editor, *The Communications Handbook*, pages 255–272. CRC Press / IEEE Press, Boca Raton, FL, 1996.

[287] M. Gerla, T. J. Kwon, and G. Pei. On Demand Routing in Large Ad Hoc Wireless Networks with Passive Clustering. In *Proceedings of the 2nd IEEE Wireless Communications and Networking Conference (WCNC)*, Chicago, IL, September 2000.

[288] M. Gerla and J. T-C. Tsai. Multicluster, Mobile, Multimedia Radio Network. *ACM/Baltzer Wireless Networks*, 1(3): 255–265, 1995.

[289] M. Gerla, K. Tang, and R. Bagrodia. TCP Performance in Wireless Multi-hop Networks. In *Proceedings of the Second IEEE Workshop on Mobile Computing Systems and Applications, 1999 (WMCSA '99)*, pages 41–50, New Orleans, LA, February 1999.

[290] E. N. Gilbert. Capacity of a Burst-Noise Channel. *Bell Systems Technical Journal*, 39: 1253–1265.

[291] L. Girod and D. Estrin. Robust Range Estimation using Acoustic and Multimodal Sensing. In *Proceedings of the IEEE/RSJ International Conference on Intelligent Robots and Systems (IROS)*, Maui, HI, October 2001.

[292] N. Glance, D. Snowdown, and J.-L. Meunier. Pollen: Using People as a Communication Medium. *Computer Networks*, 35(4): 429–442, 2001.

[293] S. Glisic and B. Vucetic. *Spread Spectrum CDMA Systems for Wireless Communications*. Artech House, Boston, MA, 1997.

[294] M. Goel and N. R. Shanbhag. Low-Power Channel Coding via Dynamic Reconfiguration. In *Proceedings of the International Conference on Acoustics, Speech and Signal Processing (ICASSP)*, Phoenix, Arizona, March 1999.

[295] S. Goel and T. Imielinski. Prediction-based Monitoring in Sensor Networks: Taking Lessons from MPEG. *ACM SIGCOMM Computer Communinication Review*, 31(5): 82–98, 2001.

[296] S. Goel, T. Imielinski, K. Ozbay, and B. Nath. Sensor on Wheels – Towards a Zero-Infrastructure Solution for Intelligent Transportation Systems. In *Proceedings of the 1st ACM International Conference on Embedded Networked Sensor Systems (SenSys)*, pages 338–339, Los Angeles, CA, November 2003.

[297] A. M. J. Goiser. *Handbuch der Spread-Spectrum Technik*. Springer Verlag, Wien, New York, 1998.

[298] A. Goldsmith and S. B. Wicker. Special Issue: Energy-Aware Ad Hoc Wireless Networks. *IEEE Wireless Communications*, 9, 2002.

[299] A. J. Goldsmith and S. B. Wicker. Design Challenges for Energy-Constrained Ad Hoc Wireless Networks. *IEEE Wireless Communications*, 9(4): 8–27, 2002.

[300] J. Gomez, A. T. Campbell, M. Naghshineh, and C. Bisdikian. Power-Aware Routing in Wireless Packet Radio. In *Proceedings of the 6th IEEE International Workshop on Mobile Multimedia Communications (MoMuC)*, San Diego, CA, November 1999. http://comet.columbia.edu/~campbell/andrew/publications/papers/momuc99c%.pdf.

[301] J. Gomez, A. T. Campbell, M. Nashineh, and C. Bisdikian. Conserving Transmission Power in Wireless Ad Hoc Networks. *Proceedings of the 9th Internationa Conference on Network Protocols (ICNP)*. PARO, November 2001.

[302] K. Govil, E. Chan, and H. Wasserman. Comparing Algorithms for Dynamic Speed-Setting of a Low-Power CPU. In *Proceedings of the 1st Conference on Mobile Computing and Networking*, pages 13–25, Berkeley, CA, November 1995.

[303] R. Govindan, J. M. Hellerstein, W. Hong, S. Madden, M. Franklin, and S. Shenker. The Sensor Network as a Database. Technical Report 02-771, USC/Information Sciences Institute, September 2002.

[304] R. Graybill and R. Melhem, editors. *Power Aware Computing*. Kluwer, 2002.

[305] B. Greenstein, D. Estrin, R. Govindan, S. Ratnasamy, and S. Shenker. DIFS: A Distributed Index for Features in Sensor Networks. In *Proceedings of the 1st IEEE International Workshop on Sensor Network Protocols and Applications (SNPA)*, Anchorage, AK, May 2003.

[306] M. Grossglauser and M. Vetterli. Locating Nodes with EASE: Last Encounter Routing in Ad Hoc Networks through Mobility Diffusion. In *Proceedings of the IEEE INFOCOM*, San Francisco, CA, March 2003.

[307] F. Gruian. Hard Real-Time Scheduling for Low Energy using Stochastic Data and DVS Processor. In *Proceedings of the International Symposium on Low Power Electronics and Design (ISLPED)*, pages 46–51, Huntington Beach, CA, August 2001.

[308] M. Grünewald, T. Lukovszki, C. Schindelhauer, and K. Volbert. Distributed Maintenance of Resource Efficient Wireless Network Topologies. In *Proceedings of the 8th International Euro-Par Conference*, pages 935–946, Paderborn, Germany, 2002.

[309] S. Guha and S. Khuller. Approximation Algorithms for Connected Dominating Set. *Algorithmica*, 20: 374–387, 1998.

[310] L. J. Guibas. Sensing, Tracking and Reasoning with Relations. *IEEE Signal Processing Magazine*, pages 73–85, March 2002.

[311] A. Chandra V. Gummalla and John O. Limb. Wireless Medium Access Control Protocols. *IEEE Communications Surveys and Tutorials*, 3(2): 2–15, 2000. http://www.comsoc.org/pubs/surveys.

[312] C. Guo, L. C. Zhong, and J. M. Rabaey. Low Power Distributed MAC for Ad Hoc Sensor Networks. In *Proceedings of the IEEE GlobeCom*, San Antonio, AZ, November 2001. http://bwrc.eecs.berkeley.edu/People/Grad_Students/czhong/documents/glo%becom2001.pdf.

[313] H. Gupta, S. Das, and Q. Gu. Connected Sensor Cover: Self-Organization of Sensor Networks for Efficient Query Execution. In *Proceedings of the 4th ACM International Symposium on Mobile Ad Hoc Networking and Computing (MobiHoc)*, Annapolis, MD, 2003.

[314] I. Gupta, R. van Renesse, and K. P. Birman. Scalable Fault-Tolerant Aggregation in Large Process Groups. In *Proceedings of the International Conference on Dependable Systems and Networks*, Goteborg, Sweden, July 2001. http://www.cs.cornell.edu/gupta/gupta_aggregn_dsn01.ps.

[315] P. Gupta and P. R. Kumar. Critical Power for Asymptotic Connectivity in Wireless Networks. In W.M. McEneany, G. Yin, and Q. Zhang, editors, *Stochastic Analysis, Control, Optimization and Applications*, pages 547–566. Birkhauser, Boston, MA, 1998.

[316] P. Gupta and P. R. Kumar. The Capacity of Wireless Networks. *IEEE Transactions on Information Theory*, 46(2): 388–404, 2000.

[317] J. A. Gutierrez, M. Naeve, E. Callaway, V. Mitter, and B. Heile. IEEE 802.15.4: A Developing Standard for Low-Power Low-Cost Wireless Personal Area Networks. *IEEE Network Magazine*, 15(5): 12–19, 2001.

[318] J. C. Haartsen. The Bluetooth Radio System. *IEEE Personal Communications*, 7(1): 28–36, 2000.

[319] J. C. Haartsen and S. Mattisson. Bluetooth – A New Low-Power Radio Interface Providing Short-Range Connectivity. *Proceedings of the IEEE*, 88(10): 1651–1661, 2000.

[320] Z. J. Haas, J. Y. Halpern, and L. Li. Gossip-Based Ad Hoc Routing. In *Proceedings of the IEEE INFOCOM*, New York, June 2002.

[321] Z. J. Haas and J. Deng. Dual Busy Tone Multiple Access (DBTMA)-Performance Evaluation. In *Proceedings of the IEEE Vehicular Technology Conference 1999 (VTC99)*, Houston, TX, May 1999.

[322] D. Haccoun and S. Pierre. Automatic Repeat Request. In Jerry D. Gibson, editor, *The Communications Handbook*, pages 181–198. CRC Press/IEEE Press, Boca Raton, FL, 1996.

[323] E. Haines. Point in Polygon Strategies. In Paul S. Heckbert, editor, *Graphics Gems IV*, pages 24–46. Academic Press, Boston, MA, 1994.

[324] M. Hajiaghayi, N. Immorlica, and V. S. Mirrokni. Power Optimization in Fault-Tolerant Topology Control Algorithms for Wireless Multi-hop Networks. *Proceedings of the 9th Annual International Conference on Mobile Computing and Networking*, pages 300–312. ACM Press, 2003.

[325] M. N. Halgamuge, S. M. Guru, and A. Jennings. Energy Efficient Cluster Formation in Wireless Sensor Networks. In *International Conference on Telecommunications (ICT)*, Papeete, Tahiti, 2003.

[326] D. L. Hall and J. Llinas. An Introduction to Multisensor Data Fusion. *Proceedings of the IEEE*, 85(1): 6–23, 1997.

[327] F. Halsall. *Data Communications, Computer Networks and Open Systems*. Addison-Wesley, Reading, MA, 1996.

[328] M. Hamdaoui and P. Ramanathan. A Dynamic Priority Assignment Technique for Streams with (m, k)-Firm Deadlines. *IEEE Transactions on Computers*, 44(12): 1443–1451, 1995.

[329] J. Handy. Energy Consumption Looms Large in Choosing Flash for Portable Applications. EEdesign, July 2001. `http://www.eedesign.com/isd/features/OEG20010711S0066`.

[330] V. Handziski, H. Karl, A. Köpke, and A. Wolisz. A Common Wireless Sensor Network Architecture? In H. Karl, editor, Proceedings 1. GI/ITG Fachgespräch "Sensornetze". Technical Report TKN-03-012 of the Telecommunications Networks Group, Technische Universität Berlin, pages 10–17, Berlin, Germany, July 2003.

[331] V. Handziski, A. Köpke, H. Karl, C. Frank, and W. Drytkiewicz. Improving the Energy Efficiency of Directed Diffusion Using Passive Clustering. In H. Karl, A. Willig, and A. Wolisz, editors, *Proceedings of the 1st European Workshop on Wireless Sensor Networks (EWSN)*, volume 2920 of *LNCS*, Springer, pages 172–187, Berlin, Germany, January 2004.

[332] S. Hara, A. Ogino, M. Araki, M. Okada, and N. Morinaga. Throughput Performance of SAW-ARQ Protocol with Adaptive Packet Length in Mobile Packet Data Transmission. *IEEE Transactions on Vehicular Technology*, 45(3): 561–569, 1996.

[333] A. Harter and A. Hopper. A Distributed Location System for the Active Office. *IEEE Network*, 8(1): 62–70, January 1994.

[334] B. A. Harvey and S. B. Wicker. Packet Combining Systems Based on the Viterbi Decoder. *IEEE Transactions on Communications*, 42(2): 1544–1557, 1994.

[335] H. Hashemi. The Indoor Radio Propagation Channel. *Proceedings of the IEEE*, 81(7): 943–968, 1993.

[336] A. T. Hayes. How Many Robots? Group Size and Efficiency in Collective Search Tasks. In *6th International Symposium on Distributed Autonomous Robotic Systems*, pages 289–298, Fukuoka, Japan, June 2002.

[337] T. W. Haynes, S. T. Hedetniemi, and P. J. Slater. *Fundamentals of Domination in Graphs*. Marcel Dekker, 1998.

[338] T. He, B. M. Blum, J. A. Stankovic, and T. F. Abdelzaher. AIDA: Adaptive Application-Independent Data Aggregation in Wireless Sensor Networks. *ACM Transactions on Embedded Computing Systems*, 3(2): 426–457, 2004.

[339] T. He, C. Huang, B. M. Blum, J. A. Stankovic, and T. Abdelzaher. Range-Free Localization Schemes for Large Scale Sensor Networks. *Proceedings of the 9th Annual International Conference on Mobile Computing and Networking*, pages 81–95. ACM Press, 2003.

[340] T. He, J. A. Stankovic, C. Lu, and T. Abdelzaher. SPEED: A Stateless Protocol for Real-Time Communication in Sensor Networks. In *Proceedings of the 23rd International Conference on Distributed Computing Systems (ICDCS'03)*, Providence, Rhode Island, May 2003.

[341] J. Heidemann, F. Silva, and D. Estrin. Matching Data Dissemination Algorithms to Application Requirements. In *Proceedings of the 1st ACM International Conference on Embedded Networked Sensor Systems (SenSys)*, pages 218–230, Los Angeles, CA, November 2003.

[342] J. Heidemann, F. Silva, C. Intanagonwiwat, R. Govindan, D. Estrin, and D. Ganesan. Building Efficient Wireless Sensor Networks with Low-Level Naming. In *Proceedings of the Symposium on Operating System Principles (SOSP 2001)*, pages 146–159, Lake Louise, Banff, Canada, October 2001.

[343] J. Heidemann, F. Silva, C. Intanagonwiwat, R. Govindan, D. Estrin, and D. Ganesan. Building Efficient Wireless Sensor Networks with Low-Level Naming. *Proceedings of the 18th ACM Symposium on Operating Systems Principles*, pages 146–159. ACM Press, 2001.

[344] W. B. Heinzelman, A. P. Chandrakasan, and H. Balakrishnan. An Application-Specific Protocol Architecture for Wireless Microsensor Networks. *IEEE Transactions on Wireless Networking*, 1(4): 660–670, 2002.

[345] W. R. Heinzelman, J. Kulik, and H. Balakrishnan. Adaptive Protocols for Information Dissemination in Wireless Sensor Networks. *Proceedings of the 5th Annual International Conference on Mobile Computing and Networking*, pages 174–185. ACM, Seattle, WA, August 1999. http://citeseer.nj.nec.com/heinzelman99adaptive.html.

[346] W. R. Heinzelman, A. Chandrakasan, and H. Balakrishnan. Energy-Efficient Communication Protocol for Wireless Microsensor Networks. In *Proceedings of the 33rd Hawaii International Conference on System Sciences*, pages 174–185, Hawaii, HI, January 2000.

[347] J. M. Hellerstein, W. Hong, S. Madden, and K. Stanek. Beyond Average: Toward Sophisticated Sensing with Queries. In *Proceedings of the 2nd International Workshop on Information Processing in Sensor Networks (IPSN)*, Palo Alto, CA, April 2003.

[348] A. Helmy. CAPTURE: Location-Free Contact-Assisted Power-Efficient Query Resolution for Sensor Networks. *ACM Mobile Computing and Communications Review*, 8(1): 27–47, 2004.

[349] J. Hightower and G. Borriello. Location Systems for Ubiquitous Computing. *IEEE Computer*, 34(8): 57–66, 2001.

[350] J. Hightower and G. Borriello. A Survey and Taxonomy of Location Systems for Ubiquitous Computing. Technical Report UW-CSE 01-08-03, University of Washington, Computer Science and Engineering, Seattle, WA, August 2001.

[351] J. Hill and D. Culler. MICA: A Wireless Platform for Deeply Embedded Networks. *IEEE Micro*, 22(6): 12–24, 2002.

[352] J. Hill, M. Horton, R. Kling, and L. Krishnamurthy. The Platform Enabling Wireless Sensor Networks. *Communication of the ACM*, 47(6): 41–46, 2004.

[353] J. Hill, R. Szewczyk, A. Woo, S. Hollar, D. E. Culler, and K. S. J. Pister. System Architecture Directions for Networked Sensors. In *Proceedings of the 9th International Conference on Architectural Support for Programming Languages and Operating Systems*, pages 93–104, Cambridge, MA, 2000.

[354] B. Hofmann-Wellenhof, H. Lichtenegger, and J. Collins. *Global Positioning System: Theory and Practice*. Springer, 4th edition, 1997.

[355] A. Honarbacht and A. Kummert. WSDP: Efficient, yet Reliable, Transmission of Real-Time Sensor Data over Wireless Networks. In *Proceedings of the Wireless Sensor Networks, First European Workshop (EWSN 2004)*, Berlin, Germany, January 2004.

[356] X. Hong, K. Xu, and M. Gerla. Scalable Routing Protocols for Mobile Ad Hoc Networks. *IEEE Network Magazine*, 16(4): 11–21, 2002.

[357] T. Hou and V. O. K. Li. Transmission Range Control in Multihop Radio Networks. *IEEE Transactions on Communications*, 34(1): 38–44, 1986.

[358] A. Howard, M. J. Mataric, and G. S. Sukhatme. An Incremental Self-Deployment Algorithm for Mobile Sensor Networks. *Autonomous Robots*, 13(2): 113–126, 2002.

[359] I. Howitt. Bluetooth Performance in the Presence of 802.11b WLAN. *IEEE Transactions on Vehicular Technology*, 51(6): 1640–1651, 2002.

[360] I. Howitt and J. A. Gutierrez. IEEE 802.15.4 Low Rate - Wireless Personal Area Network Coexistence Issues. In *Proceedings of IEEE Wireless Communications and Networking Conference 2003 (WCNC 2003)*, pages 1481–1486, New Orleans, Louisiana, March 2003.

[361] H.-Y. Hsieh and R. Sivakumar. Transport over Wireless Networks. In I. Stojmenovic, editor, *Handbook of Wireless Networks and Mobile Computing*, pages 289–308. John Wiley & Sons, New York, 2002.

[362] L. Hu. Topology Control for Multihop Packet Radio Networks. *IEEE Transactions on Communications*, 41: 1474–1481, 1993.

[363] C.-F. Huang, Y.-C. Tseng, S.-L. Wu, and J.-P. Sheu. Increasing the Throughput of Multihop Packet Radio Networks with Power Adjustment. In *Proceedings of the International Conference on Computer Communications and Networks (ICCCN)*, Scottsdale, AZ, 2001.

[364] C.-F. Huang and Y.-C. Tseng. The Coverage Problem in a Wireless Sensor Network. In *Proceedings of the Second ACM International. Workshop on Wireless Sensor Networks and Applications (WSNA'03)*, San Diego, CA, September 2003.

[365] C.-F. Huang and Y.-C. Tseng. The Coverage Problem in a Wireless Sensor Network. *MONET-Mobile Networks and Applications*, 2004. to appear.

[366] C.-F. Huang, Y.-C. Tseng, and L.-C. Lo. The Coverage Problem in Three-Dimensional Wireless Sensor Networks. In *Proceedings of IEEE Globecom*, Dallas, TX, 2004.

[367] G. T. Huang. Casting the Wireless Sensor Net. *Technology Review*, pages 51–56, July 2003. www.technologyreview.com.

[368] Q. Huang, C. Lu, and G.-C. Roman. Mobicast: Just-In-Time Multicast for Sensor Networks und Spatiotemporal Constraints. In *Proceedings of the 2nd International Workshop on Information Processing in Sensor Networks (IPSN)*, Palo Alto, CA, April 2003.

[369] Q. Huang, C. Lu, and G.-C. Romand. Spatiotemporal Multicast in Sensor Networks. *Proceedings of the 1st International Conference on Embedded Networked Sensor Systems (SenSys)*, pages 205–217. ACM, Los Angeles, CA, November 2003.

[370] B. Hull, K. Jamieson, and H. Balakrishnan. Poster Abstract: Bandwidth Management in Wireless Sensor Networks. In *Proceedings of the 1st International Conference on Embedded Networked Sensor Systems (SenSys)*, pages 306–307, Los Angeles, CA, November 2003.

[371] R. Hwang, D. Richards, and P. Winter. *The Steiner Tree Problem*, volume 53 of *Annals of Discrete Mathematics*. North-Holland, Amsterdam, The Netherlands, 1992.

[372] IEEE. *802.4 Token-passing Bus Access Method*, 1985.

[373] M. Ilyas, editor. *The Handbook of Ad Hoc Wireless Networks*. CRC Press, 2003.

[374] T. Imielinski and S. Goel. DataSpace – Querying and Monitoring Deeply Networked Collections in Physical Space. *Proceedings of ACM International Workshop on Data Engineering for Wireless and Mobile Access (MobiDE)*, pages 44–51. ACM Press, 1999.

[375] Wireless Components ASK/FSK 868 MHz Wireless Transceiver TDA 5250 D2 Version 1.6. Infineon Product data sheet, July 2002.

[376] C. Intanagonwiwat, D. Estrin, R. Govindan, and J. Heideman. Impact of Network Density on Data Aggregation in Wireless Sensor Networks. Technical Report 01-750, University of Southern California, Computer Science Department, November 2001.

[377] C. Intanagonwiwat, D. Estrin, R. Govindan, and J. Heidemann. Impact of Network Density on Data Aggregation in Wireless Sensor Networks. In *Proceedings of IEEE 22nd International Conference on Distributed Computing Systems (ICDCS)*, pages 457–458, Vienna, Austria, July 2002.

[378] C. Intanagonwiwat, R. Govindan, D. Estrin, J. Heidemann, and F. Silva. Directed Diffusion for Wireless Sensor Networks. *IEEE/ACM Transactions on Networking*, 11(1): 2–16, 2003.

[379] Intel StrongARM SA-1100 Microprocessor Brief Data Sheet. intel product documentation, August 2000.

[380] J. Ishac. Survey of Header Compression Techniques. Technical Memorandum e-13010, NASA John H. Glenn Research Center, September 2001.

[381] R. Iyer and L. Kleinrock. QoS Control for Sensor Networks. In *Proceedings of ICC'03*, pages 517–521, Anchorage, AK, May 2003.

[382] S. Jain, R. Shah, W. Brunnette, G. Borriello, and S. Roy. Exploiting Mobility for Energy Efficient Data Collection in Sensor Networks. In *Proceedings of 2nd International. Workshop on Modeling and Optimization in Mobile, Ad Hoc and Wireless Networks*, pages 292–301, Cambridge, UK, March 2004.

[383] X. Ji and H. Zha. Multidimensional Scaling Based Sensor Positioning Algorithms in Wireless Sensor Networks. In *Proceedings of the 1st ACM International Conference on Embedded Networked Sensor Systems (SenSys)*, pages 328–329, Los Angeles, CA, November 2003.

[384] G. Jiang, W. Chung, and G. Cybenko. Semantic Agent Technologies for Tactical Sensor Networks. In *Proceedings of SPIE Conference on AeroSense*, Orlando, FL, April 2003.

[385] C. E. Jones, K. M. Sivalingam, P. Agrawal, and J.-C. Chen. A Survey of Energy Efficient Network Protocols for Wireless Networks. *Wireless Networks*, 7(4): 343–358, 2001.

[386] J.-H. Ju and V. O. K. Li. An Optimal Topology-Transparent Scheduling Method in Multihop Packet Radio Networks. *IEEE/ACM Transactions on Networking*, 6(3): 298–306, 1998.

[387] H. J. Moon, H. S. Park, S. C. Ahn, and W. H. Kwon. Performance Degradation of the IEEE 802.4 Token Bus Network in a Noisy Environment. *Computer Communications*, 21: 547–557, 1998.

[388] P. Juang, H. Oki, Y. Wang, M. Martonosi, L.-S. Peh, and D. Rubenstein. Energy-Efficient Computing for Wildlife Tracking: Design Tradeoffs and Early Experiences with ZebraNet. In *Proceedings of the 10th International Conference on Architectural Support for Programming Languages and Operating Systems*, San Jose, CA, October 2002.

[389] E.-S. Jung and N. H. Vaidya. A Power Control MAC Protocol for Ad Hoc Networks. In *Proceedings of the Eighth Annual International Conference on Mobile Computing and Networking 2002 (MobiCom)*, Atlanta, Georgia, September 2002.

[390] R. Jurdak, C. V. Lopes, and P. Baldi. A Survey, Classification and Comparative Analysis of Medium Access Control Protocols for Ad Hoc Networks. *IEEE Communications Surveys and Tutorials*, 6(1), 2004. http://www.comsoc.org/livepubs/surveys.

[391] J. M. Kahn, R. H. Katz, and K. S. J. Pister. Emerging Challenges: Mobile Networking for Smart Dust. *Journal of Communications and Networks*, 2(3): 188–196, 2000.

[392] J. M. Kahn, R. H. Katz, and K. S. J. Pister. Next Century Challenges: Mobile Networking for "Smart Dust". In *Proceedings of ACM/IEEE International Conference on Mobile Computing and Networking (MobiCom 99)*, Seattle, WA, August 1999.

[393] R. Kalidindi, L. Ray, R. Kannan, and S. Iyengar. Distributed Energy Aware MAC Layer Protocol for Wireless Sensor Networks. In *Proceedings of International Conference on Wireless Networks (ICWN03)*, Las Vegas, NV, June 2003.

[394] S. Kallel. Analysis of a Type-II Hybrid ARQ Scheme with Code Combining. *IEEE Transactions on Communications*, 38(8): 1133–1137, 1990.

[395] K. Kalpakis, K. Dasgupta, and P. Namjoshi. Maximum Lifetime Data Gathering and Aggregation in Wireless Sensor Networks. In *Proceedings of the IEEE International Conference on Networking (ICN)*, pages 685–696, Atlanta, GA, August 2002.

[396] K. Kalpakis, K. Dasgupta, and P. Namjoshi. Efficient Algorithms for Maximum Lifetime Data Gathering and Aggregation in Wireless Sensor Networks. *Computer Networks*, 42: 697.

[397] R. Kannan, S. Sarangi, S. S. Iyengar, and L. Ray. Sensor-Centric Quality of Routing in Sensor Networks. In *Proceedings of IEEE INFOCOM*, San Francisco, CA, March 2003.

[398] V. Kanodia, C. Li, A. Sabharwal, B. Sadeghi, and E. Knightly. Distributed Priority Scheduling and Medium Access in Ad-Hoc Networks. *Wireless Networks*, 8(6): 455–466, 2002.

[399] A. Kansal and M.B. Srivastava. An Environmental Energy Harvesting Framework for Sensor Networks. In *Proceedings of the International Symposium on Low Power Electronics and Design (ISLPED)*, Seoul, Korea, August 2003.

[400] E. Kaplan, editor. *Understanding GPS: Principles & Applications*. Artech House, 1996.

[401] E. Kaplan, editor. *Understanding GPS: Principles and Applications*. Artech House, Boston, MA, 1996.

[402] K. Kar, M. Kodialam, T. V. Lakshman, and L. Tassiulas. Routing for Network Capacity Maximization in Energy-constrained Ad-hoc Networks. In *Proceedings of IEEE INFOCOM*, San Francisco, CA, March 2003.

[403] D. Karger, P. Klein, and R. Tarjan. A Randomized Linear-Time Algorithm to Find Minimum Spanning Trees. *Journal of the ACM*, 42: 321–328, 1995.

[404] S. Karlin and H. M. Taylor. *A First Course in Stochastic Processes*. Academic Press, San Diego, CA, second edition, 1975.

[405] S. Karlin and H. M. Taylor. *A Second Course in Stochastic Processes*. Academic Press, San Diego, CA, 1981.

[406] C. Karlof and D. Wagner. Secure Routing in Wireless Sensor Networks: Attacks and Countermeasures. *Ad Hoc Networks*, 1: 293–315, 2003.

[407] P. Karn. A New Channel Access Method for Packet Radio. In *Proceedings of the ARRL/CRRL Amateur Radio 9th Computer Networking Conference*, pages 134–140, September 1990.

[408] M. J. Karol, Z. Liu, and K. Y. Eng. An Efficient Demand-Assignment Multiple Access Protocol for Wireless (ATM) Networks. *Wireless Networks*, 1(3): 269–279, 1995.

[409] B. Karp and H. T. Kung. GPSR: Greedy Perimeter Stateless Routing for Wireless Networks. In *Proceedings of the 6th International Conference on Mobile Computing and Networking (ACM Mobicom)*, Boston, MA, 2000.

[410] R. Karp, J. Elson, D. Estrin, and S. Shenker. Optimal and Global Time Synchronization in Sensornets. CENS Technical Report Number 0012, Center for Embedded Networked Sensing (CENS), April 2003.

[411] H. Karvonen, Z. Shelby, and C. Pomalaza-Raez. Coding for Energy Efficient Wireless Embedded Networks. In *Proceedings of the International Workshop on Wireless Ad Hoc Networks (IWWAN)*, Oulu, Finland, June 2004.

[412] V. Kawadia and P. R. Kumar. A Cautionary Perspective on Cross Layer Design. `http://black.csl.uiuc.edu/~prkumar/ps_files/cross-layer-design.pdf`, July 2003. In IEEE Wireless Communication Magazine.

[413] V. Kawadia and P. R. Kumar. Power Control and Clustering in Ad Hoc Networks. In *Proceedings of IEEE INFOCOM*, San Francisco, CA, March 2003.

[414] S. M. Kay. *Fundamentals of Statistical Signal Processing: Estimation Theory*. Prentice-Hall, Upper Saddle River, NJ, 1993.

[415] C. D. Kidd, R. J. Orr, G. D. Abowrd, C. G. Atkeson, I. A. Essa, B. MacIntyre, E. Mynatt, T. E. Starner, and W. Newstetter. The Aware Home: A Living Laboratory for Ubiquitous Computing Research. In *Proceedings of the 2nd International Workshop on Cooperative Buildings*, Pittsburgh, PA, 1999.

[416] H. S. Kim, T. F. Abdelzaher, and W. H. Kwon. Minimum-Energy Asynchronous Dissemination to Mobile Sinks in Wireless Sensor Networks. In *Proceedings of the 1st ACM International Conference on Embedded Networked Sensor Systems (SenSys)*, pages 192–204, Los Angeles, CA, November 2003.

[417] M. Kim and B. Noble. Mobile Network Estimation. In *Proceedings of the Seventh Annual International Conference on Mobile Computing and Networking 2001 (MobiCom)*, Rome, GA July 2001.

[418] S. Kim, S. H. Son, J. A. Stankovic, S. Li, and Y. Choi. Safe: A Data Dissemination Protocol for Periodic Updates in Sensor Networks. In *Proceedings of the Workshop of IEEE International Conference on Distributed Computing Systems (ICDCS)*, Providence, RI, May 2003.

[419] Y. Kim, J.-J. Lee, and A. Helmy. Modeling and Analyzing the Impact of Location Inconsistencies on Geographic Routing in Wireless Networks. *ACM Mobile Computing and Communications Review*, 8(1): 48–60, 2004.

[420] M. Klein, B. Konig-Ries, and P. Obreiter. Lanes: A Lightweight Overlay for Service Discovery in Mobile ad hoc Networks. Technical Report 2003-6, Technical University Karlsruhe, May 2003. `http://citeseer.nj.nec.com/klein03lanes.html`.

[421] L. Kleinrock and J. Silvester. Optimum Transmission Radii for Packet Radio Networks or Why Six is a Magic Number. In *Proceedings of the National Telecommunications Conference.*, Birmingham, AL, December 1978.

[422] L. Kleinrock and F. A. Tobagi. Packet Switching in radio channels: Part I Carrier Sense Multiple Access Models and their Throughput-/Delay-Characteristic. *IEEE Transactions on Communications*, 23(12): 1400–1416, 1975.

[423] Y.-B. Ko and N. H. Vaidya. Location-Aided Routing (LAR) in Mobile Ad Hoc Networks. In *Proceedings of the Mobile Computing and Networking (MOBICOM)*, 66–75, Dallas, TX, 1998.

[424] Y.-B. Ko and N. H. Vaidya. GeoTORA: A Protocol for Geocasting in Mobile Ad Hoc Networks. In *Proceedings of 8th International Conference on Network Protocols (ICNP)*, 240–250, Osaka, Japan, November 2000.

[425] Y.-B. Ko and N. H. Vaidya. Flooding-Based Geocasting Protocols for Mobile Ad Hoc Networks. *Mobile Networks & Applications*, 7(6): 471–480, 2002.

[426] J. Koberstein, F. Reuter, and N. Luttenberger. The XCast Approach for Content-based Flooding Control in Distributed Virtual Shared Information Spaces – Design and Evaluation. In *Proceedings of the First European Workshop Wireless Sensor Networks, (EWSN 2004)*, Berlin, Germany, January 2004.

[427] M. Kochhal, L. Schwiebert, and S. Gupta. Role-Based Hierarchical Self-Organization for Wireless Ad hoc Sensor Networks. In *Proceedings of the 2nd ACM International Workshop on Wireless Sensor Networks and Applications (WSNA)*, San Diego, CA, September 2003.

[428] R. Koetter and M. Medard. Beyond Routing: An Algebraic Approach to Network Coding. In *Proceedings of INFOCOM*, New York, June 2002.

[429] H. Kopetz. *Real-Time Systems Design Principles for Distributed Embedded Applications*. Kluwer Academic Publishers, Dordrecht, The Netherlands, 1997.

[430] H. Kopetz and W. Schwabl. Global time in distributed real-time systems. Technical Report 15/89, Technical University Vienna, 1989.

[431] A. Köpke, V. Handziski, J.-H. Hauer, and H. Karl. Structuring the Information Flow in Component-Based Protocol Implementations for Wireless Sensor Nodes. In *Proceedings of the Work-in-Progress Session of the 1st European Workshop on Wireless Sensor Networks (EWSN)*, Technical Report TKN-04-001 of Technical University Berlin, Telecommunication Networks Group, 41–45, Berlin, January 2004.

[432] Henri Koskinen. On the Coverage of a Random Sensor Network in a bounded domain. In *Proceedings of 16th ITC Specialist Seminar*, 11–18, Antwerp, Belgium, 2004.

[433] V. A. Kottapalli, A. S. Kiremidjian, J. P. Lynch E. Carryer T. W. Kenny K. H. Law, and Y. Lei. Two-Tiered Wireless Sensor Network Architecture for Structural Health Monitoring. In *Proceedings of SPIE Annual International Symposium Smart Structures and Materials*, San Diego, CA, March 2003.

[434] L. Kou, G. Markowsky, and L. Berman. A Fast Algorithm for Steiner Trees. *Acta Informatica*, 15: 141–145, 1981.

[435] U. C. Kozat and L. Tassiulas. Network Layer Support for Service Discovery in Mobile Ad Hoc Networks. In *Proceedings of IEEE INFOCOM*, San Francisco, CA, March 2003.

[436] A. Köpke, A. Willig, and H. Karl. Chaotic Maps as Parsimonious Bit Error Models of Wireless Channels. In *Proceedings of IEEE INFOCOM 2003*, San Francisco, CA, 2003.

[437] E. Kranakis, H. Singh, and J. Urrutia. Compass Routing on Geometric Networks. In *Proceedings of the 11th Canadian Conference on Computational Geometry*, 51–54, Vancouver, BC, August 1999.

[438] P. Krishna, N. Vaidya, M. Chatterjee, and D. Pradhan. A Cluster-based Approach for Routing in Dynamic Networks. *ACM SIGCOMM Computer Communication Review*, 2: 49–65, 1997.

[439] B. Krishnamachari, D. Estrin, and S. Wicker. The Impact of Data Aggregation in Wireless Sensor Networks. In *Proceedings of the Workshops of 22nd International Conference on Distributed Computing Systems*, 575–578, Vienna, Austria, July 2002. IEEE Computer Society.

[440] B. Krishnamachari, Y. Mourtada, and S. Wicker. The Energy-Robustness Tradeoff for Routing in Wireless Sensor Networks. In *Proceedings of the International Conference on Communications (ICC)*, Anchorage, AK, May 2003.

[441] B. Krishnamachari, S. Wicker, R. Bejar, and M. Pearlman. *Advances in Coding and Information Theory*, chapter Critical Density Thresholds in Distributed Wireless Networks. Kluwer.

[442] Bhaskar Krishnamachari, Stephen B. Wicker, and Ramon Bejar. Phase Transition Phenomena in Wireless Ad-Hoc Networks. In *Proceedings of IEEE GlobeCom Symposium on Ad-Hoc Wireless Networks*, San Antonio, Texas, November 2001.

[443] B. Krishnamachari, S. B. Wicker, R. Bejar, and M. Pearlman. Critical Density Thresholds in Distributed Wireless Networks. In H. Bhargava, H. V. Poor, V. Tarokh, and S. Yoon, editors, *Communications, Information and Network Security*. Kluwer Publishers, 2002.

[444] R. Krishnan and D. Starobinski. Message-Efficient Self-Organization of Wireless Sensor Networks. In *Proceedings of IEEE Wireless Communications and Networking Conference (WCNC)*, New Orleans, LA, March 2003.

[445] C. Krsihna and Y. Lee. Voltage-Clock-Scaling adaptive Scheduling Techniques for Low Power in Hard Real-Time Systems. In *Proceedings of the 6th IEEE Real Time Technology and Applications Symposium (RTAS)*, 156–165, Washington, DC, 2000.

[446] M. Krunz, A. Muqattash, and S.-J. Lee. Transmission Power Control in Wireless Ad Hoc Networks: Challenges, Solutions, and Open Issues. *IEEE Network Magazine*, 18(5): 8–14, 2004.

[447] M. Kubisch, H. Karl, and A. Wolisz. Are Classes of Nodes with Different Power Amplifiers Good for Wireless Multi-hop Networks? In *Proceedings of the Personal Wireless Communications (Work-in-progresss session)*, Venice, Italy, September 2003.

[448] M. Kubisch, H. Karl, A. Wolisz, L. C. Zhong, and J. Rabaey. Distributed Algorithms for Transmission Power Control in Wireless Sensor Networks. In *Proceedings of IEEE Wireless Communications and Networking Conference (WCNC)*, New Orleans, LA, March 2003.

[449] M. Kubisch, S. Mengesha, D. Hollos, H. Karl, and A. Wolisz. Applying ad-hoc relaying to improve capacity, energy efficiency, and immission in infrastructure-based WLANs. In K. Irmscher, editor, *Kommunikation in Verteilten Systemen (KiVS 2003),13. ITG/GI-Fachtagung*, 195–206, Leipzig, Germany, February 2003.

[450] F. Kuhn and R. Wattenhofer. Constant-Time Distributed Dominating Set Approximation. In *Proceedings of the 22th Annual ACM Symposium on Principles of Distributed Computing (PODC)*, Boston, MA, 2003.

[451] F. Kuhn, R. Wattenhofer, Y. Zhang, and A. Zollinger. Geometric Ad-Hoc Routing: Of Theory and Practice. In *Proceedings of the 22th Annual ACM Symposium on Principles of Distributed Computing (PODC)*, Boston, MA, 2003.

[452] F. Kuhn, R. Wattenhofer, and A. Zollinger. Worst-Case Optimal and Average-Case Efficient Geometric Ad-Hoc Routing. In *Proceedings of the 4th ACM International Symposium on Mobile Ad Hoc Networking and Computing (MobiHoc)*, Annapolis, MD, 2003.

[453] Joanna Kulik, Wendy Rabiner, and Hari Balakrishnan. Adaptive Protocols for Information Dissemination in Wireless Sensor Networks. In *Proceedings of the Fifth Annual International Conference on Mobile Computing and Networks (MobiCom 1999)*, Seattle, WA, 1999.

[454] G. Kulkarni, C. Schurgers, and M. Srivastava. Dynamic Link Labels for Energy Efficient MAC Headers in Wireless Sensor Networks. In *Proceedings of IEEE International Conference on Sensors (Sensors'02)*, 1520–1525, Orlando, FL, June 2002.

[455] A. Kumar, P. Ishwar, and K. Ramchandran. On Distributed Sampling of Smooth Non-Bandlimited Fields. In *Proceedings of ICASSP 2004*, Montreal, Canada, May 2004.

[456] A. Kumar, P. Ishwar, and K. Ramchandran. On Distributed Sampling of Smooth Non-Bandlimited Fields. In *Proceedings of the Information Processing in Sensor Networks, IPSN'04*, Berkeley, CA, April 2004.

[457] R. Kumar, C. Tsiatsis, and M. Srivastava. Computation Hierarchy for In-network Processing. In *Proceedings of the 2nd ACM International Workshop on Wireless Sensor Networks and Applications (WSNA)*, San Diego, CA, September 2003.

[458] J. F. Kurose and K. W. Ross. *Computer Networking – A Top-Down Approach Featuring the Internet.* Addison-Wesley, Boston, 2001.

[459] J. Kusuma, L. Doherty, and K. Ramchandran. Distributed Compression for Sensor Networks. In *Proceedings of the International Conference on Image Processing (ICIP)*, Thessaloniki, Greece, October 2001.

[460] M. Kwon and S. Fahmy. Topology-aware overlay networks for group communication. In *Proceedings of the 12th International Workshop on Network and Operating Systems Support for Digital Audio and Video*, 127–136. ACM Press, 2002.

[461] T. J. Kwon and M. Gerla. Clustering with Power Control. In *Proceedings of MILCOM*, volume 2, 1424–1428, Atlantic City, NJ, November 1999.

[462] T. J. Kwon and M. Gerla. Efficient Flooding with Passive Clustering (PC) in Ad Hoc Networks. *ACM SIGCOMM Computer Communication Review*, 32(1): 44–56, 2002.

[463] A. Lal and J. Blanchard. The Daintiest Dynamos. *IEEE Spectrum Online*, 2004.

[464] L. Lamport. Time, clocks and the ordering of events in a distributed system. *Communications of the ACM*, 21(7): 558–565, 1978.

[465] K. Langendoen and N. Reijers. Distributed Localization in Wireless Sensor Networks: A Quantitative Comparison. *Computer Networks*, 42, August 2003. Special Issue on Wireless Sensor Networks.

[466] LAN/MAN Standards Committee of the IEEE Computer Society. *IEEE Standard for Information Technology-Telecommunications and information exchange between systems-Local and Metropolitan networks-Specific requirements-Part 11: Wireless LAN Medium Access Control (MAC) and Physical Layer (PHY) specifications: Higher speed Physical Layer (PHY) extension in the 2.4 Ghz band*, 1999.

[467] LAN/MAN Standards Committee of the IEEE Computer Society. *Information technology – Telecommunications and Information Exchange between Systems – Local and Metropolitan Area Networks – Specific Requirements – Part 11: Wireless LAN Medium Access Control (MAC) and Physical Layer (PHY) Specifications*, 1999.

[468] LAN/MAN Standards Committee of the IEEE Computer Society. *IEEE Standard for Information technology – Telecommunications and information exchange between systems – Local and metropolitan area networks – Specific requirements – Part 15.4: Wireless Medium Access Control (MAC) and Physical Layer (PHY) Specifications for Low Rate Wireless Personal Area Networks (LR-WPANs)*, October 2003.

[469] J. Lansford, A. Stephens, and R. Nevo. Wi-Fi (802.11b) and Bluetooth: enabling coexistence. *IEEE Network Magazine*, 15(5): 20–27, 2001.

[470] L. E. Larson. Radio Frequency Integrated Circuit Technology for Low-Power Wireless Communications. *IEEE Personal Communications*, 5(3): 11–19, 1998.

[471] Y. W. Law, J. Doumen, and P. Hartel. Survey and Benchmark of Block Ciphers for Wireless Sensor Networks. Technical Report TR-CTIT-04-07, University of Twente, Computer Science Department, 2004.

[472] H. Lee, B. Han, Y. Shin, and S. Im. Multipath Characteristics of Impulse Radio Channels. In *Proceedings of Vehicular Technology Conference (VTC)*, 2487–2491, Tokyo, Japan, 2000.

[473] S.-J. Lee and M. Gerla. Split Multipath Routing with Maximally Disjoint Paths in Ad hoc Networks. In *Proceedings of the IEEE International Conference on Communications (ICC)*, St. Petersburg, VA, June 2001.

[474] S.-J. Lee, W. Su, J. Hsu, M. Gerla, and R. Bagrodia. A Performance Comparison Study of Ad Hoc Wireless Multicast Protocols. In *Proceedings of the IEEE Infocom*, Tel-Aviv, Israel, March 2000. `http://www.ieee-infocom.org/2000/papers/361.ps`.

[475] S.-W. Lee and C.-S. Wu. A k-Best Paths Algorithm for Highly Reliable Communication Networks. *IEICE Transactions on Communications*, E82-B: 586–590, 1999.

[476] P. H. Lehne and M. Pettersen. An Overview of Smart Antenna Technology for Mobile Communications Systems. *IEEE Communications Surveys and Tutorials*, 2(4): 1999. http://www.comsoc.org/livepubs/surveys.

[477] M. D. Lemmon, Q. Ling, and Y. Sun. Overload Management in Sensor-Actuator Networks Used for Spatially-Distributed Control Systems. In *Proceedings of the 1st International Conference on Embedded Networked Sensor Systems (SenSys)*, pages 162–170, Los Angeles, CA, November 2003.

[478] P. Lettieri, C. Schurgers, and M. B. Srivastava. Adaptive Link Layer Strategies for Energy-Efficient Wireless Networking. *Wireless Networks*, 5(5): 339–355, 1999.

[479] P. Lettieri and M. Srivastava. Adaptive Frame Length Control for Improving Wireless Link Throughput, Range and Energy Efficiency. *Proceedings of INFOCOM 1998*, pages 564–571. IEEE, San Francisco, CA, 1998. .

[480] B. Leuf. *Peer to Peer: Collaboration and Sharing Over the Internet*. Addison-Wesley, 2002.

[481] P. Levis and D. Culler. Maté: A Tiny Virtual Machine for Sensor Networks. In *Proceedings of the 10th International Conference on Architectural Support for Programming Languages and Operating Systems*, San Jose, CA, October 2002.

[482] D. Li, K. D. Wong, Y. H. Hu, and A. M. Sayeed. Detection, Classification, and Tracking of Targets. *IEEE Signal Processing Magazine*, 19(2): 17–29, 2002.

[483] J. Li, J. Jannotti, D. S. J. De Couto, D. R. Karger, and R. Morris. A Scalable Location Service for Geographic Ad Hoc Routing. In *Proceedings of the 6th ACM International Conference on Mobile Computing and Networking*, pages 120–130, Boston, MA, August 2000.

[484] L. Li and J. Y. Halpern. Minimum-Energy Mobile Wireless Networks Revisited. In *Proceedings of IEEE International Conference on Communication (ICC)*, pages 278–283, Helsinki, Finland, June 2001.

[485] L. Li, J. Y. Halpern, P. Bahl, Y. Wang, and R. Wattenhofer. Analysis of Cone-Based Distributed Topology Control Algorithm for Wireless Multi-hop Networks. In *Proceedings of the 20th Annual ACM SIGACT-SIGOPS Symposium on Principles of Distributed Computing (PODC)*, Newport, RI, August 2001.

[486] L. Li and P. Sinha. Throughput and Energy Efficiency in Topology-Controlled Multi-hop Wireless Sensor Networks. In *Proceedings of the 2nd ACM International Workshop on Wireless Sensor Networks and Applications (WSNA)*, San Diego, CA, September 2003.

[487] N. Li, J. C. Hou, and L. Sha. Design and Analysis of an MST-Based Topology Control Algorithm. In *Proceedings of IEEE INFOCOM*, San Francisco, CA, March 2003.

[488] Q. Li, J. Aslam, and D. Rus. Hierarchical Power-Aware Routing in Sensor Networks. In *Proceedings of the DIMACS Workshop on Pervasive Networking*, Piscataway, NJ, May 2001.

[489] Q. Li, J. Aslam, and D. Rus. Online Power-Aware Routing in Ad-Hoc Networks. *Proceedings of the 7th Annual International Conference on Mobile Computing and Networking*, pages 97–107. ACM , Rome, Italy, July 2001.

[490] S. Li, S. H. Son, and J. A. Stankovic. Event Detection Services Using Data Service Middleware in Distributed Sensor Networks. In *Proceedings of the 2nd International Workshop on Information Processing in Sensor Networks (IPSN)*, Palo Alto, CA, April 2003.

[491] S.-F. Li, R. Sutton, and J. Rabaey. Low Power Operating System for Heterogeneous Wireless Communication Systems. In *Proceedings of the 10th International Conference on Parallel Architectures and Compilation Techniques (PACT 01)*, Barcelona, Spain, September 2001.

[492] S.-Y. R. Li, R. W. Yeung, and N. Cai. Linear Network Coding. *IEEE Transactions on Information Theory*, 49(2): 371–381, 2003.

[493] X. Li, Y. J. Kim, R. Govindan, and W. Hong. Multi-dimensional Range Queries in Sensor Networks. In *Proceedings of the 1st ACM International Conference on Embedded Networked Sensor Systems (SenSys)*, pages 63–75, Los Angeles, CA, November 2003.

[494] X.-Y. Li, G. Calinescu, and P.-J. Wan. Distributed Construction of a Planar Spanner and Routing for Ad Hoc Wireless Networks. In *Proceedings of IEEE INFOCOM*, New York, 2002.

[495] X. Y. Li, P.-J. Wan, Y. Wang, and C. W. Yi. Fault Tolerant Deployment and Topology Control in Wireless Networks. In *Proceedings of the 4th ACM International Symposium on Mobile Ad Hoc Networking and Computing (MobiHoc)*, Annapolis, MD, 2003.

[496] X.-Y. Li, P.-J. Wan, and O. Frieder. Coverage in Wireless Ad Hoc Sensor Networks. *IEEE Transactions on Computers*, 52(6), 2003.

[497] B. Liang and Z. J. Haas. Virtual Backbone Generation and Maintenance in Ad Hoc Network Mobility Management. In *Proceedings IEEE Infocom*, Tel-Aviv, Israel, March 2000.

[498] W. Liang. Constructing Minimum-Energy Broadcast Trees in Wireless Ad Hoc Networks. In *Proceedings of 3rd ACM International Symposium on Mobile Ad Hoc Networking and Computing (MobiHoc)*, Lausanne, Switzerland, 2002.

[499] W.-H. Liao. GeoGRID: A Geocasting Protocol for Mobile Ad Hoc Networks Based on GRID. *Journal of Internet Technology*, 1(2): 23–32, 2000.

[500] A. Lim. Distributed Services for Information Dissemination in Self-Organizing Sensor Networks. *Special Issue on Distributed Sensor Networks for Real-Time Systems with Adaptive Reconfiguration, Journal of Franklin Institute*, 388: 707–727, 2001.

[501] C. R. Lin and M. Gerla. Multimedia Transport in Multihop Dynamic Packet Radio Networks. In *Proceedings of the International Conference on Network Protocols*, pages 209–216, Tokyo, Japan, November 1995.

[502] C. R. Lin and M. Gerla. Adaptive Clustering for Mobile Wireless Networks. *IEEE Journal on Selected Areas in Communications*, 15(7): 1265–1275, 1997.

[503] E.-Y. A. Lin, J. M. Rabaey, and A. Wolisz. Power-Efficient Rendez-vous Schemes for Dense Wireless Sensor Networks. In *Proceedings of IEEE International Conference on Communications (ICC'04)*, Paris, France, June 2004.

[504] S. Lin and D. J. Costello. *Error Control Coding – Fundamentals and Applications*. Prentice-Hall, Englewood Cliffs, NJ, 1983.

[505] S. Lin, D. J. Costello, and M. J. Miller. Automatic-Repeat-Request Error-Control Schemes. *IEEE Communications Magazine*, 22(12): 5–17, 1984.

[506] T. Lin, H. Zhao, J. Wang, G. Han, and J. Wang. An Embedded Web Server for Equipments. In *Proceedings of the 7th International Symposium on Parallel Architectures, Algorithms and Networks*, pages 345–350, Hong Kong, China, May 2004.

[507] S. Lindsey and K. M. Sivalingam. Data Gathering Algorithms in Sensor Networks Using Energy Metrics. *IEEE Transactions on Parallel and Distributed Systems*, 13(9): 924–934, 2002.

[508] R. Liscano. Service Discovery in Sensor Networks: An Overview. `http://www.site.uottawa.ca/~rliscano/presentations/SDSensorNetworks.pdf%`, 2003.

[509] B. Liu and D. Towsley. On the Coverage and Detectability of Large-Scale Wireless Sensor Networks. In *Proceedings of WiOpt'03: Modeling and Optimization in Mobile, Ad Hoc and Wireless Networks*, Sophia-Antipolis, 2003.

[510] B. Liu and D. Towsley. A Study on the Coverage of Large-Scale Sensor Networks. In *Proceedings of the 1st IEEE International Conference on Mobile Ad-hoc and Sensor Systems (MASS'04)*, Fort Lauderdale, FL, 2004.

[511] H. Liu, H. Ma, M. E. Zarki, and S. Gupta. Error Control Schemes for Networks: An Overview. *MONET – Mobile Networks and Applications*, 2(2): 167–182, 1997.

[512] J. Liu and B. Li. Distributed Topology Control in Wireless Sensor Networks with Asymmetric Links. In *Proceedings of IEEE Globecom Wireless Communications symposium*, San Francisco, CA, December 2003.

[513] J. Liu, F. Zhao, and D. Petrovic. Information-Directed Routing in Ad Hoc Sensor Networks. In *Proceedings of the 2nd ACM International Workshop on Wireless Sensor Networks and Applications (WSNA)*, San Diego, CA, September 2003.

[514] J. W. S. Liu. *Real-Time Systems*. Prentice Hall, 2000.

[515] J. W. S. Liu, W.-K. Shih, K.-W. Lin, R. Bettati, and J.-Y. Chung. Imprecise Computations. *Proceedings of the IEEE*, 82(1): 83–94, 1994.

[516] J. Liu, P. Cheung, L. Guibas, and F. Zhao. A Dual-Space Approach to Tracking and Sensor Management in Wireless Sensor Networks. In *Proceedings of the First ACM International Workshop on Wireless Sensor Networks and Applications (WSNA'02)*, pages 131–139, Atlanta, GA, 2002.

[517] E. Lloyd, R. Liu, M. V. Marathe, R. Ramanathan, and S. S. Ravi. Algorithmic Aspects of Topology Control Problems for Ad Hoc Networks. In *Proceedings of the 3rd ACM International symposium on Mobile Ad Hoc Networking and Computing (MobiHoc)*, Lausanne, Switzerland, 2002.

[518] S. B. Lowen and M. C. Teich. Power-Law Shot Noise. *IEEE Transactions on Information Theory*, 36(6): 1302–1318, 1990.

[519] C. Lu, B. M. Blum, T. F. Abdelzaher, J. A. Stankovic, and T. He. RAP: A Real-Time Communication Architecture for Large-Scale Wireless Sensor Networks. In *Proceedings of the Eighth IEEE Real-Time and Embedded Technology and Applications Symposium, 2002 (RTAS 2002)*, San Jose, CA, September 2002.

[520] G. Lu, B. Krishnamachari, and C. Raghavendra. An Adaptive Energy-Efficient and Low-Latency MAC for Data Gathering in Sensor Networks. In *Proceedings of the 4th International Workshop on Algorithms for Wireless, Mobile, Ad Hoc and Sensor Networks (WMAN 04)*, Santa Fe, CA, April 2004.

[521] G. Lu, B. Krishnamachari, and C. S. Raghavendra. Performance Evaluation of the IEEE 802.15.4 MAC for Low-Rate Low-Power Wireless Networks. In *Proceedings of the 2004 IEEE International Conference on Performance, Computing, and Communications*, pages 701–706, Phoenix, AZ, April 2004.

[522] D. S. Lun, M. Medard, and T. H. R. Koetter. Network Coding with a Cost Criterion. In *Proceedings of the International symposium on Information Theory and its Applications (ISITA)*, Parma, Italy, October 2004.

[523] J. Luo, P. Th. Eugster, and J.-P. Hubaux. Route Driven Gossip: Probabilistic Reliable Multicast in Ad Hoc Networks. In *Proceedings of IEEE INFOCOM*, San Francisco, CA, March 2003.

[524] J. Luo, P T. Eugster, and J.-P. Hubaux. Pilot: Probabilistic Lightweight Group Communication System for Ad Hoc Networks. *IEEE Transactions on Mobile Computing*, 3(2): 164–179, 2004.

[525] C. Luschi, M. Sandell, P. Strauch, J.-J. Wu, C. Ilas, P.-W. Ong, R. Baeriswyl, F. Battaglia, S. Karageorgis, and R.-H. Yan. Advanced Signal-Processing Algorithms for Energy-Efficient Wireless Communications. *Proceedings of the IEEE*, 88(10): 1633–1649, 2000.

[526] N. A. Lynch. *Distributed Algorithms*. Morgan Kaufmann Publishers, San Francisco, CA, 1996.

[527] S. Madden and M. J. Franklin. Fjording the Stream: An Architecture for Queries over Streaming Sensor Data. In *Proceedings of the 18th International Conference on Data Engineering (ICDE)*, San Jose, CA, February 2002.

[528] S. Madden, M. J. Franklin, J. M. Hellerstein, and W. Hong. TAG: A Tiny Aggregation Service for Ad-Hoc Sensor Networks. *ACM SIGOPS Operating Systems Review*, 36(SI): 131–146, 2002.

[529] S. Madden, M. J. Franklin, J. M. Hellerstein, and W. Hong. The Design of an Acquisitional Query Processor For Sensor Networks. In *Proceedings of SIGMOD*, San Diego, CA, 2003.

[530] S. Madden, R. Szewczyk, M. J. Franklin, and D. Culler. Supporting Aggregate Queries Over Ad-Hoc Wireless Sensor Networks. In *Proceedings of the 4th IEEE Workshop on Mobile Computing Systems and Applications*, Callicoon, NY, June 2002.

[531] S. R. Madden, M. J. Franklin, J. M. Hellerstein, and W. Hong. TAG: a Tiny AGregation Service for Ad-Hoc Sensor Networks. In *Proceedings of OSDI*, Boston, MA, December 2002.

[532] N. P. Mahalik, editor. *Fieldbus Technology – Industrial Network Standards for Real-Time Distributed Control*. Springer, Berlin, Germany, 2003.

[533] C. Maihöfer. A Survey of Geocast Routing Protocols. *IEEE Communications Surveys & Tutorials*, 6(2): 32–42, 2004.

[534] A. Mainwaring, J. Polastre, R. Szewczyk, D. Culler, and J. Anderson. Wireless Sensor Networks for Habitat Monitoring. In *Proceedings of the 1st ACM Workshop on Wireless Sensor Networks and Applications*, Atlanta, GA, September 2002.

[535] N. Malpani, Y. Chen, N. Vaidya, and J. Welch. Distributed Token Circulation on Mobile Ad Hoc Networks. *IEEE Transactions on Mobile Computing*, 4(2): 154–165, 2004.

[536] D. Maltz. *On-Demand Routing in Multi-Hop Wireless Ad Hoc Networks*. PhD thesis, Carnegie Mellon University, Pittsburgh, PA, 2001.

[537] A. Manzak and C. Chakrabarty. Variable Voltage Task Scheduling for Minimizing Energy or Minimizing Power. In *Proceedings of the IEEE International Conference on Acoustic, Speech, and Signal Processing (ICASSP)*, pages 3239–3242, Istanbul, Turkey, June 2000.

[538] D. Marco, E. Duarte-Melo, M. Liu, and D. L. Neuhoff. On the Many-to-One Transport Capacity of a Dense Wireless Sensor Network and the Compressibility of its Data. In *Proceedings of the 2nd International Workshop on Information Processing in Sensor Networks (IPSN)*, Palo Alto, CA, April 2003.

[539] D. Marco, E. J. Duarte-Melo, M. Liu, and D. L. Neuhoff. On the Many-to-One Transport Capacity of a Dense Wireless Sensor Network and the Compressibility of its Data. In *Proceedings of the 2nd Symposium on Information Processing in Sensor Networks (IPSN '03)*, Palo Alto, CA, April 2003.

[540] I. Maric and R. D. Yates. Cooperative Multihop Broadcast for Wireless Networks. *IEEE Journal on Selected Areas in Communications*, 22(6): 1080–1088, 2004.

[541] R. J. Marks, A. K. Das, M. El-Sharkawi, P. Arabshahi, and A. Gray. Minimum Power Broadcast Trees for Wireless Networks: Optimizing Using the Viability Lemma. In *IEEE International Symposium on Circuits and Systems (ISCAS)*, Scottsdale, AZ, 2002.

[542] J. L. Massey. Information Theory Aspects of Spread-Spectrum Communications. In *Proceedings of the IEEE ISSSTA '94*, pages 16–21, Oulu, Finland, July 1994.

[543] M. Mauve, J. Widmer, and H. Hartenstein. A Survey on Position-Based Routing in Mobile Ad-Hoc Networks. *IEEE Network*, 15: 30–39, 2001.

[544] A. B. McDonald and T. Znati. A Mobility-Based Framework for Adaptive Clustering in Wireless AD-Hoc Networks. *IEEE Journal on Selected Areas in Communications*, 17(8): 1466–1487, 1999. Special Issue on Wireless Ad Hoc Networks.

[545] R. Meester and R. Roy. *Continuum Percolation*. Cambridge University Press, 1996.

[546] S. Meguerdichian, F. Koushanfar, M. Potkonjak, and M. B. Srivastava. Coverage Problems in Wireless Ad-Hoc Sensor Networks. In *Proceedings of IEEE INFOCOM 2001*, pages 1380–1387, Anchorage, AK, 2001.

[547] S. Meguerdichian, F. Koushanfar, G. Qu, and M. Potkonjak. Exposure in Wireless Ad-Hoc Sensor Networks. In *Proceedings of the 7th Annual International Conference on Mobile Computing and Networking (MobiCom '01)*, pages 139–150, Rome, Italy, July 2001.

[548] A. J. Menezes, P. C. van Oorschot, and S. A. Vanstone. *Handbook of Applied Cryptography*. CRC Press, Boca Raton, FL, 1996.

[549] S. Meninger, J. O. Mur-Miranda, R. Amirtharajah, A. P. Chandrakasan, and J. H. Lang. Vibration-to-Electric Energy Conversion. *IEEE Transactions on VLSI Systems*, 9(1): 64–76, 2001.

[550] V. Mhatre and C. Rosenberg. Design Guidelines for Wireless Sensor Networks Communication: Clustering and Aggregation. *Elsevier AdHoc Networks J. (Special Issue on Sensor Network Applications and Protocols)*, 2(1): 45–63, 2003.

[551] A. M. Michelson and A. H. Levesque. *Error-Control Techniques for Digital Communication*. John Wiley & Sons, New York, 1985.

[552] M. J. Miller and N. H. Vaidya. Minimizing Energy Consumption in Sensor Networks using a Wakeup Radio. In *Proceedings of IEEE WCNC 2004*, Atlanta, Georgia, March 2004.

[553] D. L. Mills. *Network Time Protocol (Version 3) Specification, Implementation and Analysis*. RFC 1305, 1992.

[554] D. L. Mills. Improved Algorithms for Synchronizing Computer Network Clocks. *IEEE/ACM Transactions on Networking*, 3(3): 245–254, 1995.

[555] D. L. Mills. *Simple Network Time Protocol (SNTP) Version 4 for IPv4, IPv6 and OSI*. RFC 2030, 1996.

[556] D. L. Mills. Adaptive Hybrid Clock Discipline Algorithm for the Network Time Protocol. *IEEE/ACM Transactions on Networking*, 6(5): 505–514, 1998.

[557] L. B. Milstein and M. K. Simon. Spread Spectrum Communications. In J. D. Gibson, editor, *The Communications Handbook*, pages 199–212. CRC Press/IEEE Press, Boca Raton, FL, 1996.

[558] R. Min, M. Bhardwaj, S.-H. Cho, E. Shih, A. Sinha, A. Wang, and A. Chandrakasan. Low-Power Wireless Sensor Networks. In *Proceedings of the 14th International Conference on VLSI Design (VLSID '01)*, Bangalore, India, 2001.

[559] R. Min and A. Chandrakasan. A Framework for Energy-Scalable Communication in High-Density Wireless Networks. In *Proceedings of the 2002 International Symposium on Low Power Electronics and Design*, pages 36–41. ACM Press, 2002.

[560] R. Min and A. Chandrakasan. MobiCom Poster: Top Five Myths About the Energy Consumption of Wireless Communication. *ACM SIGMOBILE Mobile Computing and Communications Review*, 7(1): 65–67, 2003.

[561] R. Min, M. Bhardwaj, S.-H. Cho, N. Ickes, E. Shih, A. Sinha, A. Wang, and A. Chandrakasan. Energy-Centric Enabling Technologies for Wireless Sensor Networks. *IEEE Wireless Communications*, 9(4): 28–39, 2002.

[562] R. Min and A. Chandrakasan. Energy-Efficient Communication for Ad-Hoc Wireless Sensor Networks. In *Proceedings of the 35th Asilomar Conference on Signals, Systems, and Computers*, pages 139–143, Pacific Grove, CA, November 2001.

[563] R. Min and A. Chandrakasan. A Framework for Energy-Scalable Communication in High-Density Wireless Networks. In *Proceedings of ISLPED '02*, pages 36–41, Monterey, CA, 2002.

[564] J. Mirkovic, G. P. Venkataramani, S. Lu, and L. Zhang. A Self-organizing approach to data forwarding in large-scale sensor networks. In *Proceedings of the IEEE International Conference on Communications (ICC)*, 5: 1357–1361, 2001.

[565] A. R. Mishra. *Fundamentals of Cellular Network Planning and Optimisation: 2G/2.5G/3G... Evolution to 4G*. John Wiley & Sons, 2004.

[566] T. Mitchell. Broad is the Way. *IEE Review*, 47(1): 35–39, 2001.

[567] S. Mitra and J. Rabek. Power Efficient Clustering for Clock Synchronization in Dynamic Multi-Hop Sensor Networks. `http://theory.lcs.mit.edu/~mitras/courses/6829/project/project_main.htm%1`, 2003.

[568] M. Mock, R. Frings, E. Nett, and S. Trikaliotis. Clock Synchronization in Wireless Local Area Networks. In *Proceedings of the 12th Euromicro Conference On Real Time Systems*, Stockholm, Sweden, June 2000.

[569] M. Mock, R. Frings, E. Nett, and S. Trikaliotis. Continuous Clock Synchronization in Wireless Real-Time Applications. In *Proceedings of the 19th IEEE Symposium on Reliable Distributed Systems (SRDS)*, Nuremberg, Germany, October 2000.

[570] E. Modiano. An Adaptive Algorithm for Optimizing the Packet Size Used in Wireless ARQ Protocols. *Wireless Networks*, 5: 279–286, 1999.

[571] J. P. Monks, J.-P. Ebert, A. Wolisz, and W. W. Hwu. A Study of the Energy Saving and Capacity Improvement Potential of Power Control in Multi-hop Wireless Networks. In *Proceedings of Workshop on Wireless Local Networks*, Tampa, FL, November 2001. Held in conjunction with Conference of Local Computer Networks (LCN).

[572] J. P. Monks, V. Bharghavan, and W.-M. Hwu. A Power Controlled Multiple Access Protocol for Wireless Packet Networks. *Proceedings of the INFOCOM Conference 2001*, pages 219–228. IEEE Press, Anchorage, AL, April 2001.

[573] J. P. Monks, J.-P. Ebert, A. Wolisz, and W.-M. Hwu. A Study of the Energy Saving and Capacity Improvement Potential of Power Control in Multi-Hop Wireless Networks. In *Proceedings of Workshop on Wireless Local Networks/Proceedings of Conference on Local Computer Networks (LCN)*, Berlin, Germany, November 2001.

[574] D. Moore and J. Hebeler. *Peer-to-Peer: Building Secure, Scalable, and Manageable Networks*. McGraw-Hill, 2001.

[575] G. Mühl, L. Fiege, and A. P. Buchmann. Filter Similarities in Content-Based Publish/Subscribe Systems. In H. Schmeck, T. Ungerer, and L. Wolf, editors, *Proceedings of the International Conference on Architecture of Computing Systems (ARCS)*, volume 2299 of *Lecture Notes in Computer Science*, pages 224–238. Springer-Verlag, Karlsruhe, Germany, 2002.

[576] G. Mühl, L. Fiege, F. C. Gartner, and A. Buchmann. Evaluating Advanced Routing Algorithms for Content-Based Publish/Subscribe Systems. *Proceedings of the 10th IEEE International symposium on Modeling, Analysis and Simulation of Computer and Telecommunications Systems (MASCOTS)*, pages 167–176. IEEE Press, 2002.

[577] A. Muqattash and M. Krunz. Power Controlled Dual Channel (PCDC) Medium Access Protocol for Wireless Ad Hoc Networks. In *Proceedings of IEEE INFOCOM*, San Francisco, CA, March 2003.

[578] A. D. Murugan, P. K. Gopala, and H. E. Gamal. Correlated Sources Over Wireless Channels: Cooperative Source-Channel Coding. *IEEE Journal on Selected Areas in Communications*, 22(6): 988–998, 2004.

[579] A. D. Myers and S. Basagni. Wireless Media Access Control. In I. Stojmenovic, editor, *Handbook of Wireless Networks and Mobile Computing*, pages 119–143. John Wiley & Sons, New York, 2002.

[580] B. A. Myers, J. B. Willingham, P. Landy, M. A. Webster, P. Frogge, and M. Fischer. Design Considerations for Minimal-Power Wireless Spread Spectrum Circuits and Systems. *Proceedings of the IEEE*, 88(10): 1598–1612, 2000.

[581] R. Nagpal, H. Shrobe, and J. Bachrach. Organizing a Global Coordinate System from Local Information on an Ad Hoc Sensor Network. In *Proceedings of the 2nd International Workshop on Information Processing in Sensor Networks (IPSN)*, Palo Alto, CA, April 2003.

[582] S. Narayanswamy, V. Kawadia, R. S. Sreenivas, and P. R. Kumar. Power Control in Ad Hoc Networks : Theory, Architecture, Algorithm and Implementation of the COMPOW Protocol. In *Proceedings of European Wireless 2002*, Florence, Italy, February 2002.

[583] B. Narendran, J. Sienicki, S. Yajnik, and P. Agrawal. Evaluation of an Adaptive Power and Error Control Algorithm for Wireless Systems. In *Proceedings of the International Conference on Communication (ICC)*, Montreal, Canada, June 1997.

[584] A. Nasipuri, R. Castaneda, and S. R. Das. Performance of Multipath Routing for On-Demand Protocols in Ad Hoc Networks. *Mobile Networks and Applications (MONET)*, 6(4): 339–349, 2002.

[585] A. Nasipuri and S. R. Das. On-Demand Multipath Routing for Mobile Ad-Hoc Networks. In *Proceedings of the 8th International Conference on Computer Communications and Networks (ICCCN)*, Boston, MA, 1999.

[586] A. Nasipuri and K. Li. A Directionality Based Location Discovery Scheme for Wireless Sensor Networks. In *Proceedings of the 1st ACM International Workshop on Sensor Networks and Applications (WSNA)*, Atlanta, GA, September 2002.

[587] B. Nath and D. Niculescu. Routing on a Curve. *ACM SIGCOMM Computer Communication Review*, 33(1): 155–160, 2003.

[588] National Semiconductors. *LMX 3162 – Single Chip Radio Transceiver*, 2000.

[589] J. C. Navas and T. Imielinski. GeoCast – Geographic Addressing and Routing. In *Proceedings of the 3rd ACM/IEEE International Conference on Mobile Computing (MobiCom)*, Budapest, Hungary, September 1997.

[590] R. Nelson and L. Kleinrock. The Spatial Capacity of a Slotted Aloha Multihop Packet Radio Network with Capture. *IEEE Transactions on Communications*, 32(6): 684–694, 1984.

[591] S. Nesargi and R. Prakash. MANETconf: Configuration of Hosts in a Mobile Ad Hoc Network. In *Proceedings of IEEE INFOCOM 2002*, pages 1587–1596, New York, June 2002.

[592] A. Neskovic, N. Neskovic, and G. Paunovic. Modern Approaches in Modeling of Mobile Radio Systems Propagation Environment. *IEEE Communications Surveys and Tutorials*, 3(3), 2000. http://www.comsoc.org/livepubs/surveys.

[593] J. Newsome and D. Song. GEM: Graph Embedding for Routing and Data-Centric Storage in Sensor Networks Without Geographic Information. In *Proceedings of the 1st ACM Conference on Embedded Networked Sensor Systems (SenSys)*, pages 76–88, Los Angeles, CA, November 2003.

[594] G. T. Nguyen, R. H. Katz, B. Noble, and M. Satyanarayanan. A Trace-Based Approach for Modeling Wireless Channel Behavior. In *Proceedings of the Winter Simulation Conference*, Coronado, CA, December 1996.

[595] S. Y. Ni, Y. C. Tseng, and J. P. Sheu. Efficient Broadcasting in a Mobile Ad Hoc Network. In *Proceedings of the IEEE International Conference on Distributed Computing and Systems*, pages 16–19, Phoenix, AZ, April 2001.

[596] S.-Y. Ni, Y.-C. Tseng, Y.-S. Chen, and J.-P. Sheu. The Broadcast Storm Problem in a Mobile Ad Hoc Network. In *Proceedings of the Fifth Annual International Conference on Mobile Computing and Networks (MobiCom 1999)*, Seattle, WA, 1999.

[597] D. Niculescu and B. Nath. Ad Hoc Positioning System (APS). In *Proceedings of IEEE GlobeCom*, San Antonio, AZ, November 2001.

[598] D. Niculescu and B. Nath. Ad Hoc Positioning System (APS) Using AOA. In *Proceedings of IEEE INFOCOM*, San Francisco, CA, March 2003.

[599] D. Niculescu and B. Nath. Localized Positioning in Ad Hoc Networks. In *Proceedings of the 1st IEEE International Workshop on Sensor Network Protocols and Applications (SNPA)*, Anchorage, AK, May 2003.

[600] D. Niculescu and B. Nath. Trajectory Based Forwarding and Its Applications. In *Proceedings of 9th International Conference on Mobile Computing and Networking (ACM MobiCom)*, San Diego, CA, 2003.

[601] R. Nowak and U. Mitra. Boundary Estimation in Sensor Networks: Theory and Methods. In *Proceedings of the 2nd Symposium on Information Processing in Sensor Networks (IPSN '03)*, Palo Alto, CA, April 2003.

[602] R. Nowak, U. Mitra, and R. Willett. Estimating Inhomogeneous Fields Using Wireless Sensor Networks. *IEEE Journal on Selected Areas in Communications*, 22(6): 999–1006, 2004.

[603] I. D. O'Donnel, M. S. W. Chen, S. B. T. Wand, and R. W. Brodersen. An Integrated, Low Power, Ultra-Wideband Transceiver Architecture for Low-Rate, Indoor Wireless Systems. In *IEEE CAS Workshop on Wireless Communications and Networking*, Pasadena, CA, September 2002.

[604] R. Ogier and N. Shacham. A Distributed Algorithm for Finding Shortest Pairs of Disjoint Paths. In *Proceedings of IEEE INFOCOM*, Ottawa, Canada, 1989.

[605] A. Okabe, B. Boots, and K. Sugihara. *Spatial Tessellations: Concepts and Applications of Voronoi Diagrams*. Wiley, 1992.

[606] C. M. Okino and M. G. Corr. Statistically Accurate Sensor Networking. In *Proceedings of IEEE Wireless Communications and Networking Conference (WCNC)*, Orlando, FL, March 2002.

[607] C. A. S. Oliveira and P. M. Pardalos. A Survey of Combinatorial Optimization Problems in Multicast Routing. To appear in *Computers and Operations Research*, 32(8): 1953–1981, 2004.

[608] A. Oram. *Peer-to-Peer: Harnessing the Power of Disruptive Technologies*. O'Reilly & Associates, 2001.

[609] J. O'Rourke. *Art Gallery Theorems and Algorithms*. Oxford University Press, New York, 1987.

[610] E. Pagani and G. P. Rossi. On the Reduction of Broadcast Redundancy in Mobile Ad Hoc Networks. *Mobile Networks and Applications*, 4: 172–192, 1999.

[611] G. A. Paleologo, L. Benini, A. Bogliolo, and G. De Micheli. Policy Optimization for Dynamic Power Management. *IEEE Transcations on CAD*, 18(6): 813–833, 1999.

[612] P. Papadimitratos, Z. Haas, and E. G. Sirer. Path-Set Selection in Mobile Ad Hoc Networks. In *Proceedings of the 3rd ACM International Symposium on Mobile Ad Hoc Networking and Computing (MobiHoc)*, Lausanne, Switzerland, 2002.

[613] I. Papadimitriou and L. Georgiadis. Minimum-Energy Broadcasting in Wireless Networks Using a Single Broadcast Tree. In *Proceedings the 2nd International Workshop on Modeling and Optimization in MObile, Ad Hoc and Wireless Networks*, pages 38–47, Cambridge, UK, March 2004.

[614] M. Papadopouli and H. Schulzrinne. Effects of Power Conservation, Wireless Coverage and Cooperation on Data Dissemination Among Mobile Devices. In *Proceedings of the 2nd ACM International Symposium on Mobile Ad Hoc Networking and Computing (MobiHoc)*, Long Beach, CA, 2001.

[615] A. Papoulis and S. Unnikrishna Pillai. *Probability, Random Variables, and Stochastic Processes*. McGraw-Hill, Boston, MA, fourth edition, 2002.

[616] A. K. Parekh. Selecting Routers in Ad-Hoc Wireless Networks. In *Proceedings of the SBT/IEEE International Telecommunications Symposium*, Rio de Janeiro, Brazil, August 1994.

[617] S.-J. Park and R. Sivakumar. Poster: Sink-to-Sensors Reliability in Sensor Networks. In *Proceedings of the 4th ACM International Symposium on Mobile Ad Hoc Networking and Computing (MOBIHOC)*, Annapolis, MD, June 2003.

[618] V. D. Park and M. S. Corson. A Highly Adaptive Distributed Routing Algorithm for Mobile Wireless Networks. In *Proceedings of INFOCOM*, Kobe, Japan, April 1997.

[619] V. D. Park and M. S. Corson. A Highly Adaptive Distributed Routing algorithm for Mobile Wireless Networks. In *Proceedings of INFOCOM*, pages 1405–1413, Kobe, Japan, April 1997.

[620] J. D. Parsons. *The Mobile Radio Propagation Channel*. Pentech Press, London, 1992.

[621] N. Passas, S. Paskalis, D. Vali, and L. Merakos. Quality-of-Service-Oriented Medium Access Control for Wireless ATM Networks. *IEEE Communications Magazine*, 35(11): 42–50, 1997.

[622] M. Patel, N. Tanna, P. Patel, and R. Banerjee. TCP over Wireless Networks: Issues, Challenges and Survey of Solutions. `citeseer.ist.psu.edu/489782.html`.

[623] S. Pattem, B. Krishnamachari, and R. Govindan. The Impact of Spatial Correlation on Routing with Compression in Wireless Sensor Networks. In *Proceedings of the 3nd International Workshop on Information Processing in Sensor Networks (IPSN)*, Berkeley, CA, April 2004.

[624] N. Patwari and A. Hero. Using Proximity and Quantized RSS for Sensor Localization in Wireless Networks. In *Proceedings of the 2nd ACM International Workshop on Wireless Sensor Networks and Applications (WSNA)*, San Diego, CA, September 2003.

[625] A. Paulraj. Diversity Techniques. In J. D. Gibson, editor, *The Communications Handbook*, pages 213–223. CRC Press/IEEE Press, Boca Raton, FL, 1996.

[626] M. Pearlman and Z. Haas. Improving the Performance of Query-Based Routing Protocols through "Diversity-Injection". In *Proceedings of the 1st IEEE Wireless Communications and Networking Conference (WCNC)*, New Orleans, LO, September 1999.

[627] M. R. Pearlman, Z. J. Haas, P. Scholander, and S. S. Tabrizi. On the Impact of Alternate Path Routing for Load Balancing in Mobile Ad Hoc Networks. In *IEEE/ACM Workshop on Mobile Ad Hoc Networking and Computing (MobiHOC)*, Boston, MA, August 2000.

[628] G. Pei, M. Gerla, X. Hong, and C.-C. Chiang. A Wireless Hierarchical Routing Protocol with Group Mobility. In *Proceedings of the 1st IEEE Wireless Communications and Networking Conference (WCNC)*, New Orleans, LO, September 1999.

[629] W. Peng and X.-C. Lu. On the Reduction of Broadcast Redundancy in Mobile Ad Hoc Networks. In *Proceedings of the 1st Annual Workshop on Mobile and Ad Hoc Networking and Computing*, pages 129–130, Boston, MA, August 2000.

[630] M. D. Penrose. On k-Connectivity for a Geometric Random Graph. *Wiley Random Structures and Algorithms*, 15(2): 145–164, 1999.

[631] M. Perillo and W. B. Heinzelman. Providing Application QoS through Intelligent Sensor Managment. In *Proceedings of the 1st IEEE International Workshop on Sensor Network Protocols and Applications (SNPA'03)*, Anchorage, AK, May 2003.

[632] M. A. Perillo and W. B. Heinzelman. Sensor Management Policies to Provide Application QoS. *Elsevier AdHoc Networks Journal (Special Issue on Sensor Network Applications and Protocols)*, 1(2–3): 235–246, 2003.

[633] C. Perkins and P. Bhagwat. Highly Dynamic Destination-Sequenced Distance-Vector Routing (DSDV) for Mobile Computers. In *Proceedings of the ACM SIGCOMM*, pages 234–244, London, UK, 1994.

[634] C. E. Perkins and E. M. Royer. Ad-Hoc On-Demand Distance Vector Routing. In *Proceedings of the 2nd IEEE Workshop on Mobile Computing Systems and Applications*, pages 90–100, New Orleans, LA, February 1999.

[635] C. E. Perkins, editor. *Ad Hoc Networking*. Addison-Wesley, Upper Saddle River, NJ, 2001.

[636] C. E. Perkins, J. T. Malinen, R. Wakikawa, E. M. Belding-Royer, and Y. Sun. *IP Address Autoconfiguration for Ad Hoc Networks*. Internet draft, IETF, November 2001. `draft-ietf-manet-autoconf-01.txt`.

[637] A. Perrig, R. Szewczyk, J. D. Tygar, V. Wen, and D. E. Culler. SPINS: Security Protocols for Sensor Networks. *Wireless Networks*, 8: 521–534, 2002.

[638] A. Perrig, R. Szewczyk, V. Wen, D. Culler, and J. D. Tygar. SPINS: Security Protocols for Sensor Networks. In *Proceedings of the 7th Annual International Conference on Mobile Computing and Networking (ACM MobiCom)*, pages 189–199, Rome, Italy, July 2001.

[639] C. Petrioli, R. R. Rao, and J. Redi. Special Issue: Energy Conserving Protocols. *ACM-Baltzer Mobile Networks and Applications Journal*, 6, 2001.

[640] D. Petrovic, R. C. Shah, K. Ramchandran, and J. Rabaey. Data Funneling: Routing with Aggregation and Compression for Sensor Networks. In *Proceedings of the 1st IEEE International Workshop on Sensor Network Protocols and Applications (SNPA)*, Anchorage, AK, May 2003.

[641] P. P. Pham and S. Perreau. Performance Analysis of Reactive Shortest Path and Multi-path Routing Mechanism With Load Balance. In *Proceedings of IEEE INFOCOM*, San Francisco, CA, March 2003.

[642] T. K. Philips, S. S. Panwar, and A. N. Tantawi. Connectivity Properties of a Packet Radio Network Model. *IEEE Transactions on Information Theory*, 35(5): 1044–1047, 1989.

[643] Physikalisch Technische Bundesanstalt, Braunschweig/Berlin, Germany. *Die gesetzlichen Einheiten in Deutschland*, 2002.

[644] J. R. Pimentel. *Communication Networks for Manufacturing*. Prentice-Hall, 1990.

[645] P. Piret. On the Connectivity of Radio Networks. *IEEE Transactions on Information Theory*, 37: 1490–1492, 1991.

[646] D. Porcino and W. Hirt. Ultra-Wideband Radio Technology: Potential and Challenges Ahead. *IEEE Communications Magazine*, 41(7): 66–74, 2003.

[647] A.-S. Porret, T. Melly, C. C. Enz, and E. A. Vittoz. A Low-Power Low-Voltage Transceiver Architecture Suitable for Wireless Distributed Sensors Network. In *IEEE International Symposium on Circuits and Systems (ISCAS)*, volume I, pages 56–58, Geneva, Switzerland, May 2000.

[648] G. J. Pottie and W. J. Kaiser. Embedding the Internet: Wireless Integrated Network Sensors. *Communications of the ACM*, 43(5): 51–58, 2000.

[649] J. Pouwelse, K. Langendoen, and H. Sips. Dynamic Voltage Scaling on a Low-Power Microprocessor. *Proceedings of the 7th Annual International Conference on Mobile Computing and Networking*, pages 251–259, ACM Press, Rome, Italy, July 2001.

[650] B. Prabhakar, E. Biyikoglu, and A. E. Gamal. Energy-Efficient Transmission Over a Wireless Link Via Lazy Packet Scheduling. In *Proceedings of IEEE INFOCOM*, pages 386–394, Anchorage, AK, April 2001.

[651] S. S. Pradhan and K. Ramchandran. Distributed Source Coding Using Syndromes (DISCUS): Design and Construction. In *Proceedings of IEEE Data Compression Conference (DCC)*, Snowbird, UT, 1999.

[652] S. S. Pradhan and K. Ramchandran. Distributed Source Coding: Symmetric Rates and Applications to Sensor Networks. In *Proceedings of the IEEE Data Compression Conference (DCC)*, Snowbird, UT, 2000.

[653] S. S. Pradhan, J. Kusuma, and K. Ramchandran. Distributed Compression in a Dense Microsensor Network. *IEEE Signal Processing Magazine*, 19(2): 51–60, 2002.

[654] S. S. Pradhan and K. Ramchandran. Distributed Source Coding Using Syndromes (DISCUS): Design and Construction. *IEEE Transactions on Information Theory*, 49(3): 626–643, 2003.

[655] L. Prasad, S. S. Iyengar, and R. L. Rao. Fault-Tolerant Sensor Integrtion using Multiresolution Decomposition. *Physical Review E*, 49(4): 3452–3461, 1994.

[656] J. Prätorius. *Discovery and Interaction with Services in a WSN via Standard User Interfaces*. Diplomarbeit, Fachgebiet Telekommunikationsnetze, Technische Universität Berlin, September 2004.

[657] R. C. Prim. Shortest Connection Networks and Some Generalizations. *Bell System Technical Journal*, 36: 1389–1401, 1957.

[658] N. B. Priyantha, H. Balakrishnan, E. Demaine, and S. Teller. Anchor-Free Distributed Localization in Sensor Networks. In *Proceedings of the 1st International Conference on Embedded Networked Sensor Systems (SenSys)*, pages 340–341, Los Angeles, CA, November 2003. ACM.

[659] N. B. Priyantha, A. Chakraborty, and H. Balakrishnan. The Cricket Location-Support System. In *Proceedings of the 6th International Conference on Mobile Computing and Networking (ACM Mobicom)*, Boston, MA, 2000.

[660] J. G. Proakis. Channel Equalization. In J. D. Gibson, editor, *The Communications Handbook*, pages 339–363. CRC Press/IEEE Press, Boca Raton, FL, 1996.

[661] J. G. Proakis. *Digital Communications*. McGraw-Hill, Boston, MA, fourth edition, 2001. International edition.

[662] B. Przysdatek, D. Song, and A. Perrig. SIA: Secure Information Aggregation in Sensor Networks. *Proceedings of the 1st International Conference on Embedded Networked Sensor Systems (SenSys)*, pages 255–265. ACM Press, Los Angeles, CA, November 2003.

[663] H. Qi, S. S. Iyengar, and K. Chakrabarty. Multiresolution Data Integration using Mobile Agents in Distributed Sensor Networks. *IEEE Transactions on Systems, Man and Cybernetics (Part C): Applications and Reviews*, 31(3): 383–391, 2001.

[664] H. Qi and F. Wang. Optimal Itineracy Analysis for Mobile Agents in Ad Hoc Wireless Sensor Networks. In *Proceedings of the International Conference on Wireless Communications*, pages 147–153, San Diego, CA, 2001.

[665] H. Qi, X. Wang, S. S. Iyengar, and K. Chakrabarty. Multisensor Data Fusion in Distributed Sensor Networks using Mobile Agents. In *Proceedings of the International Conference Information Fusion*, pages 11–16, Montreal, Canada, August 2001.

[666] H. Qi, Y. Xu, and X. Wang. Mobile-Agent-Based Collaborative Signal and Information Processing in Sensor Networks. *Proceedings of the IEEE*, 91(8): 1172–1183, 2003.

[667] J. M. Rabaey, M. J. Ammer, J. L. da Silva, D. Patel, and S. Roundy. PicoRadio Supports Ad Hoc Ultra-Low Power Wireless Networking. *IEEE Computer*, 33(7): 42–48, 2000.

[668] C. S. Raghavendra and S. Singh. PAMAS – Power Aware Multi-Access Protocol with Signalling for Ad Hoc Networks. *ACM Computer Communication Review*, 27: 5–26, 1998.

[669] V. Raghunathan, P. Spanos, and M. Srivastava. Adaptive Power-Fidelity in Energy-Aware Wireless Embedded Systems. In *Proceedings of IEEE Real Time Systems Symposium (RTSS)*, London, UK, 2001.

[670] V. Raghunathan, C. Schurgers, S. Park, and M. B. Srivastava. Energy-Aware Wireless Microsensor Networks. *IEEE Signal Processing Magazine*, 19: 40–50, 2002.

[671] R. Rajaraman. Topology Control and Routing in Ad Hoc Networks: A Survey. *ACM SIGACT News*, 33(2): 60–73, 2002.

[672] V. Rajendran, K. Obraczka, and J. J. Garcia-Luna-Aceves. Energy-Efficient, Collision-Free Medium Access Control for Wireless Sensor Networks. In *Proceedings of ACM SenSys 03*, Los Angeles, CA, November 2003.

[673] J. Raju and J. J. Garcia-Luna-Aceves. A Comparison of On-Demand and Table Driven Routing for Ad-Hoc Wireless Networks. In *Proceedings of ICC*, New Orleans, LA, June 2000.

[674] V. Ramadurai and M. L. Sichitiu. Localization in Wireless Sensor Networks: A Probabilistic Approach. In *Proceedings of 2003 International Conference on Wireless Networks (ICWN 2003)*, pages 300–305, Las Vegas, NV, June 2003.

[675] A. Ramakrishnan. 16-bit embedded Web server. In *Proceedings of ISA/IEEE Sensors for Industry Conference*, pages 187–193, New Orleans, LO, 2004.

[676] C. V. Ramamoorthy, A. Bhide, and J. Srivastava. Reliable Clustering Techniques for Large, Mobile Packet Radio Networks. In *Proceedings of Infocom*, pages 218–226, San Francisco, CA, 1987.

[677] P. Ramanathan, K. G. Shin, and R. W. Butler. Fault-Tolerant Clock Synchronization in Distributed Systems. *IEEE Computer*, 23(10): 33–42, 1990.

[678] R. Ramanathan and R. Rosales-Hain. Topology Control of Multihop Wireless Networks using Transmit Power Adjustment. In *Proceedings of IEEE Infocom*, pages 404–413, Tel-Aviv, Israel, March 2000.

[679] R. Ramanathan and M. Steenstrup. Hierarchically-Organized, Multihop Mobile Wireless Networks for Quality-of-Service Support. *ACM/Baltzer Mobile Networks & Applications (MONET)*, 3(1): 101–119, 1998.

[680] S. Ramanathan and M. Steenstrup. A Survey of Routing Techniques for Mobile Communications Networks. *ACM/Baltzer Mobile Networks and Applications*, 1: 89–104, 1996.

[681] A. Rao, S. Ratnasamy, C. Papadimitriou, S. Shenker, and I. Stoica. Geographic Routing without Location Information. In *Proceedings of the 9th ACM International Conference on Mobile Computing and Networking (MobiCom)*, San Diego, CA, 2003.

[682] T. S. Rappaport. *Wireless Communications – Principles and Practice*. Prentice Hall, Upper Saddle River, NJ, 2002.

[683] S. Ratnasamy, M. Handley, R. Karp, and S. Shenker. Topologically-Aware Overlay Construction and Server Selection. In *Proceedings of IEEE Infocom*, pages 1190–1199, New York, 2002.

[684] S. Ratnasamy, B. Karp, S. Shenker, D. Estrin, R. Govindan, L. Yin, and F. Yu. Data-Centric Storage in Sensornets with GHT, A Geographic Hash Table. *Mobile Networks and Applications (MONET)*, 8(4): 427–442, 2003. Special Issue on Wireless Sensor Networks.

[685] S. Ratnasamy, B. Karp, L. Yin, F. Yu, D. Estrin, R. Govindan, and S. Shenker. GHT: A Geographic Hash Table for Data-Centric Storage. *Proceedings of the 1st ACM International Workshop on Wireless Sensor Networks and Applications*, pages 78–87. ACM Press, 2002.

[686] S. Ratnasamy, P. Francis, M. Handley, R. Karp, and S. Shenker. A Scalable Content-Addressable Network. In *Proceedings of ACM SIGCOMM'2001 Conference*, pages 161–172, San Diego, CA, August 2001.

[687] L. Rauchhaupt. System and Device Architecture of a Radio-Based Fieldbus – The RFieldbus System. In *Proceedings of the Fourth IEEE Workshop on Factory Communication Systems 2002 (WFCS 2002)*, Vasteras, Sweden, 2002.

[688] J. M. Reason and J. M. Rabaey. A Study of Energy Consumption and Reliability in a Multi-Hop Sensor Network. *ACM Mobile Computing and Communications Review*, 8(1): 84–97, 2004.

[689] J. Redi, C. Petrioli, and I. Chlamtac. An Asymmetric, Dynamic, Energy-conserving ARQ Protocol. In *Proceedings of the 49th Annual Vehicular Technology Conference*, Houston, Texas, July 1999.

[690] R. F. Monolithics. *TR1000 916.50 MHz Hybrid Transceiver*, 2000.

[691] C. Röhl, H. Woesner, and A. Wolisz. A Short Look on Power Saving Mechanisms in the Wireless LAN Standard IEEE 802.11. In J. M. Holtzmann and M. Zorzi, editors, *Advances in Wireless Communications*, pages 219–226. Kluwer Academic Publishers, April 1998.

[692] N. Riga, I. Matta, and A. Bestavros. DIP: Density Inference Protocol for Wireless Sensor Networks and its Application to Density-Unbiased Statistics. In *Proceedings of the Second International Workshop on Sensor and Actuator Network Protocols and Applications (SANPA '04)*, Boston, MA, August 2004.

[693] J. Riihijärvi, P. Mähönen, M. J. Saaranen, J. Roivainen, and J.-P. Soininen. Providing Network Connectivity for Small Appliances: a Functionally Minimized Embedded Web Server. *IEEE Communications Magazine*, 39(10): 74–79, 2001.

[694] H. Ritter. ScatterWeb. Website `http://www.scatterweb.de/`, October 2004.

[695] K. Römer. Time Synchronization in Ad Hoc Networks. In *Proceedings of the 2nd ACM International Symposium on Mobile Ad Hoc Networking and Computing, (MobiHoc)*, Long Beach, CA, 2001.

[696] M. Robert. Discovery and Its Discontents: Discovery Protocols for Ubiquitous Computing, 2000. `http://portal.acm.org/citation.cfm?id=871253`.

[697] V. Rodoplu and T. H. Meng. Minimum Energy Mobile Wireless Networks. *IEEE Journal of Selected Areas on Communication*, 17(8): 1333–1344, 1999.

[698] K. Römer. The Lighthouse Location System for Smart Dust. In *Proceedings of ACM/USENIX International Conference on Mobile Systems, Applications, and Services (MobiSys)*, pages 15–30, San Francisco, CA, May 2003.

[699] K. Römer, O. Kasten, and F. Mattern. Middleware Challenges for Wireless Sensor Networks. *ACM Mobile Communication and Communications Review*, 6(2): 59–61, 2002.

[700] A. H. M. Ross and K. S. Gilhausen. CDMA Technology and the IS-95 North American Standard. In J. D. Gibson, editor, *The Communications Handbook*, pages 199–212. CRC Press/IEEE Press, Boca Raton, FL, 1996.

[701] S. Roundy. *Energy Scavenging for Wireless Sensor Networks*. Kluwer Academic Publishers, 2003.

[702] S. Roundy, B. Otis, Y.-H. Chee, J. Rabaey, and P. K. Wright. A 1.9 GHz Transmit Beacon using Environmentally Scavenged Energy. In *Proceedings of the IEEE International Symposium on Low Power Electronics and Devices*, Seoul, Korea, August 2003.

[703] S. Roundy, D. Steingart, L. Frechette, P. Wright, and J. Rabaey. Power Sources for Wireless Sensor Networks. In H. Karl, A. Willig, and A. Wolisz, editors, *Proceedings of 1st European Workshop on Wireless Sensor Networks (EWSN)*, pages 1-17. LNCS, Springer, Berlin, Germany, volume 2920, January 2004..

[704] A. Rowstron and P. Druschel. Pastry: Scalable, distributed object location and routing for large-scale peer-to-peer systems. In *IFIP/ACM International Conference on Distributed Systems Platforms (Middleware)*, pages 329–350, Heidelberg, Germany, November 2001.

[705] E. M. Royer, P. M. Melliar-Smith, and L. E. Moser. An analysis of the optimum node density for ad hoc mobile networks. In *Proceedings of the IEEE Intlernational Conference on Communications (ICC)*, Helsinki, Finland, June 2001.

[706] E. M. Royer and C. E. Perkins. Multicast Using Ad Hoc On-Demand Distance Vector Routing. In *Proceedings of MobiCom*, pages 207–218, Seattle, WA, August 1999.

[707] E. M. Royer and C.-K. Toh. A Review of Current Routing Protocols for Ad-Hoc Mobile Wireless Networks. *IEEE Prersonal Communications*, 6(2): 46–55, 1999.

[708] I. Rubin. Access-Control Disciplines for Multi-Access Communication Channels: Reservation and TDMA Schemes. *IEEE Transactions on Information Theory*, 25(5): 516–536, 1979.

[709] I. Rubin. Multiple Access Methods for Communications Networks. In J. D. Gibson, editor, *The Communications Handbook*, pages 622–649. CRC Press/IEEE Press, Boca Raton, FL, 1996.

[710] S. Rührup, C. Schindelhauer, K. Volbert, and M. Grünewald. Performance of Distributed Algorithms for Topology Control in Wireless Networks. In *Proceedings of the 17th International Parallel and Distributed Processing Symposium*, Nice, France, 2002.

[711] N. Sadagopan, B. Krishnamachari, and A.Helmy. The ACQUIRE Mechanism for Efficient Querying in Sensor Networks. In *Proceedings of the 1st IEEE International Workshop on Sensor Network Protocols and Applications (SNPA)*, Anchorage, AK, May 2003.

[712] N. Sadagopan, B. Krishnamachari, and A. Helmy. Active Query Forwarding in Sensor Networks (ACQUIRE), AdHoc Networks Journal, 3(1): 91–113, 2005.

[713] N. Sadagopan, B. Krishnamachari, and A. Helmy. Active Query Forwarding in Sensor Networks (ACQUIRE). In *Proceedings of the 1st IEEE International Workshop on Sensor Network Protocols and Applications (SNPA)*, Anchorage, AK, May 2003.

[714] A. Safwat, H. Hassanein, and H. Mouftah. Power-Aware Fair Infrastructure Formation for Wireless Mobile Ad hoc Communications. In *Proceedings of IEEE GlobeCom*, pages 2832–2836, San Antonio, AZ, November 2001.

[715] A. Safwat, H. Hassanein, and H. Mouftah. A MAC-based Performance Study of Energy-Aware Routing Schemes in Wireless Ad Hoc Networks. In *Proceedings of IEEE Globecom*, Taipeh, China, 2002.

[716] A. Safwat, H. Hassanein, and H. Mouftah. A Framework for Wireless Ad Hoc Networks with a Quasi-Guaranteed Minimum System Lifetime. In *Proceedings of the 8th IEEE Symposium on Computers and Communications (ISCC)*, pages 349–355, Kemer, Turkey, July 2003.

[717] A. Safwat, H. Hassanein, and H. Mouftah. Q-GSL: A Framework for Energy-Conserving Wireless Multi-Hop Ad hoc Networks. In *Proceedings of the International Conference on Communications (ICC)*, Anchorage, AK, May 2003.

[718] J. H. Saltzer. Naming and Binding of Objects. In R. Bayer, R. M. Graham, and G. Seegmüller, editors, *Operating System – An Advanced Course*, Lecture Notes in Computer Science, pages 99–208. Springer, 1978.

[719] J. H. Saltzer, D. P. Reed, and D. D. Clark. End-to-End Arguments in System Design. *ACM Transactions on Computer Systems*, 2(4): 277–288, 1984.

[720] Y. Sankarasubramaniam, O. B. Akan, and I. F. Akyildiz. ESRT: Event-to-Sink Reliable Transport in Wireless Sensor Networks. In *Proceedings of the 4th ACM International Symposium on Mobile Ad Hoc Networking and Computing (MobiHoc)*. ACM Press, Annapolis, MD, June 2003.

[721] Y. Sankarasubramaniam, I. F. Akyildiz, and S. W. McLaughlin. Energy Efficiency Based Packet Size Optimization in Wireless Sensor Networks. In *Proceedings of the 1st IEEE International Workshop on Sensor Network Protocols and Applications (SNPA)*, Anchorage, AK, May 2003.

[722] P. Santi and D. M. Blough. The Critical Transmitting Range for Connectivity in Sparse Wireless Ad Hoc Networks. *IEEE Transactions on Mobile Computing*, 2: 25–39, 2003.

[723] C. Savarese, J. M. Rabaey, and J. Beutel. Locationing in Distributed Ad-Hoc Wireless Sensor Networks. In *Proceedings of the International Conference on Acoustics, Speech and Signal Processing (ICASSP 2001)*, Salt Lake City, Utah, May 2001.

[724] C. Savarese, J. Rabay, and K. Langendoen. Robust Positioning Algorithms for Distributed Ad-Hoc Wireless Sensor Networks. In *Proceedings of the Annual USENIX Technical Conference*, Monterey, CA, 2002.

[725] A. Savvides, C.-C. Han, and M. Srivastava. Dynamic Fine-Grained Localization in Ad-Hoc Networks of Sensors. *Proceedings of the 7th Annual International Conference on Mobile Computing and Networking*, pages 166–179. ACM press, Rome, Italy, July 2001.

[726] A. Savvides, H. Park, and M. B. Srivastava. The Bits and Flops of the N-Hop Multilateration Primitive for node Localization Problems. In *Proceedings of the 1st ACM International Workshop on Sensor Networks and Applications (WSNA)*, Atlanta, GA, September 2002.

[727] A. Savvides, J. Fang, and D. Lymberopoulos. Using Mobile Sensing Nodes for Dynamic Boundary Estimation. In *Proceedings of the MobiSys 2004 Workshop on Applications of Mobile Embedded Systems (WAMES'04)*, Boston, MA, June 2004.

[728] A. Scaglione and S. D. Servetto. On the Interdependence of Routing and Data Compression in Multi-Hop Sensor Networks. *ACM/Kluwer Journal on Mobile Networks and Applications (MONET)*, 2002, http://portal.acm.org/citation.cfm?id=570663.

[729] G. Schäfer. *Security in Fixed and Wireless Networks – An Introduction to Securing Data Communications*. John Wiley & Sons, Chichester, UK, 2003.

[730] G. Schäfer. Sensor Network Security. In R. Zurawski, editor, *The Industrial Communication Technology Handbook*. CRC Press, 2004.

[731] B. Schneier. *Applied Cryptography: Protocols, Algorithms and Source Code in C*. John Wiley & Sons, second edition, 1996.

[732] D. Schonberg, S. S. Pradhan, and K. Ramchandran. Distributed Code Constructions for the Entire Slepian-Wolf Rate Region for Arbitrarily Correlated Sources. In *Proceedings of the 37th Asilomar Conference on Signals, Systems, and Computers*, Monterey, CA, November 2003.

[733] C. Schrugers, V. Tsiatsis, S. Ganeriwal, and M. Srivastava. Topology Management for Sensor Networks: Exploiting Latency and Density. In *Proceedings of the 3rd ACM International Symposium on Mobile Ad Hoc Networking and Computing (MobiHoc)*, Lausanne, Switzerland, 2002.

[734] C. Schurgers, G. Kulkarni, and M. B. Srivastava. Distributed Assignment of Encoded MAC Addresses in Sensor Networks. In *Proceedings of the Symposium on Mobile Ad Hoc Networking & Computing (MobiHoc'01)*, Long Beach, CA, October 2001.

[735] C. Schurgers, V. Raghunathan, and M. B. Srivastava. Power Management for Energy-Aware Communication Systems. *Transactions on Embedded Computing Systems*, 2(3): 431–447, 2003.

[736] C. Schurgers and M. B. Srivastava. Energy Efficient Routing in Wireless Sensor Networks. In *Proceedings of IEEE Military Communication Conference (MILCOM)*, October 2001.

[737] C. Schurgers, V. Tsiatsis, S. Ganeriwal, and M. B. Srivastava. Optimizing Sensor Networks in the Energy-Latency-Density Design Space. *IEEE Transactions on Mobile Computing*, 1(1): 70–80, 2002.

[738] C. Schurgers, O. Aberthorne, and M. B. Srivastava. Modulation Scaling for Energy Aware Communication Systems. In *Proceedings of the International Symposium on Low Power Electronics and Design (ISLPED'01)*, pages 96–99, Huntington Beach, CA, August 2001.

[739] C. Schurgers, G. Kulkarni, and M. B. Srivastava. Distributed On-Demand Address Assignment in Wireless Sensor Networks. *IEEE Transactions on Parallel and Distributed Systems*, 13(10): 1056–1065, 2002.

[740] C. Schurgers, V. Raghunathan, and M. B. Srivastava. Modulation Scaling for Real-Time Energy Aware Packet Scheduling. In *Proceedings of Global Communications Conference (GlobeCom'01)*, pages 3653–3657, San Antonio, TX, November 2001.

[741] C. Schurgers, V. Raghunathan, and M. B. Srivastava. Power Management for Energy-Aware Communication Systems. *ACM Transactions on Embedded Computing Systems*, 2(3): 431–447, 2003.

[742] C. Schurgers, V. Tsiatsis, S. Ganeriwal, and M. Srivastava. Optimizing Sensor Networks in the Energy-Latency-Density Design Space. *IEEE Transactions on Mobile Computing*, 1(1): 70–80, 2002.

[743] M. Schwartz. *Telecommunication Networks – Protocols, Modeling and Analysis*. Addison-Wesley, Reading, MA, 1988.

[744] M. Schwartz. *Mobile Wireless Communications*. Cambridge University Press, Cambridge, GB, 2005.

[745] L. Schwiebert, S. K. S. Gupta, and J. Weinmann. Research Challenges in Wireless Networks of Biomedical Sensors. In *Proceedings of the 7th International Conference on Mobile Computing and Networking (ACM Mobicom)*, pages 151–165, Rome, Italy, July 2001.

[746] K. Schwieger, H. Nuszkowski, and G. Fettweis. Analysis of Node Energy Consumption in Sensor Networks. In *Proceedings Wireless Sensor Networks, First European Workshop (EWSN 2004)*, Berlin, Germany, January 2004.

[747] K. Scott and N. Bambos. Routing and Channel Assignment for Low Power Transmission in PCS. In *Proceedings International Conference on Universal Personal Communications*, pages 469–502, Cambridge, MA, September 1996.

[748] K. Seada, A. Helmy, and R. Govindan. On the Effect of Localization Errors on Geographic Face Routing in Sensor Networks. *Proceedings of the 1st International Conference on Embedded Networked Sensor Systems (SenSys)*, pages 312–313. ACM Press, Los Angeles, CA, November 2003.

[749] Sensor Modeling Language (SensorML), 2004, http://stromboli.nsstc.uah.edu/SensorML/.

[750] S. D. Servetto and G. Barrenechea. Constrained Random Walks on Random Graphs: Routing Algorithms for Large Scale Wireless Sensor Networks. In *Proceedings of the 1st ACM International Workshop on Sensor Networks and Applications (WSNA)*, Atlanta, GA, September 2002.

[751] M. Sgroi, A. Wolisz, A. Sangiovanni-Vincentelli, and J. M. Rabaey. *A Service-Based Universal Application Interface for Ad-Hoc Wireless Sensor Networks*. White paper, private communication, November 2003.

[752] R. C. Shah and J. M. Rabaey. Energy Aware Routing for Low Energy Ad Hoc Sensor Networks. In *Proceedings of IEEE Wireless Communications and Networking Conference (WCNC)*, Orlando, FL, March 2002.

[753] R. C. Shah, S. Roy, S. Jain, and W. Brunette. Data Mules: Modeling a Three-tier Architecture for Sparse Sensor Networks. In *Proceedings of the 1st IEEE International Workshop on Sensor Network Protocols and Applications (SNPA)*, Anchorage, AK, May 2003.

[754] S. Shakkotai, R. Srikant, and N. B. Shroff. Unreliable Sensor Grids: Coverage, Connectivity and Diameter. In *Proceedings of IEEE INFOCOM 2003*, San Francisco, CA, 2003.

[755] Y. Shang, W. Ruml, Y. Zhang, and M. Fromherz. Localization from Mere Connectivity. In *Proceedings of the 4th ACM International Symposium on Mobile Ad Hoc Networking and Computing (MobiHoc)*, Annapolis, MD, 2003.

[756] C. E. Shannon. A Mathematical Theory of Communication. *Bell Systems Technical Journal*, 27: 379–423, 623–656, July, October 1948.

[757] O. Sharon and E. Altman. An Efficient Polling MAC for Wireless LANs. *IEEE/ACM Transactions on Networking*, 9(4): 439–451, 2001.

[758] C.-C. Shen, C. Srisathapornphat, and C. Jaikaeo. Sensor Information Networking Architecture and Applications. *IEEE Personal Communications*, 8(4): 52–59, 2001.

[759] N. S. Shenck and J. A. Paradiso. Energy Scavenging with Shoe-Mounted Piezolectrics. *IEEE Micro*, 21: 30–41, 2001.

[760] S. Shenker, S. Ratnasamy, B. Karp, R. Govindan, and D. Estrin. Data-Centric Storage in Sensornets. *ACM SIGCOMM Computer Communication Review*, 33(1): 137–142, 2003.

[761] E. Shih, B. H. Calhoun, S.-H. Cho, and A. P. Chandrakasan. Energy-Efficient Link Layer for Wireless Microsensor Networks. In *Proceedings of Workshop on VLSI 2001 (WVLSI '01)*, Orlando, FL, April 2001.

[762] E. Shih, S.-H. Cho, N. Ickes, R. Min, A. Sinha, A. Wang, and A. Chandrakasan. Physical Layer Driven Protocol and Algorithm Design for Energy-Efficient Wireless Sensor Networks. In *Proceedings of the Seventh Annual International Conference on Mobile Computing and Networking 2001 (MobiCom)*, pages 272–286, Rome, Italy, July 2001.

[763] Y. Shin and K. Choi. Power Concious Fixed Priority Scheduling for Hard Real-Time Systems. In *Proceedings of Design Automation Conference (DAC)*, pages 134–139, New Orleans, LA, June 1999.

[764] M. L. Sichitiu and C. Veerarittiphan. Simple, Accurate time Synchronization for Wireless Sensor Networks. In *Proceedings of Wireless Communications and Networking 2003 (WCNC)*, pages 1266–1273, New Orleans, Louisiana, March 2003.

[765] D. Sidhu, R. Nair, and S. Abdallah. Finding Disjoint Paths in Networks. In *Proceedings of SIGCOMM*, Zürich, Germany, 1991.

[766] C. K. Siew and D. J. Goodman. Packet Data Transmission over Mobile Radio Channels. *IEEE Transactions on Vehicular Technology*, 38(2): 95–101, 1989.

[767] T. Simunic, L. Benini, P. Glynn, and G. De Micheli. Dynamic Power Management for Portable Systems. In *Proceedings of the 6th International Conference on Mobile Computing and Networking (ACM Mobicom)*, pages 11–19, Boston, MA, 2000.

[768] S. Singh, M. Woo, and C. S. Raghavendra. Power-Aware Routing in Mobile Ad Hoc Networks. In *Proceedings of the 4th ACM/IEEE International Conference on Mobile Computing and Networking (MOBICOM'98)*, Dallas, TX, October 1998.

[769] A. Sinha and A. Chandrakasan. Dynamic Power Management in Wireless Sensor Networks. *IEEE Design and Test of Computers*, 18(2): 62–74, 2001.

[770] A. Sinha, A. Wang, and A. Chandrakasan. Algorithmic Transforms for Efficient Energy Scalable Computation. In *Proceedings of IEEE International Symposium on Low Power Electronics and Design*, pages 31–36, Rapallo, Italy, July 2000.

[771] R. Sivakumar, B. Das, and V. Bharghavan. The Clade Vertebrata: Spines and Routing in Ad Hoc Networks. In *Proceedings of IEEE Symposium on Computer Communications (ISCC)*, Athens, Greece, June 1998.

[772] B. Sklar. *Digital Communications – Fundamentals and Applications*. Prentice Hall, Englewood Cliffs, NJ, 1988.

[773] B. Sklar. A Primer on Turbo Code Concepts. *IEEE Communications Magazine*, 35(12): 94–102, 1997.

[774] D. Slepian and J. K. Wolf. Noiseless Coding of Correlated Information Sources. *IEEE Transactions on Information Theory*, 19(4): 471–480, 1973.

[775] S. Slijepcevic, S. Megerian, and M. Potkonjak. Location Errors in Wireless Embedded Sensor Networks: Sources, Models, and Effect on Applications. *ACM Mobile Computing and Communications Review*, 6(3): 67–78, 2002.

[776] D. Snoonian. Smart Buildings. *IEEE Spectrum*, 40(8): 18–23, 2003.

[777] J. So and N. H. Vaidya. A Distributed Self-Stabilizing Time Synchronization Protocol for Multi-Hop Wireless Networks. Technical report, Department of Electrical and Computer Engineering and Coordinated Science Laboratory, University of Illinois at Urbana-Champaign, January 2004.

[778] K. Sohrabi, J. Gao, V. Ailawadhi, and G. J. Pottie. Protocols for Self-Organization of a Wireless Sensor Network. *IEEE Personal Communications*, 7(5): 16–27, 2000.

[779] K. Sohrabi, B. Manriquez, and G. J. Pottie. Near Ground Wideband Channel Measurement in 800-1000 MHz. *Proceedings of IEEE Vehicular Technology Conference (VTC) '99*. IEEE Press, 1999.

[780] K. Sohrabi and G. J. Pottie. Performance of a Novel Self-Organization Protocol for Wireless Ad-Hoc Sensor Networks. In *Proceedings of IEEE 50th Vehicular Technology Conference (VTC)*, pages 1222–1226, Fall 1999.

[781] V. Srinivasan, C. F. Chiasserini, P. Nuggehalli, and R. R. Rao. Optimal Rate Allocation and Traffic Splits for Energy Efficient Routing in Ad Hoc Networks. In *Proceedings of IEEE Infocom*, New York, 2002.

[782] K. Sripanidkulchai, B. Maggs, and H. Zhang. Efficient Content Location Using Interest-Based Locality in Peer-to-Peer Systems. In *Proceedings of IEEE INFOCOM*, San Francisco, CA, March 2003.

[783] M. Srivastava, R. Muntz, and M. Potkonjak. Smart Kindergarten: Sensor-based Wireless Networks for Smart Developmental Problem-solving Environments (Challenge Paper). *Proceedings of the 7th Annual International Conference on Mobile Computing and Networking*, pages 132–138. ACM Press, Rome, Italy, July 2001. . `http://www.acm.org/pubs/articles/proceedings/comm/381677/p132-srivastav%a/p132-srivastava.pdf`.

[784] M. B. Srivastava, A. P. Chandrakasan, and R. W. Brodersen. Predictive System Shutdown and Other Architectural Techniques for Energy Efficient Programmable Computation. *IEEE Transactions on VLSI Systems*, 4(1): 42–55, 1996.

[785] W. Stallings. *Cryptography and Network Security: Principles and Practice*. Prentice-Hall, second edition, 1998.

[786] J. A. Stankovic, T. F. Abdelzaher, C. Lu, L. Sha, and J. C. Hou. Real-Time Communication and Coordination in Embedded Sensor Networks. *Proceedings of the IEEE*, 91(7): 1002–1022, 2003.

[787] Fred Stann and John Heidemann. RMST: Reliable Data Transport in Sensor Networks. In *Proceedings of the 1st IEEE International Workshop on Sensor Network Protocols and Applications (SNPA)*, Anchorage, Alaska, May 2003.

[788] D. C. Steere, A. Baptista, D. McNamee, C. Pu, and J. Walpole. Research Challenges in Environmental Observation and Forecasting Systems. In *Proceedings of the 6th International Conference on Mobile Computing and Networking (ACM Mobicom)*, Boston, MA, 2000.

[789] M. Stemm and R. H. Katz. Measuring and Reducing Energy Consumption of Network Interfaces in Hand-Held Devices. *IEICE Transactions on Communications, Special Issue on Mobile Computing*, E80-B(8): 1125-1131, 1997.

[790] W. R. Stevens. *TCP/IP Illustrated Volume 1 – The Protocols*. Addison-Wesley, Boston, MA, 1995.

[791] W. R. Stevens. *Unix Network Programming*, Volume 1. Prentice Hall, Upper Saddle River, NJ, second edition, 1998.

[792] I. Stoica, R. Morris, D. Karger, M. F. Kaashoek, and H. Balakrishnan. Chord: A Scalable Peer-to-peer Lookup Service for Internet Applications. In *Proceedings of ACM SIGCOMM*, pages 149–160, San Diego, CA, August 2001.

[793] I. Stojmenovic, editor. *Handbook of Wireless Networks and Mobile Computing*. Wiley, 2002.

[794] I. Stojmenovic and X. Lin. Loop-Free Hybrid Single-path/Flooding Routing Algorithms with Guaranteed Delivery for Wireless Networks. *IEEE Transactions on Parallel and Distributed Systems*, 12(10): 1023–1032, 2001.

[795] I. Stojmenovic, A. P. Rzhil, and D. K. Lobiyal. Voronoi Diagram and Convex Hull-Based Geocasting and Routing in Wireless Networks. In *Proceedings of the 6th IEEE Symposium on Computers and Communications (ISCC)*, pages 51–56, Antalya, Turkey, July 2001.

[796] I. Stojmenovic, M. Seddigh, and J. Zunic. Dominating Sets and Neighbor Elimination-Based Broadcasting Algorithms in Wireless Networks. *IEEE Transactions on Parallel and Distributed Systems*, 13(1): 14–25, 2002.

[797] J. Stone, M. Greenwald, C. Partridge, and J. Hughes. Performance of Checksums and CRC's Over Real Data. *IEEE/ACM Transactions on Networking*, 6(5): 529–543, 1998.

[798] L. Subramanian and R. H. Katz. An Architecture for Building Self-Configurable Systems. In *IEEE/ACM Workshop on Mobile Ad Hoc Networking and Computing (MobiHOC 2000)*, Boston, MA, August 2002.

[799] Y. Sun, E. M. Belding-Royer, and C. E. Perkins. Internet Connectivity for Ad Hoc Mobile Networks. *International Journal of Wireless Information Networks (special Issue on Mobile Ad hoc Networks)*, 9(2), 2002.

[800] K. Sundaresan, V. Anantharaman, H.-Y. Hsieh, and R. Sivakumar. ATP: A Reliable Transport Protocol for Ad-hoc Networks. In *Proceedings of the 4th ACM International Symposium on Mobile Ad Hoc Networking and Computing (MobiHoc) 2003*, Annapolis, MD, June 2003.

[801] A. Hu and S. D. Servetto. Algorithmic Aspects of the Time Synchronization Problem in Large-Scale Sensor Networks. *MONET - Mobile Networks and Applications*, 2003. Invited paper.

[802] A. Hu and S. D. Servetto. Asymptotically Optimal Time Synchronization in Dense Sensor Networks. In *Proceedings of the Second ACM International Workshop on Wireless Sensor Networks and Applications (WSNA'03)*, San Diego, CA, September 2003.

[803] R. Szewczyk, E. Osterweil, J. Polastre, M. Hamilton, A. Mainwaring, and D. Estrin. Habitat Monitoring with Sensor Networks. *Communication of the ACM*, 47(6): 34–40, 2004.

[804] H. Takagi and L. Kleinrock. Optimal Transmission Ranges for Randomly Distributed Packet Radio Networks. *IEEE Transactions on Communications*, COM-32: 246–257, 1984.

[805] H. Takagi. *Analysis of Polling Systems*. MIT Press, Cambridge, MA, 1986.

[806] H. Takahashi and A. Matsuyama. An Approximate Solution for Steiner Problem in Graphs. *Mathematica Japonica*, 24(6): 573–577, 1980.

[807] A. S. Tanenbaum and A. S. Woodhull. *Operating Systems: Design and Implementation*. Prentice Hall, second edition, 1997.

[808] A. S. Tanenbaum. *Computer Networks*. Prentice-Hall, Englewood Cliffs, NJ, third edition, 1997.

[809] C. Tang and C. S. Raghavendra. Correlation Analysis and Applications in Wireless Microsensor Networks. In *Proceedings of the First Annual International Conference on Mobile and Ubiquitous Systems: Networking and Services, 2004. MOBIQUITOUS 2004*, pages 184–193, Boston, MA, August 2004.

[810] C. Tang, C. S. Raghavendra, and V. K. Prasanna. An Energy Efficient Adaptive Distributed Source Coding Scheme in Wireless Sensor Networks. In *Proceedings of the IEEE International Conference on Communications (ICC'03)*, pages 732–737, Anchorage, Alaska, May 2003.

[811] K. Tang and M. Gerla. MAC Reliable Broadcast in Ad Hoc Networks. In *Proceedings of the IEEE Military Communications Conference, 2001 (MILCOM 2001)*, pages 1008–1013, October 2001.

[812] K. Tang, K. Obraczka, S.-J. Lee, and M. Gerla. Reliable Adaptive Lightweight Multicast Protocol. In *Proceedings of IEEE ICC 2003*, Anchorage, AK, May 2003.

[813] Y. C. Tay, K. Jamieson, and H. Balakrishnan. Collision-Minimizing CSMA and Its Applications to Wireless Sensor Networks. *IEEE Journal on Selected Areas in Communications*, 22(6): 1048–1057, 2004.

[814] MSP430x1xx Family User's Guide. Texas Instruments product documentation. 2004.

[815] The Editors of IEEE 802.11. *IEEE Standard for Wireless LAN Medium Access Control (MAC) and Physical Layer (PHY) specifications*, November 1997.

[816] D. Tian and N. D. Georganas. A Coverage-Preserving Node Scheduling Scheme for Large Wireless Sensor Networks. In *Proceedings of the First ACM International Workshop on Wireless Sensor Networks and Applications (WSNA)*, pages 32–41, Atlanta, GA, September 2002.

[817] D. Tian and N. D. Georganas. Energy Efficient Routing with Guaranteed Delivery in Wireless Sensor Networks. *Proceedings of IEEE Wireless Communications and Networking Conference 2003 (WCNC'03)*. Institute of Electrical and Electronics Engineers, IEEE Press, New Orleans, LA, March 2003.

[818] S. Tilak, A. Murphy, and W. Heinzelman. Non-uniform Information Dissemination for Sensor Networks. In *Proceedings of IEEE International Conference on Network Protocols (ICNP)*, Atlanta, GA, November 2003.

[819] S. Tilak, N. B. Abu-Ghazaleh, and W. Heinzelman. Infrastructure Tradeoffs for Sensor Networks. In *Proceedings of the 1st ACM International Workshop on Sensor Networks and Applications (WSNA)*, Atlanta, GA, September 2002.

[820] TinyOS Web Project. `http://webs.cs.berkeley.edu/tos/`, February 2004. (date of access).

[821] TinyOS On-line Tutorial. `http://webs.cs.berkeley.edu/tos/tinyos-1.x/doc/tutorial/index.html`, February 2004. (date of access).

[822] Tiny OS Hardware Designs. `http://webs.cs.berkeley.edu/tos/hardware/hardware.html`, January 2003. Date of access.

[823] F. A. Tobagi and L. Kleinrock. Packet Switching in Radio Channels: Part II The Hidden Terminal Problem in CSMA and Busy-Tone Solutions. *IEEE Transactions on Communications*, 23(12): 1417–1433, 1975.

[824] F. A. Tobagi and L. Kleinrock. Packet Switching in Radio Channels: Part III – Polling and (Dynamic) Split-Channel Reservation Multiple Access. *IEEE Transactions on Communications*, 24(8): 832–845, 1976.

[825] C.-K. Toh. A Novel Distributed Routing Protocol to Support Ad Hoc Mobile Computing. In *Proceedings of IEEE 15th Annual International Conference on Computers and Communications*, pages 460–486, 1996.

[826] C. K. Toh. Maximum Battery Life Routing to Support Ubiquitous Mobile Computing in Wireless Ad Hoc Networks. *IEEE Communications Magazine*, 39: 138–147, 2001.

[827] C.-K. Toh. *Ad Hoc Mobile Wireless Networks*. Prentice Hall PTR, Upper Saddle River, NJ, 2002.

[828] S. Toner and D. O'Mahony. Self-Organising Node Address Management in Ad-hoc Networks. In *Lecture Notes in Computer Science 2775*, pages 476–483. Springer, Berlin, Germany, 2003.

[829] L. Tong, Q. Zhao, and S. Adireddy. Sensor Networks with Mobile Agents. In *Proceedings of IEEE Military Communication Conference*, Boston, MA, October 2003.

[830] G. Toussaint. The Relative Neighborhood Graph of a Finite Planar Set. *Pattern Recognition*, 12: 261–268, 1980.

[831] Y. Tseng, Y. Chang, and B. Tzeng. Energy-Efficient Topology Control for Wireless Ad Hoc Sensor Networks. In *Proceedings of the International Conference Parallel and Distributed Systems (ICPADS)*, Chung-Li, 2002.

[832] Y.-C. Tseng, C.-S. Hsu, and T.-Y. Hsieh. Power-Saving Protocols for IEEE 802.11-Based Multi-Hop Ad Hoc Networks. *Proceedings of INFOCOM 2002*. IEEE Press, New York, June 2002. .

[833] Y.-C. Tseng, S.-Y. Ni, Y.-S. Chen, and J.-P. Sheu. The Broadcast Storm Problem in a Mobile Ad Hoc Network. *Wireless Networks*, 8: 153–167, 2002.

[834] W. Turin. *Digital Transmission Systems – Performance Analysis and Modeling*. McGraw-Hill Telecommunications. McGraw-Hill, New York, 1998.

[835] E. Uysal-Biyikoglu, B. Prabhakar, and A. E. Gamal. Energy-Efficient Packet Transmission Over a Wireless Link. *IEEE/ACM Transactions on Networking*, 10(4): 487–499, 2002.

[836] N. H. Vaidya. Weak Duplicate Address Detection in Mobile Ad Hoc Networks. In *Proceedings of ACM International Symposium on Mobile Ad Hoc Networking and Computing (MobiHoc)*, Lausanne, Switzerland, June 2002.

[837] A. Valera, W. K. G. Seah, and SV. Rao. Cooperative Packet Caching and Shortest Multipath Routing in Mobile Ad hoc Networks. In *Proceedings of IEEE INFOCOM*, San Francisco, CA, March 2003.

[838] T. v. Dam and K. Langendoen. An Adaptive Energy-Efficient MAC Protocol for Wireless Sensor Networks. *Proceedings of the 1st International Conference on Embedded Networked Sensor Systems (SenSys)*, pages 171–180. ACM Press, Los Angeles, CA, November 2003.

[839] J. v. Greunen and J. Rabaey. Lightweight Time Synchronization for Sensor Networks. In *Proceedings of the 2nd ACM International Workshop on Wireless Sensor Networks and Applications (WSNA)*, San Diego, CA, September 2003.

[840] G. Veltri, Q. Huang, G. Qu, and M. Potkonjak. Minimal and Maximal Exposure Path Algorithms for Wireless Embedded Sensor Networks. In *Proceedings of ACM SenSys 03*, Los Angeles, CA, November 2003.

[841] P. Verissimo, L. Rodrigues, and A. Casimiro. CESIUMSPRAY: A Precise and Accurate Global Time Service for Large-scale Systems. *Journal of Real-Time Systems*, 12(3): 243–254, 1997.

[842] D. C. Verma. *Legitimate Peer to Peer Network Applications: Beyond File and Music Swapping*. Wiley, 2004.

[843] J. R. Vig. Introduction to Quartz Frequency Standards. Technical Report SLCET-TR-92-1 (Rev. 1), Army Research Laboratory, October 1992.

[844] A. J. Viterbi. Error Bounds for Convolutional Codes and an Asymptotically Optimum Decoding Algorithm. *IEEE Transactions on Information Theory*, 13(2): 260–269, 1967.

[845] A. J. Viterbi. Convolutional Codes and Their Performance in Communication Systems. *IEEE Transactions on Communication Technology*, 19(5): 751–772, 1971.

[846] M. Waldvogel and R. Rinaldi. Efficient Topology-Aware Overlay Network. *ACM SIGCOMM Computer Communication Review*, 33(1): 101–106, 2003.

[847] B. Walke, P. Seidenberg, and M. P. Althoff. *UMTS – The Fundamentals*. John Wiley & Sons, Chichester, UK, 2003.

[848] B. Walke. *Mobile Radio Networks – Networking, Protocols and Traffic Performance*. John Wiley & Sons, Chichester, UK, 2002.

[849] C.-Y. Wan, A. T. Campbell, and L. Krishnamurthy. PSFQ: A Reliable Transport Protocol for Wireless Sensor Networks. In *Proceedings of the First ACM International Workshop on Wireless Sensor Networks and Applications (WSNA'02)*, Atlanta, GA, 2002.

[850] C.-Y. Wan, S. B. Eisenman, and A. T. Campbell. CODA: Congestion Detection and Avoidance in Sensor Networks. In *Proceedings of the First ACM Conference on Embedded Networked Sensor Systems (SenSys 2003)*, pages 266–279, Los Angeles, CA, November 2003.

[851] P. Wan, K. Alzoubi, and O. Frieder. Distributed Construction of Connected Dominating Set in Wireless Ad hoc Networks. In *Proceedings of IEEE INFOCOM*, New York, June 2002.

[852] P. Wan, G. Caliuescu, X. Li, and O. Frieder. Minimum-Energy Broadcast Routing in Static Ad Hoc Wireless Networks. In *Proceedings of IEEE Infocom*, Anchorage, AK, April 2001.

[853] P. Wan, G. Caliuescu, X. Li, and O. Frieder. Minimum-Energy Broadcast Routing in Static Ad Hoc Wireless Networks. *Wireless Networks*, 8(6): 607–617, 2002.

[854] A. Wang, W. R. Heinzelman, A. Sinha, and A. P. Chandrakasan. Energy-Scalable Protocols for Battery-Operated MicroSensor Networks. *Journals of VLSI Signal Processing*, 29: 223–237, 2001.

[855] A. Wang, S.-H. Cho, C. G. Sodini, and A. P. Chandrakasan. Energy-Efficient Modulation and MAC for Asymmetric Microsensor Systems. In *Proceedings of ISLPED 2001*, Huntington Beach, CA, August 2001.

[856] H. Wang, L. Yip, D. Maniezzo, J. C. Chen, R. E. Hudson, J. Elson, and K. Yao. A Wireless Time-Synchronized COTS Sensor Platform: Applications to Beamforming. In *Proceedings of IEEE CAS Workshop on Wireless Communications and Networking*, Pasadena, CA, September 2002.

[857] H. Wang, L. Yip, K. Yao, and D. Estrin. Lower Bounds of Localization Uncertainty in Sensor Networks. In *Proceedings of IEEE International Conference on Acoustics, Speech and Signal Processing (ICASP)*, Montreal, Canada, May 2004.

[858] H.S. Wang and N. Moayeri. Finite State Markov Channel – A Useful Model for Radio Communication Channels. *IEEE Transactions on Vehicular Technology*, 44(1): 163–171, 1995.

[859] X. Wang, G. Xing, Y. Zhang, C. Lu, R. Pless, and C. Gill. Integrated Coverage and Connectivity Configuration in Wireless Sensor Networks. *Proceedings of the 1st International Conference on Embedded Networked Sensor Systems (SenSys)*, pages 28–39. ACM Press, Los Angeles, CA, November 2003.

[860] X. Wang and M. T. Orchard. On Reducing the Rate of Retransmission in Time-Varying Channels. *IEEE Transactions on Communications*, 51(6): 900–910, 2003.

[861] Y. Wang, X.-Y. Li, P.-J. Wan, and P. Frieder. Sparse Power Efficient Topology for Wireless Networks. *Journals of Parallel and Distributed Computing*, 2002, `http://csdl.computer.org/comp/proceedings/hicss/2002/1435/09/14305296babs.htm`.

[862] Z. Wang, Y. q. Song, E.-M. Poggi, and Y. Sun. Survey of Weakly-Hard Real Time Schedule Theory and Its Application. In *Proceedings of International Symposium on Distributed Computing and Applications to Business. Engineering and Science (DCABES)*, Wuxi, Jiangsu, China, 2002.

[863] R. Want, A. Hopper, V. Falão, and J. Gibbons. The Active Badge Location System. *ACM Transactions on Information Systems*, 10(1): 91–102, 1992.

[864] A. Ward, A. Jones, and A. Hopper. A New Location Technique for the Active Office. *IEEE Personal Communications*, 4(5): 42–47, 1997.

[865] R. Wattenhofer, L. Li, P. Bahl, and Y.-M. Wang. Distributed Topology Control for Power Efficient Operation in Multihop Wireless Ad Hoc Networks. In *Proceedings of IEEE Infocom*, Anchorage, AK, April 2001.

[866] M. Weisenhorn and W. Hirt. Novel Rate-Division Multiple-Access Scheme for UWB-Radio-Based Sensor Networks. In *Proceedings of the 2004 International Zurich Seminar on Communications*, pages 76–81, Zurich, Switzerland, February 2004.

[867] M. Weiser. The Computer for the 21st Century. *Scientific American*, 43(3): 66–75, 1991.

[868] M. Weiser. Hot topic: Ubiquitous Computing. *IEEE Computer*, pages 71–72, October 1993.

[869] M. Weiser, B. Welch, A. Demers, and B. Shenker. Scheduling for Reduced CPU Energy. In *Proceedings of USENIX Symposium on Operating Systems Desing and Implementation*, pages 13–23, Monterey, CA, November 1994.

[870] K. Weniger. Passive Duplicate Address Detection in Mobile Ad Hoc Networks. In *Proceedings of IEEE WCNC 2003*, New Orleans, LA, March 2003.

[871] K. Weniger and M. Zitterbart. IPv6 Autoconfiguration in Large Scale Mobile Ad-Hoc Networks. In *Proceedings of European Wireless*, Florence, Italy, February 2002.

[872] D. D. Wentzloff, B. H. Calhoun, R. Min, A. Wang, N. Ickes, and A. P. Chandrakasan. Design Considerations for Next Generation Wireless Power-Aware Microsensor Nodes. In *Proceedings of the 17th International Conference on VLSI Design*, pages 361–367, Mumbai, India, January 2004.

[873] K. Whitehouse and D. Culler. Calibration as Parameter Estimation in Sensor Networks. In *Proceedints of the 1st ACM International Workshop on Sensor Networks and Applications (WSNA)*, Atlanta, GA, September 2002.

[874] J. E. Wieselthier, G. D. Nguyen, and A. Ephremides. On the Construction of Energy-Efficient Broadcast and Multicast Trees in Wireless Networks. In *Proceedings of IEEE Infocom*, Tel-Aviv, Israel, March 2000.

[875] J. E. Wieselthier, G. D. Nguyen, and A. Ephremides. Resource Management in Energy-Limited, Bandwidth-Limited, Transceiver-Limited Wireless Networks for Session-Based Multicasting. *Computer Networks*, 39: 113–131, 2002.

[876] B. Williams and T. Camp. Comparison of Broadcasting Techniques for Mobile Ad Hoc Networks. *Proceedings of the 3rd ACM International Symposium on Mobile Ad Hoc Networking and Computing (MobiHoc)*, pages 194–205. ACM Press, 2002.

[877] A. Willig, R. Shah, J. Rabaey, and A. Wolisz. Altruists in the PicoRadio Sensor Network. In *Proceedings of the 4th IEEE International Workshop on Factory Communication Systems*, Vasteras, Sweden, August 2002.

[878] A. Willig. Polling-Based MAC Protocols for Improving Realtime Performance in a Wireless PROFIBUS. *IEEE Transactions on Industrial Electronics*, 50(4): 806 –817, 2003.

[879] A. Willig. Intermediate Checksums for Improving Goodput over Error-Prone Links. In *Proceedings of the IEEE Vehicular Technology Conference (VTC), Fall 04*, Los Angeles, CA, September 2004.

[880] A. Willig. Some Simple Upper Bounds on the Throughput of IEEE 802.15.4 with 2.4 GHz. In *Contribution to the 3rd ESA Wireless Workgroup and Optical Onboard S/C Workshop*, Noordwijk, Netherlands, September 2004.

[881] A. Willig. Wireless LAN Technology for the Factory Floor: Challenges and Approaches. In R. Zurawski, editor, *Handbook on Industrial Communication Systems*. CRC Press, 2004.

[882] A. Willig, M. Kubisch, C. Hoene, and A. Wolisz. Measurements of a Wireless Link in an Industrial Environment Using an IEEE 802.11-Compliant Physical Layer. *IEEE Transactions on Industrial Electronics*, 49(6): 1265–1282, 2002.

[883] A. Willig and A. Wolisz. Ring Stability of the PROFIBUS Token Passing Protocol Over Error Prone Links. *IEEE Transactions on Industrial Electronics*, 48(5): 1025–1033, 2001.

[884] M. Z. Win and R. A. Scholtz. Impulse Radio: How it Works. *IEEE Communication Letters*, 2: 10–12, 1998.

[885] M. Z. Win and R. A. Scholtz. Ultra-Wide Bandwidth Time-Hopping Spread-Spectrum Impulse Radio for Wireless Multiple-Access Communications. *IEEE Transactions on Communications*, 48(4): 679–691, 2000.

[886] H. Woesner, J.-P. Ebert, M. Schlaeger, and A. Wolisz. Power-Saving Mechanisms in Emerging Standards for Wireless LAN's: The MAC-Level Perspective. *IEEE Personal Communications*, 5(3): 40–48, 1998.

[887] A. Woo, S. Madden, and R. Govindan. Networking Support for Query Processing in Sensor Networks. *Communications of the ACM*, 47(6): 47–52, 2004.

[888] A. Woo and D. Culler. A Transmission Control Scheme for Media Access in Sensor Networks. In *Proceedings of the Seventh Annual International Conference on Mobile Computing and Networking 2001 (MobiCom)*, Rome, Italy, July 2001.

[889] A. Woo and D. Culler. Evaluation of Efficient Link Reliability Estimators for Low-Power Wireless Networks. Technical Report UCB/CSD 03-1270, University of California, Berkeley, 2002.

[890] A. Woo, T. Tong, and D. Culler. Taming the Underlying Challenges of Reliable Multihop Routing in Sensor Networks. In *Proceedings of the ACM SenSys 03*, Los Angeles, CA, November 2003.

[891] A. D. Wood and J. A. Stankovic. Denial of Service in Sensor Networks. *IEEE Computer*, 35(10): 54–62, 2002.

[892] M. Woolridge. *Introduction to MultiAgent Systems*. Wiley, 2002.

[893] G. R. Wright and W. R. Stevens. *TCP/IP Illustrated Volume 2 – the Implementation*. Addison-Wesley, Reading, MA, 1995.

[894] J. Wu and H. Li. On Calculating Connected Dominating Set for Efficient Routing in Ad Hoc Wireless Networks. In *Proceedings of the 4th International Workshop on Discrete Algorithms and Methods for Mobile Computing and Communications*, Boston, MA, August 11, 2000.

[895] S.-L. Wu, Y.-C. Tseng, and J.-P. Sheu. Intelligent Medium Access for Mobiel Ad Hoc Networks with Busy Tones and Power Control. *IEEE Journal on Selected Areas in Communications*, 18(9): 1647–1657, 2000.

[896] S. Wu and C. Bonnet. A Reliable Multicasting Protocol for Ad Hoc Networks. In *Proceedings of the World Wireless Congress (WWC2004)*, San Francisco, CA, 2004.

[897] Y. Wu, P. A. Chou, and S.-Y. Kung. Network Planning in Wireless Ad Hoc Networks: A Cross-layer Approach. *IEEE Journal on Selected Areas in Communications (JSAC)*, January 2005.

[898] A. D. Wyner. Recent Results in the Shanon Theory. *IEEE Transactions on Information Theory*, IT-20(1): 2–10, 1974.

[899] A. D. Wyner and J. Ziv. The Rate-Distortion Function for Source Coding with Side Information at the Decoder. *IEEE Transactions on Information Theory*, IT-22: 1–10, 1976.

[900] A. D. Wyner. On Source Coding with Side Information at the Decoder. *IEEE Transactions on Information Theory*, IT-21: 294–300, 1975.

[901] Z. Xiong, A. Liveris, and S. Cheng. Distributed Source Coding for Sensor Networks. *IEEE Signal Processing Magazine (Special Issue on Signal Processing for Networks)*, 21: 80–94.

[902] Y. Xu, S. Bien, Y. Mori, J. Heidemann, and D. Estrin. Topology Control Protocols to Conserve Energy in Wireless Ad Hoc Networks. Technical Report CENS Technical Report 0006, University of California at Los Angeles, CENS, January 2003.

[903] Y. Xu, J. Heidemann, and D. Estrin. Geography-Informed Energy Conservation for Ad Hoc Routing. In *Proceedings of the 7th Annual International Conference on Mobile Computing and Networking (MobiCom)*, pages 70–84, Rome, Italy, July 2001. ACM.

[904] F. Xue and P. R. Kumar. The Number of Neighbors Needed for Connectivity of Wireless Networks. *Wireless Networks*, 10(2): 169–181, 2004.

[905] A. Yao. On Constructing Minimum Spanning Trees in k-Dimensional Spaces and Related Problems. *SIAM Journal on Computing*, 11(4): 721–736, 1982.

[906] F. Yao, A. Demers, and S. Shenker. A Scheduling Model for Reduced CPU Energy. In *Proceedings of the 36th Annual Symposium on Foundations of Computer Science (FOCS)*, pages 374–385, Milwaukee, WI, October 1995.

[907] K. Yao, R. E. Hudson, C. W. Reed, D. Chen, and F. Lorenzelli. Blind Beamforming on a Randomly Distributed Sensor Array System. *IEEE Journal on Selected Areas in Communications*, 16(8): 1555–1567, 1998.

[908] Y. Yao and J. Gehrke. The COUGAR Approach to In-Network Query Processing in Sensor Networks. *ACM SIGMOD Record*, 31(3): 9–18, 2002.

[909] Y. Yao and J. Gehrke. Query Processing for Sensor Networks. In *First Biennial Conference on Innovative Data Systems Research(CIDR 2003)*, Asilomar, January 2003.

[910] F. Ye, A. Chen, S. Lu, and L. Zhang. A Scalable Solution to Minimum Cost Forwarding in Large Scale Sensor Networks. In *Proceedings of the International Conference on Computer Communications and Networks (ICCCN)*, Scottsdale, AZ, 2001.

[911] F. Ye, H. Luo, J. Cheng, S. Lu, and L. Zhang. A Two-Tier Data Dissemination Model for Large-Scale Wireless Sensor Networks. *Proceedings of the 8th ACM Annual International Conference on Mobile Computing and Networking (MobiCom)*, pages 148–159. ACM Press, 2002.

[912] F. Ye, G. Zhong, S. Lu, and L. Zhang. A Robust Data Delivery Protocol for Large Scale Sensor Networks. In *Proceedings of the 2nd International Workshop on Information Processing in Sensor Networks (IPSN)*, Palo Alto, CA, April 2003.

[913] F. Ye, G. Zhong, S. Lu, and L. Zhang. PEAS: A Robust Energy Conserving Protocol for Long-lived Sensor Networks. In *Proceedings of the 23rd International Conference on Distributed Computing Systems (IEEE ICDCS)*, Providence, RI, 2003.

[914] W. Ye, J. Heidemann, and D. Estrin. An Energy-Efficient MAC Protocol for Wireless Sensor Networks. *Proceedings of INFOCOM 2002*. IEEE Press, New York, June 2002.

[915] W. Ye, J. Heidemann, and D. Estrin. Medium Access Control with Coordinated, Adaptive Sleeping for Wireless Sensor Networks. *IEEE/ACM Transactions on Networking*, 2004. `http://portal.acm.org/citation.cfm?id=1008463.1008471`.

[916] H. Y. Youn, C. Yu, and B. Lee. *The Handbook of Ad Hoc Wireless Networks*, chapter Routing Algorithms for Balanced Energy Consumption in Ad Hoc Networks, pages 25-1–25-14. CRC Press, 2003.

[917] H. Y. Youn, C. Yu, B. Lee, and S. Moh. *The Handbook of Ad Hoc Wireless Networks*, chapter Energy Efficient Multicast in Ad Hoc Networks, pages 23-1–23-12. CRC Press, 2003.

[918] O. Younis and S. Fahmy. Distributed Clustering in Ad-hoc Sensor Networks: A Hybrid, Energy-Efficient Approach. In *Proceedings of IEEE INFOCOM*, Hong Kong, March 2004.

[919] Y. Yu, R. Govindan, and D. Estrin. Geographical and Energy Aware Routing: A Recursive Data Dissemination Protocol for Wireless Sensor Networks. Technical Report UCLA/CSD-TR-01-0023, University of California at Los Angeles, May 2001.

[920] W. Yuan, S. Krishnamurthy, and S. K. Tripathi. Synchronization of Multiple Levels of Data Fusion in Wireless Sensor Networks. In *Proceedings of IEEE Globecom*, San Francisco, CA, December 2003.

[921] H. Zhang and A. Arora. GS3: Scalable Self-Configuration and Self-Healing in Wireless Networks. In *Proceedings of the 21st Annual Symposium on Principles of Distributed Computing*, Monterey, CA, 2002.

[922] B. Y. Zhao, J. D. Kubiatowicz, and A. D. Joseph. Tapestry: A Fault-Tolerant Wide-Area Application Infrastructure. *ACM SIGCOMM Computer Communication Review*, 32(1): 81–81, 2002.

[923] F. Zhao, J. Shin, and J. Reich. Information-Driven Dynamic Sensor Collaboration for Tracking Applications. *IEEE Signal Processing Magazine*, 19(2): 61–72, 2002.

[924] F. Zhao and L. Guibas. *Wireless Sensor Networks – An Information Processing Approach*. Elsevier/Morgan-Kaufman, Amsterdam, NY, 2004.

[925] F. Zhao, J. Liu, J. Liu, L. Guibas, and J. Reich. Collaborative Signal and Information Processing: An Information Directed Approach. *Proceedings of the IEEE*, 91(8): 1199–1209, 2003.

[926] F. Zhao, J. Shin, and J. Reich. Information-Driven Dynamic Sensor Collaboration. *IEEE Signal Processing Magazine*, 19(2): 61–72, 2002.

[927] J. Zhao, R. Govindan, and D. Estrin. Computing Aggregates for Monitoring Wireless Sensor Networks. In *Proceedings of the 1st IEEE International Workshop on Sensor Network Protocols and Applications (SNPA)*, Anchorage, AK, May 2003.

[928] Q. Zhao and M. Effros. Optimal Code Design for Lossless and Near Lossless Source Coding in Multiple Access Networks. In *Proceedings of IEEE Data Compression Conference (DCC)*, Snowbird, UT, March 2001.

[929] J. Zheng and M. J. Lee. Will IEEE 802.15.4 Make Ubiquitous Networking a Reality?: A Discussion on a Potential Low Power, Low Bit Rate Standard. *IEEE Communications Magazine*, 42(6): 140–146, 2004.

[930] L. C. Zhong, J. Rabaey, C. Guo, and R. Shah. Data Link Layer Design for Wireless Sensor Networks. In *Proceedings of IEEE MILCOM 2001*, Washington, DC, October 2001.

[931] L. C. Zhong, R. C. Shah, C. Guo, and J. M. Rabaey. An Ultra-Low Power and Distributed Access Protocol for Broadband Wireless Sensor Networks. In *IEEE Broadband Wireless Summit*, Las Vegas, NV, May 2001.

[932] L. C. Zhong, J. M. Rabaey, and A. Wolisz. An Integrated Data-Link Energy Model for Wireless Sensor Networks. In *IEEE International Conference on Communications (ICC)*, Paris, France, June 2004.

[933] C. Zhou and B. Krishnamachari. Localized Topology Generation Mechanisms for Self-Configuring Sensor Networks. In *Proceedings of IEEE Globecom*, San Francisco, CA, December 2003.

[934] G. Zhou, T. He, S. Krishnamurthy, and J. Stankovic. Impact of Radio Asymmetry on Wireless Sensor Networks. In *Proceedings of ACM/USENIX International Conference on Mobile Systems, Applications, and Services (MobiSys)*, Boston, MA, June 2004.

[935] H. Zhou and S. Singh. Content-based Multicast CBM for Ad hoc Networks. In *Proceedings of the 1st ACM International Symposium on Mobile Ad Hoc Networking and Computing (MOBIHOC)*, Boston, MA, June 2000.

[936] H. Zhou, L. M. Ni, and M. W. Mutka. Prophet Address Allocation for Large Scale MANETs. In *Proceedings of IEEE INFOCOM*, San Francisco, CA, March 2003.

[937] S. Zhu, S. Setia, and S. Jajodia. LEAP: Efficient Security Mechanisms for Large-Scale Distributed Sensor Networks. In *Proceedings of the 10th ACM Conference on Computer and Communication Security, CCS'03*, pages 62–72, Washington, DC, 2003.

[938] R. E. Ziemer and W. H. Tranter. *Principles of Communications*. Wiley, 2002.

[939] R. Ziemer, M. Wickert, and T. Williams. A comparison between UWB and DSSS for use in a multiple access secure wireless sensor network. In *Proceedings of the 2003 IEEE Conference on Ultra Wideband Systems and Technologies*, pages 428 – 432, Reston, VA, November 2003.

[940] M. Zorzi and R. R. Rao. Geographic Random Forwarding (GeRaF) foR Ad Hoc and Sensor Networks: Energy and Latency Performance. *IEEE Transactions on Mobile Computing*, 2: 337–347, 2003.

[941] M. Zorzi and R. R. Rao. Geographic Random Forwarding (GeRaF) for Ad Hoc and Sensor Networks: Multihop Performance. *IEEE Transactions on Mobile Computing*, 2: 349–364, 2003.

[942] M. Zorzi and R. R. Rao. Error Control and Energy Consumption in Communications for Nomadic Computing. *IEEE Transactions on Computers*, 46(3): 279–289, 1997.

[943] M. Zorzi and R. R. Rao. Coding Tradeoffs for Reduced Energy Consumption in Sensor Networks. In *Proceedings of PIMRC 04*, Barcelona, Spain, 2004.

[944] Y. Zou and K. Charkrabarty. Sensor Deployment and Target Localization Based on Virtual Forces. In *Proceedings of IEEE INFOCOM*, San Francisco, CA, March 2003.

[945] Y. Zou and K. Chakrabarty. Sensor Deployment and Target Localization in Distributed Sensor Networks. *ACM Transaction on Embedded Computing Systems*, 2003, `http://portal.acm.org/citation.cfm?id=972627.972631`. accepted for publication.

[946] M. Zuniga and B. Krishnamachari. Integrating Future Large Scale Sensor Networks with the Internet. Technical Report CS 03-792, University of Southern California, Department Computer Science, 2003.

[947] M. Zuniga and B. Krishnamachari. Optimal Transmission Radius for Flooding in Large Scale Sensor Networks. In *Proceedings of Workshop on Mobile and Wireless Networks (MWN)*, May 2003. In conjunction with 23rd IEEE International Conference on Distributed Computing Systems (ICDCS).

[948] G. Zussman and Adrian Segall. Energy Efficient Routing Ad Hoc Disaster Recovery Networks. In *Proceedings of IEEE INFOCOM*, San Francisco, CA, March 2003.

Index